Springer Handbook of Auditory Research

Series Editors: Richard R. Fay and Arthur N. Popper

SPRINGER HANDBOOK OF AUDITORY RESEARCH

Volume 1: The Mammalian Auditory Pathway: Neuroanatomy
Edited by Douglas B. Webster, Arthur N. Popper, and Richard R. Fay

Volume 2: The Mammalian Auditory Pathway: Neurophysiology
Edited by Arthur N. Popper and Richard R. Fay

Volume 3: Human Psychophysics
Edited by William Yost, Arthur N. Popper, and Richard R. Fay

Volume 4: Comparative Hearing: Mammals
Edited by Richard R. Fay and Arthur N. Popper

Volume 5: Hearing by Bats
Edited by Arthur N. Popper and Richard R. Fay

Volume 6: Auditory Computation
Edited by Harold L. Hawkins, Teresa A. McMullen, Arthur N. Popper, and Richard R. Fay

Volume 7: Clinical Aspects of Hearing
Edited by Thomas R. Van De Water, Arthur N. Popper, and Richard R. Fay

Volume 8: The Cochlea
Edited by Peter Dallos, Arthur N. Popper, and Richard R. Fay

Volume 9: Development of the Auditory System
Edited by Edwin W Rubel, Arthur N. Popper, and Richard R. Fay

Volume 10: Comparative Hearing: Insects
Edited by Ronald Hoy, Arthur N. Popper, and Richard R. Fay

Volume 11: Comparative Hearing: Fish and Amphibians
Edited by Richard R. Fay and Arthur N. Popper

Volume 12: Hearing by Whales and Dolphins
Edited by Whitlow W.L. Au, Arthur N. Popper, and Richard R. Fay

Volume 13: Comparative Hearing: Birds and Reptiles
Edited by Robert Dooling, Arthur N. Popper, and Richard R. Fay

Volume 14: Genetics and Auditory Disorders
Edited by Bronya J.B. Keats, Arthur N. Popper, and Richard R. Fay

Volume 15: Integrative Functions in the Mammalian Auditory Pathway
Edited by Donata Oertel, Richard R. Fay, and Arthur N. Popper

Volume 16: Acoustic Communication
Edited by Andrea Simmons, Arthur N. Popper, and Richard R. Fay

Volume 17: Compression: From Cochlea to Cochlear Implants
Edited by Sid P. Bacon, Richard R. Fay, and Arthur N. Popper

Volume 18: Speech Processing in the Auditory System
Edited by Steven Greenberg, William Ainsworth, Arthur N. Popper, and Richard R. Fay

Volume 19: The Vestibular System
Edited by Stephen M. Highstein, Richard R. Fay, and Arthur N. Popper

Volume 20: Cochlear Implants: Auditory Prostheses and Electric Hearing
Edited by Fan-Gang Zeng, Arthur N. Popper, and Richard R. Fay

Volume 21: Electroreception
Edited by Theodore H. Bullock, Carl D. Hopkins, Arthur N. Popper, and Richard R. Fay

Continued after index

Ruth Anne Eatock
Richard R. Fay
Arthur N. Popper
Editors

Vertebrate Hair Cells

With 84 illustrations

Ruth Anne Eatock
Department of Otorhinolaryngology
Baylor College of Medicine
Houston, TX 77030
USA
eatock@bcm.edu

Richard R. Fay
Parmly Hearing Institute and Department
 of Psychology
Loyola University of Chicago
Chicago, IL 60626
USA
rfay@wpo.it.luc.edu

Arthur N. Popper
Department of Biology
University of Maryland
College Park, MD 20742
USA
apopper@umd.edu

Series Editors:
Richard R. Fay
Parmly Hearing Institute and Department
 of Psychology
Loyola University of Chicago
Chicago, IL 60626
USA

Arthur N. Popper
Department of Biology
University of Maryland
College Park, MD 20742
USA

Cover illustration: Electron microscopy of hair cells and their mechanosensitive hair bundles. *Left*, An outer hair cell from the mammalian cochlea is crowned by the hair bundle, comprising rows of stereocilia. *Middle*, Hair bundle on a mammalian vestibular hair cell. *Right*, Scanning (top) and transmission (bottom) electron micrographs showing the filamentous tip links connecting rows of stereocilia. Mechanosensitive ion channels are located near the tip links' insertions into the stereocilia. From David Furness and Carole Hackney (Chapter 3).

Library of Congress Control Number: 2005925504

ISBN 10: 0-387-95202-0   Printed on acid-free paper.
ISBN 13: 978-0387-95202-4

© 2006 Springer Science+Business Media, Inc.
All rights reserved. This work may not be translated or copied in whole or in part without the written permission of the publisher (Springer Science+Business Media, Inc., 233 Spring Street, New York, NY 10013, USA), except for brief excerpts in connection with reviews or scholarly analysis. Use in connection with any form of information storage and retrieval, electronic adaptation, computer soft-ware, or by similar or dissimilar methodology now known or hereafter developed is forbidden. The use in this publication of trade names, trademarks, service marks, and similar terms, even if they are not identified as such, is not to be taken as an expression of opinion as to whether or not they are subject to proprietary rights.

Printed in the United States of America. (SHER)

9 8 7 6 5 4 3 2 1

springeronline.com

We dedicate this book to Alfons Rüsch (1959–2002), whose work touched on almost every aspect of function in vertebrate hair cells. A true naturalist, Alfons investigated hair cells from diverse organs. A pragmatic experimentalist, he was at the leading edge in development of elegant physiological preparations and assays. His results enriched our understanding of transduction and the changes in auditory and vestibular hair cells that accompany the maturation of hearing and balance.

Series Preface

The *Springer Handbook of Auditory Research* presents a series of comprehensive and synthetic reviews of the fundamental topics in modern auditory research. The volumes are aimed at all individuals with interests in hearing research including advanced graduate students, postdoctoral researchers, and clinical investigators. The volumes are intended to introduce new investigators to important aspects of hearing science and to help established investigators to better understand the fundamental theories and data in fields of hearing that they may not normally follow closely.

Each volume presents a particular topic comprehensively, and each serves as a synthetic overview and guide to the literature. As such, the chapters present neither exhaustive data reviews nor original research that has not yet appeared in peer-reviewed journals. The volumes focus on topics that have developed a solid data and conceptual foundation rather than on those for which a literature is only beginning to develop. New research areas will be covered on a timely basis in the series as they begin to mature.

Each volume in the series consists of a few substantial chapters on a particular topic. In some cases, the topics will be ones of traditional interest for which there is a substantial body of data and theory, such as auditory neuroanatomy (Vol. 1) and neurophysiology (Vol. 2). Other volumes in the series deal with topics that have begun to mature more recently, such as development, plasticity, and computational models of neural processing. In many cases, the series editors are joined by a co-editor having special expertise in the topic of the volume.

RICHARD R. FAY, Chicago, Illinois
ARTHUR N. POPPER, College Park, Maryland

Volume Preface

Vertebrate hair cells have the attractive ability to represent key problems in diverse disciplines. For example, developmental biologists see orderly programs of development, from the formation of the mechanosensitive hair bundle to the organization of the hair cell and supporting cell mosaic. Biophysicists analyze hair cell receptor potentials into contributions from sets of specific mechanosensory, voltage- and ligand-gated ion channels. Synaptologists see hair cells as relatively accessible and high-functioning presynaptic terminals. Each chapter in this volume summarizes the hair cell from a distinct research perspective, with the overall goal of providing a comprehensive picture of hair cell function.

In Chapter 1, Eatock introduces the modern view of hair cells as represented in this volume and then provides a historic review of experimental milestones in hair cell study. Chapter 2 follows with a comprehensive synthesis of the rapidly expanding field of hair cell development by Goodyear, Kros, and Richardson.

The next several chapters advance from top to bottom through the afferent cascade of sensory processing, from transduction at the apical pole of the hair cell through the effect of basolateral channels on the receptor potential, concluding with synaptic transmission at the basal pole. Furness and Hackney (Chapter 3) describe the beautifully diverse and ordered structures of hair bundles at the light, ultrastructural, and molecular levels. In Chapter 4, Fettiplace and Ricci discuss mechanoelectrical transduction, the remarkable process by which bundle deflection gates channels in the hair bundle, and associated feedback processes that shape the transduction current.

The transduction current changes the membrane potential, modulating voltage-gated ion channels in the hair cell's basolateral membrane. In Chapter 5, Art and Fettiplace review elegant experimental series showing how the biophysical properties of basolateral channels tune the receptor potentials of turtle and chick cochlear hair cells. In Chapter 6, Fuchs and Parsons describe novel mechanisms supporting afferent transmission from hair cells to eighth-nerve fibers and efferent transmission from nerve fibers originating in the brainstem.

The volume concludes with two chapters on specialist hair cells. In Chapter 7, Brownell discusses an electromechanical transduction process unique to the highly specialized outer hair cells of mammalian cochleas and fundamental to mammalian hearing sensitivity. In Chapter 8, Eatock and Lysakowski review transduction and transmission in mammalian vestibular hair cells, with special focus on the unusual properties and possible functions of the type I hair cell and its calyceal afferent ending.

Complementary chapters on hair cells can be found in several other volumes in the Springer Handbook of Auditory Research series. Volume 8 (*The Cochlea*) covers the structure and function of the inner ear, including physiology and other aspects of hair cells. Vestibular hair cells are discussed in Volume 19 (*The Vestibular System*) in chapters covering afferent morphophysiology (Lysakowski and Goldberg) and basolateral hair cell currents (Steinacker). The evolution of hair cells is considered in a chapter by Coffin et al. in *Evolution of the Vertebrate Auditory System* while the molecular biology of ear (and hair cell) development is a topic of *Development of the Inner Ear*.

RUTH ANNE EATOCK, Houston, Texas
RICHARD R. FAY, Chicago, Illionois
ARTHUR N. POPPER, College Park, Maryland

Contents

Series Preface vii
Volume Preface ix
Contributors xiii

Chapter 1   Vertebrate Hair Cells: Modern and Historic
            Perspectives. 1
            RUTH ANNE EATOCK

Chapter 2   The Development of Hair Cells in the Inner Ear. 20
            RICHARD J. GOODYEAR, CORNÉ J. KROS, AND
            GUY P. RICHARDSON

Chapter 3   The Structure and Composition of the Stereociliary
            Bundle of Vertebrate Hair Cells 95
            DAVID N. FURNESS AND CAROLE M. HACKNEY

Chapter 4   Mechanoelectrical Transduction in Auditory Hair Cells .. 154
            ROBERT FETTIPLACE AND ANTHONY J. RICCI

Chapter 5   Contribution of Ionic Currents to Tuning in
            Auditory Hair Cells. 204
            JONATHAN J. ART AND ROBERT FETTIPLACE

Chapter 6   The Synaptic Physiology of Hair Cells. 249
            PAUL A. FUCHS AND THOMAS D. PARSONS

Chapter 7   The Piezoelectric Outer Hair Cell 313
            WILLIAM E. BROWNELL

Chapter 8   Mammalian Vestibular Hair Cells. 348
            RUTH ANNE EATOCK AND ANNA LYSAKOWSKI

Index 443

Contributors

JONATHAN J. ART
Department of Anatomy and Cell Biology, University of Illinois College of Medicine, Chicago, IL 60612, USA

WILLIAM E. BROWNELL
Department of Otorhinolarynglogy, Baylor College of Medicine, Houston, TX 77030, USA

RUTH ANNE EATOCK
Department of Otorhinolaryngology, Baylor College of Medicine, Houston, TX 77030, USA

ROBERT FETTIPLACE
Department of Physiology, University of Wisconsin Medical School, Madison, WI 53706, USA

PAUL A. FUCHS
The Cellular Neurotransmission Laboratory, Center for Hearing and Balance, The Johns Hopkins University School of Medicine, Baltimore, MD 21205, USA

DAVID N. FURNESS
MacKay Institute of Communication and Neuroscience, School of Life Sciences, Keele University, Keele, Staffordshire ST5 5BG, United Kingdom

RICHARD J. GOODYEAR
School of Life Sciences, University of Sussex, Falmer, Brighton BN1 9QG, United Kingdom

CAROLE M. HACKNEY
MacKay Institute of Communication and Neuroscience, School of Life Sciences, Keele University, Keele, Staffordshire ST5 5BG, United Kingdom

CORNÉ J. KROS
School of Life Sciences, University of Sussex, Falmer, Brighton BN1 9QG, United Kingdom

ANNA LYSAKOWSKI
Department of Anatomy and Cell Biology, University of Illinois College of Medicine, Chicago, IL 60612, USA

THOMAS D. PARSONS
School of Veterinary Medicine, University of Pennsylvania, Kennett Square, PA 19348, USA

ANTHONY J. RICCI
Neuroscience Center, Louisiana Sate University Health Science Center, New Orleans, LA 70112, USA

GUY P. RICHARDSON
School of Life Sciences, University of Sussex, Falmer, Brighton BN1 9QG, United Kingdom

1
Vertebrate Hair Cells: Modern and Historic Perspectives

RUTH ANNE EATOCK

1. Introduction to the Volume

Vertebrate hair cells share many fundamental features, reflecting their common origin in vertebrate ancestors (chordates; Coffin et al. 2004). Conversely, the intervening half-billion years have allowed hair cells to diversify across species and organs. In their well-ordered diversity, hair cells afford natural experiments in biomechanics, development, sensory transduction, excitability, and synaptic transmission. Each chapter in this book represents hair cells from a distinct research perspective, so that the ensemble presents a comprehensive summary of their development and function.

The volume begins with a chapter by Richard Goodyear, Corné Kros, and Guy Richardson on hair cell development (Chapter 2; Fig. 1.1A). The fast growth of research into hair cell development reflects its ability to attract investigators from outside fields: Hair cell organs are choice tissues in which to examine how complex three-dimensional structures, such as the vestibular labyrinth, are carved out during development (Chang et al. 2004); how highly ordered two-dimensional arrays of cells achieve their mature pattern (Bryant et al. 2002); how individual cells regulate the growth of specialized processes such as stereocilia (Tilney et al. 1988; Fig. 1.1A); and how and why ion channel expression changes with development (Marcotti et al. 2003). Such routine processes as bundling of filamentous actin by cross-linking proteins acquire panache when the bundling is essential for hearing and balance (Zheng et al. 2000b). Developmental changes in ion channel expression are well known throughout the brain, but are precisely anchored in cochlear hair cells by the onset of hearing, at which time a set of ion channel transformations appears to switch the hair cells from spontaneously spiking promoters of synaptogenesis to proper sensory cells.

The next several chapters advance from top to bottom through the hair cell and the afferent cascade of sensory processing. In Chapter 3, David Furness and Carole Hackney describe the structure of hair bundles. These arrays of specialized microvilli are organized in a staircase of rows behind a single true cilium and are cross-linked internally and externally to form a stiff structure that

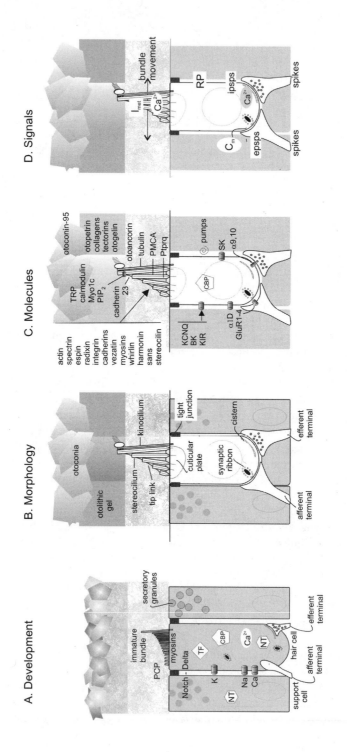

bends about its insertion into a dense apical web (the cuticular plate) (Fig. 1.1B). New data from genetics and molecular biology (Fig. 1.1C) are rapidly expanding our knowledge of the component molecules and how they may link to produce an intact transducing bundle.

In Chapter 4, Robert Fettiplace and Tony Ricci describe our unfolding understanding of how the mechanosensitive apparatus transduces hair bundle deflection into current and how the properties of its component parts, especially the transduction channels, shape the transduction current. It has long been ap-

FIGURE 1.1. Levels of study of inner ear hair cells. (**A**) Investigations of hair cell development (Chapter 2) have identified proteins involved in: setting hair cell fate, such as the transcription factor (TF) Math1 and the ligand-receptor combination Delta–Notch; establishing the supporting cell/hair cell mosaic (Delta–Notch); making hair bundles (myosins VI, VIIa, and XV) and orienting them (planar cell polarity, PCP, proteins); making accessory tectorial and otolithic gels, in part via secretion of such component proteins as tectorins and otogelin from supporting cells; expression of age-specific calcium-binding proteins (CBPs); and promoting synapses via neurotrophin (NT) release and action potentials (spikes) generated by Ca, Na, and K channels. (**B**) Hair cell morphology (Chapters 3, 6, 7, and 8). Hair cells are epithelial cells, with an apical bundle of specialized microvilli (stereocilia) and a basolateral soma contacted by neural endings. Presynaptic to each afferent contact is a synaptic ribbon and postsynaptic to each efferent contact (on hair cells) is a cistern. (**C**) A sampling of molecules revealed by genetic, molecular, immunocytochemical and biophysical approaches (Chapters 2–8). Congenital inner ear defects frequently reflect mutations in components of hair bundles or extracellular matrices. Thus, genetic approaches have populated the hair bundle with numerous cytoskeletal (actin, fimbrin, espin), linkage (vezatin, cadherins), and motor (myosin) proteins (Chapters 2 and 3). There are strong candidates for several structures key to mechanotransduction, including TRP channels and Myo1c adaptation motors (Chapter 4). Plasma membrane ATPases, PMCA, maintain trans-membrane Ca^{2+} gradients and are required for transduction and normal otoconia (Chapters 4 and 8). The enzyme PTPRQ at the base of the bundle may be responsible for apical segregation of phosphatidylinositol 4,5-bisphosphate (PIP_2), where it modulates adaptation (Hirono et al. 2004). Extracellular gels include the secreted collagens, tectorins and otogelin (Chapter 8). Otoconia are largely calcareous crystals, but some have a proteinaceous core that includes otoconin-95, and otopetrin may organize their formation (Chapter 8). Basolateral ion channels include the K channels KCNQ, BK, and IR; the SK channel, which generates the efferent-evoked inhibitory postsynaptic potential; the presynaptic Ca^{2+} channel, α_{1D}; the neurotransmitter receptor channels $\alpha_{9,10}$; and, on the afferent terminal, glutamate receptors 1–4 (Chapters 2, 5, 6, and 8). CBPs such as parvalbumin, calretinin and calbindin are present at high levels and differentially expressed (Chapters 4 and 8). (**D**) Hair cell signals, monitored as changes in bundle stiffness (Chapter 4), intracellular free Ca^{2+} (Chapters 4–6), capacitance (a measure of exocytosis; Chapter 6) or electrical activity in the form of transduction currents (I_{met}; Chapters 4 and 8), receptor potential (RP; Chapters 5 and 8), inhibitory postsynaptic potentials (IPSPs, Chapter 6), or excitatory postsynaptic potentials (EPSPs) from afferent nerve fibers (Chapter 6) and spikes.

preciated that in the cochlea, narrow-band tuning enhances the sensitivity of hair cells (Block 1992), and for just as long arguments have raged over the site(s) of such tuning. For the mammalian cochlea, which hears at frequencies well above the upper limit in other species (10 kHz), current debate centers on whether amplification depends more on processes localized in the bundle (Chapter 4) or in the basolateral membrane of outer hair cells (Chapter 7).

The transduction current changes membrane potential, secondarily modulating ion channels in the basolateral hair cell membrane. In Chapter 5, Jonathan Art and Robert Fettiplace review how the basolateral channels condition the receptor potentials of cochlear hair cells. In non-mammals, hair cells tuned to stimuli below several kilohertz often show electrical tuning: amplification of the receptor potential at the acoustic or vibratory best frequency by the electrical properties of the membrane. In the afferent sequence, electrical tuning follows tuning of accessory structures such as tectorial membranes and of the transduction apparatus (the stereociliary amplifier). Electrical tuning is achieved with specific complements of basolateral ion channels such that current flowing through the channels enhances the receptor potential at the best frequency. This virtuoso display of ion channel engineering has made its way into most neuroscience textbooks. Current research is focused on the molecular mechanisms underlying this tuning—how are gradients in tuning achieved, such that one cell is tuned to a slightly higher frequency than its apical neighbor?

In Chapter 6, Paul Fuchs and Tom Parsons consider the synaptic pole, at which the hair cell transmits its signal to afferent nerve terminals and receives input from efferent fibers originating in the brainstem. Vertebrate hair cells are excellent models for studies of excitatory transmission. Like excitatory presynaptic terminals in the brain, hair cell afferent synapses release glutamate, but they offer several experimental advantages. First, they are relatively large and accessible, so that it is possible to monitor capacitance changes as a measure of exocytosis (Parsons et al. 1994; Moser and Beutner 2000). Second, hair cell synapses are "souped-up," delivering transmitter at remarkably high peak and tonic rates. Third, because we know a great deal about the sensory signals being transmitted, we can relate performance (e.g., speed of transmission) to function (e.g., temporal representation of sound frequencies). Fourth, vesicles are organized around intriguing electron-dense structures called presynaptic ribbons or dense bodies, also found in the high-functioning synapses of photoreceptors and bipolar cells.

Vertebrate hair cells, unlike other sensory cells, receive direct feedback from brain neurons. As described in Chapter 6, we now have a good idea of the physiological effect of efferents on afferent auditory responses (de-tuning) and of the underlying molecular mechanisms. Unraveling the latter involved discovering a new kind of cholinergic receptor and novel coupling to a special class of potassium channels (Fuchs and Murrow 1992). These mechanisms have been worked out in avian and mammalian cochleas, in which efferent feedback is strongly directed to hair cells that are not the principal sources of afferent information. Somehow, efferent feedback to one class of hair cells (outer hair cells in mammals and short hair cells in chicks) is communicated to a second class of hair cell (inner and tall hair cells, respectively).

In the mammalian cochlea, this occurs via a special electromechanical transduction process—somatic "electromotility"—in outer hair cells (Brownell et al. 1985). Electrically evoked changes in force generation by outer hair cells feed into cochlear mechanics, affecting the afferent signals produced by the inner hair cells. The somatic electromotility of outer hair cells, although sensitive to efferent-mediated voltage changes in the hair cells, is also driven by the outer hair cell's own receptor potential. Indeed, it has been argued that positive feedback from this mechanism onto cochlear mechanics is critical to the unique sensitivity of mammalian cochleas to high sound frequencies (> 10 kHz). A membrane protein essential for the electromotility has been identified, but, as Bill Brownell describes in Chapter 7, how that protein works in its lipid environment and how the hair cell overcomes the effects of low-pass filtering of membrane voltage at high sound frequencies remain mysterious.

In Chapter 8, Eatock and Lysakowski turn to the low end of the frequency spectrum, occupied by hair cells in vestibular organs. Vestibular hair cells provide input about head movements to brainstem reflexes that control eye, head, and body position. Chapter 8 presents evidence that mammalian vestibular epithelia represent head movement frequencies in a coarse zonal map, quite different from the finely graded representations of sound frequency along cochleas. A prominent feature of amniote vestibular epithelia is the type I hair cell, which has at least two highly unusual properties whose significance is not understood. It is enveloped by a large cup-shaped afferent ending (Wersäll 1956)—the only postsynaptic calyx known—and it expresses an enormous potassium conductance that is significantly activated at resting potential (Correia and Lang 1990).

In summary, this volume reviews our current understanding of hair cell development, mechanoelectrical and electromechanical transduction, ion channel signaling, excitatory synaptic transmission, and efferent feedback. The next section provides context for contemporary results with a selective review of historical developments.

2. A Brief History of Research on Vertebrate Hair Cells

The earliest anatomical studies of hair cells coincide with the development of histological methods for light microscopy in the 1800s and are linked to studies of the mammalian cochlea, which received then, as now, the lion's share of the attention devoted to inner ear organs. Hair cells were apparently first described by Alfonso Corti in 1851 in his famous light microscopic observations of the sensory epithelium (now the "organ of Corti") of the mammalian cochlea, although he did not call them hair cells (Hawkins 2004). Leydig may have been the first to describe stereocilia (Gitter and Preyer 1991). Deiters (1860, cited by Schacht and Hawkins 2004) actually referred to cochlear hair cells as "Corti cells" and called the support cells that now bear his name "hair cells." Between

1860 and the early 1880s, hair cells acquired their name, appearing as "Haarzellen" in Gustaf Retzius' remarkable illustrations of inner ear anatomy (Retzius 1881, 1884).

The first recordings of hair cell activity were made inadvertently by Wever and Bray (1930), who were attempting to record auditory nerve activity but picked up instead an extracellular trace of the transduction currents flowing across many hair cells, which became known as the cochlear "microphonic"[1] potential. Through the 1960s, microphonic potentials were recorded from preparations as diverse as lizard auditory organs and salamander lateral line organs (reviewed by Flock 1971). The electron microscope came into widespread use and the orderly beauty of hair cell epithelia made them favored subjects of study (Spoendlin 1970). Tract-tracing methods revealed the complex innervation pattern of the mammalian cochlea, with most afferent fibers arising from the single row of inner hair cells (Spoendlin 1972) and a strong efferent projection onto the three rows of outer hair cells (Warr 1978).

Key findings established by these pioneering experiments include that hair cells are directional, responding to the stimulus component parallel to the bundle's axis of bilateral symmetry (its "orientation"), and that the receptor potential is a saturating, asymmetric function of bundle displacement or deflection (Flock 1971). From the 1970s to 1990s, the advance of single-cell recording and electron microscopic methods permitted experiments that established hair cells as model cells for studies of mechanotransduction and the roles of voltage- and Ca^{2+}-gated ion channels in signaling, synaptic release, and efferent feedback.

2.1 Hair Cells as Models of Mechanotransduction

Beginning in 1970, investigators succeeded in using sharp micropipettes to record intracellular receptor potentials from hair cells in the lateral lines of amphibians and fish (Harris et al. 1970; Flock and Russell 1976), the auditory papilla of the alligator lizard (Mulroy et al. 1974), and the cochlea of the guinea pig (Russell and Sellick 1977). The latter experiments demonstrated sharp frequency selectivity in auditory hair cells. In the mid-to-late 1970s, these in vivo approaches to hair cell transduction began to give way to in vitro experiments. Breakthroughs were made by investigators trained in phototransduction, where in vitro methods for stimulating and recording from isolated sensory cells were far advanced (Hudspeth and Corey 1977; Fettiplace and Crawford 1978). Fettiplace and Crawford developed a turtle preparation that maintains the integrity of the sound transmission pathway while providing microelectrode access to auditory afferents and hair cells. They showed that turtle auditory hair cells have sharp electrical tuning that accounts for much of their acoustic tuning (Crawford and Fettiplace 1981).

---

[1]See a biographical memoir of Wever by J. Vernon at www.nap.edu/readingroom/books/biomems.ewever.html for a description of how a comment by E.D. Adrian led to this nomenclature.

Hudspeth and Corey (1977) went more fully in vitro, excising the frog saccular epithelium, pinning it out in an experimental chamber and viewing it with high-power light microscopy. Because hair cells, unlike photoreceptors, are exquisitely mechanosensitive, many within the cochlear field were skeptical that in vitro approaches could work; surely the large inadvertent mechanical stimuli during preparation would destroy the mechanotransduction process. But hair cells have proven remarkably resilient. Figure 1 in Shepherd and Corey (1994) demonstrates how transduction and adaptation survive a bundle deflection many times the size of the cell's instantaneous operating range. With the otolithic gel layer in place, extracellular recordings could be made of summed hair cell responses to deflections of the gel layer (Corey and Hudspeth 1979, 1983). For intracellular recordings, judicious use of protease facilitated removal of the gel to expose the apical hair-bundle-bearing surface of the epithelium. Now it was possible to bring in a fine glass probe to deflect the hair bundle and an electrolyte-filled micropipette for recording from the hair cell.

The development of experiments on semi-intact epithelia permitted rapid advances in our documentation of mechanotransduction, so that our knowledge of the biophysics of hair cell transduction vastly outstrips our understanding of mechanotransduction in vertebrate somatosensors or any invertebrate mechanosensors. Only the bacterial mechanosensing channel, MscL, for which there is crystal structure, is better understood (Sukharev and Corey 2004). Several technical aspects were important to the success of early in vitro hair cell experiments. The large exposed apical surface of hair cells in the pinned-out frog saccular macula and the use of bent microelectrodes (Hudspeth and Corey 1978) permitted penetration of the hair cell with minimal damage. To deflect hair bundles or otolithic gels in a controlled, calibrated fashion, Corey and Hudspeth (1980) affixed glass pipettes to piezoelectric wafers that moved rapidly and linearly with applied voltage; probe motion was monitored with photodiodes. These advances made it possible to measure hair cell transducer functions (membrane voltage as a function of bundle displacement) and the latency of transduction (tens of microseconds), and to show that the receptor potential adapts during maintained bundle deflections. From the very short latency of transduction and the shape of the transducer function, the gating spring model of mechanotransduction was developed (Corey and Hudspeth 1983). This model continues to dominate our thinking about how hair bundle deflection produces a transducer current and has strongly influenced models of mechanotransduction in other specialist mechanosensors, such as the fly's insect bristle organ (Gillespie and Walker 2001) and the nematode's soft-touch sensors (Chalfie 1997).

In vitro stimulating methods were adapted in the 1980s to test the mechanical properties of hair bundles. Bundle motions were monitored optically and probes were made wispier so that their stiffness was comparable to or less than that of the bundle. The bundle's motion in response to a force delivered by a probe of known stiffness yields, by Hooke's law, the bundle's stiffness. Such experiments on frog saccular and turtle cochlear hair cells provided substantial insights into the hair cell's mechanosensitivity. It was directly shown that bundles transduce

Brownian motions of endolymph (Denk et al. 1989). The transduction apparatus was found to be a major component of total bundle stiffness. Remarkably, gating of the transduction channels substantially softens the bundle, producing a large "gating compliance" analogous to the gating current of voltage-gated channels (Howard and Hudspeth 1988). Fast and slow phases of adaptation were shown to have opposite mechanical correlates, with fast adaptation stiffening the bundle and slow adaptation relaxing it (Howard and Hudspeth 1987). When calcium levels are matched to in vivo levels, both adaptation phases can produce resonances of bundle motion (Benser et al. 1996; Ricci et al. 2000). Thus, what appears to be adaptation in some conditions may actually reflect tuning processes that amplify bundle motion over a restricted frequency range (i.e., stereociliary amplifiers).

From the 1970s to the 1990s, when biochemical approaches were cracking open the molecular cascade of phototransduction in vertebrate photoreceptors, the molecules of mechanotransduction in hair cells remained mostly biophysical constructs. Unlike photoreceptors, hair cells have a direct transduction mechanism that relies on relatively few molecules and is therefore resistant to large-scale biochemistry. Compare, for example, the numbers of functioning transduction channels per sensory cell: at most hundreds per hair bundle versus up to a million cyclic-nucleotide-gated transduction channels per rod outer segment (Yau and Baylor 1989). Thus, known molecular constituents of vertebrate hair cells were confined to abundant cytoskeletal elements that could be recognized by immunocytochemical and biochemical methods (Drenckhahn et al. 1985). Investigators speculated about the identity of just a handful of transduction molecules. Were the transduction channels members of the DEG/ENaC family of amiloride-sensitive Na^+ channels, based on genetic work implicating such channels in soft-touch sensation in the nematode (Chalfie and Sulston 1981)? Did a complex of unconventional myosin molecules connect transduction channels to the actin core of the stereocilium (Gillespie and Hudspeth 1994)?

Beginning in the 1990s, however, genetic studies began identifying bundle proteins at an impressive rate (Fig. 1.1C). Such studies begin with animals with spontaneous or experimentally induced genetic mutations that have a mechanosensory phenotype. Methods include behavioral screens of randomly mutagenized animals (Kernan and Zuker 1995) and genetic and anatomical analysis of spontaneous hearing and balance mutants in animals and humans (Brown and Steel 1994). A screen in fruitfly identified the probable mechanotransduction channel for the fly bristle organ (Walker et al. 2000) as NOMPC, a member of the TRP superfamily of ion channels. The founding member of this family is the "transient receptor potential" channel, isolated in a screen of mutagenized fruitflies for vision problems (Hardie and Minke 1993). Now, six TRP subfamilies are known and most have channel representatives implicated in cellular stretch sensitivity or in thermo-, chemo-, and noci-sensing (Moran et al. 2004; Maroto et al. 2005). TRP channels are permeable to monovalent cations and Ca^{2+}, allowing them to carry large currents and to use Ca^{2+} as a feedback signal.

Inspired by the fruitfly work, investigators have proposed a zebrafish orthologue of NOMPC and a related channel called TRPA1 as candidates for hair cell transduction channel(s) (Sidi et al. 2003; Corey et al. 2004). Biophysical studies on turtle cochlear hair cells show variations in transduction channels that correlate with best acoustic frequency (Ricci et al. 2003). With the recent identification of TRP candidates for these channels, we can anticipate molecular studies to identify variants that affect transduction channel tuning. Such variants might affect the channel's Ca^{2+} sensitivity; both the frequency and quality of stereociliary tuning depend strongly on Ca^{2+}.

Our view of gating of the transduction channel is similarly being refined by genetic approaches. Until very recently, the tip link, an extracellular link between stereocilia in adjacent rows, was a favored candidate for the gating spring (Pickles and Corey 1992). But a behavioral screen of mutagenized zebrafish and a spontaneously occurring mouse mutation independently led to the identification of the tip link protein as a Ca^{2+}-dependent adhesion molecule, cadherin 23 (Siemens et al. 2004; Söllner et al. 2004), which may be too stiff to be the elastic gating element (Corey and Sotomayor 2004). An alternative suggested by the structure of NOMPC and TRPA1 channels is that the gating spring comprises a helical concatenation of ankyrin moieties on cytoplasmic N termini of the putative transduction channels (Howard and Bechstedt 2004).

Another example of how genetic approaches can refine or transform the view painted by biophysical experiments concerns the kinocilium, the single true cilium at the tall edge of many hair bundles (Fig. 1.1). The nonessentiality of the kinocilium for transduction is evident from its resorption during maturation of mammalian and avian cochleas (Engström and Engström 1978; Hirokawa 1978). Even in vestibular bundles, where kinocilia persist throughout life, transduction appears to be relatively unaffected by extirpation of the kinocilium (Hudspeth and Jacobs 1979). But new results locate putative mechanosensitive channels in kinocilia as well as stereocilia, suggesting that kinocilia do transduce (Corey et al. 2004). This is an attractive possibility: cilia with sensory functions include photoreceptor outer segments, olfactory cilia, and the mechanosensitive cilia of invertebrate hair cells. There is even evidence for a mechanosensory function in the primary cilia of cells in nonmechanosensory epithelia (Pazour and Whitman 2003). Kinocilia originate as centrally located primary cilia in developing hair cells (Tilney et al. 1992); could cilial mechanosensitivity play a role in bundle development?

A mouse mutation in the tip link protein, cadherin 23, produces disheveled hair bundles and the charmingly named balance disorder, the waltzer (Di Palma et al. 2001). A number of other hearing and balance disorders arise from defects in hair bundles, accessory structures, and their connections. Analyses of mouse and human mutants have implicated more than a dozen proteins in hair bundle formation, including cytoskeletal proteins, adhesion proteins (including cadherin 23), scaffolding proteins, and intracellular motors (myosins), leading to complex cartoons (Müller and Littlewood-Evans 2001; Boeda et al. 2002; Frolenkov et al. 2004). Once a protein is identified by genetic analysis of an animal with a hair bundle defect, placement of the protein in the cartoon depends on its homology with proteins of known function, at least until ultrastructural immunocytochemistry pinpoints its location. For many bundle proteins, it has been difficult to separate possible functions in the mature bundle from developmental functions, because failure to form a perfect bundle may produce complete loss of transduction capability. Two exceptional studies, however, have shown specific and interesting effects of myosin mutations. Naturally occurring myosin VII mutations produce bundles that are severely disheveled but can transduce, at least early in postnatal development (Kros et al. 2002). The operating range is strongly shifted relative to normal hair bundles and adaptation is exaggerated. In experiments on myosin 1c, the putative motor molecule of slow adaptation, Gillespie and colleagues designed a transgenic mouse expressly for patch-clamp experiments that could functionally knock out myosin 1c during recording (Holt et al. 2002). The manipulation eliminated the slow component of adaptation, confirming an essential role for myosin 1c.

Genetic analyses have also, unsurprisingly, uncovered mutations that affect accessory structures, such as tectorial membranes and the calcareous otoconia that move during linear accelerations (Jones et al. 2004). This work is renewing interest in these heretofore underappreciated components and their role in shaping the input to transduction.

2.2 Hair Cells as Models of Voltage-Dependent Signaling

In 1981, the first tight-seal patch-clamp recordings of membrane current were reported (Hamill et al. 1981). Hair cells, unlike many mechanosensors, are compact epithelial cells and well suited to the whole-cell mode of patch clamping in which total membrane voltage is controlled while current is recorded (voltage clamp recordings) or total membrane current is controlled while voltage is recorded (current clamp recordings). Patch-clamp methods were quickly adapted to hair cells (Lewis and Hudspeth 1983; Ohmori 1984; Art and Fettiplace 1987). This approach revealed that both frog saccular and turtle cochlear hair cells have large complements of voltage- and Ca^{2+}-gated ion channels that work in concert to tune the receptor potential (Hudspeth and Lewis 1988a,b; Wu et al. 1995). With such observations, the hair cell field took the lead in showing the importance of secondarily activated (non-transduction) ion channels for sensory signaling. This development owes much to the inspired choices of preparations made by Jim Hudspeth and Robert Fettiplace and colleagues. Frog saccular hair cells and turtle cochlear hair cells are particularly robust, not just in semi-intact preparations, as originally intended, but also as dissociated, solitary cells. Moreover, both epithelia are tuned to frequencies between 10 and 1000 Hz, a range suited for electrical tuning. Chick auditory hair cells also show electrical tuning (Fuchs et al. 1988), probably up to several kilohertz after correction for temperature effects. Experiments in chick and turtle auditory papillae continue to explore the general question of how to achieve tonotopic electrical tuning, that is, graded variation in tuning frequency with position along the organ. The answer is not fully in but involves variation in both channel number and in the Ca^{2+} sensitivity of the K^+ channels, the latter mediated partly by splice variants in the pore-forming α subunit and partly by differential expression of β subunits (Fettiplace and Fuchs 1999).

Modulation of voltage- and Ca^{2+}-gated conductances is too slow to work in the frequency range above 10 kHz, to which mammals alone are sensitive. In the late 1970s, receptor potential recordings from guinea pig inner hair cells (Russell and Sellick 1977) showed that they are sharply tuned. This tuning, unlike electrical tuning, was not private to the inner hair cell but rather depended on intact outer hair cells (Ryan et al. 1979). The unknown process by which outer hair cells remotely boost the tuning of inner hair cells was labeled the "cochlear amplifier" (Davis 1983). A potential amplifier emerged from early experiments looking at voltage-dependent behavior of ultralong outer hair cells isolated from the guinea pig cochlea (Brownell et al. 1985). Remarkably, the hair cells changed shape as membrane potential was changed. (Again, the choice of preparation was critical, in that the smaller movements of the shorter outer hair cells of other species might have gone unnoticed.) Was this "electromotility" the cochlear amplifier? Gold (1948) had proposed that a piezoelectric-like feedback could help overcome viscous damping in the cochlea at high frequencies.

The observation by Brownell and colleagues has stimulated many investigations, some incorporating outer hair cell electromotility into models of cochlear mechanics and others aimed at the underlying biophysical mechanism. Drawing from models of voltage sensing by voltage-gated channels, Dallos et al. (1993) suggested that electromotility reflects shape changes in a voltage-sensitive protein expressed at high density in the long basolateral membrane. This hypothesis received strong support with the isolation of prestin, an outer hair cell protein that can induce electromotility when transfected into other cell types (Zheng et al. 2000a) and whose expression is required for high auditory sensitivity (Liberman et al. 2002).

How does electromotility, which depends on membrane voltage, work up to frequencies so high that membrane voltage should hardly change at all? There are two classes of proposed solution. Investigators of outer hair cell electromotility tend to propose mechanisms to counter its expected high-frequency roll-off (Brownell, Chapter 7, this volume). Investigators of mechanoelectrical transduction argue that somatic electromotility can work only at low frequencies, with Ca^{2+}-driven feedback intrinsic to transduction (the stereociliary amplifier) dominating at high frequencies (Jaramillo et al. 1993; Chan and Hudspeth 2005; Kennedy et al. 2005).

2.3 Hair Cells as Models of Synaptic Transmission

Early on, the patch-clamp method was adapted to measure membrane capacitance changes that reflect fusion of large numbers of vesicles in secretory chromaffin cells (Lindau and Neher 1988). The technique was not easily applied to neurons, in which vesicle fusion usually occurs at small terminals that are remote from the site of recording in the soma. In hair cells, however, the soma is actually a compact presynaptic terminal. In the early 1990s, the capacitance recording method was applied simultaneously to frog saccular hair cells (Parsons et al. 1994) and retinal bipolar cells (von Gersdorff and Matthews 1994). The method has since been used profitably on mouse and chick cochlear hair cells to probe the dynamics of Ca^{2+}-mediated events in vesicle fusion at ribbon synapses (Moser and Beutner 2000; Eisen et al. 2004). Capacitance changes measured in this way are averaged over all the active zones of the hair cell. Recently, Glowatzki and Fuchs (2002) developed methods to study transmission at a single ribbon synapse by recording excitatory postsynaptic currents from individual afferent terminals on inner hair cells.

Neurochemical studies have suggested that efferent neurons release multiple transmitters in the inner ear. In the best-studied effect, efferents release acetylcholine (ACh) onto hair cells, evoking a fast inhibitory postsynaptic potential (IPSP). This was shown in the 1970s and early 1980s in lateral line hair cells (Flock and Russell 1976) and in guinea pig and turtle cochlear hair cells (Brown et al. 1983; Art et al. 1984). In the latter, the IPSPs reduce the sensitive response at best frequency—that is, they de-tune the cell. But the underlying mechanism was a puzzle: ACh was known to gate receptor-channels that should depolarize the cell, not hyperpolarize it. The puzzle was solved a decade later by experiments on short hair cells isolated from the chick cochlea (Fuchs and Murrow 1992) which, like mammalian outer hair cells, receive a strong cholinergic efferent input. ACh puffed onto hair cells while they were patch clamped was shown to open a pore-forming ACh receptor with high Ca^{2+} permeability and odd pharmacology, now known to be of unique molecular composition (Chapter 6). The incoming ions, including Ca^{2+}, briefly depolarize the cell before activating Ca^{2+}-dependent K^+ channels, which produce a much larger IPSP.

3. Summary and Future

By the mid-1990s, microelectrode and patch-clamp methods, biochemical and anatomical experiments had revealed first-order physiological properties of hair cells, from transduction through voltage-dependent processing to afferent transmission and efferent modulation. These studies established that transduction channels are located near the tips of stereocilia, that their gating is fast and direct, and that myosin is involved in a major Ca^{2+}-mediated feedback mechanism. A number of post-transduction specializations important for different aspects of sensory signaling by the inner ear had been identified. In several model hair cells, a voltage-gated Ca^{2+} conductance and a large Ca^{2+}-activated K^+ conductance were known to work together to create electrical tuning. The Ca^{2+} conductance had been shown to be well suited to rapid and sustained transmitter release, even near resting potential. A second Ca^{2+}-activated K^+ conductance was known to couple to special ACh receptors to create the efferent-evoked IPSP.

Since 1990; increases in the numbers and kinds of hair cell preparations (and investigators) have revealed greater diversity than was originally appreciated. For example, mammalian hair cells have substantial outwardly rectifying basolateral K^+ conductances at resting potential (Correia and Lang 1990; Housley and Ashmore 1992), unlike the turtle and chick cochlear and frog saccular hair cells that dominated early investigations. Also, technical improvements have substantially altered the biophysical picture. Both the stereociliary amplifier and electrical tuning have been particularly vulnerable to recording conditions. In early microelectrode recordings, adaptation was not always seen and when it was, the use of high extracellular Ca^{2+}, a trick to improve membrane stability, made it both faster and more damped than is likely to be the case in vivo. Slight changes in resting potential can eliminate spontaneous evidence of electrical tuning, while enzymes used to dissociate hair cells can alter the properties of basolateral ion channels, modifying the electrical tuning frequency (Armstrong and Roberts 1998).

The transforming technical developments of the past decade, however, have been molecular and genetic. These can be most immediately productive when they identify candidate proteins for known functions, such as ion channels carrying voltage-gated, neurotransmitter-gated, and transduction currents; motor proteins that mediate adaptation or somatic electromotility; and proteins that make cytoskeletal or extracellular structures. Genetic approaches often yield proteins with a phenotype but not a clearly identified cellular function. Investigators first localize expression of the proteins and develop ideas by inference and analogy with related proteins, then make null mutants to which they apply behavioral, anatomical, and physiological assays. Null mutant animals can show complex effects reflecting developmental processes as well as mature functions of the targeted molecule (e.g., the ACh receptor channels of outer hair cells; Vetter et al. 1999). To understand the adult function of the protein may require more pointed and transient knockdowns, achievable by combining the cre-recombinase method with hair cell-specific promoters (Li et al. 2004), or by adenoviral transfections, small interfering RNAs or morpholinos (Sidi et al. 2003; Corey et al. 2004).

One future direction is toward a refined understanding of developmental acquisition of sensory properties. Once the key biophysical and molecular differences between hair cells of distinct tuning frequency are characterized, a natural question is how such distinctions are achieved at the level of gene translation. How does a hair bundle know how to be a micrometer taller than its neighbor? How do K^+ channels know to be slightly faster than those next door? What property of a type I vestibular hair cell encourages calyx formation by the primary afferent?

An equally important direction is toward a comprehensive understanding of how hair cell properties shape higher level sensory processing. Because of technical differences in preparations (in vitro vs. in vivo), and in part because of cultural gaps between molecular, cellular, and systems-oriented scientists, we are rather poorly informed about the connection between hair cell properties and behavior, except at the crudest level. For example, in the mammalian vestibular system, comprehensive understanding of the organization of hair cell epithelia is hampered by a lack of such basic anatomical data as the central targets of different epithelial zones. Also, we know little of how hair bundles move in vivo, an essential link between known hair cell properties and the well-documented responses of eighth-nerve afferents to sounds or complex head movements. Recent technical achievements in this direction include the preservation of accessory structures in experiments on hair cells in excised inner ear organs (He et al. 2004; Chan and Hudspeth 2005) and increasingly sensitive measures of the in vivo performance of inner ear organs (Fridberger et al. 2004).

References

Armstrong CE, Roberts WM (1998) Electrical properties of frog saccular hair cells: distortion by enzymatic dissociation. J Neurosci 18:2962–2973.

Art JJ, Fettiplace R (1987) Variation of membrane properties in hair cells isolated from the turtle cochlea. J Physiol 385: 207–242.

Art JJ, Fettiplace R, Fuchs PA (1984) Synaptic hyperpolarization and inhibition of turtle cochlear hair cells. J Physiol 356:525–550.

Benser ME, Marquis RE, Hudspeth AJ (1996) Rapid, active hair bundle movements in hair cells from the bullfrog's sacculus. J Neurosci 16:5629–5643.

Block SM (1992) Biophysical principles of sensory transduction. In: Corey DP, Roper SD (eds), Sensory Transduction. New York: The Rockefeller University Press, pp. 1–17.

Boeda B, El Amraoui A, Bahloul A, Goodyear R, Daviet L, Blanchard S, Perfettini I, Fath KR, Shorte S, Reiners J, Houdusse A, Legrain P, Wolfrum U, Richardson G, Petit C (2002) Myosin VIIa, harmonin and cadherin 23, three Usher I gene products that cooperate to shape the sensory hair cell bundle. EMBO J 21:6689–6699.

Brown MC, Nuttall AL, Masta RI (1983) Intracellular recordings from cochlear inner hair cells: effects of stimulation of the crossed olivocochlear efferents. Science 222: 69–72.

Brown SD, Steel KP (1994) Genetic deafness—progress with mouse models. Hum Mol Genet 3:1453–1456.

Brownell WE, Bader CR, Bertrand D, de Ribaupierre Y (1985) Evoked mechanical responses of isolated cochlear outer hair cells. Science 227:194–196.

Bryant J, Goodyear RJ, Richardson GP (2002) Sensory organ development in the inner ear: molecular and cellular mechanisms. Br Med Bull 63:39–57.

Chalfie M (1997) A molecular model for mechanosensation in *Caenorhabditis elegans*. Biol Bull 192:125.

Chalfie M, Sulston J (1981) Developmental genetics of the mechanosensory neurons of *Caenorhabditis elegans*. Dev Biol 82:358–370.

Chan DK, Hudspeth AJ (2005) Ca^{2+} current-driven nonlinear amplification by the mammalian cochlea in vitro. Nat Neurosci 8:149–155.

Chang W, Brigande JV, Fekete DM, Wu DK (2004) The development of semicircular canals in the inner ear: role of FGFs in sensory cristae. Development 131:4201–4211.

Coffin A, Kelley MW, Manley GA, Popper AN (2004) Evolution of sensory hair cells.

In: Manley GA, Popper AN, and Fay RR (eds), Evolution of the Vertebrate Auditory System. New York: Springer-Verlag, pp. 55–94.

Corey DP, Hudspeth AJ (1979) Response latency of vertebrate hair cells. Biophys J 26: 499–506.

Corey DP, Hudspeth AJ (1980) Mechanical stimulation and micromanipulation with piezoelectric bimorph elements. J Neurosci Methods 3:183–202.

Corey DP, Hudspeth AJ (1983) Kinetics of the receptor current in bullfrog saccular hair cells. J Neurosci 3:962–976.

Corey DP, Sotomayor M (2004) Hearing: tightrope act. Nature 428:901–903.

Corey DP, Garcia-Añoveros J, Holt JR, Kwan KY, Lin SY, Vollrath MA, Amalfitano A, Cheung EL, Derfler BH, Duggan A, Géléoc GS, Gray PA, Hoffman MP, Rehm HL, Tamasauskas D, Zhang DS (2004) TRPA1 is a candidate for the mechanosensitive transduction channel of vertebrate hair cells. Nature 432:723–730.

Correia MJ, Lang DG (1990) An electrophysiological comparison of solitary type I and type II vestibular hair cells. Neurosci Lett 116:106–111.

Crawford AC, Fettiplace R (1981) An electrical tuning mechanism in turtle cochlear hair cells. J Physiol 312:377–422.

Dallos P, Hallworth R, Evans BN (1993) Theory of electrically driven shape changes of cochlear outer hair cells. J Neurophysiol 70:299–323.

Davis H (1983) An active process in cochlear mechanics. Hear Res 9:79–90.

Denk W, Webb WW, Hudspeth AJ (1989) Mechanical properties of sensory hair bundles are reflected in their Brownian motion measured with a laser differential interferometer. Proc Natl Acad Sci USA 86:5371–5375.

Di Palma F, Holme RH, Bryda EC, Belyantseva IA, Pellegrino R, Kachar B, Steel KP, Noben-Trauth K (2001) Mutations in Cdh23, encoding a new type of cadherin, cause stereocilia disorganization in waltzer, the mouse model for Usher syndrome type 1D. Nat Genet 27:103–107.

Drenckhahn D, Schäfer T, Prínz M (1985) Actin, myosin, and associated proteins in the vertebrate auditory and vestibular organs. Immunochemical and biochemical studies. In: Drescher DG (ed), Auditory Biochemistry. Springfield, IL: Charles C. Thomas, pp. 312–335.

Eisen MD, Spassova M, Parsons TD (2004) Large releasable pool of synaptic vesicles in chick cochlear hair cells. J Neurophysiol 91:2422–2428.

Engström H, Engström B (1978) Structure of the hairs on cochlear sensory cells. Hear Res 1:49–66.

Fettiplace R, Crawford AC (1978) The coding of sound pressure and frequency in cochlear hair cells of the terrapin. Proc R Soc Lond B 203:209–218.

Fettiplace R, Fuchs PA (1999) Mechanisms of hair cell tuning. Annu Rev Physiol 61: 809–834.

Flock A (1971) Sensory transduction in hair cells. In: Loewenstein WR (ed), Handbook of Sensory Physiology, Vol. 1. Berlin: Springer-Verlag, pp. 396–441.

Flock A, Russell I (1976) Inhibition by efferent nerve fibres: action on hair cells and afferent synaptic transmission in the lateral line canal organ of the burbot *Lota lota*. J Physiol 257:45–62.

Fridberger A, de Monvel JB, Zheng J, Hu N, Zou Y, Ren T, Nuttall A (2004) Organ of corti potentials and the motion of the basilar membrane. J Neurosci 24:10057–10063.

Frolenkov GI, Belyantseva IA, Friedman TB, Griffith AJ (2004) Genetic insights into the morphogenesis of inner ear hair cells. Nat Rev Genet 5:489–498.

Fuchs PA, Murrow BW (1992) Cholinergic inhibition of short (outer) hair cells of the chick's cochlea. J Neurosci 12:800–809.

Fuchs PA, Nagai T, Evans MG (1988) Electrical tuning in hair cells isolated from the chick cochlea. J Neurosci 8:2460–2467.

Gillespie PG, Hudspeth AJ (1994) Pulling springs to tune transduction: adaptation by hair cells. Neuron 12:1–9.

Gillespie PG, Walker RG (2001) Molecular basis of mechanosensory transduction. Nature 413:194–202.

Gitter AH, Preyer S (1991) [A brief history of hearing research. III. icroscopic anatomy]. Laryngorhinootologie 70:417–421.

Glowatzki E, Fuchs PA (2002) Transmitter release at the hair cell ribbon synapse. Nat Neurosci 5:147–154.

Gold T (1948) Hearing. ii. The physical basis of the action of the cochlea. Proc R Soc Lond [Biol] 135:492–498.

Hamill OP, Marty A, Neher E, Sakmann B, Sigworth FJ (1981) Improved patch-clamp techniques for high-resolution current recording from cells and cell-free membrane patches. Pflugers Arch 391:85–100.

Hardie RC, Minke B (1993) Novel Ca^{2+} channels underlying transduction in *Drosophila* photoreceptors: implications for phosphoinositide-mediated Ca^{2+} mobilization. Trends Neurosci 16:371–376.

Harris GG, Frishkopf LS, Flock A (1970) Receptor potentials from hair cells of the lateral line. Sci 167:76–79.

Hawkins JE (2004) Sketches of otohistory. Part 3: Alfonso Corti. Audiol Neurootol 9:259–264.

He DZ, Jia S, Dallos P (2004) Mechanoelectrical transduction of adult outer hair cells studied in a gerbil hemicochlea. Nature 429:766–770.

Hirokawa N (1978) The ultrastructure of the basilar papilla of the chick. J Comp Neurol 181:361–374.

Hirono M, Denis CS, Richardson GP, Gillespie PG (2004) Hair cells require phosphatidylinositol 4,5-bisphosphate for mechanical transduction and adaptation. Neuron 44:309–320.

Holt JR, Gillespie SK, Provance DW, Shah K, Shokat KM, Corey DP, Mercer JA, Gillespie PG (2002) A chemical-genetic strategy implicates myosin-1c in adaptation by hair cells. Cell 108:371–381.

Housley GD, Ashmore JF (1992) Ionic currents of outer hair cells isolated from the guinea-pig cochlea. J Physiol 448:73–98.

Howard J, Bechstedt S (2004) Hypothesis: a helix of ankyrin repeats of the NOMPC-TRP ion channel is the gating spring of mechanoreceptors. Curr Biol 14:R224–R226.

Howard J, Hudspeth AJ (1987) Mechanical relaxation of the hair bundle mediates adaptation in mechanoelectrical transduction by the bullfrog's saccular hair cell. Proc Natl Acad Sci USA 84:3064–3068.

Howard J, Hudspeth AJ (1988) Compliance of the hair bundle associated with gating of mechanoelectrical transduction channels in the bullfrog's saccular hair cell. Neuron 1:189–199.

Hudspeth AJ, Corey DP (1977) Sensitivity, polarity, and conductance change in the response of vertebrate hair cells to controlled mechanical stimuli. Proc Natl Acad Sci USA 74:2407–2411.

Hudspeth AJ, Corey DP (1978) Controlled bending of high-resistance glass microelectrodes. Am J Physiol 234:C56–C57.

Hudspeth AJ, Jacobs R (1979) Stereocilia mediate transduction in vertebrate hair cells. Proc Natl Acad Sci USA 76:1506–1509.

Hudspeth AJ, Lewis RS (1988a) A model for electrical resonance and frequency tuning in saccular hair cells of the bull-frog, *Rana catesbeiana*. J Physiol 400:275–297.

Hudspeth AJ, Lewis RS (1988b) Kinetic analysis of voltage- and ion-dependent conductances in saccular hair cells of the bull-frog, *Rana catesbeiana*. J Physiol 400:237–274.

Jaramillo F, Markin VS, Hudspeth AJ (1993)Auditory illusions and the single hair cell. Nature 364:527–529.

Jones SM, Erway LC, Johnson KR, Yu H, Jones TA (2004) Gravity receptor function in mice with graded otoconial deficiencies. Hear Res 191:34–40.

Kennedy HJ, Crawford AC, Fettiplace R (2005) Force generation by mammalian hair bundles supports a role in cochlear amplification. Nature 433:880–883.

Kernan M, Zuker CS (1995) Genetic approaches to mechanosensory transduction. Curr Opin Neurobiol 5:443–448.

Kros CJ, Marcotti W, van Netten SM, Self TJ, Libby RT, Brown SD, Richardson GP, Steel KP (2002) Reduced climbing and increased slipping adaptation in cochlear hair cells of mice with Myo7a mutations. Nat Neurosci 5:41–47.

Lewis RS, Hudspeth AJ (1983) Voltage- and ion-dependent conductances in solitary vertebrate hair cells. Nature 304:538–541.

Li M, Tian Y, Fritzsch B, Gao J, Wu X, Zuo J (2004) Inner hair cell Cre-expressing transgenic mouse. Genesis 39:173–177.

Liberman MC, Gao J, He DZ, Wu X, Jia S, Zuo J (2002) Prestin is required for electromotility of the outer hair cell and for the cochlear amplifier. Nature 419:300–304.

Lindau M, Neher E (1988) Patch-clamp techniques for time-resolved capacitance measurements in single cells. Pflügers Arch 411:137–146.

Marcotti W, Johnson SL, Holley MC, Kros CJ (2003) Developmental changes in the expression of potassium currents of embryonic, neonatal and mature mouse inner hair cells. J Physiol 548:383–400.

Maroto R, Raso A, Wood TG, Kurosky A, Martinac B, Hamill OP (2005) TRPC1 forms the stretch-activated cation channel in vertebrate cells. Nat Cell Biol 7:179–185.

Moran MM, Xu H, Clapham DE (2004) TRP ion channels in the nervous system. Curr Opin Neurobiol 14:362–369.

Moser T, Beutner D (2000) Kinetics of exocytosis and endocytosis at the cochlear inner hair cell afferent synapse of the mouse. Proc Natl Acad Sci U S A 97:883–888.

Müller U, Littlewood-Evans A (2001) Mechanisms that regulate mechanosensory hair cell differentiation. Trends Cell Biol 11:334–342.

Mulroy MJ, Altmann DW, Weiss TF, Peake WT (1974) Intracellular electric responses to sound in a vertebrate cochlea. Nature 249:482–485.

Ohmori H (1984) Studies of ionic currents in the isolated vestibular hair cell of the chick. J Physiol 350:561–581.

Parsons TD, Lenzi D, Almers W, Roberts WM (1994) Calcium-triggered exocytosis and endocytosis in an isolated presynaptic cell: capacitance measurements in saccular hair cells. Neuron 13:875–883.

Pazour GJ, Witman GB (2003) The vertebrate primary cilium is a sensory organelle. Curr Opin Cell Biol 15:105–110.

Pickles JO, Corey DP (1992) Mechanoelectrical transduction by hair cells. Trends Neurosci 15:254–259.

Retzius G (1881) Das Gehörorgan der Wirbelthiere. Vol. 1. Stockholm: Samson and Wallin.

Retzius G (1884) Das Gehörorgan der Wirbelthiere. Vol. 2. Stockholm: Samson and Wallin.

Ricci AJ, Crawford AC, Fettiplace R (2000) Active hair bundle motion linked to fast transducer adaptation in auditory hair cells. J Neurosci 20:7131–7142.

Ricci AJ, Crawford AC, Fettiplace R (2003) Tonotopic variation in the conductance of the hair cell mechanotransducer channel. Neuron 40:983–990.

Russell IJ, Sellick PM (1977) Tuning properties of cochlear hair cells. Nature 267:858–860.

Ryan A, Dallos P, McGee T (1979) Psychophysical tuning curves and auditory thresholds after hair cell damage in the chinchilla. J Acoust Soc Am 66:370–378.

Schacht J, Hawkins JE (2004) Sketches of otohistory part 4: a cell by any other name: cochlear eponyms. Audiol Neurootol 9:317–327.

Shepherd GM, Corey DP (1994) The extent of adaptation in bullfrog saccular hair cells. J Neurosci 14:6217–6229.

Sidi S, Friedrich RW, Nicolson T (2003) NompC TRP channel required for vertebrate sensory hair cell mechanotransduction. Sci 301:96–99.

Siemens J, Lillo C, Dumont RA, Reynolds A, Williams DS, Gillespie PG, Müller U (2004) Cadherin 23 is a component of the tip link in hair-cell stereocilia. Nature 428: 950–955.

Söllner C, Rauch GJ, Siemens J, Geisler R, Schuster SC, Muller U, Nicolson T (2004) Mutations in cadherin 23 affect tip links in zebrafish sensory hair cells. Nature 428: 955–959.

Spoendlin H (1970) Vestibular labyrinth. In: Bischoff A (ed), Ultrastructure of the Peripheral Nervous System and Sense Organs. Saint Louis, MO: C.V. Mosby Company.

Spoendlin H (1972) Innervation densities of the cochlea. Acta Otolaryngol (Stockh) 73: 235–248.

Sukharev S, Corey DP (2004) Mechanosensitive channels: multiplicity of families and gating paradigms. Sci STKE 219:re4.

Tilney LG, Tilney MS, Cotanche DA (1988) Actin filaments, stereocilia, and hair cells of the bird cochlea. V. How the staircase pattern of stereociliary lengths is generated. J Cell Biol 106:355–365.

Tilney LG, Tilney MS, DeRosier DJ (1992) Actin filaments, stereocilia, and hair cells: how cells count and measure. Annu Rev Cell Biol 8:257–274.

Vetter DE, Liberman MC, Mann J, Barhanin J, Boulter J, Brown MC, Saffiote-Kolman J, Heinemann SF, Elgoyhen AB (1999) Role of alpha9 nicotinic ACh receptor subunits in the development and function of cochlear efferent innervation. Neuron 23:93–103.

von Gersdorff H, Matthews G (1994) Dynamics of synaptic vesicle fusion and membrane retrieval in synaptic terminals. Nature 367:735–739.

Walker RG, Willingham AT, Zuker CS (2000) A *Drosophila* mechanosensory transduction channel. Sci 287:2229–2243.

Warr WB (1978) The olivocochlear bundle: its origins and terminations in the cat. In: Naunton RF, Fernández C (eds), Evoked Electrical Activity in the Auditory Nervous System. New York: Academic Press, pp. 43–65.

Wersäll J (1956) Studies on the structure and innervation of the sensory epithelium of the crista ampullares in the guinea pig. Acta Otolaryngol 126 (Suppl.):1–85.

Wever EG, Bray CW (1930) Action currents in the auditory nerve in response to acoustical stimulation. Proc Natl Acad Sci USA 16:344–350.

Wu Y-C, Art JJ, Goodman MB, Fettiplace R (1995) A kinetic description of the calcium-activated potassium channel and its application to electrical tuning of hair cells. Prog Biophys Mol Biol 63:131–158.

Yau K-W, Baylor DA (1989) Cyclic GMP-activated conductance of retinal photoreceptor cells. Annu Rev Neurosci 12:289–327.

Zheng J, Shen W, He DZ, Long KB, Madison LD, Dallos P (2000a) Prestin is the motor protein of cochlear outer hair cells. Nature 405:149–155.

Zheng L, Sekerkova G, Vranich K, Tilney LG, Mugnaini E, Bartles JR (2000b) The deaf jerker mouse has a mutation in the gene encoding the espin actin-bundling proteins of hair cell stereocilia and lacks espins. Cell 102:377–385.

2
The Development of Hair Cells in the Inner Ear

RICHARD J. GOODYEAR, CORNÉ J. KROS, AND GUY P. RICHARDSON

1. Introduction

Hair cells are mechanosensitive cells found in the sensory organs of the inner ear and lateral line organs of vertebrates. There are up to eight discrete sensory organs in the inner ears of some species. In the mouse there are six, the three cristae in the ampullae of the semicircular canals that react to rotation, the maculae of the utricle and saccule that detect gravity or linear acceleration, and, in the cochlea, the organ of Corti, a receptor that responds to sound. In the bird, there are two other mechanosensitive epithelia, the lagena macula and the macula neglecta, making a total of eight sensory organs.

Despite considerable variety in both form and function, the different mechanosensory organs of the vertebrate inner ear all conform to the same basic structural plan (see Fig. 2.1). They are relatively simple epithelia of ectodermal origin that rest on an underlying basal lamina. They contain two cell types, sensory hair cells and nonsensory supporting cells, both of which are polarized epithelial cells. The hair cells have a stereociliary bundle, or hair bundle, on their apical surface that is required for stimulus reception and mechanoelectrical transduction, and a basolateral membrane that has a number of different functions, including shaping the receptor potential and controlling neurotransmitter release. The basolateral membranes of the hair cells do not reach the basal lamina, and are contacted by afferent and efferent nerve fibres that penetrate the epithelium through the basal lamina. The supporting cells surround the hair cells, effectively isolating the hair cells from each other. The basal membranes of supporting cells sit upon the basal lamina, and their lateral membranes envelop the hair cells and the axons and their nerve terminals. Apically, the lateral membranes of the supporting cells form adherens and tight junctions with the membranes of adjacent hair cells and other supporting cells. The apical membranes of the hair and supporting cells therefore form a single continuous surface sealed by tight junctions, the apical surface of the sensory epithelium. Most mechanosensory organs have a prominent piece of extracellular matrix contacting this apical surface, either a cupula, an otoconial (or otolithic) membrane, or a tectorial membrane. These matrices are usually attached to both the hair

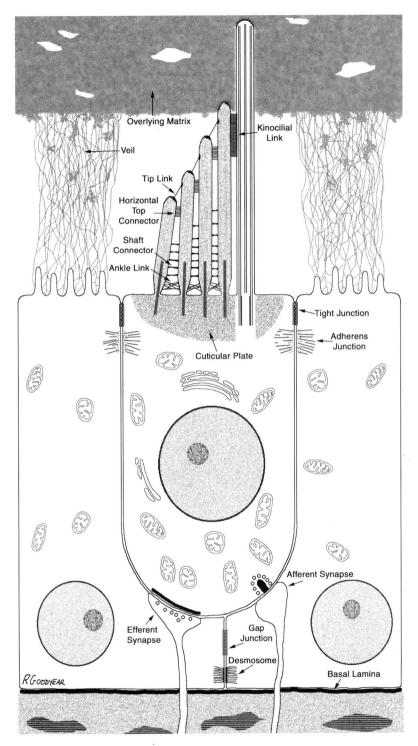

FIGURE 2.1. A diagram illustrating the basic organization of a typical sensory organ in the inner ear, and some of the key cytological features of hair and supporting cells.

bundles and the microvilli on the apical surface of the supporting cells. They serve to deliver the stimulus to the hair bundle, or act as a structure against which it can react.

The inner ear develops from the otic placode, a thickening of the head ectoderm that forms adjacent to the rhombencephalon during the early stages of embryonic development (see Fig. 2.2). Head ectoderm is a simple cuboidal epithelium that sits, like all epithelia, on a basal lamina. As the head ectoderm thickens to form a placode it becomes pseudostratified. The cells become elongated and spindle-shaped but retain contact with both the basal and apical surfaces of the epithelium, with their nuclei staggered at different levels within the thickness of the epithelium. The placode invaginates to form a depression, the otic pit, that sinks in further and eventually pinches off from the overlying ectoderm to form the otic vesicle. The otic vesicle is a hollow, pear-shaped epithelial cyst. It is surrounded by a basal lamina and the apical surfaces of the epithelial cells face the lumen of the cyst. The different sensory organs of the inner ear, and the neurons of the VIIIth ganglion that innervate them, all develop from the pseudostratified epithelial wall of the otocyst. In the first part of this chapter we consider when, where, and how various regions of the otocyst wall become specified to form mechanosensory organs.

Once various regions of the inner ear are specified to become sensory organs, the epithelial cells withdraw from the cell cycle and differentiate along different pathways to form hair cells or supporting cells. In the auditory organs of birds and mammals, the patterns formed by these two cell types are remarkably precise and ordered. In the second part of this chapter we discuss the mechanisms that influence whether a postmitotic cell becomes a hair cell or a supporting cell, and how the patterns observed in the sensory organs are generated.

The hair bundle is a structure located at the apical end of the hair cell that is essential for mechanotransduction. Hair bundles are composed of modified microvilli, actin-packed finger-like projections known as stereocilia that are arranged in rows of increasing height toward one side of the hair cell (see Furness and Hackney, chapter 3). A single, microtubule-containing kinocilium is found in many types of mature hair bundle and, when present, it is located adjacent to, and central with respect to, the tallest row of stereocilia. The actin filaments

FIGURE 2.2. Schematic diagram illustrating some of the key steps in inner ear development and comparing the embryonic age at which they occur in the chick (*left*) and the mouse (*right*). For the chick, the first day of incubation is E0. For the mouse, midday on the day of the vaginal plug is E0.5. The exact timing of events in the chick depends on the temperature at which the eggs are incubated, and the ages quoted are therefore only approximate. Hatching occurs between E20 and E22 in the chick, and mice are born between E18 and E21. Cartoons for the chick (E4 to E9) and mouse (E11.5 to E15) are based on data from inner-ear paint fills published by Bissonnette and Fekete (1996) and Morsli et al. (1998) and are drawn to scale.

2. The Development of Hair Cells in the Inner Ear   23

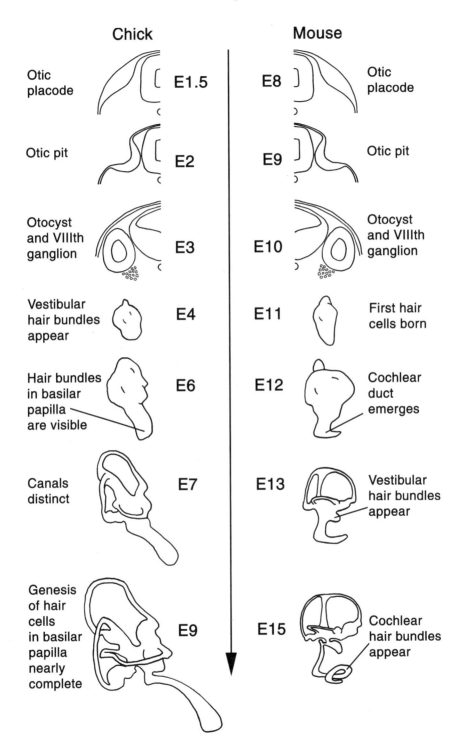

within each stereocilium are packed in a semicrystalline array and are crosslinked by two actin bundling proteins, fimbrin and espin. The actin filaments are all oriented in the same direction with the barbed or + end of each filament (the preferred end for actin monomer addition) located at the tip of each stereocilium. The stereocilia are connected to one another by a number of different link types. Up to four morphologically distinct link types can be recognized on the hair bundles of fish and bird hair cells (see Fig. 2.1). Working from the top of the bundle down there are tip links, horizontal top connectors, shaft connectors, and finally, around the base of the bundle, ankle links. Additional links, kinocilial links, are also present between the kinocilium and the tallest row of stereocilia. Not all hair cells have this entire complement of link types, although all that have been studied to date have tip links. Hair bundles are polarized, bilaterally symmetrical structures, and exhibit maximal sensitivity to deflections parallel to the bundle's plane of mirror symmetry. Deflecting the hair bundle toward the tallest row of stereocilia (or the kinocilium when present) increases the probability of mechanotransducer channel opening, whereas deflections in the opposite direction reduce it. As tip links are oriented along the hair bundle's axis of sensitivity, and are located such that they would apply tension to an attached channel only when the bundle is displaced toward the tallest row of stereocilia, they are thought to gate the mechanotransducer channel. Hair bundles differ in their size and shape, and in the numbers and packing arrangements of their constituent stereocilia, both from organ to organ and within any one organ in a precisely determined fashion. Furthermore, their orientation within any organ is critical for that organ to function correctly. The way in which such a complex piece of cellular machinery as the hair bundle is generated by the hair cell, and how the desired orientation is achieved, is considered in the third section of this chapter.

In addition to mechanotransducer channels, hair cells require a number of other membrane channels and receptors to perform as efficient transducers and transmitters of sensory information to the central nervous system (CNS). Furthermore, outer hair cells in the mammalian cochlea have electromotile properties and are able to generate axial movements of their elongated cell bodies that are thought to amplify the motion of the basilar membrane. In the final section we review what is known about the development of mechanotransduction, basolateral membrane currents, transmitter receptors, and electromotility in hair cells. Although hair cells develop in close association with the axons that innervate them, there is thus far little evidence that neural input influences the development of hair cells. We therefore focus specifically on how hair cells develop. For a recent review on the subject of synaptogenesis and the development of innervation patterns in the inner ear the reader is referred to Rubel and Fritzsch (2002).

Most information concerning the development of hair cells comes from studies that have been done in the mouse and the chick. As a reference point for the descriptions that follow, the developmental stages at which key events in ear development occur during embryogenesis in the two species are compared in Figure 2.2.

2. The Formation of Sensory Patches in the Otocyst

The mechanisms that determine which regions of the otocyst will develop as sensory patches, and what type of sensory organ each patch will become, are still far from being understood. Nonetheless, a potentially testable theory has been proposed (Fekete 1996), and a number of molecular markers are now known that demarcate the regions that will become sensory organs in advance of the birth and overt cytodifferentiation of hair and supporting cells (von Bartheld et al. 1991; Ekker et al. 1992; Oh et al. 1996; Wu and Oh 1996; Morsli et al. 1998; Cole et al. 2000; Goodyear et al. 2001; Mowbray et al. 2001).

Bone morphogenetic protein 4 (BMP4) is a member of the transforming growth factor beta family of secreted signaling molecules known to be involved in interactions between epithelial and mesenchymal tissues. It is a homologue of the *Drosophila* protein decapentaplegic, a molecule involved in many aspects of fly development. In the chick, the expression patterns of the *BMP4* gene provide the earliest indication that discrete sensory patches have appeared within the otocyst (Wu and Oh 1996). *BMP4* is first detected in the medial and posterior margins of the placode, as it begins to invaginate at embryonic day (E)1.5 (stage 11). The *BMP4* expression pattern rapidly changes over the next few stages but it becomes concentrated in two discrete foci at the anterior and posterior poles of the otocyst by E3 (stage 19). The gene for the low-affinity nerve growth factor receptor, p75NGFR, is also expressed in anterior and posterior poles of the chick otocyst at E3 (von Bartheld et al. 1991). The two foci of *BMP4* expression correspond to the future anterior (superior) and posterior cristae, and by E4 (stage 24) most of the eight different organs in the chick inner ear can be defined as discrete sites of *BMP4* expression (Wu and Oh 1996). For some, but not all of the organs, *p75NGFR* and the homeobox-containing gene *Msx-1* are simultaneously co-expressed in the prospective sensory patches at the same early stage as *BMP4*, although the expression domains of *p75NGFR* are usually broader than those observed for *BMP4*, and *Msx-1* is also expressed in nonsensory regions (Wu and Oh 1996). Likewise, although *BMP4* appears relatively specific for prospective sensory patches when first expressed in the chick otocyst, at later stages it is expressed in a number of regions that will develop into nonsensory structures. The expression patterns of *BMP5* and *BMP7* have also been examined in the developing chick inner ear (Oh et al. 1996). *BMP5* is transiently expressed in a pattern similar to that of *BMP4*, but only during the very early stages of otocyst formation. *BMP7* has a more widespread and less discrete expression pattern than *BMP4*, although almost all *BMP4*-expressing prospective sensory patches lie within *BMP7* expression domains. Within the avian auditory organ, the basilar papilla, *BMP4* expression finally becomes restricted to the hair cells, whereas that of *BMP7* becomes localized to the supporting cells. In the vestibular organs, *BMP4* expression is finally restricted to the supporting cells, whereas *BMP7* expression becomes excluded from the sensory epithelium and only continues in the flanking nonsensory epithelial regions.

In the zebrafish, *bmp2b* and *bmp4* mRNAs are concentrated laterally at the anterior and posterior ends of the otocyst by 24 hours post fertilization (hpf) (Mowbray et al. 2001). These *bmp* expression domains may, as in the chick, demarcate the cristae, as the first hair cells generated in the zebrafish inner ear are found at the anterior and posterior ends of the otocyst at 24 hpf (Haddon and Lewis 1996), where the foci of *bmp* expression abut a medial domain of *pax2.1* expression (Mowbray et al. 2001). By 27 hpf, the expression of *bmp2b* and *bmp4* demarcates a third domain that is coincident with that of *msxC*, a homeobox gene that is expressed within the lateral crista (Ekker et al. 1992). Although the *bmp2* and *bmp4* expression patterns mark the developing cristae in the zebrafish as in the chick, these genes are not expressed in the developing maculae of the zebrafish otocyst (Mowbray et al. 2001).

In the mouse, *BMP4* expression also appears to serve as an early marker for the three cristae, but not for the maculae or the organ of Corti (Morsli et al. 1998). For the organ of Corti and the maculae, *Lunatic fringe* (*Lfng*), a modulator of the Notch signaling pathway, is a better marker (Morsli et al. 1998). In the chick, *Serrate1* (*Ser1*), a gene encoding a Notch ligand, and *Lfng* are both specifically expressed within the sensory patches by the time hair-cell differentiation is beginning (E5, stage 26; Adam et al. 1998). At earlier stages (from E2.5+, stage 18), *Lfng* and *Ser1* are expressed within a single, large, ventromedial region of the otocyst that extends from the anterior to the posterior pole of the otocyst (Cole et al. 2000). All of the discrete patches of *BMP4* expression that mark the sites of sensory organ generation fall within or largely overlap with this broad expression domain of *Lfng* and *Ser1*, indicating these latter two molecules may define a ventromedial prosensory patch from which the sensory organs develop (Cole et al. 2000). This prosensory ventromedial patch defined by the expression of *Lfng* and *Ser1* at stage 18 may also be identical to that delineated by the distribution of the cell-cell adhesion molecule BEN (Goodyear et al. 2001). At earlier stages, whilst the placode is invaginating to form the pit, although the expression patterns of *BMP4* and *Ser1* may be quite similar, those of *Lfng* and BEN may be different, possibly from each other and from those of *BMP4* and *Ser1*. However, once the otocyst has formed, the expression patterns of *Ser1*, *Lfng*, and BEN together define a prosensory patch from which the different sensory organs are generated. These recent molecular observations confirm the suggestion (Knowlton 1967) that all sensory organs develop from a *macula communis*—a common ventromedial region of the otocyst.

These studies do not reveal why it is that sensory organs form at the sites that they do, or why each organ finally has a different identity. A boundary model of sensory organ specification has been formulated (Fekete 1996). This model proposes that the otocyst is divided into a relatively small number of compartments, each defined by the differential expression of a few genes and within which the cells remain restricted by lineage and are prevented from mixing with cells in adjacent compartments. Signaling across the boundaries induces the formation of a sensory patch. The location of a sensory patch would depend on its proximity to different boundaries, and its identity on the combination of interacting compartments. The initial compartments may correspond to different parts of the otocyst, like its dorsal or its ventral segments, and the signals that specify these compartments may be derived from adjacent tissues such as the rhombencephalon and the surface ectoderm, or more distant structures such as the notochord. Genes encoding a variety of transcription factors (e.g., *Pax2*, *Dlx5*, *Otx1*, *Eya1*, *Six1*) are expressed within the otocyst in domains consistent with the existence of a number of compartments, and the phenotypes of mice that are null mutants for some of these genes are also consistent with the hypothesis. For example *Pax2*$^{-/-}$ mice lack a cochlear duct (Torres et al. 1996), and the anterior and posterior cristae are missing in *Dlx5*$^{-/-}$ mice, along with their associated ampullae and canals (Acampora et al. 1999). In *Otx1*$^{-/-}$ mice, the lateral crista is missing, also along with its associated ampulla and canal (Acampora et al. 1996; Morsli et al. 1998). The transcription factor eyes absent (Eya1) is expressed in the ventromedial region of the mouse otic vesicle (Kalatzis et al. 1998), and the development of the inner ear of Eya1 mutant mice arrests at the otic vesicle stage, with no sensory structures being formed (Xu et al. 1999). Expression of the homeobox gene *Six1*, a homologue of *Drosophila* sine oculis (*so*), is dependent on *Eya1* expression, while that of *Pax2* is not (Xu et al. 1999), suggesting some conservation of the *Pax–Eya–Six* regulatory hierarchy that is seen in *Drosophila* eye development.

Six1 expression first becomes prominent around E8.75 in the ventral region of the early mouse otic pit (Zheng et al. 2003; Ozaki et al. 2004), and is seen in all sensory epithelia of the inner ear by E15.5 (Zheng et al. 2003). In *Six1* mutant mice, sensory organs do not develop and the inner ear arrests at the otocyst stage, forming simply an enlarged endolymphatic sac and some dorsal canal structures (Li et al. 2003; Zheng et al. 2003; Ozaki et al. 2004). In *Six1* mutant mice, no expression of ventral otocyst markers is seen, while the expression of dorsal otocyst markers expands ventrally, indicating that *Six1* is important for establishing the correct patterning of the otocyst (Ozaki et al. 2003; Zheng et al. 2003).

A recent study has shown that the expression patterns of some of these genes within the otocyst are regulated by Sonic hedgehog, derived either from the notochord or floor plate of the adjacent neural tube (Riccomagno et al. 2002). The *Six1* mutant phenotype closely resembles that seen in Shh deficient mice, however, the expression levels of *Six1* and *Shh* are mutually independent (Ozaki et al. 2003), indicating the two genes interact to properly pattern the developing otocyst. *Six1* is required for the normal expression of a number of genes expressed in the sensory patches, including those encoding BMP4, FGF10, and Lnfg.

The role of BMP4 in the developing sensory organs has been studied by blocking its function with the antagonist Noggin, introduced by infecting the chick otocyst with retroviral vectors encoding *Noggin* cDNA (Chang et al. 1999), or by implanting beads impregnated with Noggin or coated with cells expressing *Noggin* into the adjacent periotic mesenchyme (Gerlach et al. 2000). The formation and growth of the semicircular canals were found to be readily blocked by Noggin treatment (Chang et al. 1999; Gerlach et al. 2000), and this may be due to a decrease in cell proliferation followed by an increase in cell death within the otic epithelium (Chang et al. 1999). Sensory structures were found to be more resistant to Noggin treatment. Although a decrease in the expression of both *p75NGFR* and *Msx1*, but not of *BMP4* or *Lfng*, was observed in the anterior crista in response to Noggin treatment, most of the sensory organs were still present and only misshapen, and the differentiation of hair cells, as assessed by the expression of the hair-cell antigen (HCA, Richardson et al. 1990), was unaffected. Neither Noggin treatment, nor the introduction of beads coated with *BMP4* expressing cells led to the formation of ectopic or supernumerary hair cells (Chang et al. 1999; Gerlach et al. 2000). Overall these results suggest that BMP4 is most likely to control the development of adjacent accessory structures, rather than acting in a cell-autonomous fashion to regulate hair- and supporting-cell development within the sensory patch.

Fibroblast growth factor (FGF) 10 is one of the 19 known members of the FGF family. The genes for FGF10, neurotrophin-3 (NT-3) and brain derived neurotrophic factor (BDNF) are expressed in the early developing mouse otocyst in the presumptive sensory epithelia (Pirvola et al. 1992, 1994, 2000). The precise relationship between the sites of expression of these molecules and other molecular markers for sensory patches in the developing otocyst has not been fully elucidated, although the expression patterns of *Lfng* and *NT-3* mRNAs do overlap in the developing mouse cochlea at E12 (Morsli et al. 1998). *FGF10* expression overlaps with that of both *NT-3* and *BDNF*, and the gene encoding the receptor for *FGF10, Fgfr2(IIIb)*, is expressed in a complementary pattern in adjacent areas of the epithelium. Like BMP4, FGF10 produced by the sensory patches may operate in a paracrine fashion and control the development of adjacent nonsensory tissues (Pirvola et al. 2000). *Fgfr1* is expressed in the ventromedial wall of the otocyst at E10.5 and its expression becomes restricted to the outer hair cells of the cochlear duct by E16.5 (Pirvola et al. 2002). Loss of function mutations in the *Fgfr1* gene reveal that it is required for the development of the organ of Corti, but not the vestibular epithelia, and it has been suggested that it may control the proliferation of a pool of sensory precursor cells, possibly via an interaction with FGF10 (Pirvola et al. 2002).

The picture that emerges is one in which the different sensory organs of the inner ear develop within the ventromedial prosensory patch at discrete sites according to their proximity to the boundaries formed between the expression domains of genes encoding transcription factors. The prosensory patch defined by the expression of *Ser1*, *Lfng*, and BEN may itself be induced by an interaction between a dorsal and a ventral compartment. As components of the Notch signaling pathway, Serrate1 and Lunatic fringe may maintain Notch in an activated state within the patch, thereby maintaining sensory competence (Eddison et al. 2000). BEN, as a cell–cell adhesion molecule, may reinforce the boundaries and prevent the cells of the prosensory patch from mixing with those in adjacent compartments (Goodyear et al. 2001). The molecules that signal across the compartment boundaries and induce sensory organs of different identity have yet to be elucidated, although a recent study indicates that components of the wingless type (Wnt) signaling pathway may be involved (Stevens et al. 2003).

3. The Production of Hair and Supporting Cells

Hair cells, like neurons in the CNS, are terminally differentiated postmitotic cells. Supporting cells in the sensory organs of the mammalian inner ear and avian basilar papilla are mitotically quiescent (Jørgensen and Mathiesen 1988), and those in the avian papilla retain the capacity to reenter the cell cycle (Corwin and Cotanche 1988; Ryals and Rubel 1988). In the mammalian inner ear and the avian basilar papilla, hair and supporting cells are born during a brief time window during early development (Ruben 1967; Katayama and Corwin 1989). However, in the inner ears of fishes (Corwin 1981, 1983; Popper and Hoxter 1984, 1990; Lanford et al. 1996) and frogs (Corwin 1985), and in the vestibular organs of birds (Jørgensen and Mathiesen 1988; Roberson et al. 1992), hair cells may be produced throughout the lifetime of the animal.

3.1 Timing of Hair-Cell Production in Birds and Mammals

In the mouse, pulse labelling with [^{3}H]thymidine has been used to study the time at which hair and supporting cells undergo terminal mitosis (Ruben 1967). In the cochlea, the largest percentage of labeled hair and supporting cells (approximately 70% of the total observed at all time points) is observed following a pulse at E13.5, and 90% of the cells are produced between E12.5 and E14.5. In the maculae and cristae, most of the hair and supporting cells are produced over a broader time frame (95% of the total between E13.5 and E17.5). Labeled hair and supporting cells ($< 10\%$ of the total) are first observed in both the cochlea and the vestibular system after a pulse on E11.5, and not following a pulse on E9.5, indicating that the first hair cells to be born are produced as early as E11.5, probably coincident with the first expression of a number of molecular markers (e.g., *NT3*, *BDNF*, *FGF10*). Pulse labeling with [^{3}H]thymidine is not possible with chick embryos in ovo as the label remains continuously available. However, a negative labeling protocol can be employed, where cells that are not labeled following injection of [^{3}H]thymidine can be considered to have undergone terminal mitosis prior to administration of the label. Using this approach, it has been shown that the first hair and supporting cells to be born in the basilar papilla become postmitotic between E4 and E5 (stages 23 to 26) (Katayama and Corwin 1989). Birthdates for the first hair and supporting cells to be produced in the vestibular organs of the chick have not been reported. As described above, the prospective basilar papilla can be distinguished as a discrete sensory organ by the expression of *BMP4* and *Msx1* by E4 (Wu and Oh 1996), so the birth of hair and supporting cells in the basilar papilla appears just subsequent to the onset of expression of these two molecular markers. By E8, almost all hair and supporting cells in the basilar papilla, except a few at the distal end of the organ, have been produced (Katayama and Corwin 1989). The production of hair cells in the avian basilar papilla is therefore limited to a short time period in development, as it is in all sensory organs of the mouse inner ear. However, in the vestibular organs of the bird inner ear, hair, and supporting cells are produced over a considerably extended period, including the first few weeks after hatching and probably throughout life (Jørgensen and Mathiesen 1988; Roberson et al. 1992; Goodyear et al. 1999). As in fishes and frogs, this postembryonic production can lead to an increase in hair-cell numbers in the organs in which it occurs, as well as serving to replace hair cells that have been lost.

3.2 Patterns of Hair-Cell Production

The pattern of hair- and supporting-cell production in the avian papilla indicates that the first cells to be born are produced in a strip that runs along most of the length of the organ, with the cells generated later being added around the edge of this strip (Katayama and Corwin 1989). This is similar to the situation in postembryonic toads and elasmobranch fishes in which new hair cells are added around the periphery of the growing maculae (Corwin 1981, 1983, 1985). It is also similar to that in the developing vestibular system of the rat, where the oldest hair cells in the lateral crista and the utricular macula are located centrally, with new hair cells being added peripherally (Sans and Chat 1982). However, in the postembryonic maculae of birds and teleost fishes, it appears that new hair cells are distributed fairly uniformly throughout the maculae, and not preferentially around the periphery (Popper and Hoxter 1990; Roberson et al. 1992; Lanford et al. 1996; Bang et al. 2001). In the cochlea of the mouse, the first hair cells to be born are those at the apex, and the last are those at the base (Ruben 1967). This is surprising as almost all observations show that hair cells differentiate first at the base and last at the apex (or centrifugally from a point in the mid-basal coil) and suggests that the first hair cells to be born are the last to differentiate. As Ruben (1967) pointed out, the data showing that cells in the apical end of the organ of Corti are born before those in the base need to be interpreted with caution as the distributions of labelled cells along the cochlea are based on a very small average number of total labeled cells (e.g., fewer than seven per cochlea for outer hair cells at E11.5). However, a similar apex-to-base trend for birthdates was observed for both inner hair cells (IHCs) and outer hair cells (OHCs), and most of their immediately adjacent supporting cells, suggesting it is correct. Clearly this is an aspect of hair-cell development in the cochlea that needs to be reexamined, possibly using a continuous supply of BrdU and a negative labeling technique similar to that employed for the chick basilar papilla.

Cyclin-dependent kinases are proteins that influence whether a mitotically active cell reenters the cell cycle or withdraws from the cell cycle and becomes quiescent, and the activity of these kinases is regulated by two families of cyclin-dependent kinase inhibitor (CKI) proteins, the CIP/KIP and INK families. A member of the CIP/KIP family, $p27^{Kip1}$, is specifically expressed within the developing organ of Corti by E14 (Chen and Segil 1999). Expression is not observed at E12, but by E14 the expression pattern of $p27^{Kip1}$ has tightly defined boundaries that precisely delineate the region of the cochlear duct within which the hair and supporting cells will differentiate. Differentiated supporting cells continue to express $p27^{Kip1}$ into adulthood, but the expression of $p27^{Kip1}$ is down-regulated in hair cells as they begin to differentiate morphologically and start to express the hair-cell marker myosin VIIa. A targeted deletion of *$p27^{Kip1}$* leads to the appearance of supernumerary hair and supporting cells within the organ of Corti and continued cell proliferation (Chen and Segil 1999; Lowenheim et al. 1999). This indicates that $p27^{Kip1}$ normally negatively regulates cell division within the organ of Corti and controls the final number of cells. However, $p27^{Kip1}$ cannot be the only cell-cycle control molecule operating, as hair and supporting cells do withdraw from the cell cycle and differentiate in *$p27^{Kip1}$* null mutant mice. $P27^{Kip1}$ is also expressed in vestibular organs from E14 onward (Chen and Segil 1999) and becomes restricted to the supporting cells by P6, but its expression levels are low in these organs and it is not known whether excess numbers of hair and supporting cells are produced in the cristae and maculae of *$p27^{Kip1}$* null mutants. Other members of the CIP/KIP family of cyclin-dependent kinase inhibitors may play a more prominent role in the vestibular system.

3.3 Lineage Relationships

In the mammalian cochlea (Ruben 1967) and the avian basilar papilla (Katayama and Corwin 1989) hair and supporting cells at any one point in the organ are born simultaneously, indicating that they may share a common lineage. Tracing studies with replication-deficient retroviral vectors have been used to demonstrate that hair and supporting cells are indeed related by lineage in the sensory organs of the avian inner ear (Fekete et al. 1998; Lang and Fekete 2001). Following injection of virus into the otocyst between E3.5 and E5, individual clones of marked cells were observed in basilar papillae (harvested just before hatching) that were composed of either just hair cells, just supporting cells, or a combination of both hair and supporting cells. Hair and supporting cells therefore share a common progenitor. Clones within the papillae varied in size, containing a single marked cell of either type, or a mixture of up to 17 cells of both types. The average ratio of hair cells to supporting cells in these latter clones containing both cell types was 1.6:1 (Fekete et al. 1998). Importantly, two-cell clones were found that contained either two supporting cells or one hair cell and one supporting cell, indicating that a progenitor cell can retain the potential to become either a hair cell or a supporting cell up to its final mitotic division. Time-lapse studies of living cells in the lateral line organs of axolotls have also clearly demonstrated that a single progenitor cell can give rise to both a hair cell and a supporting cell (Jones and Corwin 1996).

4. Differentiation of Hair and Supporting Cells

The mechanism by which cells of common origin adopt different fates during development has been thoroughly characterized, most notably in the nervous system of *Drosophila*, where it involves an interaction of the products of the proneural and neurogenic genes. Proneural genes like *atonal* and those of the *achaete–scute* complex encode bHLH transcription factors and their expression defines groups of cells, called proneural clusters or equivalence groups, that all have the potential to become neurons. The neurogenic genes *Notch* and *Delta* encode the transmembrane receptor Notch and its transmembrane ligand Delta, respectively, and a process of lateral inhibition mediated by these two proteins singles out which cell in the proneural cluster adopts the primary, neural fate and prevents its neighbours from doing likewise, forcing them to adopt the secondary fate. Activation of Notch by its ligand Delta induces the expression of genes of the *Enhancer-of-split* complex, and the products of this complex in turn repress the expression of the proneural genes of the *achaete–scute* complex. The products of the *achaete–scute* complex induce the expression of neuron-specific genes and *Delta*. Activation of Notch in any one cell thereby reduces the expression of Delta in that cell, and in a system composed of two or more cells this generates a feedback loop, as a cell with reduced Delta expression will be less able to activate Notch in its neighbors, which in turn will thereby express more Delta and reciprocally activate more Notch. Cells that express more Delta and have low levels of Notch activation adopt the primary fate and become neurons, those with high levels of Notch activation adopt the secondary fate. Mutations in neurogenic genes usually lead to the overproduction of neurons in the CNS (Lewis 1996). In a population of equivalent cells that initially express uniform levels of Notch and Delta, a stochastic fluctuation in Notch activation (or Delta expression) can theoretically, as a consequence of lateral inhibition with feedback, lead to the production of uniform mosaics of two cell types, similar to those observed in the sensory organs of the inner ear (Collier et al. 1996; Pickles and van Heumen, 2000).

It was suggested more than a decade ago that a similar process involving lateral inhibition underlies cell fate determination in the developing inner ear (Corwin et al. 1991; Lewis 1991). These suggestions were inspired to a large degree by structural homologies between the sensory organs of the vertebrate inner ear and the mechanosensory bristle organs of *Drosophila*, a system in which many aspects of lateral inhibition were originally elucidated. It has been shown since that *Notch* and the genes encoding its ligands Delta and Serrate, vertebrate homologues of the Drosophila neurogenic genes, are expressed in the inner ear and experimental studies have shown that they play a key role in mediating the differentiation of hair and supporting cells.

Transcription factors are another key class of molecules that regulate developmental processes, acting either up or downstream of signaling molecules, and a number are now known to be involved in the differentiation of hair and supporting cells, including a vertebrate homologue of a proneural gene required for sense organ development in *Drosophila*. The expression patterns of these transcription factors and the neurogenic genes in the developing inner ear, and the effects of manipulating their expression, will now be discussed in detail.

4.1 The Transcription Factor Math1

Math1 is a mouse homologue of the *Drosophila* proneural gene *atonal*. It is expressed in the developing vestibular system by E12.5 and first appears in a narrow strip of cells in the basal end of the cochlea between E13.5 and E14.5 (Bermingham et al. 1999; Shailam et al. 1999; Lanford et al. 2000; Chen et al. 2002). A careful comparison using tissues from transgenic *EGFP/Math1* mice clearly indicates that the onset of $p27^{Kip1}$ expression in the basal end of the cochlea (between E13 and E14) precedes the onset of Math1 expression (between E13.5 and E14.5) (Chen et al. 2002). Math1 expression is initially observed to extend from the basal to the lumenal surface of the sensory epithelia and becomes restricted to the apical cell layer as the hair cells overtly differentiate (Lanford et al. 2000). These vertical bands of *Math1*-expressing cells may be bipotential progenitors, a subset of which become hair cells (Lanford et al. 2000). Alternatively, they may be vertical stacks of hair cells that then spread out along the length of the cochlea as it elongates, by a process that has been referred to as radial intercalation (Chen et al. 2002). After birth, the expression of *Math1* decreases, in both the cochlea and the vestibular system (Shailam et al. 1999; Lanford et al. 2000).

Math1 null mutant mice have been produced by replacing the coding region of *Math1* with the gene encoding β-galactosidase. Hair cells are completely absent from the inner ears of these mice (Bermingham et al. 1999). The sensory epithelia are thinner than normal and lack nuclear stratification, but retain their overlying extracellular matrices, suggesting the supporting cells, the cells that normally produce these structures, have differentiated. Myosin VI, an unconventional myosin specifically expressed by hair cells in the inner ear, is not expressed in the inner ear of *Math1* null mutants, nor is another hair-cell marker, calretinin. It was originally reported that apoptosis was not apparent in the developing epithelia of the null mutant mice, and that there was an overcrowding of nuclei in the basal layer. These observations suggested that there may have been a fate switch in the null mutants leading to the overproduction of supporting cells, and it was proposed (Bermingham et al. 1999) that *Math1* acts as a "pro-hair cell" gene required for the specification of hair cells. However, a more recent analysis has indicated that apoptosis does indeed occur in the organ of Corti of *Math1* null mutants, and that the appearance of dying cells follows the normal base-apex sequence of hair-cell differentiation. It may therefore be the case that hair cells are produced in *Math1* null mutant mice, never express any of the known markers for hair cells, and then die (Chen et al. 2002).

In cochlear cultures prepared from early postnatal rats, the ectopic overexpression of *Math1* in the cells of the greater epithelial ridge lying adjacent to the organ of Corti leads to the production of additional hair cells (Zheng and Gao 2000). The cells of the greater epithelial ridge cells do not normally express *Math1* and they recede and disappear as the internal sulcus forms after birth. When transfected with *Math1*, they detach their basally directed radial cell process from the underlying basal lamina, express myosin VIIA, and a well-defined stereociliary bundle forms on the apical surface. Furthermore, if the supporting cells of rat utricle cultures are transfected with *Math1* their conversion into hair cells is facilitated. These results have led to the suggestion that Math1 acts as a positive regulator of hair-cell differentiation, and antagonizes the effects of negative regulators that lie downstream of Notch signaling (Zheng and Gao 2000).

4.2 The Transcription Factors Brn3.1 and Gfi1

POU-domain transcription factors are known to be involved in the development and survival of many neuronal cell types. Brn3.1 (also known as Brn3c) is a POU-domain transcription factor that is specifically expressed by the hair cells of the adult rat and mouse inner ear (Erkman et al. 1996; Xiang et al. 1997). Brn3.0 (Brn3a) and Brn3.2 (Brn3b) are expressed by the neurons of the spiral and vestibular ganglia but not by hair cells (Xiang et al. 1997). In the developing vestibular sensory patches, postmitotic cells express Brn3.1 as early as E12.5, one day before the appearance of the hair-cell markers myosin VI and VIIa (Xiang et al. 1998). In the developing cochlea, Brn3.1 is expressed by E14.5, also one day before the appearance of myosin VI and myosin VIIa (Xiang et al. 1998).

In *Brn3.1* null mutant mice, all sensory hair cells are lost from the inner ear by P14 (Erkman et al. 1996; Xiang et al. 1997). Myosin VI, myosin VIIa, and calretinin are all expressed by hair cells in *Brn3.1* null mutants between E15.5 and P8, indicating hair cells undergo the initial stages of differentiation (Xiang et al. 1998). However, the hair cells do not form stereociliary bundles, their nuclei do not segregate to a plane above that of the supporting cell nuclei, and they die by apoptosis (Xiang et al. 1998). The ectopic overexpression of Brn3.1 in rat cochlear cultures fails to induce hair-cell differentiation (Zheng and Gao 2000). These results all indicate that Brn3.1 is not required for the determination of hair cells, but is required for their maintenance and maturation.

Gfi1, encoding a zinc finger transcriptional repressor, is a vertebrate homologue of *Drosophila senseless,* a gene that is required for the survival of neurons in the embryonic and adult peripheral nervous system, and is also both necessary and sufficient for sensory organ formation (Nolo et al. 2000). In *Drosophila*, the expression of *senseless* depends on the expression of the proneural genes, and the maintained expression of the proneural genes is, in turn, dependent on the expression of *senseless*. Gfi1 protein is expressed in the mouse inner ear from E12.5 onward and is clearly restricted to the hair cells in the utricle and saccule by E14.5, and those in the cochlea by E16.5. In *Gfi1* null mutant mice, hair cells undergo the early stages of differentiation, and express Math1, Brn3.1, and myosin VI/VIIa. However, the hair cells in the organ of Corti begin to degenerate between E15.5 and P0 and are completely lost by P14. In the vestibular system, the hair cells do not degenerate but the sensory epithelia are disorganized, with some hair cells remaining within the basal supporting-cell nuclear layer rather than being apically located within the epithelium (Wallis et al. 2003). The phenotype of *Gfi1* null mutants is similar in certain respects to that of the *Brn3.1* null mutants, although *Gfi1* is not required for the long-term survival of vestibular hair cells.

4.3 The Transcription Factors GATA3 and Pax2

GATA3 is one of a family of six transcription factors that are involved in the differentiation and development of a number of different cell types. In the mouse, GATA3 is expressed by all of the epithelial cells that form the dorsal floor of the cochlear duct at E14 (Rivolta and Holley 1998; Lawoko-Kerali et al. 2002). It is also expressed by the neurons of the spiral ganglion, but not by any of the vestibular epithelia or other cell types in the ear at this stage. As hair and supporting cells differentiate in the cochlea, GATA3 expression selectively decreases, first in the hair cells and then in the supporting cells. The decrease in GATA3 expression progresses in the usual base to apex fashion, and it has been suggested that it may act as a negative regulator of hair and supporting cell differentiation that ensures controlled differentiation along the length of the cochlea (Rivolta and Holley 1998). The paired-box gene *Pax2* is expressed throughout the early otic placode in mouse and chick and becomes restricted to the medial regions as the otocyst develops (Hutson et al. 1999; Groves and Bronner-Fraser 2000; Lawoko-Kerali et al. 2002). Pax2 is expressed by hair cells in all of the sensory patches until at least E16.5 in the developing mouse inner ear (Lawoko-Kerali et al. 2002), and in zebrafish the closely related homologue *pax2.1* is expressed by differentiating hair cells (Riley et al. 1999).

4.4 The Notch Signaling Pathway Molecules

4.4.1 Expression Patterns

Expression of the Notch receptor and its ligands Delta and Serrate (the latter is called Jagged in mammals) has been studied in the developing inner ears of chicks, fishes, mice, and rats (Adam et al. 1998; Haddon et al. 1998; Lewis et al. 1998; Lanford et al. 1999; Morrison et al. 1999; Riley et al. 1999; Shailam et al. 1999; Stone and Rubel 1999; Zine et al. 2000). In chick, *C-Notch1* has a widespread distribution in the developing otocyst and is present throughout the sensory patches until at least E12 (Adam et al. 1998). In the mitotically quiescent basilar papilla of the early posthatch chick, *C-Notch1* expression is restricted to the supporting-cell layer (Stone and Rubel 1999). In mouse and rat the situation is probably very similar. In the mouse, in situ hybridization shows that *Notch1* is expressed throughout the vestibular sensory patches and the entire dorsal floor of the cochlear duct at E14.5 (Lanford et al. 1999; Shailam et al. 1999). *Notch1* expression decreases in the hair cells as they differentiate, while it continues in the surrounding supporting cells, at least until birth. In the rat, antibody staining also indicates that hair and supporting cells both initially express Notch1, at least by E18 although possibly not at earlier stages, and that hair cells, but not supporting cells, stop expressing Notch1 as they differentiate (Zine et al. 2000). In surprising contradiction, there is one claim, based on the use of a GFP-reporter construct for the *Notch1* promoter, that *Notch1* is exclusively expressed by hair cells in the developing mouse inner ear (Lewis et al. 1998).

In the chick otocyst, *C-Delta1* is expressed in sensory patches from E3.5 onward until at least E12 (Adam et al. 1998). It is expressed exclusively in the sensory patches and, within these, in a scattered subset of cells that eventually become located in the apical, upper layer of the epithelium and are almost certainly nascent, differentiating hair cells. In the inner ear of the early posthatch chick, *C-Delta1* is not expressed by either the hair cells or the supporting cells of the basilar papilla, but it is expressed by a subpopulation of cells in the utricle where hair-cell production is still an ongoing event (Stone and Rubel 1999). In situ hybridization studies and transgenic mice expressing a lacZ reporter construct for *cis*-acting regulatory sequences of the *Delta1* gene both indicate that hair cells in the mouse express *Delta1* (Morrison et al. 1999). In the vestibular system, *Delta1* expression is seen from E12.5, and in the cochlear duct it is seen from E14.5 onward, initially in the inner hair cells and then in the outer hair cells. Cells surrounding the hair cells do not express detectable levels of *Delta1* at any stage. Nascent hair cells most likely use Delta1 to laterally inhibit their surrounding supporting cells. In zebrafish, expression of *deltaA*, *deltaB*, and *deltaD* is observed in small patches at the anterior and posterior ends of the otocyst at 14 hpf (Haddon et al. 1998), foreshadowing the appearance of morphologically identifiable hair cells at 24 hpf (Haddon and Lewis 1996).

In the mouse, the expression of *Jagged2* (*Serrate2*) is first detected at E14.5, in a narrow strip of cells that extends along the basal coil of the cochlea (Lanford et al. 1999). This strip is initially the width of a single cell. By E18 it expands to a width of four to five cell diameters, extends along the full length of the cochlea, and is clearly restricted to the developing hair cells. *Jagged2* is expressed in all the sensory organs of the vestibular system (Lanford et al. 1999). It is detected in the cristae of the ampullae as early as E13.5 and is restricted to cells that lie close to the lumenal border of the epithelium and are presumably hair cells. In the cochlea of the rat, antibody staining indicates Jagged2 is initially expressed by the IHCs at E18, and by both the IHCs and the OHCs by E20 (Zine et al. 2000). *Serrate2* expression has not been described in the chick inner ear, but in zebrafish the expression of *serrateB* (which is more closely related to *C-Serrate2* than to *C-Serrate1*) closely resembles that of *deltaA*, *deltaB*, and *deltaD*, although its expression persists for longer than that of the *delta* genes in the differentiated hair cells (Haddon et al. 1998).

In the chick, as described earlier, *C-Serrate1* expression is restricted to the developing sensory patches from the very early stages of development (Adam et al. 1998). *C-Serrate1* mRNA is observed throughout the thickness of the developing epithelium and its expression is observed in both the hair and supporting cells of the early posthatch basilar papilla (Stone and Rubel 1999). In the mouse, *Jagged1* (*Serrate1*) is also expressed early, by E12.5, and its distribution demarcates the future sensory patches (Morrison et al. 1999). Expression of *Serrate1* in the mouse otocyst is initially observed throughout the thickness of the epithelium but becomes restricted to the supporting-cell layer as the hair cells differentiate (Morrison et al. 1999). In the rat cochlea, Jagged1 (Serrate1) protein, as detected by immunofluorescence, becomes restricted to the supporting cells in the organ of Corti by E20 (Zine et al. 2000). The expression of *Lunatic fringe* (*Lfng*), a gene encoding a modulator of the Notch signalling pathway is, as described above, initially similar to that of *Serrate1*, demarcating the future sensory patches and eventually becoming restricted to the supporting cells (Morsli et al. 1998; Cole et al. 2000). In the mouse cochlea, *Lfng* is expressed in the cells that immediately surround the IHCs and the OHCs at E18, that is, by the inner phalangeal cells, Deiters cells and the outer pillar cells, but not by the inner pillar cells (Zhang et al. 2000).

4.4.2 Mutant Phenotypes and Experimental Observations

Homozygous *Notch1* null mutants die by E9.5, before the appearance of sensory patches in the inner ear, so it has not been possible to assess hair-cell differentiation in these mice. However, heterozygous *Notch1*$^{+/-}$ mice are viable and a significant increase is observed in the numbers of regions along the cochlea where there are four instead of three rows of outer hair cells (Zhang et al. 2000). A reduced level of *Notch1* therefore has an effect on the numbers of outer hair cells that differentiate and their patterning. Treating late embryonic (E16 to E20) rat cochlear cultures with anti-sense *Notch1* oligonucleotides leads to the formation of extra rows of both inner and outer hair cells (Zine et al. 2000). Up to eight rows of OHCs and 3 rows of IHCs are observed in cultures treated with anti-sense *Notch1* oligonucleotides, and the number of hair cells per unit length of cochlea is significantly increased. The effect declines with the age at which treatment is initiated. Fewer extra rows are observed in antisense-treated explants prepared from the early postnatal stages (P0 to P3), and these are found in the more apical, less differentiated end of the cochlea. In regions where antisense *Notch1* oligonucleotide treatment caused formation of additional rows of hair cells, hair-bundle polarity is disrupted and there is some evidence that hair cells may be in direct contact with one another. While this suggests additional hair cells may have formed at the expense of intervening supporting cells in response to antisense *Notch1* oligonucleotide treatment, a definitive analysis of supporting-cell numbers in these cultures has yet to be performed.

In zebrafish that are homozygous for a dominant negative allele of the *deltaA* gene, *deltaA$^{dx2/dx2}$*, there is a five to sixfold fold increase in the numbers of hair cells that form (Riley et al. 1999). The maculae are enlarged, supporting cells are almost totally missing, and otolith formation is delayed. The zebrafish *pax2.1* gene is expressed by developing and mature hair cells, but not by supporting cells. In *no isthmus* mutants with a mutation (*noi^{tb21}*) in the C-terminus of pax2.1, the expression levels of *deltaA* and *deltaB* in nascent hair cells are reduced and nearly twice the normal number of hair cells is produced. The expression of *pax2.1* precedes the expression of *delta* genes in hair cells and the analysis of double mutants indicates that *pax2.1* is required for normal levels of *delta* gene expression in developing hair cells (Riley et al. 1999). In the zebrafish *mindbomb* mutant, which has a neurogenic phenotype in the CNS, the expression of *delta* genes and *serrateB* is increased and there is a nearly 15-fold increase in the number of hair cells that form by 36 hpf, accompanied by a total absence of supporting cells (Haddon et al. 1998). Recent studies have identified mindbomb as a ubiquitin ligase that interacts with the intracellular domain of delta, promoting its internalization (Itoh et al. 2003). In the chick, retrovirally mediated expression of dominant-negative forms of either *C-Delta1* (*Dll1^{dn}*) or suppressor of hairless (a transcription factor that acts downstream of Notch) (*Su(H)^{dn}*) within the developing sensory patches does not lead to an increase in the linear density of hair cells, nor does overexpression of full length *C-Delta1* cDNA inhibit hair-cell production (Eddison et al. 2000). This latter, unexpected observation may be explained by the fact that differentiated hair cells in the chick inner ear express Numb (Eddison et al. 2000), a protein that interacts with Notch and blocks its activity. Differentiating hair cells may therefore be immune to the effects of Notch activation caused by the overexpression of Delta1 in surrounding cells. Although blocking Notch signaling by expressing dominant negative Delta1 or Su(H) does not lead to an increase in hair-cell numbers, it does down-regulate the expression of Serrate1 in the sensory patches (Eddison et al. 2000). Notch activation therefore positively regulates the expression of its ligand, Serrate 1, in sensory patches via lateral induction.

Mice that are homozygous null mutants for the *Jagged2* (*Jag2*) gene have nearly complete duplication of the normal single row of inner hair cells, and regions along the length of the cochlea that contain four rather than three rows of outer hair cells (Lanford et al. 1999). The number of inner hair cells present per unit length is increased by approximately 50%, and the number of OHCs by nearly 20%. Hair-bundle polarity is also disrupted in the *Jag2* null mutants, as in cultures treated with antisense Notch oligonuleotides. No effects are observed in heterozygotes. A null mutant for the *Lfng* gene has no effect on hair-cell patterning or the numbers of hair cells that are produced in the cochlea, but the appearance of extra rows of inner hair cells in *Jag2* null mutant mice is suppressed on a *Lfng* null background (Zhang et al. 2000). This suppression of the *Jag2* null mutant phenotype is restricted to the inner hair cells. Extra rows of outer hair cells and a general disorganization of hair-bundle polarity in outer hair cells are still observed in mice that are double homozygous $Jag2^{-/-}$, $Lfng^{-/-}$ mutants. It has been suggested that the Lfng protein may normally block the inhibitory signal delivered by Delta1, but not that delivered by Jag2, and that this inhibitory signal from Delta1 can function in the absence of Lfng to suppress the overproduction of inner hair cells caused by the loss of Jag2 signaling in a *Jag2* null mutant (Zhang et al. 2000). An alternative explanation is that Lfng normally weakly inhibits the Notch-activating effect of Ser1, and that an increase in Notch activation is revealed as a decrease in hair cell production in a *Jag2* mutant background where the action of Ser1 becomes more critical (Eddison et al. 2000). Neither hypothesis explains why the *Lfng* mutation does not suppress the phenotype observed with OHCs in the *Jag2* null mutant mouse.

Treating cochlear cultures prepared from E16 to P3 rats with *Jag1* antisense oligonucleotides leads to the formation of extra rows of both inner and outer hair cells, and some disruption in hair-bundle polarity (Zine et al. 2000). This effect is similar, although not as dramatic, to that observed with antisense oligonucleotides for *Notch1* (see above). Dominant missense mutations in the mouse *Jag1* gene, assumed to be dominant negative mutations, result in extra hair cells, both normal and atypical, in the inner hair cell row of the cochlea, but also a loss of most third-row outer hair cells (Kiernan et al. 2001; Tsai et al. 2001). In addition, these mutations cause a complete loss of the posterior ampulla, and sometimes the anterior ampulla, along with their cristae, and a truncation of the associated canals. Jag1 may play two roles in ear development (Kiernan et al. 2001): an early one in specifying sensory patches and a later one in regulating hair development within the patches.

4.5 The Transcription Factors Hes1 and Hes5

Homologues of the *Drosophila* genes *Hairy* and *Enhancer-of-split*, *Hes1* and *Hes5*, encode bHLH transcription factors that are known to negatively regulate neurogenesis in vertebrates (Ohtsuka et al. 1999). In the cochlea of the rat at E17.5, *Hes1* is expressed by the cells of the greater epithelial ridge that lie adjacent to the developing inner hair cells, by the cells of the lesser epithelial ridge, and by supporting cells throughout the utricle. At the same stage, *Hes5* is expressed by the cells of the lesser epithelial ridge that surround the emerging hair cells, and by supporting cells in the striolar region of the utricular macula (Zheng JL et al. 2000). In the mouse, *Hes5* is first detected in the basal end of the cochlea at E15, 2 days after the onset of *Math1* expression, throughout the thickness of the epithelium and in a thin strip of cells that is narrower than that defined by the expression pattern of *Math1* at this stage. By E17, *Hes5* expression is observed throughout the length of the mouse cochlea and is restricted to Deiters cells in the basal coils (Lanford et al. 2000). In homozygous *Hes1* null mutant mice, additional inner hair cells but not outer hair cells are formed (Zheng JL et al. 2000). The effect is dose dependent with some but fewer additional inner hair cells observed in heterozygotes. In homozygous *Hes5* null mutant mice additional outer, but not inner, hair cells are produced (Zine et al. 2001). Additional hair cells are found in the utricles of both *Hes1* and *Hes5* null mutants, and to a lesser extent in those of double heterozygotes. Cotransfecting cells in the greater epithelial ridge of rat cochlear cultures with a mixture of plasmids expressing both *Math1* and *Hes1* leads to a reduction, relative to cultures transfected with *Math1* alone, in the number of GER cells that can be induced to become ectopic hair cells (Zheng JL et al. 2000). *Hes1*, and probably also *Hes5*, can therefore antagonize the action of *Math1* and negatively regulate hair-cell differentiation. In *Jag2* null mutants, the expression of *Hes5* is markedly reduced, suggesting its expression is positively regulated by Notch activation (Lanford et al. 2000).

4.6 Models for Hair- and Supporting-Cell Differentiation

The expression patterns of the genes described above, their mutant phenotypes, and the experimental findings are all generally consistent with the hypothesis that lateral inhibition mediated via Notch signaling controls the differentiation of hair and supporting cells in the vertebrate inner ear. The situation is clearly more complex than originally conceived, however, and it seems unlikely that lateral inhibition with feedback is all that is required to generate organs composed of hair and supporting cells. Several Notch ligands are operating, and the system can be modulated at different levels. Furthermore, although models incorporating lateral inhibition with feedback can successfully generate regular patterns of two cell types similar to those seen in the sensory organs of the ear, observations in the developing avian basilar papilla have shown that a rearrangement of cells with respect to one another contributes to the formation of a regular cellular mosaic of hair and supporting cells (Goodyear and Richardson 1997).

J. Lewis and colleagues have recently proposed a system of regulatory interactions (see Fig. 2.3) between Notch and its ligands that accounts for the observations described above (Eddison et al. 2000). In this system, newly differentiating hair cells express Delta1, Serrate2, and Numb, and newly differentiating supporting cells express *Serrate1*. Notch activation positively regulates the expression of *Serrate1*, and negatively regulates the expression of *Delta1* and *Serrate2*. Delta1 and Serrate 2 that are expressed by the hair cells activate Notch1 in the supporting cells, thereby preventing supporting cells from expressing *Delta1* and *Serrate2*, inhibiting their differentiation as hair cells, and promoting the expression of *Serrate1*. Serrate1 that is expressed by supporting cells does not activate Notch1 in the neighboring hair cells because the hair cells express Numb, but it will further activate Notch1 in adjacent supporting cells. A low level of Notch activation in hair cells ensures they maintain a high level of expression of *Delta1* and *Serrate2*, and means that they express a low

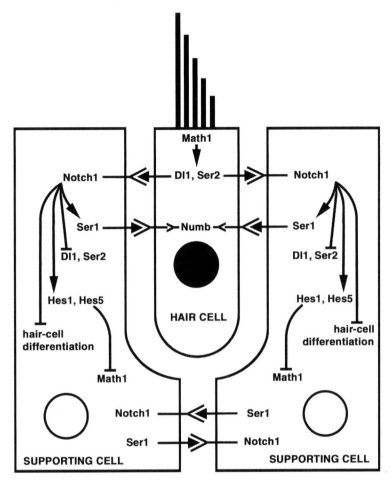

FIGURE 2.3. A model for the interactions occurring between components of the Notch signaling pathway in a differentiating sensory organ of the inner ear. Hair cells express Delta1 (Dl1) and Serrate2 (Ser2), thereby activating Notch1 in the supporting cells. This prevents the supporting cells from expressing Delta1 and Serrate2, and positively regulates the expression of both Serrate1 (Ser1) and Hes1/5, with Hes1/5 inhibiting hair-cell differentiation. Hair cells express Numb and are therefore immune to Serrate1 signaling from adjacent supporting cells. Cells with a low level of Notch activation become hair cells, while cells with a high level of Notch activation become supporting cells. Math1 may lie upstream of Delta1 and Serrate2, and Hes1/5 may inhibit the expression of Math1. Modified with permission from Eddison et al. (2000).

level of *Serrate1*. The system operates to maintain a high level of Notch activation in supporting cells, and a low level of Notch activation in hair cells. Any perturbation that reduces levels of Notch activation should cause an overproduction of hair cells, and in most cases (e.g., in $Notch^{+/-}$, $Jag2^{-/-}$, and dominant negative *Delta* mutants, and in cultures treated with antisense *Jag1* and *Notch1* oligonucleotides) this has been shown to be the case. Increased activation of Notch should lead to a decrease in hair-cell numbers, but in the one instance in which this experiment has been attempted (Eddison et al. 2000), the retroviral overexpression of *Delta1* in chick ears did not inhibit hair-cell production, possibly because Notch cannot be activated in cells that already express Numb.

The model (Fig. 2.3) can be further elaborated by incorporating *Math1* upstream of the genes encoding the Notch ligands *Delta* and *Serrate2*, and *Hes1/5* downstream of activated Notch, with *Hes1/5* repressing the expression of *Math1*. Clearly there is some evidence for this. *Math1* is probably expressed before *Delta1* and *Jagged2*, *Hes5* is up-regulated in supporting cells as they differentiate, and the ectopic coexpression of *Hes1* with *Math1* blocks the effects of *Math1*. What is not clear is whether hair cells are selected from a population of bipotential progenitor cells that are expressing *Math1*, or whether the cells that express *Math1* have already been selected by some as yet unidentified signal/process to be hair cells, or at least to be strongly biased in that direction, and then reinforce that decision and maintain it by lateral inhibition. Clearly *Math1* is not acting as a proneural gene like its *Drosophila* homologue *atonal*, and a bHLH transcription factor that specifies the prosensory patch, as defined by the zone of nonproliferating cells that are expressing $p27^{Kip1}$ remains to be identified.

5. Roles of Retinoic Acid and Thyroid Hormone

Retinoic acid and thyroid hormone are known to be involved in a number of different aspects of embryonic development. Both operate through nuclear receptors, transcription factors with a zinc finger DNA binding domain and a domain for ligand binding and dimerization. There are six receptors that mediate the effects of retinoic acid, three retinoic acid receptors (RARα, β and γ) and three retinoid X receptors (RXRα, β, and γ), each the product of separate genes. There are three T3-responsive thyroid hormone receptors, TRα1, TRβ1, and TRβ2, derived from two genes, *Thra* and *Thrb*. The roles of these receptors in inner ear development have been reviewed fairly recently (Raz and Kelley 1997), so only key observations and more recent findings related specifically to the development and differentiation of hair cells will be discussed.

Exposing chick otocysts to retinoic acid in vitro causes a decrease in cell proliferation, a decrease in mitogen-induced c-fos expression and the precocious differentiation of hair cells (Represa et al. 1990; Leon et al. 1995). Treating mouse cochlear cultures derived from early (E12.5 to E16), but not later, stage embryos with retinoic acid leads to the production of regions with extra rows of both inner and outer hair cells, along with additional accompanying supporting cells (Kelley et al. 1993). In the mouse, retinoic acid is produced by the embryonic organ of Corti, but not by the adult organ (Kelley et al. 1993). In the rat, cellular retinol binding protein I (CRBPI) is expressed transiently by Deiters cells and the outer pillar cells between E15 and P15 (Ylikoski et al. 1994). CRBPI is thought to be involved in controlling the availability or synthesis of retinoic acid, so these supporting cells may be the source of retinoic acid within the organ of Corti. However, in situ hybridization studies suggest CRBPI is not expressed in the developing mouse cochlea, and the development of the organ of Corti appears normal in CRBPI null mutant mice (Romand et al. 2000). The cellular retinoic acid binding proteins I and II (CRABPI and CRABPII) may also control retinoic acid availability. They are expressed in the greater epithelial ridge adjacent to the developing organ of Corti, but defects are not observed in the morphological and physiological development of the organ of Corti in CRABPI/CRABPII double null mutant mice (Romand et al. 2000). RARα, RXRα, and RXRβ mRNAs are expressed in the developing cochlear epithelium by E13, and transcripts for all three receptors become concentrated to a certain extent in hair cells as development proceeds (Raz and Kelley 1999). RXRα mRNA expression decreases dramatically between E17 and P3, correlating with the observed decrease in sensitivity to exogenously supplied retinoic acid (Kelley et al. 1993). As active retinoic acid receptors are usually heterodimers composed of one RAR and one RXR subunit, and as RXRβ/RXRγ compound null mutants are viable and have normal hearing (Barros et al. 1998), it seems that RARα/RXRα is the most likely hair-cell receptor for retinoic acid, although the organ of Corti develops normally in RARα null mutant mice (Romand et al. 2002). Nonetheless, treating E12.5 cochlear cultures with the RARα antagonist Ro-41-5253 for 2 days followed by cultivation for a further 5 days leads to reductions in both the number of hair cells and the numbers of hair-cell rows that develop (Raz and Kelley 1999). If cochlear cultures are established at E13 and treated with Ro-41-5253 for 48 h prior to fixation at the equivalent of E15.5 (i.e., after a total of 2.5 days in vitro), cells expressing myosin VI, but not myosin VIIA or Brn3.1, are observed along the entire length of the cochlea. These results indicate that the expression of myosin VI and the initial differentiation of hair cells are unlikely to require retinoic acid, and have been interpreted as supporting the hypothesis that retinoic acid plays a role in certain specific aspects of the differentiation of progenitor cells as hair cells, rather than a role in determining the size of the prosensory cell population for the organ of Corti (Raz and Kelley 1999).

Thyroid hormone receptors are expressed in the rat inner ear from E12.5 onward (Bradley et al. 1994). TRα1 and TRα2 are expressed throughout most of the sensory epithelia of the inner ear, while TRβ1 and TRβ2 are expressed only in the cochlear epithelium, by cells of both the greater and lesser epithelial ridges including the hair cells. Chemically induced hypothyroidism causes a general delay in the maturation of many components of the organ of Corti, but most structures, with the exception of the tectorial membrane, eventually develop normally (Uziel et al. 1981). A similar effect is observed in compound null mutant mice that lack all known thyroid hormone receptors (Rüsch et al. 2001). *Thrb* is essential for the maturation of auditory function (Forrest et al. 1996), and in *Thrb* null mutant mice, development of the fast potassium conductance, $I_{K,f}$, in IHCs is specifically delayed, but the onset of outer hair cell motility is unaffected (Rüsch et al. 1998a). In $Thra^{-/-}/Thrb^{-/-}$ double mutant mice, the onset of $I_{K,f}$ expression in IHCs and the onset of OHC motility are both affected (Rüsch et al. 2001). A recent study has shown that there are thyroid response elements in the promotor for the gene encoding prestin, the OHC motor (Zheng J et al. 2000), and that expression of this gene is indeed regulated by thyroid hormone (Weber et al. 2002).

6. Hair-Bundle Development

Scanning and transmission electron microscopy have been used to describe the formation and development of sensory hair bundles in a number of different mammals (Li and Ruben 1979; Anniko 1983 a, b; Lim and Anniko 1985; Mbiene and Sans 1986; Lavigne-Rebillard and Pujol 1986; Lenoir et al. 1987; Lim and Rueda 1992; Kaltenbach et al. 1994; Zine and Romand 1996). Studies on the chick auditory organ, the basilar papilla, have provided the most thorough and complete description of hair-bundle development (Cotanche and Sulik 1983, 1984; Tilney and DeRosier 1986; Tilney et al. 1986, 1988a, 1992). These papers will therefore be summarized in some detail, and the extent to which hair-bundle development in the chick auditory organ is representative or atypical of that in different organs and species will be considered. Despite good morphological descriptions, the basic molecular mechanisms underlying hair-bundle development remain largely unknown. Hair bundles come in a wide range of very precisely determined shapes and sizes, even within one organ, and the control systems that orchestrate this variety of cellular form are even less well understood. Nonetheless studies of mouse and zebrafish mutants, and human deafness genes, have recently revealed a number of molecules that are required for the development and maintenance of hair-bundle structure and polarity, and these will be the focus of the second half of this section.

6.1 Appearance and Growth of Hair Bundles

6.1.1 The Chick Basilar Papilla

Hair bundles can be first identified by scanning electron microscopy (SEM) or with antibodies directed against the hair-cell antigen (HCA) at E6.5 (stage 29) in the distal, low-frequency end of the developing basilar papilla (Cotanche and Sulik 1983; 1984; Bartolami et al. 1991; Goodyear and Richardson 1997). A single kinocilium and numerous microvilli are present on the apical, lumenal surfaces of all the epithelial cells of the papilla at this stage (Fig. 2.4). The emerging hair bundles seen in the distal end of the papilla consist of 10 to 15 microvilli that are are clustered together, slightly taller and thicker than the microvilli on adjacent cells, and are grouped around a kinocilium that is centrally located on an acutely domed cell surface (Cotanche and Sulik 1984). Over the next 4 days, the number of hair bundles increases rapidly, and the final adult complement is nearly present by E10 (Tilney et al. 1986; Goodyear and Richardson 1997). Both SEM and immunofluorescence microscopy indicate that hair bundles first appear at the distal end of the papilla, with the distal patch of new hair bundles expanding proximally as the papilla elongates. A few scattered hair bundles are, however, observed in the proximal end of the papilla before the distal patch has expanded that far (Goodyear and Richardson 1997). In contrast, the first hair cells to be born are found in a long narrow strip that extends along the superior margin of the papilla, and new hair cells are added peripherally to this strip (see above; Katayama and Corwin 1989), so hair bundles do not simply appear on hair cells at a fixed time after they have been born.

The stereocilia of the newly emerging hair bundles are of fairly uniform height, and contain only a few peripherally located actin filaments. Between E8 and E12 the following events occur (see Fig. 2.4): the kinocilium migrates to one side of the cell, defining the polarity of the bundle (Cotanche and Sulik 1984; Tilney et al. 1992); the hair bundles reorient and become aligned with those of neighboring hair cells (Cotanche and Corwin 1991); the number of actin filaments within each stereocilium increases and the actin filaments become increasingly crossbridged (Tilney and DeRosier 1986); the stereocilia nearest the kinocilium begin to elongate, followed by those progressively further away from the kinocilium, thereby generating a hair bundle with rows of stereocilia of increasing height, the "stereocilial staircase" (Tilney et al. 1986). The initiation of stereocilia elongation occurs synchronously throughout the length and width of the papilla (Tilney et al. 1986). During this first phase of bundle differentiation, the hair cell surface is continuing to grow in area and new stereocilia are being added, filling the entire apical surface, to the extent that by E12 each hair cell has 1.5 to 2.0 times the number of stereocilia than that present on mature hair cells (Tilney et al. 1992).

During the next phase, from E13 to E15, there is a pause in stereociliary elongation, and the stereocilia begin to increase in width in those cells in which stereociliary thickness is not yet at adult values. Simultaneously the stereocilia begin to taper at their bases, rootlets form and project down into the developing cuticular plate, and the excess stereocilia are resorbed from the surface. Subsequently, from E17 onward, the elongation of stereocilia is reinitiated in those bundles that have not yet attained their final height. Cessation of stereociliary elongation occurs first in the row of shortest stereocilia, and last in the tallest row, thereby increasing the difference in height-ranking between the rows.

These studies have indicated that the growth in length and thickness of a chick stereocilium are processes that are separated in time (Tilney et al. 1986). Furthermore, it has been suggested that while the actin filaments in stereocilia

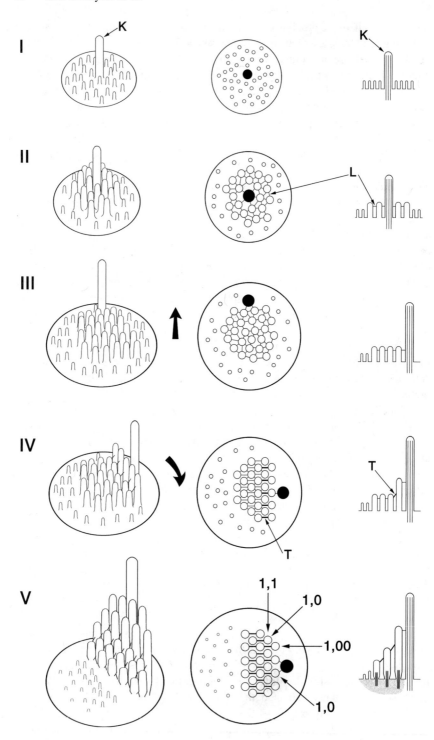

elongate via the addition of actin monomers to the most distal, barbed end at the tip of the stereocilium, the growth of rootlets into the cuticular plate that occurs while the stereocilia are not actively elongating is due to the addition of actin monomers to the proximal, nonpreferred, pointed ends of the few filaments that pass through the constricting taper developing at the stereocilial base (Tilney and DeRosier 1986). An analysis of the site of radiolabeled amino acid incorporation during the final growth phase of stereocilia in the chick basilar papilla (E17) failed to provide evidence for the hypothesis that stereocilia grow at their tips (Pickles et al. 1996). However, recent experimental studies in cultures of the early postnatal rat cochlea, using hair cells transfected with GFP-actin, have shown that new actin is continuosly incorporated at the tip of each stereocilium at this stage of development (Schneider et al. 2002; Rzadzinska et al. 2004).

6.1.2 The Mammalian Cochlea

Despite its importance as a model system, good ultrastructural descriptions of the normal process of hair-bundle development in the embryonic mouse cochlea are lacking. Evidence from transmission electron microscopy indicates hair bundles are distinguishable in the basal region of the mouse cochlea at E15 (Anniko 1983a). A monoclonal antibody (SC1) that recognizes stereocilia labels a few hair cells in the mouse cochlea at E16 (Holley and Nishida 1995). In the rat cochlea, stereociliary bundles can be first recognized by SEM on IHCs in the base of the cochlea at E18 (Zine and Romand 1996). As in the chick basilar papilla, these early rat hair bundles consist of densely packed stereocilia of approximately the same height surrounding a more or less centrally positioned kinocilium. Newly emerged hair bundles in the cochlea of the neonatal golden

FIGURE 2.4. A diagram illustrating how a generic sensory hair bundle with tip links develops. The images depict the apical surface of a hair cell at different stages of its formation as viewed from above at an oblique angle (*left*), from directly overhead (*center*), or as sectioned transversely in a plane passing through the kinocilium (*right*). The kinocilium is depicted in black in the overhead view. Stage I: Hair cell with a single central kinocilium (K), and a few stubby microvilli on its apical surface. Stage II: Hair bundle with a central kinocilium surrounded by short stereocilia of uniform height. The tips of the stereocilia are interconnected by lateral links (L) running in all directions (shown center and right). Stage III: Kinocilium migrates to the periphery of the cell (*bold vertical arrow*). Stage IV: Kinocilium repositions defining hair-bundle polarity (*bold curved arrow*). Stereocilia immediately next to the kinocilium begin to increase in height, pulling the lateral links that are directed along the 1,00 lattice plane up to form oblique tip links (T, links shown in bold center and right). Stage V: Successive rows of stereocilia become incorporated into the hair bundle, with tip links forming between new rows along the 1,00 lattice line. Excess stereocilia begin to be resorbed. Rootlets and the cuticular plate begin to form. The scheme for link formation is based on Pickles et al. (1991). Nomenclature for lattice planes is from Tilney et al. (1988b).

hamster have stereocilia that are undifferentiated in height and can be distinguished from the microvilli on surrounding supporting cells by their relative shortness and clustered appearance (Kaltenbach et al. 1994).

In the hamster, differences in the height of the stereocilia bundles from the base to the apex of the cochlea (longest bundles at the apex, shortest at the base) are generated by differences in the growth rate of stereocilia (Kaltenbach et al. 1994). This is in contrast to the scenario proposed for the chick (Tilney et al. 1986), where it is the time period over which growth occurs (at a presumably constant rate) that determines overall bundle height as well as differences in the height of stereocilia between the rows in the bundle. Furthermore, during postnatal development in the hamster cochlea, increases in the width and height of stereocilia occur simultaneously (Kaltenbach et al. 1994), rather than being separated in time as they are in the chick (Tilney et al. 1986). In the rat cochlea, there is some evidence that the stereocilia bundles on basal-coil outer hair cells may actually shorten between 4 and 24 days after birth (Roth and Bruns 1992). In all mammalian species studied thus far, hair-bundle appearance and differentiation in the cochlea proceeds in a base-to-apex direction. Hair cells in the apex begin to differentiate after those in the base, although hair cells in the apex may be born before those in the base (see above; Ruben 1967). This is in contrast to the chick basilar papilla, in which hair bundles in the distal, low-frequency end appear before those in the proximal, high-frequency end (see Section 6.1.1) and where the different phases of early stereociliary growth that lead to differentiation of a bundle with rows of ranked stereocilia are initiated simultaneously through the length of the papilla (Tilney et al. 1986). In both birds and mammals, hair bundles do not simply appear at a fixed time point after the cells have withdrawn from the cell cycle.

6.1.3 The Vestibular System

Hair bundles can be first identified in the cristae of the chick inner ear at E4 (stage 26) using the anti-HCA monoclonal antibody (Bartolami et al. 1991). Scanning electron microscope studies have unambiguously identified hair bundles in the cristae (Mbiene and Sans 1986) and maculae (Denman-Johnson and Forge 1999) of the mouse inner ear at E13.5, within 2 days of the birth of the first hair cells in these organs (see above; Ruben 1967). As in the auditory organs, the newly emerging hair bundles consist of short stereocilia of relatively uniform length clumped around a central kinocilium of similar height (Mbiene and Sans 1986; Denman Johnson and Forge 1999). At an earlier stage of development (E12.5), cells tentatively identified as immature hair cells have a single central kinocilium, very few microvilli relative to surrounding cells, and just a few small dimples on their apical surface (Denman-Johnson and Forge 1999). Bundles with eccentrically placed kinocilia and stereocilia of either uniform or staggered height are observed by E14.5 in the cristae, and as early as E13.5 in the maculae, alongside bundles with a centrally placed kinocilium, suggesting that bundle polarization and formation of the stereocilial staircase occur very rapidly in these organs.

In the anterior crista of the mouse ear, a central-to-peripheral gradient of hair-bundle differentiation is observed. Mature bundles first appear in central regions at the apex of the crest, and the less mature bundles are located peripherally around the sloping flanks (Mbiene and Sans 1986). This gradient of hair-bundle differentiation roughly matches that described for hair-cell birthdates in the lateral crista, where the oldest cells are located at the top of the ridge (see above; Sans and Chat 1982). However, despite a general gradient of differentiation in the anterior crista, from center to periphery, immature bundles are also observed interspersed among more mature bundles (Mbiene and Sans 1986). A similar situation has been described in the developing mouse maculae, where hair bundles of differing maturity are found intermingled, suggesting there may be two or more waves of differentiation occurring at different times (Denman-Johnson and Forge 1999).

6.2 Determination of Hair-Bundle Polarity

The mechanisms that determine the orientation, or planar polarity, of hair bundles in vertebrate mechanosensory organs are not yet understood, although they may well turn out to be be similar to those that determine the planar polarity of sensory bristles in *Drosophila* (Lewis and Davies 2002). Hair bundles all have a similar polarity in the auditory organs and the cristae of the inner ear. In contrast, hair bundles in the maculae are oriented toward (in the utricle) or away (in the saccule) from a curved line running through a strip of the epithelium known as the striola, and are therefore of opposite polarity on different sides of the organ. A single kinocilium is a feature of developing hair bundles in all types of hair cell, although it is lost during the later stages of hair-bundle maturation in the auditory organs of many species. Migration of the kinocilium to one side of the cell surface is one of the earliest events occurring during hair-bundle development (Fig. 2.4). In the chick basilar papilla, this initial movement of the kinocilium produces fields of adjacent hair bundles with a unimodal but very broad distribution of polarities (Cotanche and Corwin 1991). Subsequently, this variation in distribution decreases considerably and bundle orientation on neighboring cells becomes more uniform. In some areas of the papilla this occurs very rapidly, before significant growth of the hair bundle, while in other areas it occurs over a much slower time frame. The degree to which this reorientation is due to rotation of the entire cell, or movement of the bundle within the cell is unknown. In the maculae of the mouse, the adult-like pattern of hair bundle orientation characteristic of these organs is already clearly apparent by E15.5, before the birth of many hair cells and well before the maturation of most hair bundles (Denman-Johnson and Forge 1999). A systematic variation in orientation across the epithelium is even observed as early as E13.5, the stage at which polarized bundles can be first identified. In the cochleae of both rats and hamsters, unlike the situation in the basilar papilla, hair-bundle polarity appears to be precisely determined from the outset (Kaltenbach et al. 1994; Zine and Romand 1996). As soon as a hair bundle is observed with an eccentrically placed kinocilium, its polarity is as it will be in the adult. In a more recent study of the developing mouse cochlea, however, although hair-bundle orientation is initially nonrandom it is not as precise as it is in the adult, and there is good evidence for hair-bundle reorientation occurring in both inner and outer hair cells between E17 and P10 (Dabdoub et al. 2003).

It has been suggested that the differential growth of layers of matrix within the overlying tectorial membrane provides a traction force that aligns hair bundles in the developing basilar papilla (Cotanche and Corwin 1991). Data from SEM studies have been used to argue that this hypothesis is unlikely to work in mammals. In the hamster cochlea, hair bundles of correct orientation appear in advance of visible tectorial membrane coverage (Kaltenbach et al. 1994). In the mouse maculae, correctly oriented bundles are observed before significant production of a thickened otoconial membrane, and differences in the sites of otoconial membrane production that could account for the opposing hair-bundle polarities seen in these organs have not been observed (Denman-Johnson and Forge 1999). However, it should be noted that mRNAs for the tectorins, proteins that are major components of the tectorial and otoconial membranes, begin to be expressed by cells directly adjacent to the sensory primordium of the mouse cochlea at E12.5, well before the appearance of hair bundles (Rau et al. 1999), and that β-tectorin mRNA begins to be expressed exclusively within the striolar region of the maculae coincident with the appearance of correctly oriented bundles. It is not certain to what extent the immature otoconial and tectorial membranes survive the preparation procedures used for SEM, and their artefactual loss could lead to erroneous conclusions concerning their function.

Within the hair-cell body, microtubules tend to be arranged predominantly parallel to the long axis of the cell as hair-bundle polarity is being first established, suggesting they are unlikely to play a role in determining orientation (Troutt et al. 1994; Denman-Johnson and Forge 1999). However, organelles known as striated rootlets tend to lie parallel to the apical surface, and are seen to run between the basal body that lies beneath the kinocilium and the circumferential adherens junction and may therefore play a role in positioning the kinocilium (Denman-Johnson and Forge 1999). Aligned microfilament assemblies that are associated with the adherens junctions and lie parallel to the apical epithelial surface are prominent components of supporting cells in mature and developing maculae. These are also present to a certain extent in differentiating hair cells, and it has been suggested that a network of such microfilament assemblies might provide a signaling framework through which hair-bundle polarity is coordinated throughout the epithelium (Denman-Johnson and Forge 1999). In the zebrafish *mindbomb* mutant, however, in which supporting cells are completely absent, hair bundles develop the appropriate orientation for their location in the epithelium and are aligned fairly well with their neighbors (Haddon et al. 1999). Supporting cells are therefore not required for determining hair-cell polarity, nor for transmitting any signal through the epithelium, and any signals that are transmitted through the epithelium, as opposed to external influences, can presumably be transmitted directly from hair cell to hair cell if necessary.

The primary cell polarity genes of *Drosophila* encode a number of proteins including Frizzled, a member of the Wingless/Wnt receptor family; Dishevelled, a PDZ domain protein recruited to the membrane by Frizzled; two members of the cadherin superfamily, Dachsous and Flamingo; and a protein with a PDZ-domain binding interface known as Strabismus or Van Gogh. Recent studies have revealed that two mouse mutants derived from an ENU screen, spin cycle and crash, have mutations in Celsr1, a mammalian homologue of *Drosophila* frizzled (Curtin et al. 2003). In mice heterozygous for these dominant mutations there is a three-to sixfold increase in the number of misoriented hair bundles found on the outer, but not inner, hair cells in the adult cochlea. Mice homozygous for these mutations are not viable because of neural tube defects, but have large numbers of severely misoriented outer hair cells at E18.5. The degree of misorientation observed in these mutants is substantially greater (up to 180°) than the nonrandom orientation (up to 35°) observed in wild type mice at E17, indicating the defect is not simply due to a failure in the normal process of reorientation that refines planar polarity in the developing organ of Corti, unless substantial reorientation normally occurs between E15 and E17.

In loop-tail mice homozygous for a mutation in the gene encoding a mammalian homologue of *Drosophila* Van Gogh, *Vangl2*, the hair bundles of both inner and outer hair cells are misoriented by up to 180° and analysis at E16.5 indicates that this mutation most likely affects the initial biased movement of the kinocilium to the abneural edge of the hair cell (Montcouquiol et al. 2003). Although Frizzled and family members are cell-surface receptors for the Wingless/Wnt family of secreted signaling molecules, there is no evidence that these ligands are required for planar polarity in *Drosophila*. Cochlear hair-bundle orientation is normal in Wnt7a mutant mice, but experimental evidence from mouse cochlear cultures indicates that Wnt7a-conditioned media and inhibitors of Wnt signaling can block hair-bundle reorientation that is seen in the basal end of the cochlea in vitro, and in situ studies suggest the pillar cells lying between the inner and outer hair cells may act as a line source of Wnt7a that provides the gradient for hair cells to respond to when reorienting during development (Dabdoub et al. 2003). Presumably, other signaling pathways may compensate for the loss of Wnt7a signaling in the mutants.

As described earlier, hair-bundle polarity is also disrupted in the hair cells that are overproduced in *Notch*$^{+/-}$ and *Jag2*$^{-/-}$ mutant mice, and in rat cochlear cultures treated with antisense oligonucleotides for *Notch1* and *Jag1* (Lanford et al. 1999; Zhang et al. 2000; Zine et al. 2000). These data suggest Notch signaling may play some role in this aspect of hair-cell development, in addition to determining the cell-fate choice, although the *mindbomb* mutant (Haddon et al. 1999) suggests otherwise, and the polarity defects observed in response to altered Notch signaling may be a secondary effect resulting from overcrowding.

Mutations in cytoskeletal and cell-surface molecules can also cause alterations in hair-bundle polarity (see Section 6.6).

6.3 Formation of Hair Bundle Shape

Emerging stereociliary bundles in every organ, when viewed from above, all initially have a circular shape. As development proceeds they acquire their characteristic adult shape. In the mammalian cochlea, the stereociliary bundles of inner hair cells develop a straight or slightly curved crescent-like profile, and those of outer hair cells become V- or W-shaped. In the chick basilar papilla, the stereociliary bundles on proximal hair cells develop a rectangular profile, while those at the distal end retain a circular shape. In the vestibular system most bundles have a circular profile, although some are more box-shaped. The development of bundle shapes other than circular involves at least two processes, a straightening of the rows of stereocilia that are recruited into the elongating staircase (Tilney et al. 1992) and a resorption of both the excess stereocilia that are initially produced and those that appear as the apical surface area increases. Resorption of excess stereocilia is a feature that has been described in the auditory organs of mouse (Lim and Anniko 1985; Furness et al. 1989), rat (Lenoir et al. 1980; Roth and Bruns 1992), hamster (Kaltenbach et al. 1994), and chick (Tilney et al. 1986), but not in studies of vestibular organ development (Mbiene and Sans 1986; Denman-Johnson and Forge 1999), suggesting it may be important for generating the more complex, location-dependent arrays of stereocilia that are found in the auditory hair cells.

6.4 Development of the Cuticular Plate

The cuticular plate is a dense meshwork of actin filaments and other cytoskeletal proteins including α-actinin, spectrin, tropomyosin, myosin Ie, and myosin VI into which the stereociliary rootlets project (Drenkhahn et al. 1985; Slepecky 1996; Hasson et al. 1997; Dumont et al. 2002). It is located just beneath the hair bundle and presumably forms a rigid intracellular structure to which the bundle is anchored. In the chick basilar papilla, transmission electron microscope (TEM) studies indicate that the cuticular plate begins to form at E11, after the initial polarity has been established, and just as the first phase of stereociliary elongation is ending and the onset of rootlet formation is starting (Tilney and DeRosier 1986). In the mouse, TEM observations show cuticular plates are present by E15.5 in macular hair cells with polarized bundles (Denman-Johnson and Forge 1999). Structurally distinct cuticular plates are visible in mouse cochlear hair cells at E17 (Anniko 1983a,b), and are well defined by P2 (Furness et al. 1989). A monoclonal antibody that recognizes an unidentified antigen associated with the cuticular plates of mature hair cells stains the apical pole of macular hair cells by E14, and that of basal coil inner hair cells by E15, indicating that some molecular components of the cuticular plate may be in place before a distinct plate is discernable by purely morphological criteria (Nishida et al. 1998).

6.5 Development of Links

Despite the likely importance of interstereociliary links for mechanotransduction and the development and maintenance of hair-bundle integrity, the appearance and distribution of these membrane specializations on the hair-bundle surface has received little systematic attention. Tip links have been described on mouse macular hair cells at E15.5 (Denman-Johnson and Forge 1999), rat inner hair cells at birth (Zine and Romand 1996), gerbil outer hair cells by P2 (Souter et al. 1995), and mouse outer hair cells in vitro at the equivalent of P3 (Furness et al. 1989). In the chick basilar papilla, structures that are probably tip links, but do not have associated submembranous densities, can be observed by TEM as soon as a gradation in stereociliary height is apparent (at E9, Pickles et al. 1991). Links between adjacent rows of stereocilia can be observed running up the 1,00 lattice plane (see Fig. 2.4, stages IV and V) of the hair bundle (i.e., like tip links in the adult) in the basilar papilla of E11 chick embryos (Tilney et al. 1992). Submembranous densities are first observed at the insertion points of the nascent tip links sporadically at E13, and reliably by E16. The emerging tip links seen at E9 appear to develop from a network of fine filamentous strands that is concentrated at the top of the bundle at earlier stages (E7.5, Pickles et al. 1991). These filaments run along all lattice planes and interconnect all the nascent stereocilia within a new bundle (Fig. 2.4, stages II and III). Similar spoke-like arrangements of lateral links interconnecting all neighboring stereocilia have been described at the tops of newly emerging hair bundles in the mouse macula at E13.5 (Forge et al. 1997), and in early postnatal mouse cochlear hair cells between all stereocilia, including those that will eventually be resorbed (Furness et al. 1989). A protein antigen associated with both tip links and kinocilial links, the tip-link antigen (TLA, Goodyear and Richardson 2003) and cadherin 23, another component of tip and kinocilial links (see Section 6.6.3, Siemens at al. 2004), are expressed at high levels in embryonic hair bundles, relative to those observed in the adult (Goodyear and Richardson 2003; Boeda et al. 2002) consistent with the observation that tip links are generated from a population of filaments that are initially more abundant and less restricted in their orientation during the early developmental stages of hair-bundle development.

A model for tip link formation suggests that a subset of the links in these arrays of lateral links are pulled upward to a more vertical inclination as stereociliary elongation is initiated and become tip links, while the others remain in a basal location and contribute to the ankle links (Pickles et al. 1991). In an interesting model for hair-bundle growth, Tilney et al. (1988a), proposed that ion flow through the transducer channels that are gated by the tip links may uncap the barbed ends of the actin filaments at the stereocilial tips and thereby initiate stereociliary elongation. A system such as this—stereocilial elongation increasing tip-link stretching, tip-link stretching increasing channel opening, ion influx increasing actin uncapping and therefore further elongation—requires some triggering stimulus, perhaps traction caused by upward growth of the kinocilium. This remains an interesting model for hair-bundle growth, although recent evidence from both the chick (Si et al. 2003) and the mouse (Geleoc and Holt 2003) suggest the onset of mechanotransduction (see below) occurs after the ranking of stereocilia has been initiated.

Far less attention has been paid to the development of the other interstereociliary link types. It is not known when the horizontal top connectors described in certain types of hair cell in the frog (Jacobs and Hudspeth 1990), fish (Neugebauer and Thurm 1985), and chick (Goodyear and Richardson 1992) inner ear are first formed. The development of linkages has been examined in the utricle and saccule of the European roach by comparing hair bundles on marginally located, apparently immature hair cells with those on central, presumably more mature hair cells (Neugebauer 1986). Peripheral bundles have a dense collection of linkages connecting their entire surface. In addition, spherical, membrane-bound structures are associated with the links and decorate the entire stereociliary surface. In central bundles with a more mature conformation, these spherical structures are not observed, but connectors around the base of the bundle are enhanced, connectors further up the bundle are reduced in number, and tip links are more apparent (Neugebauer 1986). In the chick inner ear, components of the shaft connectors (the 275 kDa HCA, Richardson et al. 1990), ankle links (the ankle-link antigen, Goodyear and Richardson 1999) and tip links (the tip-link antigen, Goodyear and Richardson 2003) are expressed over the entire surface of the hair bundle as it first forms (Bartolami et al. 1991; Goodyear and Richardson 1999; 2003). The ankle-link antigen eventually becomes restricted to a narrow zone on the hair bundle surface that lies just above the level of the cuticular plate, but is also transiently concentrated at the very tip of the developing hair bundle for a brief period (Goodyear and Richardson 1999). The tip-link antigen finally becomes restricted to the top of the mature hair bundle, where it is associated with tip links and kinocilial links (Goodyear and Richardson 2003).

Various aspects of the process of hair-bundle development (Fig. 2.4), like the changes observed in the position of the kinocilium and the formation of tip links from arrays of lateral top links, may be common to all hair-cell types and species, while others, such as the sculpting of a hair bundle from a proportion of the stereocilia that are initially produced and the subsequent resorption of the unwanted excess numbers, may occur only in auditory as opposed to vestibular hair cells. There are a limited number of studies to draw upon, and those that are available are usually restricted to one organ from one species, so the common, and therefore presumably essential, features of the process are still hard to determine with certainty.

6.6 Role of Cytoskeletal and Cell-Surface Proteins

The mechanisms that control the growth of actin filaments in a hair cell must, to a large extent, determine the final shape and form of the stereociliary bundle, and the potential role of molecules that cap the ends of actin filaments has already been discussed above. Rho-GTPases, members of the Ras family of small G-proteins including RhoA, Rac1 and Cdc42, play a pivotal role in directing the assembly and polarity of the actin cytoskeleton in a variety of cell types, and many upstream effectors and downstream targets of these molecules are now known (Hall 1998). This knowledge has been used to suggest how the formation of actin filament bundles is regulated during hair bundle development (Kollmar 1999; Muller and Littlewood-Evans 2001) but thus far there is little firm evidence for these various hypotheses. The human homologue of the *Drosophila* gene *diaphanous*, *HDIA1*, encoding a downstream target of RhoA, is mutated in an autosomal dominant form of hereditary deafness, *DFNA1*, but this is a late onset progressive form of deafness, the cellular phenotype in the ear is unknown, and it is not clear whether *HDIA1* is required for normal development of the hair bundle (Lynch et al. 1997). However, there are several lines of evidence indicating that unconventional myosins, actin-bundling proteins and cytoplasmic scaffolding proteins, and a variety of cell-surface molecules all play a role in the process of hair-bundle development.

6.6.1 Unconventional Myosins

Myosins are actin-binding molecular motors that form a large superfamily currently comprising 17 unconventional types of myosin as well as the conventional type II myosins (for a recent review see Berg et al. 2001). Myosins are defined by the presence of a highly conserved actin- and ATP-binding head domain that in most cases is followed by a light-chain binding region comprising one or more IQ motifs, and a C-terminal, class-specific tail. Tail domains vary greatly and can contain, for example, FERM domains (e.g., myosin VIIa, myosin XVa), thought to be involved in linking cell membranes with the cytoskeleton, and SH3 domains that are associated with roles in endocytosis and protein localization. Myosins are likely to have a wide range of important roles in cells, including roles in motility, organelle trafficking, and maintenance of cell shape. More recently described roles in yeast include establishment of polarity by myosin V (Yin et al. 2000) and actin polymerization by myosin I (Berg et al. 2001). Whether these or other myosins play a similar role in hair cells has yet to be determined. However, unconventional myosins VI, VIIa, and XV are all specifically expressed by hair cells in the cochlea, and mutations in the genes encoding these proteins cause disruption of the hair bundle and result in profound deafness. Myosin IIIa, a vertebrate homologue of *Drospohila* ninac—an unusual myosin with an N-terminal kinase domain, is also expressed by hair cells in the inner ear, and mutations in human myosin IIIa cause a progressive form of nonsyndromic hearing loss (Walsh et al. 2002).

Myosin VI localizes to the cuticular plates of hair cells, where it is associated with the stereociliary rootlets, and to the pericuticular necklace, a vesicle-rich zone that surrounds the cuticular plate (Hasson et al. 1997). The *Snell's waltzer* mouse exhibits circling behavior and is deaf, has a deletion in the *Myo 6* gene (Avraham et al. 1995), and does not produce any detectable myosin VI protein. At P1, the hair bundles in the cochleae of *Snell's waltzer* mutant mice are polarized and have ranked stereocilia. However, the outer hair cell bundle lacks the clear V-shape characteristic of wild-type hair bundles at P1, the excess stereocilia that are normally reabsorbed from the apical surface are more apparent, and some of the stereocilia are fused (Self et al. 1999). By P7, severe fusion of stereocilia is evident leading to many giant stereocilia. This degenerative pathology progresses as maturation continues and only a few hair cells are left by P20. Myosin VI is unique among myosins in that it moves toward the minus ends of actin filaments (Wells et al. 1999), and it has been suggested that it tethers the plasma membrane at the base of each stereocilium to the rootlet (Self et al. 1999; Cramer 2000). In the absence of myosin VI the plasma membrane may slide up the actin cores resulting in fusion of the stereocilia. Although myosin VI is important for apical endocytosis in polarized epithelial cells (Buss et al. 2001, Biemesderfer et al. 2002), the phenotype observed in *Snell's waltzer* mice is thought not to result from a defect in membrane retrieval. Coated pits are associated with the apical membranes of hair cells in *Snell's waltzer* mutants (Self et al. 1999), and the fusion of stereocilia apparently does not involve an increase in the area of membrane covering the stereocilia (Cramer 2000).

Myosin VIIa is highly abundant in hair cells, being present in the stereocilia, cuticular plate, and pericuticular necklace (Hasson et al. 1997), and generally distributed throughout the cell soma. *Shaker1* mice have recessive mutations in the *Myo7a* gene that cause deafness and vestibular dysfunction (Gibson et al. 1995). There are several different alleles of *shaker1*. The original *shaker1* mouse, *Myo7a^{sh1}*, has a missense mutation in a poorly conserved region of the motor head of myosin VIIa (Gibson et al. 1995). In *Myo7a^{sh1}* mutants, inner hair cell stereocilia are somewhat irregularly organized whereas the stereocilia of outer hair cells, although well organized even at P15, tend to possess two rows of stereocilia instead of the normal three (Self et al. 1998). The *Myo7a^{6J}* and *Myo7a^{816SB}* alleles of *shaker1* have mutations in the conserved core region of the myosin VIIa head, and the phenotype is more severe than that in the *Myo7a^{sh1}* mutant. Hair bundles can be clearly distinguished by SEM at E18, and in both wild type and mutant mice the hair bundles are already polarized with rows of stereocilia of ranked height situated on one side of the cell surface. However, in comparison to those of wild type animals, the shapes of the *Myo7a^{6J}* and *Myo7a^{816SB}* mutant bundles are rather irregular, the kinocilia are not always in their normal position with respect to the stereocilial arrays, and the hair bundles are poorly oriented with respect to one another (Self et al. 1998). By P3, severe disruption of hair bundles is seen in mutants. The hair bundles are often split into multiple small clumps of stereocilia that are randomly arranged with respect to one another, although the stereocilia within each clump are ranked in height and show a typical staircase pattern of organization. The excess stereocilia appear to be resorbed on schedule in these *Myo7a* mutants, and endocytotic activity at the apical surface is, as judged by cationic ferritin uptake, normal (Richardson et al. 1997). The cuticular plate also forms on time in these mutants and is visible by E18, but becomes interrupted by regions of vesicle-rich cytoplasm forming smaller islands with associated clumps of stereocilia. As the mouse matures many stereocilia are lost and the hair cells finally degenerate and are lost from the sensory epithelium (Self et al. 1998). The phenotype of these mutants indicates *Myo7a* may be required to retain the kinocilium, the cuticular plate and its associated stereocilia associated as a single coherent assembly. In the zebrafish circler mutant *mariner*, mutations in the myosin VIIa gene also cause splaying of the sensory hair bundles (Nicolson et al. 1998; Ernest et al. 2000). Myosin VIIa binds via its FERM domain with vezatin, a protein that is a ubiquitous component of adherens junctions and interacts with the adherens cadherin/catenin complex. Vezatin has been shown to localize to the ankle-link region of hair bundles in the chick basilar papilla and the early postnatal mouse cochlea (Kussel-Andermann et al. 2000), a region of the hair bundle to which myosin VIIa localizes in the frog saccule (Hasson et al. 1997). The abnormal adaptation behavior and the reduced stiffness of the hair bundles of *Myo7a* mutant mice (Kros et al. 2002) provide additional evidence for myosin VIIa's role in maintaining structural coherence in the hair bundle. By connecting the stereociliary membrane to its actin core it may contribute to anchoring and tensioning membrane-bound elements such as transducer channels, tip links, and other interstereociliary links.

Myosin XVa is present in the stereocilia, the cuticular plate, and cell body of wild-type cochlear hair cells (Anderson et al. 2000). *Shaker2* (*Sh2*) mice have recessive mutations in the *Myo15* gene. The *Sh2* mouse has a missense mutation in the head domain of myosin XVa, and the *Sh2^J* mouse has a truncation of the tail domain. Significant amounts of myosin XVa are observed in the cell bodies of *Sh2* and *Sh2^J* mutant hair cells, but the stereocilia do not contain myosin XVa and are abnormally short by 1 month after birth (Probst et al. 1998; Anderson et al. 2000). It is not known whether the bundles in these *Myo15* mutant mice initially grow to the correct height during development and then shrink, or whether they never attain the correct height. The hair bundles are of normal shape and orientation, and the rows of stereocilia are ranked in height, despite being very short. In inner, but not outer, hair cells and in vestibular hair cells of *Sh2* mutants an extremely long, abnormal bundle of actin filaments called a cytocaud (Beyer et al. 2000) is observed that projects basally from a region just below the cuticular plate and can extend downward for up to 50 μm from the base of the cell. If there is a limited pool of actin monomers available in each hair cell, the generation of these aberrant, actin-based cytocauds in the mutants may limit the amount of actin available for the normal growth of stereocilia. Recent studies have shown that myosin XVa localizes to the extreme tip of each stereocilium, and that the amount of myosin XVa at the tip correlates with stereocilial length, with the longest stereocilia having the highest concentration and the shortest the least (Belyantseva et al. 2003; Rzadzinska et al. 2004). In the inner ear of the mouse, myosin XVa expression is detected from E14.5 onward in the vestibule and from E18.5 in the cochlea (Belyantseva et al. 2003). As actin flux rates are highest in the tallest stereocilia, and as *Sh2* mice have very short stereocilia, it has been suggested that actin polymerization may be regulated by myosin XVa, possibly acting in conjunction with other capping and scaffolding proteins (Rzadzinska et al. 2004).

6.6.2 Scaffolding and Actin-Bundling Proteins

There are seven loci (A to G) responsible for Usher syndrome type I, a human disorder characterized by severe to profound, congenital nonprogressive hearing impairment, vestibular dysfunction, and progressive retinitis pigmentosa. The *USH1B* locus encodes myosin VIIA, and the *USH1C* locus encodes a protein called harmonin (Verpy et al. 2000), a PDZ domain protein. PDZ domains are found in a large number of different proteins and are thought to organize multiprotein complexes at the plasma membrane. There are three classes of harmonin isoforms that can be generated by alternative splicing, harmonin a, b, and c, and harmonin b is expressed in the hair bundles of the inner ear as soon as they appear (Verpy et al. 2000; Boeda et al. 2002). Harmonin b has three PDZ domains, two coiled-coil domains and a PST (proline/serine/threonine rich) domain. Harmonin b is an actin-binding protein that bundles actin filaments in vitro, and interacts via its first PDZ domain with the FERM and Myth4 tail domains of myosin VIIa (Boeda et al. 2002). The deaf circler and deaf circler 2 Jackson mice each have deletions in the murine orthologue of the *USH1C* gene, both of which affect harmonin b transcripts (Johnson et al. 2003). At 3 weeks of age, the only stage thus far described, the hair bundles in these two mouse mutants have height-ranked rows of stereocilia but exhibit a splayed and disorganized morphology (Johnson et al. 2003). The product of the *USH1G* locus, a protein known as SANS (nested acronym for *s*caffolding protein containing three *an*kyrin repeats and a *S*AM [*s*terile *a*lpha *m*otif] domain), also interacts with harmonin b (Weil et al. 2003), and mutations in *Sans* cause defects in hair-bundle development in the Jackson shaker mouse (Kikkawa et al. 2003).

The whirler mutant mouse has a large deletion in the gene encoding whirlin, a protein with three PDZ domains and a proline-rich region (Mburu et al. 2003) that is expressed in hair bundles of the mouse inner ear from E18.5 onward. In whirler mice, the stereocilia of inner hair cells are, like those in *Sh2* mice, considerably shorter than normal, and the outer hair cell bundles form a more rounded U-shape as opposed to the typical V- or W-shape (Holme et al. 2002). Defects in *whirler* also underlie DFNB31 (Mburu et al. 2003).

Espins are a recently characterized family of calcium-independent actin bundling proteins (Bartles 2000). A 110-kDa espin isoform is expressed in the Sertoli cell-spermatid junction, and a small 30-kDa isoform is expressed in the brush border microvilli of the intestine and kidney. A novel 45 to 50-kDa isoform of espin is expressed in the auditory organs of chicks and mammals and is a component of the hair bundle. Recent experiments with cultured brush-border cells transfected with different levels of espin have shown that the length of microvilli is dependent on espin concentration, with the highest levels of espin generating the longest microvilli (Loomis et al. 2003). In the cochlea of the rat, there is also correlation between levels of espin expression in hair bundles and their length, with the longest hair bundles at the apex expressing ninefold more espin than those at the basal end (Loomis et al. 2003). The deaf *jerker* mouse has a missense mutation in the region of the espin gene encoding the actin bundling module of the protein, the region that is necessary and sufficient to cause the elongation of microvilli in vitro. Although espin is expressed while stereociliary elongation is occuring in the developing chick inner ear (Li et al. 2004), hair bundles in both the auditory and vestibular system of the *jerker* mouse develop relatively normally and only begin to degenerate by P11 with the stereocilia shortening and sometimes fusing (Zheng L et al. 2000). Other actin bundling proteins, fimbrin (I-plastin) and T-plastin, a plastin isoform that is only expressed transiently during the development of auditory hair cells (Daudet and Lebart 2002), may play a greater role in regulating stereocilia length during early development, and espin may only become critical for maintaining stereociliary length at a later stage, possibly as a consequence of changes in actin-treadmilling rates (Loomis et al. 2003).

6.6.3 Molecules of the Apical Cell Surface

Cadherins comprise a large family of membrane proteins that usually mediate homotypic and heterotypic adhesion of cells (for recent review see Angst et al. 2001). Cadherins typically have an extracellular domain composed of several cadherin repeats (a negatively charged sequence motif that is stabilized by calcium), a transmembrane domain, and an intracellular domain that that can interact with cytoskeletal and other intracellular proteins. Two members of the cadherin family have recently been identified as components of the hair bundle, cadherin 23, the product of the *USH1D* locus (Bolz et al. 2001), and protocadherin 15, product of the *USH1F* locus (Ahmed et al. 2001).

Cadherin 23 (Cdh23), also termed otocadherin, is a large, single-pass, transmembrane protein with an ectodomain containing 27 cadherin repeats and a cytoplasmic domain with no significant homology to any other known proteins (Di Palma et al. 2001). Cdh23 is expressed in a wide variety of adult tissues and, within the cochlea, mRNA for Cdh23 is localized to inner and outer hair cells, and Reissner's membrane (Di Palma et al. 2001; Wilson et al. 2001). Recent studies (Siemens et al. 2004; Söllner et al. 2004) have suggested that cadherin 23 is a component of the tip links. A splice variant containing an additional exon, exon 68, may be specifically expressed by hair cells (Siemens et al. 2004). Cdh23 is mutated in the *waltzer* mutant mouse (Di Palma et al. 2001), and in alleles of *waltzer* predicted to cause loss of function, hair-bundle defects are seen very early in development. At E18.5 cochlear bundles have a particularly immature appearance, with less prominent stereocilia than normal.

In addition, the kinocilium is often misplaced. By P4, the bundles of IHCs are somewhat irregular in form, while stereocilia of OHCs are arranged in disorganized clumps, a phenotype that closely resembles that seen in *shaker1* mice with mutations in the *Myo7a* gene. In the utricle, bundles are highly splayed and the stereocilia often appear to be thicker than those in normal mice, and are sometimes fused together (Di Palma et al. 2001). Cdh23 is found on hair bundles from the onset of their appearance in both the vestibule and the cochlea of the mouse inner ear, coincident with the expression of harmonin b. The cytoplasmic domain of Cdh23 is also know to bind to the PDZ2 domain of harmonin b (Boeda et al. 2002).

The gene for protocadherin 15 (Pcdh15) predicts a protein with 11 extracellular cadherin repeats, a single-pass transmembrane domain, and an intracellular domain that has two proline-rich regions but no homology to other known proteins (Alagraman et al. 2001a). Two isoforms of Pcdh15, a long full-length form and a short isoform with a truncated ectodomain, are expressed in a variety of different tissues including brain, retina, liver, spleen, and cochlea (Ahmed et al. 2003). Pcdh15 is expressed by hair bundles from E16 in the mouse cochlea, and becomes restricted to the tallest stereocilia at later stages of development (Ahmed et al. 2003). Pcdh15 is mutated in the *Ames waltzer* mouse, a mouse in which extensive degeneration occurs in the sensory organs in the cochlea and saccule, but not in the utricle or ampullae. Four mutant alleles of *Ames waltzer* exist, encoding truncated proteins, or proteins with deletions in either the extracellular or cytoplasmic domain of Pcdh15. Although hair-cell degeneration and loss is observed in mature mice, IHCs and OHCs are still present at P10. At this stage ranked stereocilia are visible but the bundles of both IHC and OHCs are fragmented and disorganized, and are not of the correct shape. The stereocilia of OHCs are shorter than those of wild type littermates and are sometimes arranged in a circular pattern around the periphery of the cell surface (Alagramam et al. 2000; 2001a; Raphael et al. 2001). Hair-bundle defects have been described as early as P0 in a mouse with a recently characterized spontaneous missense mutation in Pcdh15 (Hampton et al. 2003). It has been suggested that Pcdh15 acts as a stereociliary cross-linker and therefore has a role in the development of hair bundles (Alagraman et al. 2001a).

Athough it is not known yet with which hair-bundle proteins Pcdh15 interacts, the evidence described above suggests that signaling from the hair-cell surface to the cytoskeleton through protein complexes containing varying combinations of Cdh23, Pcdh15, myosin VIIa, harmonin, vezatin, and Sans may orchestrate hair-bundle development. Such signaling could occur via the heterotypic adherens junctions that exist between hair and supporting cells, and/or via orthotypic junctions between the adjacent stereocilia within a hair bundle. Other cell surface molecules may also be required for the development and maintenance of hair-bundle structure; these include one of the integrin family and a receptor-like lipid phosphatase, Ptprq.

Integrins are cell surface receptors that interact with extracellular matrix molecules and are involved in regulating actin cytoskeleton dynamics via focal adhesion kinase (FAK) (Giancotti and Ruoslahti 1999). Integrins are heterodimers, formed from the association of different α and β subunits in various combinations. The α8 integrin subunit is expressed by the hair cells in all organs of the mouse inner ear by E16, where it localizes to the apical pole of the cell including the hair bundle. FAK is also localized at the apical pole of developing hair cells, although it is not obviously present on the surface of the hair bundle, and antibodies to the extracellular matrix molecules fibronectin and collagen type IV stain the apical surface of the developing sensory epithelium. A targeted mutation of the mouse α8 integrin subunit, which inactivates the α8β1 receptor, causes the malformation or loss of some, but not all, hair bundles within the utricle, whereas the other organs of the inner ear are unaffected (Littlewood Evans and Müller 2000). FAK is not recruited to the apical end of the utricular hair cells in the absence of α8β1 integrin, fibronectin is no longer detected on the apical surface of the utricular epithelium, and the distribution of type IV collagen on the epithelial surface is perturbed. Obvious hair bundle defects are first observed at E18 in α8β1-deficient mice, several days after the onset of hair-bundle differentiation at E13 (see above), although FAK recruitment to the apical end of hair cells is affected as early as E16. Reciprocal interactions between the hair cell cytoskeleton and the overlying matrix molecules mediated by α8β1 integrin may therefore be required for the growth and/or maintenance of a subset of hair bundles in the utricle, but not their initial formation. As not every utricular hair bundle is affected by the loss of the α8β1 integrin receptor, and as other sensory organs are unaffected, other integrins may also play a similar role in the inner ear.

Ptprq is a receptor-like protein phosphatase with an ectodomain containing 18 fibronectin type III repeats, a single-pass transmembrane domain, and an intracellular domain that has activity against a broad spectrum of inositol phospholipid substrates (Oganesian et al. 2003). Ptprq is almost certainly the mammalian orthologue of the chicken HCA and is a component of the hair bundle's shaft connectors (Goodyear et al. 2003). Ptprq is expressed on vestibular hair bundles of the mouse inner ear from E13.5 onward, but begins to be expressed by cochlear hair cells only at E17.5, some 2 days after the first emergence of hair bundles. In transgenic mice lacking Ptprq, hair bundles in both the vestibule and the cochlea begin to develop normally but those in the cochlea show signs of disorganization and shortening by P1. By P8 there is considerable disruption of the hair bundle, stereocilia are missing or fused, and the hair cells eventually degenerate. Hair bundles in the vestibule are apparently unaffected by the loss of Ptprq, although shaft connectors are absent from the surface of the vestibular hair bundles (Goodyear et al. 2003). These studies indicate that shaft connectors are not necessary to hold the stereocilia in the bundle together, but are required, at least by cochlear hair cells, for the correct maturation and maintenance of hair bundle structure in the cochlea. As an inositol–lipid phosphatase, Ptprq may regulate PIP2 levels in the hair bundle membrane and thereby control the assembly/disassembly of actin filaments in the stereocilia. Other phosphatases may compensate for the loss of Ptprq in vestibular hair bundles.

6.6.4 Other Mutations that Affect Hair-Bundle Formation

An intriguing phenotype, in which the development of stereocilia is disrupted, is seen in the *tailchaser* mutant mouse that has progressive, dominant non-syndromic deafness and vestibular defects. In this mouse, outer hair cell bundles lack the V-shape normally visible at P0, and the stereocilia of both inner and outer hair cells are somewhat splayed. By P3, further abnormalities are evident, bundles are misoriented with misplaced kinocilia, and spurious patches of stereocilia are seen on the hair-cell apical surface. In the vestibular system, no clear phenotype is seen at the late embryonic or early postnatal stages. However, as *tailchaser* mice age, hair cells of both the cochlear and vestibular epithelia become substantially reduced in number. The gene affected in *tailchaser* is unknown and, although the early stereociliary phenotypes resemble those caused by mutations in the myosin VIIa gene, partial mapping of *tailchaser* suggests it is a novel locus involved in stereociliary development (Kiernan et al., 1999).

A mouse mutant called Tasmanian devil (*tde*) has a transgene inserted into a locus in the middle of chromosome 5 that appears to regulate the diameter of stereocilia (Erven et al. 2002). A significant decrease in the diameter of stereocilia is observed for all three rows of stereocilia in the early postnatal stages, with the tallest row of stereocilia on IHCs showing the greatest difference, a 50% decrease in diameter, from 0.35 to 0.175 μm. The hair bundles become progressively disorganized with age, cytocauds resembling those seen in *shaker2* mice are apparent in cochlear hair cells by P7, and the hair cells eventually degenerate.

The *Spinner* mouse has a recessive mutation in a novel gene predicted to encode a small transmembrane protein called Tmie (for *trans*membrane *i*nner *e*ar) that is expressed in a number of tissues including the inner ear (Mitchem et al. 2002). The mutation results in progressive hair-cell degeneration in the cochlea from P25. Hair-bundle defects are observed by P15, with a shortening or loss of a small number of stereocilia, often around the periphery of the bundle. In addition, kinocilia are seen on IHCs at P15 in spinner mutants but not on wild type mice at this time, suggesting that the *spinner* gene is needed for normal maturation of the organ of Corti (Mitchem et al. 2002).

The mucolipins are a family of proteins that share sequence similarities with members of the transient receptor potential (TRP) ion channel family. The gene for mucolipin 3, *Mcoln3*, encodes a putative six-transmembrane domain protein that localises to the stereocilial membrane and cytoplasmic vesicles distributed throughout hair cells. In the *varitint-waddler* (*Va*) mice, mutations in *Mcoln3* cause abnormalities in cochlear hair-bundle structure that can be detected as early as E17.5 (Di Palma et al. 2002). These abnormalities include irregularly arranged stereocilia that eventually become clumped and show signs of fusion, and may result from defects in intracellular vesicle traffic or membrane turnover at the hair cell's apical membrane.

7. Development of Electrophysiological Properties

Hair cells transduce sound, linear acceleration or rotation into receptor potentials, which ultimately lead to release of neurotransmitter. To do this they need mechanoelectrical transducer channels in the hair bundle, and potassium and calcium channels with appropriate kinetics in the basolateral cell membrane (Kros 1996). Before sensory function commences there is no a priori reason why any of these specific ion channels should be functioning or indeed present. In this section we examine what is currently known about the developmental acquisition of ion channels and other transmembrane proteins that mature hair cells need to function appropriately. For reference, the time course for the acquisition of a number of ionic currents and other membrane properties by mammalian cochlear hair cells is summarized in Figure 2.5. The purpose of ion

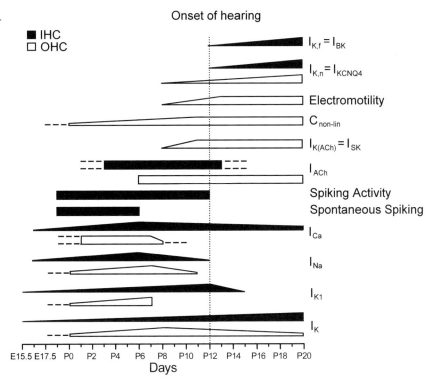

FIGURE 2.5. Diagrams showing the expression patterns in cochlear hair cells of ionic currents and other functional characteristics during maturation, as discussed in the text. Data are from small rodents (mouse, rat, gerbil) with broadly similar time courses of auditory development, and timings refer to apical-coil hair cells. All results at embryonic stages are from mouse, with E19.5 equivalent to P0. *Dashed lines* indicate uncertainty about the exact timing. Where known, developmental increases or decreases in size are indicated by a ramp. I_K represents the total classical delayed rectifier K$^+$ current ($I_{K,emb}$, $I_{K,neo}$ and $I_{K,s}$ in IHCs; $I_{K,emb}$, $I_{K,neo}$ and I_K in OHCs).

channels present in the hair cells before they start working as sensory transducers is also considered. For comparison with the timing of ion channel expression discussed below, auditory function, as judged from compound action potential recordings and/or behaviorally, commences postnatally between P10 and P12 in mice and rats (Romand 1983) and around P14 in the gerbil (Romand 1983). In the guinea pig, which has an average gestation of 68 days, function commences in utero at about E52 to E53 (Pujol et al. 1991). In the chick, auditory function develops more gradually, but a notable improvement occurs quite suddenly at E19, 2 days before hatching (Saunders et al. 1973). Less is known about the development of vestibular function, but in rats semicircular canal afferents are already responsive to angular acceleration at P0, the earliest tested, and adult sensitivity is reached by P4 (Curthoys 1979). It is likely that a similar time scale applies to mice.

7.1 Development of Mechanoelectrical Transduction

7.1.1 Cochlear Hair Cells

Hair bundles can first be distinguished at E15 in the mouse cochlea (Anniko 1983a), but it is not entirely clear at what point the tip links that are believed to gate the transducer channels first appear in these cells (see Section 6.5). Many aspects of mechanoelectrical transduction seem already mature in neonatal mice. In cochlear cultures that have been prepared from P1 or P2 mouse pups and kept in vitro for 1 to 4 days, large transducer currents can be recorded from IHCs and OHCs when the hair bundle is stimulated by a fluid jet (Kros et al. 1992). The external ear canal and middle ear are not patent until about P8 (e.g., Mikaelian and Ruben 1965), so the transducer channels are already operational at a time where they cannot possibly be stimulated by external sound. The OHC bundles from which transducer currents were recorded were between 3 and 5 μm tall (Kros et al. 1995), and moving their tips by 150 nm in the excitatory direction activates 90% of the maximum transducer current (Géléoc et al. 1997). This corresponds to an operating range of 2 degrees of rotation, exactly the same range as that estimated for adult OHCs stimulated by basilar membrane vibration in a gerbil hemicochlea (He et al. 2004). With a maximum transducer conductance of 9.2 nS and a single-channel conductance of 112 pS, up to about 80 functional transducer channels are estimated to be present in apical-coil OHCs at P1 to P2 (Kros et al. 1992; Géléoc et al. 1997). In the adult gerbil cochlea the total conductance is up to 15 nS in the apex and 35 nS in the base (He et al. 2004). This may not simply reflect a larger number of transducer channels in basal hair cells, in line with the larger number of stereocilia. In turtle hair cells, the single-channel conductance has been reported to increase tonotopically toward the base (Ricci et al. 2003) and it is currently not known whether this is also the case in the mammalian cochlea.

The currents in the immature cochlear cultures can adapt rapidly during small bundle displacements, with time constants of about 4 ms (Kros et al. 1992).

Two components of adaptation, a fast one with time constants under 1 ms and a slower one with time constants between 10 and 20 ms, have been observed in somewhat older OHCs (P4 to P7) stimulated by a fluid jet in the acutely isolated mouse cochlea (Kros et al. 2002; Goodyear et al. 2003). The fast, submillisecond adaptation process, first described in auditory hair cells of the turtle, is likely to be due to Ca^{2+} acting directly on the transducer channels, while the slower process may be regulated by myosin motors (Wu et al. 1999; Holt and Corey 2000). The apparent absence of submillisecond adaptation in the P1 to P2 cochlear cultures may reflect a developmental change or may be a technical issue due to the slower stimulus kinetics of the fluid jet used for these experiments (Wu et al. 1999; Vollrath and Eatock 2003). The first recordings of transducer currents from turtle hair cells (Crawford et al. 1989) using a stiff glass probe, also reported slower "fast" adaptation time constants (3 to 5 ms) than those reported later. In rat OHCs stimulated by a stiff glass probe fast adaptation time constants of about 0.2 ms were recorded between P3 and P18, demonstrating that this aspect of transduction is acquired early and remains after the onset of hearing (Kennedy et al. 2003). The adaptation process ensures that only a small fraction of 5% to 10% of the transducer channels are open when the hair bundle is not stimulated, as is already the case in neonatal mouse cochlear hair cells at P1 to P2 (Kros et al. 1992; Géléoc et al. 1997).

The styryl dye FM1-43, which enters hair cells through the mechanoelectrical transducer channels, is taken up by hair cells in mouse cochlear cultures prepared from P1 or P2 mice, with stronger labeling observed in the basal than in the apical coil (Gale et al. 2001). Labeling of hair cells in the apical coil increases with the time the cultures are maintained in vitro, suggesting the usual positional gradient in the maturation of hair cells along the cochlea. The exact timing of the acquisition of mechanoelectrical transducer channels in the cochlea, which may well be before birth in mice, and whether any transduction properties change before about P3, remain to be explored with electrophysiological and/or optical techniques. In the chick basilar papilla hair bundles can be identified at E6.5, with transducer currents and FM1-43 dye loading occuring from E12 onward (Si et al. 2003), slightly after the first appearance of tip links (Pickles et al. 1991; Tilney et al. 1992); see Section 6.5 above. At E12 the transducer currents were smaller in size than later on in development, had a broader operating range (i.e., were less sensitive to displacement) and showed little or no adaptation. Between E12 and the day of hatching (E21) the fraction of the transducer channels open at rest decreased from about 0.5 to 0.15, again indicative of a lack of adaptation early in development.

Little is known at present about the activation kinetics of mammalian transducer channels, and whether these kinetics vary with development, with tonotopic position along the cochlea, or between cochlear and vestibular hair cells. The fastest activation of OHC bundles reported so far has been achieved with stiff probes, but the measured transducer-current activation time constant τ of about 50 µs (corresponding to a corner frequency $f_{3dB} = 1/2\pi\tau$ of 3.2 kHz) is limited by the experimental conditions used and not by the transducer channels

(Kennedy et al. 2003). Other techniques such as laser tweezers may be required to probe the true limits of transducer current activation in mammals. For IHCs and OHCs at the basal end of the cochlea at least, the gating kinetics must be sufficiently fast to support time-varying transducer currents with frequencies as high as the upper limits of mammalian hearing (100 kHz or so).

7.1.2 Vestibular Hair Cells

In the murine vestibular organs hair bundles can be identified earlier than in the cochlea, at E13.5 (see Section 6.1.3). In the mouse utricle FM1-43 labeling indicative of functional transducer channels (Gale et al. 2001) was seen from E17, as were indeed transducer currents (Géléoc and Holt 2003). Tip links were evident at E15.5 (Denman-Johnson and Forge 1999). In contrast to the chick basilar papilla where size, operating range, and adaptation properties of the transducer current developed slowly over a period of 5 to 10 days (Si et al. 2003), mechanotransduction in the mouse utricle acquired mature characteristics within 24 hr (Géléoc and Holt 2003). Over this brief period only the size of the transducer conductance changed, increasing to 2.7 nS by E17, similar to the conductance reported for postnatal vestibular hair cells (VHCs) (Lennan et al. 1996; Géléoc et al. 1997; Holt et al. 1997; Vollrath et al. 2003). The values for fast (though slower than in the cochlea, τ about 3 ms) and slow adaptation (τ about 50 ms), the fraction of the transducer channels open at rest, and the operating range were all similar to those reported for more mature VHCs up to P10 (Holt et al. 1997; Vollrath and Eatock 2003). Transducer current adaptation in type I and type II hair cells was similar and also independent of whether they were positioned in the striolar or the extrastriolar region in the postnatal mouse utriculus (Vollrath and Eatock 2003). Regional variations in vestibular afferent responses, which can be tonic or phasic, may thus be due to the basolateral currents (see below) shaping the receptor potentials: for example, an inactivating K^+ current could counteract adaptation (Vollrath and Eatock 2003).

7.2 Development of Basolateral Membrane Properties

7.2.1 Inner Hair Cells: Basolateral Voltage-Gated Currents, Spiking, and Neurotransmitter Release

Hair-cell receptor potentials are likely to have a much lower upper frequency limit than transducer currents, because of the low-pass filtering effect of the membrane time constant τ; the corner frequency $f_{3dB} = 1/2\pi\tau$, where $\tau = R_m C_m$, the product of membrane resistance and capacitance. Nevertheless, auditory nerve fibres are able to fire action potentials that are locked onto the phase of sound waves up to several kilohertz. This phase locking, which helps with locating sound sources in space and speech processing, depends on IHCs being able to release neurotransmitter periodically at the same frequencies, necessitating a τ of the order of a few tenths of a millisecond near the cell's resting potential (Palmer and Russell 1986; Kros and Crawford 1990). Before the onset of hearing, IHCs have membrane time constants more than 100 times slower than this (Kros et al. 1998), ruling out any phase locking at audible frequencies. This developmental decrease in membrane time constant is mostly due to differences in the composition of the K^+ currents present in the basolateral membrane of IHCs before and after the onset of hearing.

Before the onset of hearing IHCs express a delayed-rectifier potassium current, called $I_{K,emb}$ and $I_{K,neo}$ in embryonic and immature postnatal hair cells, respectively (Marcotti et al. 2003a). The current's activation kinetics are slow (Fig. 2.6A), considerably slower than the voltage-gated inward currents also present in immature IHCs: a Ca^{2+} current found in all the cells from E16.5 onward and a Na^+ current evident in about 70% the cells between E16.5 and P11 (Marcotti et al. 2003b). Both Ca^{2+} and Na^+ currents activate quite close to the resting potential. The net effect of this particular mix of rapid and negatively activating inward currents and slow, more positively activating outward currents is that the membrane potential of immature IHCs is unstable. As a consequence IHC generate spontaneous and depolarization-induced action potentials from 1 to 2 days before birth until P6 (Marcotti et al. 2003a). The spikes have the character of Ca^{2+} action potentials, relatively broad and small (Fig. 2.6C) with I_{Na} increasing the frequency of the action potentials but having only a modest effect on their shape (Marcotti et al. 2003b). Another ingredient in the mix of basolateral membrane currents in developing IHCs is a strong inward rectifier, I_{K1}, that is already present at E15.5 and increases in size up to the onset of hearing, at which point it disappears (Marcotti et al. 1999, 2003a). This current is the main determinant of the resting potential and contributes to shaping the action potentials. From P7 until the onset of hearing IHCs still spike vigorously in response to depolarizing current injection, but no longer spontaneously, as the resting potential becomes more negative (about -75 mV). Likely causes include a reduction in size of I_{Ca} (Beutner and Moser 2001; Marcotti et al. 2003b), the increasing size of I_{K1} and $I_{K,neo}$ and the hyperpolarizing shift in activation of $I_{K,neo}$ (Marcotti et al. 2003a).

Such slow action potentials would be severely inadequate for encoding the frequency and intensity of sound within the range of mammalian hearing, but several lower vertebrates have regions within their mature auditory organs that encode extremely low-frequency sounds by means of slow action potentials firing at up to 20 Hz or so (e.g., Fuchs and Evans 1988, 1990; Fettiplace and Fuchs 1999). Interestingly, these spiking hair cells of lower vertebrates often also express I_{K1}, pointing to its importance in the mix of currents supporting slow action potentials.

As auditory function commences an outward potassium current, $I_{K,f}$, is expressed quite suddenly, increasing the total outward potassium current (measured at -25 mV) sevenfold between P12 and P16 (Kros et al. 1998). This current (Fig. 2.6B), with much more rapid kinetics than $I_{K,neo}$ and activating more negatively, close to -85 mV in the mouse (Oliver et al. 2003; Marcotti et al. 2004), was first described in mature guinea pig IHCs (Kros and Crawford 1990). The effect of its appearance is to disrupt the interplay of inward and outward currents

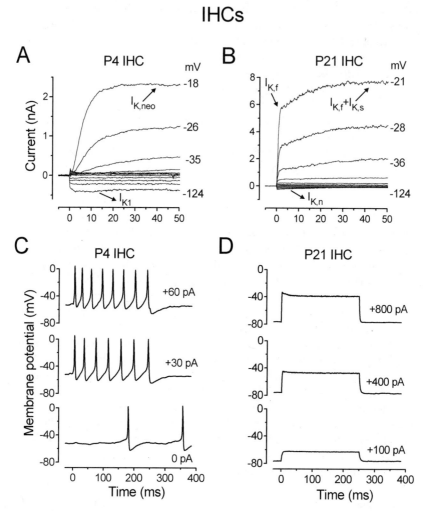

FIGURE 2.6. Electrophysiological changes during development of IHCs. (**A, B**). Current responses under voltage clamp of an immature (A) and a mature (B) IHC. The cells were held at −84 mV and depolarizing and hyperpolarizing voltage steps were applied in 10-mV increments to the potentials shown by some of the current recordings. In A, depolarizing voltage steps elicit the slowly activating delayed-rectifier $I_{K,neo}$, whereas hyperpolarizing voltage steps activate the fast inward rectifier I_{K1}. In B, depolarization activates the two main outward currents: the rapidly activating $I_{K,f}$ and the more slowly activating $I_{K,s}$. I_{K1} is no longer present and hyperpolarizing voltage steps result in deactivation of the small $I_{K,n}$, which is about 50% activated at the holding potential. Small linear leak currents have been subtracted. (**C, D**). Voltage responses under current clamp. The immature cell (C) fires spontaneous action potentials. When 250 ms depolarizing current steps are injected (size next to the traces) the spike frequency increases. Same cell as in A. A mature IHC (D) responds with fast, graded receptor potentials to depolarizing current steps. Recordings are at body temperature (except B) and are single traces.

that makes the cells spike, at the same time enabling them to encode sound stimuli in graded receptor potentials with kinetics sufficiently rapid to support phase locking (Fig. 2.6D). A slow K$^+$ current called $I_{K,s}$ is also present in mature IHCs (Fig. 2.6B), but it differs from $I_{K,neo}$. It activates more negatively, with 5% of the maximum conductance available at -70 mV, is less sensitive to tetraethylammonium (TEA) ions and it shows less inactivation (Kros and Crawford 1990; Kros et al. 1998). There are no sharp transitions between the classical delayed-rectifier currents $I_{K,emb}$, $I_{K,neo}$, and $I_{K,s}$—these names merely relate to developmental stages at which the currents are observed. All are in fact mixtures of two different conductances: one sensitive to block by 4-aminopyridine (4-AP), the other one not. During development, the 4-AP sensitive conductance gradually becomes the larger of the two (Marcotti et al. 2003a). On maturation, the IHC resting potential is no longer set by the inward rectifier I_{K1} because it disappears after P14. Another current, $I_{K,n}$, first seen at P12, takes over this role, keeping the resting potential around -75 mV (Marcotti et al. 2003a; Oliver et al. 2003). $I_{K,n}$ is an unusual delayed-rectifier K$^+$ current which is already half-activated at -85 mV and does not inactivate. It is much more prominent in OHCs (see Section 7.2.2) and in IHCs it is dwarfed by the other outward K$^+$ currents (Fig. 2.6B). However, because of its negative activation range and lack of inactivation the conductance of $I_{K,n}$ available at the resting potential (about 5.5 nS: Marcotti et al. 2003a) is comparable to that of $I_{K,f}$ (about 4 nS: Marcotti et al. 2004). The size of $I_{K,s}$ at the resting potential is smaller at about 1.5 nS (Marcotti, Johnson, and Kros, unpublished).

In contrast to the transducer channels, $I_{K,f}$ is thus expressed only when necessary to support auditory function (Kros et al. 1998). Similarly, chick cochlear hair cells acquire I_{BK} (with which $I_{K,f}$ shares many features) at E19 (Fuchs and Sokolowski 1990), exactly at the time auditory function improves rapidly (Saunders et al. 1973). Developmental changes in the pharmacology and inactivation of the slow delayed rectifier K$^+$ current of chick tall hair cells also show remarkable similarities with the transition from $I_{K,neo}$ to $I_{K,s}$ (Griguer and Fuchs 1996). Evoked (but not spontaneous) Ca^{2+} spikes were recorded from embryonic chick hair cells between E12 and E14 (Fuchs and Sokolowski 1990).

Whether immature IHCs and tall hair cells in the chick generate action potentials spontaneously in vivo, and what the precise function of this activity may be, remain to be established. Eventual degeneration of IHCs that lack the L-type Ca^{2+} current suggests that the spiking activity may be important for the normal development of these cells (Platzer et al. 2000; Glueckert et al. 2003). Regenerative spiking behavior in neonatal hair cells might also control the maturation of synaptic connections in the immature auditory system by initiating propagated activity, analogous to events in the immature retina (Maffei and Galli-Resta 1990; Meister et al. 1991; Copenhagen 1996) that occur well before the eyes become responsive to light (Ratto et al. 1991). Rhythmic spontaneous activity has been observed in neurons at various levels along the auditory brainstem of prehatch chicks (E13 to E18), including the auditory nerve (Lippe 1994; Jones et al. 2001); by E19 any spontaneous activity became unpatterned. Lippe

(1994) suggested that the rhythmic activity in the auditory brainstem might be of cochlear origin, because it disappeared on removal of both basilar papillae. Less is known about spontaneous activity in the auditory nerve and brainstem of mammals before the onset of hearing, but there are a few reports of spontaneous rhythmic discharges (reviewed by Rübsamen and Lippe 1998). In the first postnatal week, IHCs are already able to release neurotransmitter, as judged by increases in membrane capacitance indicative of fusion of synaptic vesicles, by postnatal day 1 (Marcotti et al. 2003b). Spiral ganglion neurons isolated from P0 to P5 mice are also spontaneously active, but not in a rhythmic manner, and the spikes are Na^+ dependent (Lin and Chen 2000). Thus the rhythmicity of early postnatal afferents may derive from the slow Ca^{2+} spikes of the IHCs, rather than from intrinsic properties of the afferents. To complicate the puzzle further, the spiking in the IHCs may in turn be modulated by the central nervous system (CNS), as the medial efferent system forms transient synapses on IHCs between P3 and P10 while en route to the OHCs (rat: Lenoir et al. 1980; mouse: Shnerson et al. 1982). Between P7 and P13 (the age range tested) rat IHCs do indeed respond to acetylcholine (ACh). Moreover, the spontaneous IPSPs that were occasionally observed were found to interfere with the generation of action potentials in immature IHCs (Glowatzki and Fuchs 2000).

7.2.2 Outer Hair Cells: Electromotility, Basolateral Voltage-Gated Currents and ACh Receptors

The distinguishing feature of OHCs is electromotility. Uniquely among hair cells, the length of these tall cylindrical cells varies with the receptor potentials set up by the transducer current, such that depolarization shortens and hyperpolarization lengthens them. These changes have been measured at frequencies up to 100 kHz, a limit set by the recording systems, not by the OHCs (Frank et al. 1999), and are induced by conformational changes in the transmembrane protein prestin (Zheng et al. 2000; Oliver et al. 2001). The purpose of these movements appears to be to feed mechanical forces back into the cochlea on a cycle-by-cycle basis at acoustic frequencies, to boost sensitivity and frequency selectivity. As for IHCs, this would require the OHCs (particularly those near the base of the cochlea) to have extremely fast membrane time constants—orders of magnitude faster than measured to date (Mammano and Ashmore 1996)—unless one assumes that the time-constant limitations can somehow be circumvented (e.g., Dallos and Evans 1995a,b; Kolston 1995). This electromotility can be altered by activity in the CNS. Efferent nerve fibers synapse directly onto the OHCs, and postsynaptic ACh receptors are required in the OHC membrane. It seems reasonable to expect that electromotility and its supporting features (i.e., K^+ channels speeding up the membrane time constant, responsiveness to ACh) would develop together at the onset of hearing.

Guinea pig OHCs do indeed acquire electromotility exactly at the onset of auditory function, at E52 to E53 for cells from the basal coil, with apical cells lagging about 3 days behind (Pujol et al. 1991). In other species electromotility is first observed well before the onset of hearing, at P7 in the base and at P8 in the apex of the gerbil cochlea (He et al. 1994), and at P8 and P5 in the apical coils of the mouse and rat cochlea, respectively (Marcotti and Kros 1999; Belyantseva et al. 2000). The maximum amplitude of electromotile responses, expressed as percentage length change of the cell, reached adult levels at the onset of hearing (He et al. 1994). Two lines of evidence show that prestin is already incorporated in the OHC membrane well before electromotility can be measured. Faint immunolabeling with anti-prestin antibodies was already noticed at P0 in OHCs all along the gerbil cochlea (Belyantseva et al. 2000). Also, the first signs of the nonlinear membrane capacitance caused by conformational changes of the prestin molecules were observed at P0 in apical rat OHCs, with basal OHCs surprisingly lagging behind somewhat (Oliver and Fakler 1999). The development of the size of the nonlinear capacitance and of the immunolabeling intensity shows that the density of prestin molecules in the OHC membrane reaches mature levels near the onset of hearing. The delay of about 1 week between the time at which prestin is first inserted in the membrane and measurable electromotility is most likely due to a minimum density of prestin molecules being required before their movements can be coupled together in series. Alternative possibilities include a lower sensitivity of the optical motility measurements compared to the capacitance measurements or the need for other structural molecules that might appear later in or near the OHC membrane. The latter are unlikely to include GLUT-5, proposed as a candidate for the OHC motor just before the discovery of prestin (Géléoc et al. 1999), as immunolabeling was not seen before P15 in the OHC membrane (Belyantseva et al. 2000).

The potassium current $I_{K,n}$ (Fig. 2.7B) is the dominant K^+ current of mature OHCs, which also express a small delayed rectifier, I_K. $I_{K,n}$ is more than 50% activated at the OHCs' resting potential, thus reducing the membrane time constant (Fig. 2.7D), if not sufficiently to fully explain high-frequency electromotility (Housley and Ashmore 1992; Mammano and Ashmore 1996; Marcotti and Kros 1999). In the guinea pig it is tonotopically expressed with larger current densities for the smaller, more basal cells. The ion channel underlying this current contains the KCNQ4 subunit (Kubisch et al. 1999; Marcotti and Kros 1999) and is a member of the M-current family (Brown and Adams 1980; Selyanko et al. 2000). In mice, $I_{K,n}$ is expressed from P8 onward in apical OHCs, at exactly the time that electromotility can first be detected and increasing to half the maximum adult size by P12 (Marcotti and Kros 1999). Its expression has a huge effect on the resting potential, bringing it from -55 mV before $I_{K,n}$ is present to -75 mV in mature OHCs. Mutations in KCNQ4 lead to a dominant form of slowly progressive hearing loss (DFNA2) (Kubisch et al. 1999). $I_{K,n}$ may thus be necessary for the survival of hair cells and auditory nerve fibers, by providing an efficient exit path for K^+ ions entering through the transducer channels (OHCs and IHCs) and by maintaining a low submembrane Ca^{2+} concentration (IHCs) (Oliver et al. 2003) which in turn ensures a low resting level of neurotransmitter release, preventing excitotoxic damage to the type I auditory nerve afferents. Before their functional maturation at P8, mouse and rat OHCs

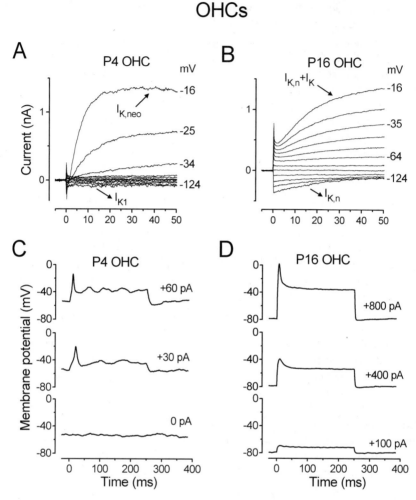

FIGURE 2.7. Electrophysiological changes during OHC development. (**A, B**). Current responses under voltage clamp of an immature (A) and a mature (B) OHC. Recording conditions are as in Fig. 2.6. In A, depolarizing voltage steps elicit the slowly activating delayed rectifier $I_{K,neo}$. Hyperpolarizing voltage steps activate the fast inward rectifier I_{K1}. Both currents are smaller than in IHCs of the same age (Fig. 2.6A). In B, the dominant current $I_{K,n}$ is already partially activated at the holding potential of -84 mV. Depolarizing steps further activate $I_{K,n}$ with slow kinetics (as well as a small I_K), whereas hyperpolarizing steps slowly turn the I_{Kn} current off. Linear leak currents subtracted. (**C, D**) Voltage responses under current clamp. The immature cell (C) fires a single action potential upon depolarization. Same cell as in A. The mature OHC (D) responds with graded receptor potentials to depolarizing current. Same cells as in B. All recordings are at body temperature and are single traces.

have a mix of membrane currents, including $I_{K,emb}/I_{K,neo}$ (Marcotti and Kros 1999; Helyer et al. 2005), a Ca^{2+} current (Michna et al. 2003) and a Na^+ current (Oliver et al. 1997), that appear similar to those in IHCs, but the currents are smaller in size (Fig. 2.7A). Under current clamp (Fig. 2.7C) immature OHCs do not fire action potentials spontaneously. On injection of depolarizing current from the resting potential only a single spike occurs followed by smaller oscillations (Marcotti and Kros 1999). OHCs in an L-type Ca^{2+} channel knockout mouse degenerate even faster than IHCs, suggesting spiking activity may occur in vivo and may be important for their development (Platzer et al. 2000; Glueckert et al. 2003). The steep decline in Ca^{2+} current density of mouse OHCs at P8 (Michna et al. 2003) and the fact that mature OHCs retain only a sparse afferent innervation support the idea that this current may have mainly a developmental role in these cells. The Na^+ current also declines sharply around this time (Oliver et al. 1997).

Currents in response to ACh can be recorded from P6 on in rat OHCs (Dulon and Lenoir 1996), exactly the age at which medial efferent synaptic connections are established on OHCs (Lenoir et al. 1980). At this age the currents are inward at negative potentials and reverse near 0 mV, which would lead to cholinergic excitation instead of the usual inhibition (Housley and Ashmore 1991). These currents are likely to flow through nicotinic ACh (nACh) receptors, which are nonselective cation channels, made up of α9 and α10 subunits (Elgoyhen et al. 2001; Lustig et al. 2001; Sgard et al. 2002). In adult OHCs and in the functionally similar short hair cells of the chick (Fuchs and Murrow 1992), this inward current is often barely visible as the Ca^{2+} ions entering through the nACh receptors open nearby SK channels that carry larger outward currents at physiological potentials positive to the K^+ equilibrium potential (Blanchet et al. 1996; Evans 1996, Nenov et al. 1996; Dulon et al. 1998; Oliver et al. 2000). In rats, responses characteristic of secondary SK channel activity were first seen at P8, suggesting that the SK channels at P6 are not yet functional or not yet properly colocalized with the nACh receptors (Dulon and Lenoir 1996). In an immortalized cell line prepared from mouse cochlea at E13 putative hair-cell precursors also respond to ACh only with inward currents, again showing a lack of coupling between ACh receptors and SK channels (Rivolta et al. 1998; Jagger et al. 2000). In gerbils, responses to ACh are also first observed at P6 in basal OHCs, but right from the start currents are outward indicative of a functional nAChR–SK complex (He and Dallos 1999). The responses in gerbil OHCs reach adult size by P11.

ACh responses can thus be elicited in OHCs at the time efferent synapses are established and just before the onset of OHC motility, raising the question of whether any of these events depend on one another. Using cochlear cultures made from P0 gerbil pups, He (1997) found that electromotility develops nearly as normal in the absence of efferent innervation. Functional ACh receptors and their associated SK channels also develop normally under the same conditions (He et al. 2001).

7.2.3 Vestibular Hair Cells: Development of Basolateral Voltage-Gated Currents

Hair cells in the mouse utricle show electrophysiological signs of maturation starting at around P4 (Rüsch et al. 1998b). Between P4 and P7, type I cells acquire their characteristic I_{KI} or $I_{K,L}$, a very large potassium current that is substantially activated at the resting potential with a number of similarities to the (much smaller) $I_{K,n}$ of cochlear hair cells (Rennie and Correia 1994; Rüsch and Eatock 1996), but some differences, including considerably lower sensitivity to block by linopirdine (Rennie et al. 2001). Another sign of maturation of both type I and type II cells is the appearance, also around P4, in a fraction of the cells of the weak inward rectifier I_h. Upon maturation, I_h was present in almost all type II cells but in only one third of type I cells. Between birth and P4, immature type I and type II VHCs can sometimes be distinguished by their morphology (including the presence of developing calyces) but not by their basolateral membrane currents: a classical delayed rectifier ($I_{DR,N}$ for neonatal), a classical fast inward rectifier (I_{K1}), a Ca^{2+} current (Rüsch et al. 1998b), as well as a large Na^+ current (Rüsch and Eatock 1997; Lennan et al. 1999; Chabbert et al. 2003). This complement of channels is, in fact, not very different from that in neonatal hair cells of the mouse cochlea. In contrast to cochlear hair cells (Marcotti et al. 1999), I_{K1} persists on maturation, certainly in type II and possibly in type I cells (Rüsch et al. 1998b). Broadly the same mix of currents, with regional variations depending on the cell's position within the ampulla, is found in developing semicircular canal hair cells of the chick embryo. Type I cells become identifiable morphologically by E15 and acquire $I_{K,L}$ by E17 (Masetto et al. 2000).

Perhaps surprisingly, given the similarities with developing cochlear hair cells, Ca^{2+} spikes, either spontaneous or induced, have not been described in developing VHCs where they have been studied under current clamp (rat utricle: Lennan et al. 1999; chick semicircular canal: Masetto et al. 2000). Instead, quite small and slow Na^+ action potentials could be evoked in developing rat utricular hair cells between the ages of P0 and P2 (Chabbert et al. 2003). Like OHC Ca^{2+} spikes (Marcotti and Kros 1999), only a single action potential could be elicited on depolarizing current injection and spontaneous action potentials did not occur. After P4 the Na^+ current declines in type II cells and disappears altogether in type I cells (Chabbert et al. 2003). It is presently not clear whether the classical delayed rectifier in mature type II cells ($I_{DR,II}$) is pharmacologically or kinetically different from $I_{DR,N}$ and there is some controversy over the nature of a possible K^+ current that may remain in mature type I cells after $I_{K,L}$ is blocked (Rennie and Correia 1994; Rüsch and Eatock 1996), no doubt in part because the large $I_{K,L}$ is very difficult to block completely (Eatock, personal communication).

7.3 Functions of Conductances in Immature Hair Cells

The IHCs' ability to spike spontaneously (Kros et al. 1998) and release neurotransmitter (Beutner and Moser 2001; Marcotti et al. 2003b) long before the onset of hearing suggests they may actively contribute to the development of neuronal connections in the peripheral and central auditory system. Spontaneous spiking is also likely to be important for the development of the IHCs themselves, as IHCs without L-type Ca^{2+} currents degenerate rapidly (Platzer et al. 2000; Glueckert et al. 2003). As in many other developing cell types, spiking and associated intracellular Ca^{2+} transients may be required for the expression of the mature complement of ion channels (e.g., Moody 1998; Spitzer et al. 2002) and altering the spike pattern may even affect the type of neurotransmitter being released (Borodinsky et al. 2004). In IHCs the L-type Ca^{2+} channel or the spiking activity is required for the acquisition of $I_{K,f}$ (Brandt et al. 2003). The generation of action potentials in immature IHCs is made possible by the particular combination of currents present during early development: a mixture of large and fast Ca^{2+} and Na^+ currents, and small and slow K^+ currents. Expression of the large and fast $I_{K,f}$ prevents spiking and turns the IHCs into high-frequency sensory receptors (Kros et al. 1998). Trains of action potentials can be evoked early in development in the tall hair cells of the chick basilar papilla (Fuchs and Sokolowski 1990), although spontaneous spiking has not been observed. Immature OHCs express voltage-gated ion channels similar to those of immature IHCs and can fire a single spike on depolarization (Marcotti and Kros 1999). OHCs also degenerate in the absence of L-type Ca^{2+} channels (Platzer et al. 2000; Glueckert et al. 2003), suggesting they may spike spontaneously in vivo. In OHCs, induced spiking is inhibited by the expression of $I_{K,n}$ at the time the cells become electromotile (Marcotti and Kros 1999). Depolarization results in a single Na^+ spike in immature vestibular hair cells, with size and kinetics much like the Ca^{2+} spikes of auditory hair cells (Chabbert et al. 2003). It is perhaps surprising that the transducer channels are already working well before the onset of auditory or vestibular function (Kros et al. 1992; Géléoc et al. 2003; Kennedy et al. 2003). This indicates a possible developmental function. Possibilities include the regulation of developmental processes within the hair bundle, perhaps triggered by Ca^{2+} entry, and the provision of a source of inward current to facilitate spontaneous spiking (Kros et al. 1998).

8. Conclusion

Considerable advances have been made over the last decade in our understanding of the molecular processes underlying the development of hair cells in the inner ear. The sequence of events leading to the formation of a sensory organ containing a patterned mixture of hair and supporting cells is summarized in Figure 2.8. Levels of Notch activation may play a critical role in this process. Initially,

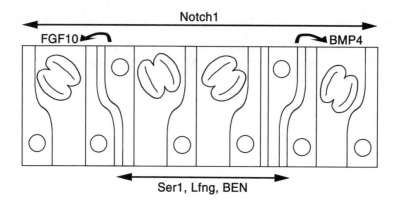
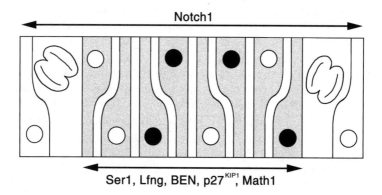
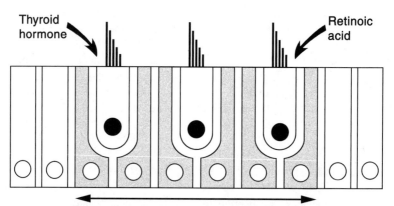

Notch activation (by *Serrate1/Jagged1*) may confer sensory competence upon a patch of cells in the otocyst. Subsequently levels of Notch activation determine whether a cell within the patch will become a hair cell or a supporting cell, with high levels of activation (by *Delta1* and *Serrate2/Jagged2*) leading to cells becoming supporting cells and low levels leading to cells differentiating as hair cells. A number of aspects nonetheless remain to be resolved in more detail. Our knowledge of the signals that determine the formation of the ventromedial prosensory region of the otocyst and their source, or what determines whether a sensory patch will become a crista, a macula, or an organ of Corti, is still rudimentary. A true proneural gene for sensory patches has yet to be found, and the current data indicate the decision to become a hair cell may be made before the hair cells express detectable levels of *Delta1* and *Serrate2*, implying that lateral inhibition may really only serve to reinforce a decision that has already been made at some earlier unknown step.

Our understanding of the molecular mechanisms underlying hair-bundle development remains very much in its infancy. Potential players, for example, unconventional myosins, PDZ domain proteins, cadherins, Wnt signalling molecules and matrix proteins are emerging on the scene, but it will be a challenge to discover how the different pathways involving these molecules are coordinated in space and time to produce a functional hair bundle of the desired form and orientation. Another vitally important aspect of hair-cell development that has yet to be touched on is how gradients in the morphological, molecular, and physiological properties of hair cells that are found within and along the sensory organs are generated and maintained. As discussed above, much more has yet to be learned about the significance of the developmental changes observed in

FIGURE 2.8. A diagram illustrating how hair and supporting cells are produced during development of the inner ear. (*Top*) Notch1 is expressed throughout the wall of the otocyst. The prosensory region can be defined as a region where the Notch ligand Serrate1, the modulator of Notch signaling Lfng, and the cell adhesion molecule BEN are expressed. Serrate 1 may activate Notch1 within the prosensory region, endowing it with the ability to produce hair and supporting cells. BMP4 and FGF10 may operate in a paracrine fashion to influence development of surrounding tissues. (*Middle*) Cells withdraw from the cell cycle and express the cyclin-dependent kinase inhibitor p27^{KIP1} (gray shading). A subset of these cells then express the transcription factor Math1 (*black nuclei*). (*Bottom*) Hair cells that express Math1 also express Numb, rendering them insensitive to Notch1 activation. They express the Notch ligands Delta1 and Serrate2, thereby activating Notch1 in the surrounding cells. Activation of Notch1 in these surrounding cells stimulates the production of Serrate1, and activates the expression of Hes1/5 thereby forcing them to differentiate as supporting cells. Hair cells then stop expressing p27^{KIP1}, express Brn3.1 and a number of hair-cell markers, including myosin VIIa, myosin VI, and calretinin. Expression of Notch1 and Serrate1 becomes restricted to supporting cells as they differentiate. Retinoic acid may influence the early stages of hair-cell differentiation. Thyroid hormone may control the timing of hair-cell differentiation.

the ion channel complement of hair cells, and the functional consequences of the spontaneous spiking observed in inner hair cells during the early stages of their life. Clearly, the developing inner ear is still hiding many an interesting secret from us, so the challenge is there.

Acknowledgments. The authors would like to acknowledge funding from The Wellcome Trust (R.G. and G.R., grant ref: 057410/Z/99/Z, C.K., grant ref: 04065/Z/96/Z), The Medical Research Council (C.K., grant ref: G0100798), and Defeating Deafness (formerly The Hearing Research Trust). The authors would to thank Stuart Johnson for his help with the preparation of Figure 2.5, Walter Marcotti for his help with the preparation of Figures 2.6 and 2.7, and Tao Kwan for editorial assistance.

References

Acampora D, Mazan S, Anataggaiato V, Barone P, Tuorto F, Lallemand Y, Brulet P, Simeone A (1996) Epilepsy and brain abnormalities in mice lacking the *Otx1* gene. Nat Genet 14:218–222.

Acampora D, Merlo GR, Paleari L, Zerega B, Postiglione MP, Mantero S, Bober E, Barbieri O, Simeone A, Levi G (1999) Craniofacial, vestibular and bone defects in mice lacking the *Distal-less*-related gene *Dlx5*. Development 126:3795–3809.

Adam J, Myat A, Le Roux I, Eddison M, Henrique D, Ish-Horowicw, Lewis J (1998) Cell fate choices and the expression of Notch, Delta and Serrrate homologues in the chick inner ear: parallels with *Drosophila* sense-organ development. Development 125:4645–4654.

Ahmed ZM, Riazuddin S, Bernstein SL, Ahmed Z, Khan S, Griffith AJ, Morell RJ, Friedman TB, Riazuddin S, Wilcox ER (2001) Mutations of the protocadherin gene *PCDH15* cause Usher syndrome type 1F. Am J Hum Genet 69:25–34.

Ahmed ZM, Riazuddin S, Ahmad J, Bernstein SL, Guo Y, Sabar MF, Sieving P, Riazuddin S, Griffith AJ, Friedman TB, Belyantseva IA, Wilcox ER (2003) PCDH15 is expressed in the neurosensory epithelium of the eye and ear and mutant alleles are responsible for both USH1F and DFNB23. Hum Mol Genet 12:3215–3223.

Alagramam KN, Zahorsky-Reeves J, Wright CG, Pawlowski KS, Erway LC, Stubbs L, Woychik RP (2000) Neuroepithelial defects of the inner ear in a new allele of the mouse mutation Ames waltzer. Hear Res 148:181–191.

Alagramam KN, Murcia CL, Kwon HY, Pawlowski KS, Wright CG, Woychik RP (2001a) The mouse Ames waltzer hearing-loss mutant is caused by a mutation of *Pdch15*, a novel protocadherin gene. Nat Genet 27: 99–102.

Alagramam KN, Yuan H, Kuehn MH, Murcia CL, Wayne S, Srisailpathy CR, Lowry RB, Knaus R, Van Laer L, Bernier FP, Schwartz S, Lee C, Morton CC, Mullins RF, Ramesh A, Van Camp G, Hageman GS, Woychik RP, Smith RJ, Hagemen GS (2001b) Mutations in the novel protocadherin *PCDH15* cause Usher syndrome type 1F. Hum Mol Genet 10:1709–1718.

Anderson DW, Probst FJ, Belyantseva IA, Fridell RA, Beyer L, Martin DM, Wu D, Kachar B, Friedman TB, Raphael Y, Camper SA (2000) The motor and tail regions of

myosin XV are critical for normal structure and function of auditory and vestibular hair cells. Hum Mol Genet 9:1729–1738.
Angst BD, Marcozzi C, Magee AI (2001) The cadherin superfamily: diversity in form and function. J Cell Sci 114:629–641.
Anniko M (1983a) Cytodifferentiation of cochlear hair cells. Am J Otolaryngol 4:375–388.
Anniko M (1983b) Postnatal maturation of cochlear sensory hairs in the mouse. Anat Embryol 166:355–368.
Avraham KB, Hasson T, Steel KP, Kingsley DM, Russell LB, Mooseker MS, Copeland NG, Jenkins NA (1995) The mouse *Snell's waltzer* deafness gene encodes an unconventional myosin required for structural integrity of the inner ear hair cells. Nat Genet 11:369–375.
Bang PI, Sewell WF, Malicki JJ (2001) Morphology and cell type heterogeneities of the inner ear epithelia in adult and juvenile zebrafish (*Danio rerio*). J Comp Neurol 438:173–190.
Barros AC, Erway LC, Krezel W, Curran T, Kastenr P, Chambon P, Forrest D (1998) Absence of thyroid hormone receptor beta-retinoid X receptor interactions in auditory function and in the pituiatry-thyroid axis. NeuroReport 9:2933–2937.
Bartles JR (2000) Parallel actin bundles and their multiple actin-bundling proteins. Curr Opin Cell Biol 12:72–78.
Bartolami S, Goodyear R, Richardson G (1991) Appearance and distribution of the 275 kD hair-cell antigen during development of the avian inner ear. J Comp Neurol 314:777–788.
Belyantseva IA, Adler HJ, Curi R, Frolenkov GI, Kachar B (2000) Expression and localization of prestin and the sugar transporter GLUT-5 during development of electromotility in cochlear outer hair cells. J Neurosci 20:RC116 (1–5).
Belyantseva IA, Boger ET, Friedman TB (2003) Myosin XVa localizes to the tips of inner ear sensory cell stereocilia and is essential for staircase formation of the hair bundle. Proc Natl Acad Sci USA 100:13958–13963.
Berg JS, Powell BC, Cheney RE (2001) A millenial myosin census. Mol Biol Cell 12:780–794.
Bermingham NA, Hassan BA, Price SD, Vollrath MA, Ben-Arie N, Eatock RA, Bellen, HJ, Lysakowski A, Zoghbi HY (1999) *Math1*: an essential gene for the generation of inner ear hair cells. Science 284:1837–1841.
Beutner D, Moser T (2001) The presynaptic function of mouse cochlear hair cells during development of hearing. J Neurosci 21:4593–4599.
Beyer LA, Odeh H, Probst FJ, Lambert EH, Dolan DF, Camper SA, Kohrman DC, Raphael Y (2000) Hair cells in the inner ear of the pirouette and shaker 2 mutant mice. J Neurocytol 29:227–239.
Biemesderfer D, Mentone SA, Mooseker M, Hasson T (2002) Expression of myosin VI within the early endocytotic pathway in adult and developing proximal tubules. Am J Physiol Renal Physiol 282:F785–F794.
Bissonnette JP, Fekete DM (1996) Standard atlas of the gross anatomy of the developing inner ear of the chicken. J Comp Neurol 368:620–630.
Blanchet C, Erostegui C, Sugasawa M, Dulon D (1996) Acetylcholine-induced potassium current of guinea pig outer hair cells: its dependence on a calcium influx through nicotinic-like receptors. J Neurosci 16:2574–2584.
Boeda B, El-Amraoui A, Bahloul A, Goodyear R, Daviet L, Blanchard S, Perfettini I,

Fath KR, Shorte S, Reiners J, Houddusse A, Legrain P, Wolfrum U, Richardson G, Petit C (2002) Myosin VIIA, harmonin, and cadherin 23, three Usher I gene products that cooperate to shape the sensory hair bundle. EMBO J 21:6689–6699.

Bolz H, von Brederlow B, Ramirez A, Bryda EC, Kutsche K, Nothwang HG, Seeliger M, del C-Salcedo Caberera, Vila MC, Molina OP, Gal A, Kubisch C (2001) Mutation of *CDH23*, encoding a new member of the cadherin gene family, causes Usher syndrome type 1D. Nat Genet 27:108–112.

Borodinsky LN, Root CM, Cronin JA, Sann SB, Gu X, Spitzer NC (2004) Activity-dependent homeostatic specification of transmitter expression in embryonic neurons. Nature 429:523–530.

Bradley DJ, Towle HC, Young III WS (1994) α and β thyroid hormone receptor (TR) gene expression during auditory neurogenesis: evidence for TR isoform-specific transcriptional regulation *in vivo*. Proc Natl Acad Sci USA 91:439–443.

Brandt A, Striessnig J, Moser T (2003) $Ca_v1.3$ channels are essential for development and presynaptic activity of cochlear inner hair cells. J Neurosci 23:10832–10840.

Brown DA, Adams PR (1980) Muscarinic suppression of a novel voltage-sensitive K^+-current in a vertebrate neurone. Nature 283:673–676.

Buss F, Arden SD, Lindsay M, Luzio JP, Kendrick-Jones J (2001) Myosin VI isoform localized to clathrin-coated vesicles with a role in clathrin-mediated endocytosis. EMBO J 20:3676–3684.

Chabbert C, Mechaly I, Sieso V, Giraud P, Brugeaud A, Lehouelleur J, Couraud F, Valmier J, Sans A (2003) Voltage-gated Na^+ channel activation induces both action potentials in utricular hair cells and brain-derived neurotrophic factor release in the rat utricle during a restricted period of development. J Physiol 553:113–123.

Chang W, Numes FD, De Jesus-Escobar JM, Harland R, Wu DK (1999) Ectopic Noggin blocks sensory and nonsensory organ morphogenesis in the chicken inner ear. Dev Biol 216:369–381.

Chen P, Segil N (1999) p27[Kip1] links cell proliferation to morphogenesis in the developing organ of Corti. Development 126:1581–1590.

Chen P, Johnson JE, Zoghbi HY, Segil N (2002) The role of Math1 in inner ear development: uncoupling the establishment of the sensory primordium from hair cell fate determination. Development 129:2495–2505.

Cole LK, Le Roux I, Nunes, Laufer E, Lewis J, Wu DK (2000) Sensory organ generation in the chick inner ear: contributions of Bone Morphogenetic Protein 4, Serrate1, and Lunatic Fringe. J Comp Neurol 424:509–520.

Collier JR, Monk NA, Maini PK, Lewis JA (1996) Pattern formation by lateral inhibition: A mathematical model of Delta–Notch intercellular signalling. J Theor Biol 183:429–446.

Copenhagen DR (1996) Retinal development: on the crest of an exciting wave. Curr Biol 6:1368–1370.

Corwin JT (1981) Postembryonic production and aging of inner ear hair cells in sharks. J Comp Neurol 201:541–553.

Corwin JT (1983) Postembryonic growth of the macula neglecta auditory detector in the ray, *Raja clavata*: continual increases in hair cell number, neural convergence, and physiological sensitivity. J Comp Neurol 217:345–356.

Corwin JT (1985) Perpetual production of hair cells and maturational changes in hair cell ultrastructure accompany postembryonic growth in an amphibian ear. Proc Nat Acad Sci USA 82:3911–3915.

Corwin JT, Cotanche DA (1988) Regeneration of sensory hair cells after acoustic trauma. Science 240:1772–1774.

Corwin JT, Jones JE, Katayama A, Kelley MW, Warchol ME (1991) Hair cell regeneration: the identities of progenitor cells, potential triggers and instructive cues. Ciba Found Symp 160:103–120.

Cotanche DA, Sulik KK (1983) Early differentiation of hair cells in the embryonic chick basilar papilla. Arch Otorhinolarngol 237:191–195.

Cotanche DA, Sulik KK (1984) The development of stereociliary bundles in the cochlear duct of chick embryos. Dev Brain Res 16:181–193.

Cotanche DA, Corwin JT (1991) Stereociliary bundles reorient during hair cell development and regeneration in the chick cochlea. Hear Res 52:379–402.

Cramer LP (2000) Myosin VI: roles for a minus end-directed actin motor in cells. J Cell Biol 150:F121–F126.

Crawford AC, Evans MG, Fettiplace R (1989) Activation and adaptation of transducer currents in turtle hair cells. J Physiol 419:405–434.

Curthoys IS (1979) The development of function of horizontal semicircular canal primary neurons in the rat. Brain Res 167:41–52.

Curtin JA, Quint E, Tsipouri V, Arkell RM, Cattanach B, Copp AJ, Henderson DJ, Spurr N, Stanier P, Fisher EM, Nolan PM, Steel KP, Brown SDM, Gray IC, Murdoch JN (2003) Mutation of *Clesr1* disrupts planar polarity of inner ear hair cells and causes severe neural tube defects in the mouse. Curr Biol 13:1129–1133.

Dabdoub A, Donohue MJ, Brennan A, Wolf V, Montcouquiol M, Sassoon DA, Hseih J-C, Rubin JS, Salinas PC, Kelley MW (2003) Wnt signalling mediates reorientation of outer hair cell stereociliary bundles in the mammalian cochlea. Development 130: 2375–2384.

Dallos P, Evans BN (1995a) High-frequency motility of outer hair cells and the cochlear amplifier. Science 267:2006–2009.

Dallos P, Evans BN (1995b) High-frequency outer hair cell motility: corrections and addendum. Science 268:1420–1421.

Daudet N, Lebart M-C (2002) Transient expression of the T-isoform of plastins/fimbrin in the stereocilia of developing auditory hair cells. Cell Motil Cytoskel 53:326–336.

Denman-Johnson K, Forge A (1999) Establishment of hair bundle polarity and orientation in the developing vestibular system of the mouse. J Neurocytol 28:821–835.

Di Palma F, Holme RH, Bryda EC, Belyantseva IA, Pellegrino R, Kachar B, Steel KP, Noben-Trauth K (2001) Mutations in *Cdh23*, encoding a new type of cadherin, cause stereocilia disorganization in waltzer, the mouse model for Usher syndrome type 1D. Nat Genet 27:103–107.

Di Palma F, Belyantseva IA, Kim HJ, Vogt TF, Kachar B, Noben-Trauth K (2002) Mutations in *Mcoln3* associataed with deafness and pigmentation defects in varitint-waddler (*Va*) mice. Proc Natl Acad Sci USA 99:14994–14999.

Dulon D, Lenoir M (1996) Cholinergic responses in developing outer hair cells of the rat cochlea. Eur J Neurosci 8:1945–1952.

Dulon D, Luo L, Zhang C, Ryan AF (1998) Expression of small-conductance calcium-activated potassium channels (SK) in outer hair cells of the rat cochlea. Eur J Neurosci 10:907–915.

Dumont R, Zhao Y, Holt JR, Bähler M, Gillespie PG (2002) Myosin-I isozymes in neonatal rodent audiorty epithelia. JARO 3:375–389.

Eddison M, Le Roux I, Lewis J (2000) Notch signalling in the development of the inner ear: lessons from *Drosophila*. Proc Natl Acad Sci USA 97:11692–11699.

Ekker M, Akimenko MA, Bremiller R, Westerfield M (1992) Regional expression of three homeobox transcripts in the inner ear of zebrafish embryos. Neuron 9: 27–35.

Elgoyhen AB, Vetter DE, Katz E, Rothlin CV, Heinemann SF, Boulter J (2001) α10: a determinant of nicotinic cholinergic receptor function in mammalian vestibular and cochlear mechanosensory hair cells. Proc Natl Acad Sci USA 98:3501–3506.

Erkman L, McEvilly RJ, Luo L, Ryan AK, Hooshmand F, O'Connell, Keithley EM, Rapaport DH, Ryan AF, Rosenfeld MG (1996) Role of transcription factors Brn-3.1 and Brn-3.2 in auditory and visual system development. Nature 381:603–606.

Ernest S, Rauch GJ, Haffter P, Geisler R, Petit C, Nicolson T (2000) Mariner is defective in myosin VIIA: a zebrafish model for human hereditary deafness. Hum Mol Genet 9:2189–2196.

Erven A, Skynner MJ, Okumura K, Takebayashi, Brown SDM, Stell KP, Allen ND (2002) A novel stereocilia defect in sensory hair cells of the deaf mutant Tasmanian devil. Eur J Neurosci 16:1433–1441.

Evans MG (1996) Acetylcholine activates two currents in guinea-pig outer hair cells. J Physiol 491:563–578.

Fekete DM (1996) Cell fate specification in the inner ear. Curr Opin Neurobiol 6:533–541.

Fekete DM, Muthukumar S, Karagogeos D (1998) Hair and supporting cells share a common progenitor in the avian inner ear. J Neurosci 18:7811–7821.

Fettiplace R, Fuchs PA (1999) Mechanisms of hair cell tuning. Annu Rev Physiol 61: 809–834.

Forge A, Souter M, Denman-Johnson K (1997) Structural development of sensory hair cells in the ear. Semin Cell Dev Biol 8:225–237.

Forrest D, Erway LC, Ng L, Altschuler R, Curran T (1996) Thyroid hormone receptor β is essential for development of auditory function. Nat Genet 13:354–357.

Frank G, Hemmert W, Gummer AW (1999) Limiting dynamics of high-frequency electromechanical transduction of outer hair cells. Proc Natl Acad Sci USA 96:4420–4425.

Fuchs PA, Evans MG (1988) Voltage oscillations and ionic conductances in hair cells isolated from the alligator cochlea. J Comp Physiol A 164:151–163.

Fuchs PA, Evans MG (1990) Potassium currents in hair cells isolated from the cochlea of the chick. J Physiol 429:529–551.

Fuchs PA, Murrow BW (1992) Cholinergic inhibition of short (outer) hair cells of the chick's cochlea. J Neurosci 12:800–809.

Fuchs PA, Sokolowski BHA (1990) The acquisition during development of Ca-activated potassium currents by cochlear hair cells of the chick. Proc R Soc Lond B 241:122–126.

Furness DN, Richardson GP, Russell IJ (1989) Stereociliary bundle morphology in organotypic cultures of the mouse cochlea. Hear Res 38:95–109.

Gale JE, Marcotti W, Kennedy HJ, Kros CJ, Richardson GP (2001) FM1-43 dye behaves as a permeant blocker of the hair-cell mechanotransducer channel. J Neurosci 21: 7013–7025.

Géléoc GW, Holt JR (2003) Developmental acquisition of sensory transduction in hair cells of the mouse inner ear. Nat Neurosci 6:1019–1020.

Géléoc GSG, Lennan GWT, Richardson GP, Kros CJ (1997) A quantitative comparison of mechanoelectrical transduction in vestibular and auditory hair cells of neonatal mice. Proc R Soc Lond B 264:611–621.

Géléoc GSG, Casalotti SO, Forge A, Ashmore JF (1999) A sugar transporter as a candidate for the outer hair cell motor. Nat Neurosci 2:713–719.

Gerlach LM, Hutson MR, Germiller JA, Nguyen-Luu D, Victor JC, Barald KF (2000) Addition of the BMP4 antagonist, noggin, disrupts avian inner ear development. Development 127:45–54.

Giancotti FG, Ruoslahti E (1999) Integrin signaling. Science 285:1028–1032.

Gibson F, Walsh J, Mburu P, Varela A, Brown KA, Antonio M, Beisel KW, Steel KP, Brown SD (1995) A type VII myosin encoded by the mouse deafness gene shaker-1. Nature 374:62–64.

Glowatzki E, Fuchs PA (2000) Cholinergic synaptic inhibition of inner hair cells in the neonatal mammalian cochlea. Science 288:2366–2368.

Glueckert R, Wietzorrek G, Kammen-Jolly K, Scholtz A, Stephan K, Striessnig J, Schrott-Fischer A (2003) Role of class D L-type Ca^{2+} channels for cochlear morphology. Hear Res 178:95–105.

Goodyear R, Richardson G (1992) Distribution of the 275 kD hair cell antigen and cell surface specialisations on auditory and vestibular hair bundles in the chick inner ear. J Comp Neurol 3256:243–256.

Goodyear R, Richardson G (1997) Pattern formation in the basilar papilla: evidence for cell rearrangement. J Neurosci 17:6289–6301.

Goodyear R, Richardson G (1999) The ankle-link antigen: an epitope sensitive to calcium chelation associated with the hair-cell surface and the calycal processes of photoreceptors. J Neurosci 19:3761–3772.

Goodyear R, Richardson G (2003) A novel antigen that is associated with the tip links and kinocilial links of sensory hair bundles. J Neurosci 23:4878–4887.

Goodyear RJ, Gates R, Lukashkin AN, Richardson GP (1999) Hair-cell numbers continue to increase in the utricular macula of the early posthatch chick. J Neurocytol 28: 851–861.

Goodyear RJ, Kwan T, Oh S-H, Raphael Y, Richardson GP (2001) The cell adhesion molecule BEN defines a prosensory patch in the developing avian otocyst. J Comp Neurol 434:275–288.

Goodyear RJ, Legan PK, Wright MB, Marcotti W, Oganesian A, Coats SA, Booth CJ, Kros CJ, Seifert RA, Bowen-Pope DF, Richardson GP (2003) A receptor-like inositol lipid phosphatase is required for the maturation of developing cochlear hair bundles. J Neurosci 23:9208–9219.

Griguer C, Fuchs PA (1996) Voltage-dependent potassium currents in cochlear hair cells of the embryonic chick. J Neurophysiol 75:508–513.

Groves AK, Bronner-Faser M (2000) Competence, specification and commitment in otic placode induction. Development 127:3489–3499.

Haddon C, Lewis J (1996) Early ear development in the embryo of the zebrafish, *Danio rerio*. J Comp Neurol 365:113–128.

Haddon C, Jiang Y-J, Smithers L, Lewis J (1998) Delta–Notch signalling and the patterning of sensory cell differentiation in the zebrafish ear: evidence from the mind bomb mutant. Development 125:4637–4644.

Haddon C, Mowbray C, Whitfield T, Jones D, Gschmeissner S, Lewis J (1999) Hair cells without supporting cells: further studies in the ear of the zebrafish *mind bomb* mutant. J Neurocytol 28:837–850.

Hall A (1998) Rho GTPases and the actin cytoskeleton. Science 279:509–514.

Hampton LL, Wright CG, Alagramam KN, Battey JF, Noben-Truth K (2003) A new spontaneous mutation in the Ames waltzer gene, *Pcdh15*. Hear Res 180:65–75.

Hasson T, Gillespie PG, Garcia JA, MacDonald RB, Zhao Y, Yee AG, Mooseker MS, Corey DP (1997) Unconventional myosins in inner-ear sensory epithelia. J Cell Biol 137:1287–1307.

He DZZ (1997) Relationship between the development of outer hair cell electromotility and efferent innervation: a study in cultured organ of Corti of neonatal gerbils. J. Neurosci 17:3634–3643.

He DZZ, Dallos P (1999) Development of acetylcholine-induced responses in neonatal gerbil outer hair cells. J Neurophysiol 81:1162–1170.

He DZZ, Evans BN, Dallos P (1994) First appearance and development of electromotility in neonatal gerbil outer hair cells. Hear Res 78:77–90.

He DZZ, Zheng J, Dallos P (2001) Development of acetylcholine receptors in cultured outer hair cells. Hear Res 162:113–125.

He DZZ, Jia S, Dallos P (2004) Mechanoelectrical transduction of adult outer hair cells studied in a gerbil hemicochlea. Nature 429:766–770.

Helyer RJ, Kennedy HJ, Davies D, Holley MC, Kros CJ (2005) Development of outward potassium currents in inner and outer hair cells from the embryonic mouse cochlea. Audiol Neuro-Otol 10:22–34.

Holley MC, Nishida Y (1995) Monoclonal antibody markers for early development of the stereociliary bundles of mammalian hair cells. J Neurocytol 24:853–864.

Holme RH, Kiernan BW, Brown SDM, Steel KP (2002) Elongation of hair cell stereocilia is defective in the mouse mutant whirler. J Comp Neurol 450:94–102.

Holt JR, Corey DP (2000) Two mechanisms for transducer adaptation in vertebrate hair cells. Proc Natl Acad Sci USA 97:11730–11735.

Holt JR, Corey DP, Eatock RA (1997) Mechanoelectrical transduction and adaptation in hair cells of the mouse utricle, a low-frequency vestibular organ. J Neurosci 17:8739–8748.

Housley GD, Ashmore JF (1991) Direct measurement of the action of acetylcholine on isolated outer hair cells of the guinea-pig cochlea. Proc R Soc Lond B 244:161–167.

Housley GD, Ashmore JF (1992) Ionic currents of outer hair cells isolated from the guinea-pig cochlea. J Physiol 448:73–98.

Hutson MR, Lewis JE, Nguyen-Luu D, Lindberg KH, Barald KF (1999) Expression of Pax2 and patterning of the chick inner ear. J Neurocytol 28:795–807.

Itoh M, Kim CH, Parlardy G, Oda T, Jiang YJ, Maust D, Yeo SY, Lorick K, Wright GJ, Ariza-McNaughton L, Weissman AM, Lewis J, Chandrasekharappa SC, Chitnis AB (2003) Mind bomb is a ubiquitin ligase that is essential for efficient activation of Notch signaling by Delta. Dev Cell 4:67–82.

Jacobs RA, Hudspeth AJ (1990) Ultrastructural correlates of mechanoelectrical transduction in hair cells of the bullfrog's internal ear. Cold Spring Harbor Symp Quant Biol 55:547–561.

Jagger DJ, Griesinger CB, Rivolta MN, Holley MC, Ashmore JF (2000) Calcium signalling mediated by the α9 acetylcholine receptor in a cochlear cell line from the Immortomouse. J Physiol 527:49–54.

Johnson KR, Gagnon LH, Webb LS, Peters LL, Hawes NL, Chang B, Zheng QY (2003) Mouse models of USH1C and DFNB18: phenotypic and molecular analyses of two new spontaneous mutations of the *Ush1c* gene. Hum Mol Genet 12:3075–3086.

Jones JE, Corwin JT (1996) Regeneration of sensory cells after laser ablation in the lateral line system: hair cell lineage and macrophage behaviour revealed by time-lapse video microscopy. J Neurosci 16:649–662.

Jones TA, Jones SM, Paggett KC (2001) Primordial rhythmic bursting in embryonic cochlear ganglion cells. J Neurosci 21:8129–8135.

Jørgensen JM, Mathiesen C (1988) The avian inner ear: continuous production of hair cells in the vestibular organs but not in the auditory papilla. Naturewissenschaften 75: 319–320.

Kalatzis V, Sahly I, El-Amraoui A, Petit C (1998) *Eya1* expression in the developing ear and kidney: towards the understanding of the pathogenesis of branchio-oto-renal (BOR) syndrome. Dev Dyn 213:486–499.

Kaltenbach JA, Falzarano PR, Simpson TH (1994) Postnatal development of the hamster cochlea. II. Growth and differentiation of stereocilia bundles. J Comp Neurol 350: 187–198.

Katayama A, Corwin JT (1989) Cell production in the chicken cochlea. J Comp Neurol 281:129–135.

Kelley MW, Xu XM, Wagner MA, Warchol ME, Corwin JT (1993) The developing organ of Corti contains retinoic acid and forms supernumerary hair cells in response to exogenous retinoic acid in culture. Development 119:1041–1053.

Kennedy HJ, Evans MG, Crawford AC, Fettiplace R (2003) Fast adaptation of mechanoelectrical transducer channels in mammalian cochlear hair cells. Nat Neurosci 6: 832–836.

Kiernan AE, Zalzman M, Fuchs H, Harabe de Angelis M, Balling R, Steel KP, Avraham KB (1999) Tailchaser (*Tlc*): a new mouse mutation affecting hair bundle differentiation and hair cell survival. J Neurocytol 28:969–985.

Kiernan AE, Ahituv N, Fuchs H, Balling R, Avraham KB, Steel KP, Hrabé de Angelis M (2001) The Notch ligand *Jagged1* is required for inner ear sensory development Proc Natl Acad Sci USA 98:3873–3878.

Kikkawa Y, Shitara H, Wakan S, Kohara Y, Takada T, Okamoto M, Taya C, Kamiya K, Yoshikawa Y, Tokano H, Kitamura K, Shimizu K, Wakabayashi Y, Shiroishi T, Kominami R, Yonekawa H (2003) Mutations in a new scaffold protein Sans cause deafness in Jackson shaker mice. Hum Mol Genet 12:453–461.

Knowlton VY (1967) Correlation of the development of membranous and bony labyrinths, acoustic ganglia, nerves, and brain centers of chick embryos. J Morphol 121: 179–208.

Kollmar R (1999) Who does the hair cell's do? Rho GTPase and hair-bundle morphogenesis. Curr Opin Neurobiol 9:394–398.

Kolston PJ (1995) A faster transduction mechanism for the cochlear amplifier? Trends Neurosci 18:427–429.

Kros CJ (1996) Physiology of mammalian cochlear hair cells. In: Dallos P, Popper AN, Fay RR (eds), The Cochlea. New York: Springer-Verlag, pp. 318–385.

Kros CJ, Crawford AC (1990) Potassium currents in inner hair cells isolated from the guinea-pig cochlea. J Physiol 421:263–291.

Kros CJ, Rüsch A, Richardson GP (1992) Mechano-electrical transducer currents in hair cells of the cultured mouse cochlea. Proc R Soc Lond B 249:185–193.

Kros CJ, Lennan GWT, Richardson GP (1995) Transducer currents and bundle movements in outer hair cells of neonatal mice. In: Flock Å, Ottoson D, Ulfendahl M (eds), Active Hearing. Amsterdam: Elsevier, pp. 113–125.

Kros CJ, Ruppersberg P, Rüsch A (1998) Expression of a potassium current in inner hair cells during development of hearing in mice. Nature 394:281–284.

Kros CJ, Marcotti W, van Netten SM, Self TJ, Libby RT, Brown SDM, Richardson GP,

Steel KP (2002) Reduced climbing and increased slipping adaptation in cochlear hair cells of mice with *Myo7a* mutations. Nat Neurosci 5:41–47.
Kubisch C, Schroeder BC, Friedrich T, Lütjohann B, El-Amoraoui A, Marlin S, Petit C, Jentsch TJ (1999) KCNQ4, a novel potassium channel expressed in sensory outer hair cells, is mutated in dominant deafness. Cell 96:437–446.
Kussel-Andermann P, El-Amraoui A, Safieddine S, Nouaille S, Perfettini I, Lecuit M, Cossart P, Wolfrum U, Petit C (2000) Vezatin, a novel transmembrane protein, bridges myosin VIIA to the cadherin–catenins complex. EMBO J 19:6020–6029.
Lanford PJ, Presson JC, Popper AN (1996) Cell proliferation and hair cell addition in the ear of the goldfish, *Carssius auratus*. Hear Res 100:1–9.
Lanford PJ, Lan Y, Jiang R, Lindsell C, Weinmaster G, Gridley T, Kelley MW (1999) Notch signalling pathway mediates hair cell development in mammalian cochlea. Nat Genet 21:289–292.
Lanford PJ, Shailam R, Norton CR, Gridley T, Kelley MW (2000) Expression of *Math1* and *Hes5* in the cochlea of wild type and Jag2 mutant mice. JARO 1:161–171.
Lang H, Fekete DM (2001) Lineage analysis in the chicken inner ear shows differences in clonal dispersion for epithelial, neuronal, and mesenchymal cells. Dev Biol 234: 120–137.
Lawoko-Kerali G, Rivolta MN, Holley M (2002) Expression of the transcription factors GATA3 and Pax2 during development of the mammalian inner ear. J Comp Neurol 442:378–391.
Lavigne-Rebillard M, Pujol R (1986) Development of the auditory hair cell surface in human fetuses. A scanning electron microscope study. Anat Embryol 174:369–377.
Lenoir M, Shnerson A, Pujol R (1980) Cochlear receptor development in the rat with emphasis on synaptogenesis. Anat Embryol 160:253–262.
Lenoir M, Puel J-L, Pujol R (1987) Stereocilia and tectorial membrane development in the rat cochlea. A SEM study. Anat Embryol 175:477–487.
Lennan GWT, Géléoc GSG, Kros CJ (1996) Displacement sensitivity of mammalian vestibular transducers. Ann NY Acad Sci 781:650–652.
Lennan GWT, Steinacker A, Lehouelleur J, Sans A (1999) Ionic currents and current-clamp depolarisations of type I and type II hair cells from the developing rat utricle. Pflügers Arch 438:40–46.
Leon Y, Sanchez JA, Miner C, Ariza-McNaughton L, Repress JJ, Giraldez F (1995) Developmental regulation of Fos-protein during proliferative growth of the otic vesicle and its relation to differentaition induced by retinoic acid. Dev Biol 167: 75–86.
Lewis AK, Frantz GD, Carpenter DA, de Sauvage FJ, Gao W-Q (1998) Distinct expression patterns of notch family receptors and ligands during development of the mammalian inner ear. Mech Dev 78:159–163.
Lewis J (1991) Rules for the production of sensory hair cells. Ciba Found Symp 160: 25–39.
Lewis J (1996) Neurogenic genes and vertebrate neurogenesis. Curr Opin Neurobiol 1: 3–10.
Lewis J, Davies A (2002) Planar cell polarity in the inner ear: how do hair cells acquire their oriented structure? J Neurobiol 53:190–201.
Li CW, Ruben RJ (1979) Further study of the surface morphology of the embryonic mouse cochlear sensory epithelia. Otolaryngol Head Neck Surg 87:479–485.
Li H, Liu H, Balt S, Mann S, Corrales CE, Heller S (2004) Correlation of expression of

the actin filament-bundling protein espin with stereociliary bundle formation in the developing inner ear. J Comp Neurol 468:125–134.

Li X, Oghi KA, Zhang J, Krones A, Bush KT, Glass CK, Nigam SK, Aggarwal AK, Maas R, Rose DW, Rosenfeld MG (2003) Eya protein phosphatase activity regulates *Six1-Dach-Eya* transcriptional effects in mammalian organogenesis. Nature 426:247–254.

Lim DJ, Anniko M (1985) Developmental morphology of the mouse inner ear. A scanning electron microscope observation. Acta Otolaryngol 422:1–69.

Lim DJ, Rueda J (1992) Structural development of the cochlea. In Development of Auditory and Vestibular Systems, Vol. 2. Amsterdam: Elsevier, pp. 33–58.

Lin X, Chen S (2000) Endogenously generated spontaneous spiking activities recorded from postnatal spiral ganglion neurons in vitro. Dev Brain Res 119:297–305.

Lippe WR (1994) Rhythmic spontaneous activity in the developing avian auditory system. J Neurosci 14:1486–1495.

Littlewood-Evans A, Müller U (2000) Stereocilia defects in the sensory hair cells of the inner ear in mice deficient in integrin α8β1. Nat Genet 24:424–428.

Loomis PA, Zheng L, Sekerkova G, Chnagyaleket B, Mugnaini E, Bartles JR (2003) Espin cross-links cause the elongation of microvillus-type parallel actin bundles in vivo. J Cell Biol 163:1045–1055.

Lowenheim H, Furness DN, Kil J, Zinn C, Gultig K, Fero ML, Frost D, Gummer AW, Roberts AW, Roberts JM, Rubel EW, Hackney CM, Zenner H-P (1999) Gene disruption of p27^{Kip1} allows cell proliferation in the postnatal and adult organ of Corti. Proc Natl Acad Sci USA 96:4084–4088.

Lustig LR, Peng H, Hiel H, Yamamoto T, Fuchs PA (2001) Molecular cloning and mapping of the human nicotinic acetylcholine receptor alpha10 (CHRNA10). Genomics 73:272–283.

Lynch ED, Lee MK, Morrow JE, Welsch PL, Leon PE, King MC (1997) Nonsyndromic deafness DFNA1 associated with mutation of a human homolog of the *Drosophila* gene *diaphanous*. Science 278:1223–1224.

Maffei L, Galli-Resta L (1990) Correlation in the discharges of neighboring rat retinal ganglion cells during prenatal life. Proc Natl Acad Sci USA 87:2861–2864.

Mammano F, Ashmore JF (1996) Differential expression of outer hair cell potassium currents in the isolated cochlea of the guinea-pig. J Physiol 496:639–646.

Marcotti W, Kros, CJ (1999) Developmental expression of the potassium current $I_{K,n}$ contributes to maturation of mouse outer hair cells. J Physiol 520:653–660.

Marcotti W, Géléoc GSG, Lennan GWT, Kros, CJ (1999) Transient expression of an inwardly rectifying potassium conductance in developing inner and outer hair cells along the mouse cochlea. Pflügers Arch 439:113–122.

Marcotti W, Johnson SL, Holley MC, Kros CJ (2003a) Developmental changes in the expression of potassium currents of embryonic, neonatal and mature mouse inner hair cells. J Physiol 548:383–400.

Marcotti W, Johnson SL, Rüsch A, Kros CJ (2003b) Sodium and calcium currents shape action potentials in immature mouse inner hair cells. J Physiol 552:743–761.

Marcotti W, Johnson SL, Kros CJ (2004) Effects of intracellular stores and extracellular Ca^{2+} on Ca^{2+}-activated K^+ currents in mature mouse inner hair cells. J Physiol 557:613–633.

Masetto S, Perin P, Malusa A, Zucca G, Valli P (2000) Membrane properties of chick semicircular canal hair cells in situ during embryonic development. J Neurophysiol 83:2740–2756.

Mbiene J-P, Sans A (1986) Differentiation and maturation of the sensory hair bundles in the fetal and postnatal vestibular receptors of the mouse: a scanning electron microscope study. J Comp Neurol 254:271–278.

Mburu P, Mustapha M, Varela A, Weil D, El-Amraoui A, Holme RH, Rump A, Hardisty RE, Blanchard S, Coimbra RS, Perfettini I, Parkinson N, Mallon A-M, Glenister P, Rogers MJ, Paige AJ, Moir L, Rosental A, Liu XZ, Blanco G, steel KP, Petit C, Brown SDM (2003) Defects in whirlin, a PDZ domain molecule involved in stereocilia elongation, cause deafness in the whirler mouse and families with DFNB31. Nat Genet 34:421–428.

Meister M, Wong ROL, Baylor D.A., Shatz CJ (1991) Synchronous bursts of action potentials in ganglion cells of the developing mammalian retina. Science 252:939–943.

Michna M, Knirsch M, Hoda J-C, Muenkner S, Langer P, Platzer J, Striessnig J, Engel J (2003) $Ca_v1.3$ ($\alpha1D$) Ca^{2+} currents in neonatal outer hair cells of mice. J Physiol 553:747–758.

Mikaelian D, Ruben RJ (1965) Development of hearing in the normal CBA-J mouse. Acta Otolaryngol 59:451–461.

Mitchem K, Hibbard E, Beyer LA, Bosom K, Dootz GA, Dolan DF, Johnson KR, Raphael Y, Kohrman DC (2002) Mutation of the novel gene *Tmie* results in sensory cell defects in the inner ear of spinner, a mouse model of human hearing loss DFNB6. Hum Mol Genet 11:1887–1898.

Montcouquiol M, Rachel RA, Landford PJ, Copeland NG, Jenkins NA, Kelley MW (2003) Identification of *Vangl2* and *Scrb1* as planar polarity genes in mammals. Nature 423:173–177.

Moody WJ (1998) The development of voltage-gated ion channels and its relation to activity-dependent development events. Curr Top Dev Biol 39:159–185.

Morrison A, Hodgetts C, Gossler A, Hrabé de Angelis M, Lewis J (1999) Expression of *Delta1* and *Serrate1 (Jagged1)* in the mouse inner ear. Mech Dev 84:169–172.

Morsli H, Choo D, Ryan A, Johnson R, Wu DK (1998) Development of the mouse inner ear and origin of its sensory organs. J Neurosci 18:3327–3335.

Mowbray C, Hammerschmidt M, Whitfield TT (2001) Expression of BMP signalling pathway members in the developing zebrafish inner ear and lateral line. Mech Dev 108:179–184.

Muller U, Littlewood-Evans A (2001) Mechanisms that regulate mechanosensory hair cell differentiation. Trends Cell Biol 11:334–342.

Nenov AP, Norris C, Bobbin RP (1996) Acetylcholine response in guinea pig outer hair cells. II. Activation of a small conductance Ca^{2+}-activated K^+ channel. Hear Res 101:149–172.

Neugebauer DC (1986) Interconnections between the stereovilli of the fish inner ear. III. Indications for developmental changes. Cell Tissue Res 246:447–453.

Neugebauer D-Ch, Thurm U (1985) Interconnections between the stereovilli of the fish inner ear. Cell Tissue Res 20:449–453.

Nicolson T, Rüsch A, Friedrich RW, Granato M, Ruppersberg JP, Nusslein-Volhard C (1998) Genetic analysis of vertebrate sensory hair cell mechanosensation: the zebrafish circler mutants. Neuron 20:271–283.

Nishida Y, Rivolta MN, Holley MC (1998) Timed markers for the differentiation of the cuticular plate and stereocilia in hair cells from the mouse inner ear. J Comp Neurol 395:18–28.

Nolo R, Abbot LA, Bellen HJ (2000) Senseless, a Zn finger transcription factor, is necessary and sufficient for sensory organ development in *Drosophila*. Cell 102:349–362.

Oganesian A, Poot M, Daum G, Coats SA, Wright MB, Seifert RA, Bowen-Pope DF (2003) Protein tyrosine phsophatse RG is a phosphatidylinositol phosphate that can regulate cell survival and proliferation. Proc Natl Acad Sci USA 100:7563–7568.

Oh S-H, Johnson R, Wu DK (1996) Differential expression of bone morphogenetic proteins in the developing vestibular and auditory sensory organs. J Neurosci 16:6463–6475.

Ohtsuka T, Ishibashi M, Gradwohl G, Nakanishi S, Guillemot F, Kageyama R (1999) *Hes1* and *Hes5* as Notch effectors in mammalian neuronal differentiation. EMBO J 18:2196–2207.

Oliver D, Fakler B (1999) Expression density and functional characteristics of the outer hair cell motor protein are regulated during postnatal development in rat. J Physiol 519:791–800.

Oliver D, Plinkert P, Zenner HP, Ruppersberg JP (1997) Sodium current expression during postnatal development of rat outer hair cells. Pflügers Arch 434:772–778.

Oliver D, Klöcker N, Schuck J, Baukrowitz T, Ruppersberg JP, Fakler B (2000) Gating of Ca^{2+}-activated K^+ channels controls fast inhibitory synaptic transmission at auditory outer hair cells. Neuron 26:595–601.

Oliver D, He DZZ, Klöcker N, Ludwig J, Schulte U, Waldegger S, Ruppersberg JP, Dallos P, Fakler B (2001) Intracellular anions as the voltage sensor of prestin, the outer hair cell motor protein. Science 292:2340–2343.

Oliver D, Knipper M, Derst C, Fakler B (2003) Resting potential and submembrane calcium concentration of inner hair cells in the isolated mouse cochlea are set by KCNQ-type potassium channels. J Neurosci 23:2141–2149.

Ozaki H, Nakamura K, Funahashi J, Ikeda K, Yamada G, Tokano H, Okamura H, Kitamura K, Muto S, Kotaki H, Sudo K, Horai R, Iwakura Y, Kawakami K (2004) *Six1* controls patterning of the mouse otic vesicle. Development 131:551–562.

Palmer AR, Russell IJ (1986) Phase-locking in the cochlear nerve of the guinea-pig and its relation to the receptor potential of inner hair cells. Hear Res 24:1–15.

Pickles JO, van Heumen WRA (2000) Lateral interactions account for the pattern of the hair cell array in the chick basilar papilla. Hear Res 145:65–74.

Pickles JO, von Perger M, Rouse GW, Brix J (1991) The development of links between stereocilia in hair cells of the chick basilar papilla. Hear Res 54:153–163.

Pickles JO, Billieux-Hawkins DA, Rouse GW (1996) The incorporation and turnover of radiolabelled amino acids in developing stereocilia of the chick cochlea. Hear Res 101:45–54.

Pirvola U, Ylikoski J, Palgi J, Lehtonen E, Arumae U, Saarma M (1992) Brain-derived neurotrophic factor and neurotrophin-3 mRNAs in the peripheral fields of developing inner ear ganglia. Proc Natl Acad Sci USA 89:9915–9919.

Pirvola U, Arumae U, Moshnyakov M, Palgi J, Saarma M, Ylikoski J (1994) Coordinated expression and function of neurotrophins and their receptors in the rat inner ear during target innervation. Hear Res 75:131–144.

Pirvola U, Spencer-Dene B, Xing-Qun L, Kettunen P, Thesleff I, Fritzsch B, Dickson C, Ylikoski J (2000) FGF/FGFR-2(IIIb) signalling is essential for inner ear morphogenesis. J Neurosci 20:6125–6134.

Pirvola U, Ylikoski Y, Trokovic R, Hébert JM, McConnell SK, Partanen J (2002) FGFR1 is required for the development of the auditory sensory epithelium. Neuron 35: 670–680.

Platzer J, Engel J, Schrott-Fischer A, Stephan K, Bova S, Chen H, Zheng H, Striessnig J (2000) Congenital deafness and sinoatrial node dysfunction in mice lacking class D L-type Ca^{2+} channels. Cell 102:89–97.

Popper AN, Hoxter B (1984) Growth of a fish ear: I. Quantitative analysis of hair and ganglion cell proliferation. Hear Res 15:133–142.

Popper AN, Hoxter B (1990) Growth of a fish ear II. Locations of newly proliferated sensory hair cells in the saccular epithelium of *Astronotus ocellatus*. Hear Res 45:35–40.

Probst FJ, Fridell RA, Raphael Y, Saunders TL, Wang A, Liang Y, Morell RJ, Touchman JW, Lyons RH, Noben-Trauth K, Friedman TB, Camper SA (1998) Correction of deafness in shaker-2 mice by an unconventional myosin in a BAC transgene. Science 280:1444–1447.

Pujol R, Zajic G, Dulon D, Raphael Y, Altschuler RA, Schacht J (1991) First appearance and development of motile properties in outer hair cells isolated from guinea-pig cochlea. Hear Res 57:129–141.

Raphael Y, Kobayashi KN, Dootz GA, Beyer LA, Dolan DF, Burmeister M (2001) Severe vestibular and auditory impairment of three alleles of Ames waltzer (*av*) mice. Hear Res 151:237–249.

Ratto GM, Robinson DW, Yan B, McNaughton PA (1991) Development of the light response in neonatal mammalian rods. Nature 351, 654–657.

Rau A, Legan KP, Richardson GP (1999) Tectorin mRNA expression is spatially and temporally restricted during mouse inner ear development. J Comp Neurol 405:271–280.

Raz Y, Kelley MW (1997) Effects of retinoic and thyroid hormone receptors during development of the inner ear. Semin Cell Dev Biol 8:257–264.

Raz Y, Kelley MW (1999) Retinoic acid signaling is necessary for the development of the organ of Corti. Dev Biol 213:180–193.

Rennie KJ, Correia MJ (1994) Potassium currents in mammalian and avian isolated type I semicircular canal hair cells. J Neurophysiol 71:317–329.

Rennie KJ, Weng T, Correia MJ (2001) Effects of KCNQ channel blockers on K^+ currents in vestibular hair cells. Am J Physiol Cell Physiol 280:C473–C480.

Represa J, Sanchez A, Milner C, Lewis J, Giraldez F (1990) Retinoic acid modulation of the early development of the inner ear is associated with the control of c-fos expression. Development 110:1081–1090.

Ricci AJ, Crawford AC, Fettiplace R (2003) Tonotopic variation in the conductance of the hair cell mechanotransducer channel. Neuron 40:983–990.

Riccomagno MM, Martinu L, Mulheisen M, Wu DK, Epstein DJ (2002) Specification of the mammalian cochlea is dependent on Sonic hedgehog. Genes Dev 16:2365–2378.

Richardson GP, Bartolami S, Russell IJ (1990) Identification of a 275 kDa protein associated with the apical surfaces of sensory hair cells in the avian inner ear. J Cell Biol 110:1055–1066.

Richardson GP, Forge A, Kros CJ, Fleming J, Brown SDM, Steel KP (1997) Myosin VIIA is required for aminoglycoside accumulation in cochlear hair cells J Neurosci 17:9506–9519.

Riley B, Chiang M-Y, Farmer L, Heck R (1999) The *deltaA* gene of zebrafish mediates lateral inhibition of hair cells in the inner ear and is regulated by *pax2.1*. Development 126:5669–5678.

Rivolta MN, Holley MC (1998) GATA3 is downregulated during hair cell differentiation in the mouse cochlea. J Neurocytol 27:637–647.

Rivolta MN, Grix N, Lawlor P, Ashmore JF, Jagger DJ, Holley MC (1998) Auditory hair cell precursors immortalized from the mammalian inner ear. Proc R Soc Lond B 265: 1595–1603.

Roberson DF, Weisleder P, Bohrer PS, Rubel EW (1992) Ongoing production of sensory cells in the vestibular epithelium of the chick. Hear Res 57:166–174.

Romand R (1983) Development of the cochlea. In: Romand R (ed), Development of Auditory and Vestibular Systems. New York: Academic Press, pp. 47–88.

Romand R, Sapin V, Ghyselinck NB, Avan P, Le Calvez S, Dolle P, Chambon P, Mark M (2000) Spatio-temporal distribution of cellular retinoid binding protein gene transcripts in the developing and the adult cochlea. Morphological and functional consequences in CRABP- and CRBPI-null mutant mice. Eur J Neurosci 12:2793–2804.

Romand R, Hashino E, Dolle P, Vonesch J-L, Chambon P, Ghyselinck NB (2002) The retinoic acid receptors RARα and RARγ are required for inner ear development. Mech Dev 119:213–223.

Roth B, Bruns V (1992) Postnatal development of the rat organ of Corti. II. Hair cells receptors and their supporting elements. Anat Embryol 185:571–581.

Rubel EW, Fritzsch B (2002) Auditory system development: primary auditory neurons and their targets. Annu Rev Neurosci 25:51–101.

Ruben RJ (1967) Development of the inner ear of the mouse: a radioautographic study of terminal mitoses. Acta Otolaryngol Suppl 220:1–44.

Rübsamen R, Lippe WR (1998) The development of cochlear function. In: Rubel EW, Popper AN, Fay RR (eds), Development of the Auditory System. New York: Springer-Verlag, pp. 193–270.

Rüsch A, Eatock RA (1996) A delayed rectifier conductance in type I hair cells of the mouse utricle. J Neurophysiol 76:995–1004.

Rüsch A, Eatock RA (1997) Sodium currents in hair cells of the mouse utricle. In: Lewis ER, Long GR, Lyon RF, Steele CR, Narins PM, Hecht-Poinar E (eds), Diversity in Auditory Mechanics. Singapore: World Scientific, pp. 549–555.

Rüsch A, Erway LC, Oliver D, Vennstrom B, Forrest D (1998a) Thyroid hormone receptor β-dependent expression of a potassium conductance in inner hair cells at the onset of hearing. Proc Natl Acad Sci USA 95:15758–15762.

Rüsch A, Lysakowski A, Eatock RA (1998b) Postnatal development of type I and type II hair cells in the mouse utricle: acquisition of voltage-gated conductances and differentiated morphology. J Neurosci 18:7487–7501.

Rüsch A, Ng L, Goodyear R, Oliver D, Lisoukov I, Vennstrom B, Richardson G, Kelley MW, Forrest D (2001) Retardation of cochlear maturation and impaired hair cell function caused by deletion of all known thyroid hormone receptors. J Neurosci 21: 9792–9800.

Ryals BM, Rubel EW (1988) Hair cell regeneration after acoustic trauma in adult Coturnix quail. Science 240:1774–1776.

Rzadzinska AK, Schneider ME, Davies C, Riordan GP, Kachar B (2004) An actin treadmill and myosins maintain stereocilia functional architecture and self-renewal. J Cell Biol 164:887–897.

Sans A, Chat M (1982) Analysis of temporal and spatial patterns of rat vestibular hair cell differentiation by tritiated thymidine radioautography. J Comp Neurol 206:1–8.

Saunders JC, Coles RB, Gates GR (1973) The development of auditory evoked responses in the cochlea and cochlear nuclei of the chick. Brain Res 63:59–74.

Schneider ME, Belyantseva IA, Azevedo RB, Kachar B (2002) Rapid renewal of auditory hair bundles. Nature 418:837–838.

Self T, Mahony M, Fleming J, Walsh J, Brown SDM, Steel KP (1998) Shaker-1 mutations reveal roles for myosin VIIA in both development and function of cochlear hair cells. Development 125:557–566.

Self T, Sobe T, Copeland NG, Jenkins NA, Avraham KB, Steel KP (1999) Role of myosin VI in the differentiation of cochlear hair cells. Dev Biol 214:331–341.

Selyanko AA, Hadley JK, Wood IC, Abogadie FC, Jentsch TJ, Brown DA (2000) Inhibition of KCNQ1-4 potassium channels expressed in mammalian cells via M_1 muscarinic acetylcholine receptors. J Physiol 522:349–355.

Sgard F, Charpantier E, Bertrand S, Walker N, Caput D, Graham D, Bertrand D, Besnard F (2002) A novel human nicotinic receptor subunit, α 10, that confers functionality to the α9-subunit. Mol Pharmacol 61:150–159.

Shailam R, Lanford PJ, Dolinsky CM, Norton C, Gridley T, Kelley MW (1999) Expression of proneural and neurogenic genes in the embryonic mammalian vestibular system. J Neurocytol 28:809–819.

Shnerson A, Devigne C, Pujol R (1982) Age-related changes in the C57BL/6J mouse cochlea. II. Ultrastructural findings. Dev Brain Res 9:305–315.

Si F, Brodie H, Gillespie PG, Vazquez AE, Yamoah EN (2003) Developmental assembly of transduction apparatus in chick basilar papilla. J Neurosci 23:10815–10826.

Siemens J, Lillo C, Dumont RA, Reynolds A, Williams DS, Gillespie PG, Muller U (2004) cadherin 23 is a component of the tip link in hair-cell stereocilia. Nature 428:955–959.

Slepecky NB (1996) Structure of the mammalian cochlea. In: Dallos P, Popper AN, Fay RR (eds), The Cochlea. New York: Springer-Verlag, pp. 44–129.

Söllner C, Rauch G-J, Siemens J, Geisler R, Schuster SC, the Tubingen 2000 Screen Consortium, Müller U, Nicolson T (2004) Mutations in *cadherin 23* affect tip links in zebrafish sensory hair cells. Nature 428:955–959.

Souter M, Nevill G, Forge A (1995) Postnatal development of membrane specialisations of gerbil outer hair cells. Hear Res 91:43–62.

Spitzer NC, Kingston PA, Manning TJ Jr, Conklin MW (2002) Outside and in: development of neuronal excitability. Curr Opin Neurobiol 12:315–323.

Stevens CB, Davies AL, Battista S, Lewis JH, Fekete DM (2003) Forced activation of Wnt signaling alters morphogenesis and sensory organ identity in the chicken inner ear. Dev Biol 261:149–164.

Stone JS, Rubel EW (1999) Delta1 expression during avian hair cell regeneration. Development 126:961–973.

Tilney LG, DeRosier DJ (1986) Actin filaments, stereocilia, and hair cells of the bird cochlea. IV. How the actin filaments become organized in developing stereocilia and in the cuticular plate. Dev Biol 116:119–129.

Tilney LG, Tilney MS, Saunders JS, DeRosier DJ (1986) Actin filaments, stereocilia, and hair cells of the bird cochlea. III. The development and differentiation of hair cells and stereocilia. Dev Biol 116:100–118.

Tilney LG, Tilney MS, Cotanche DA (1988a) Actin filaments, stereocilia, and hair cells of the bird cochlea. V. How the staircase pattern of stereociliary lengths is generated. J Cell Biol 106:355–365.

Tilney LG, Tilney MS, Cotanche DA (1988b) New observations on the stereocilia of hair cells of the chick cochlea. Hear Res 37:71–82.

Tilney LG, Cotanche DA, Tilney MS (1992) Actin filaments, stereocilia, and hair cells of the bird cochlea. VI. How the number and arrangement of stereocilia are determined. Development 116:213–226.

Torres M, Gomez-Pardo E, Gruss P (1996) Pax2 contributes to inner ear patterning and optic nerve trajectory. Development 122:3381–3391.

Troutt LL, van Heumen WRA, Pickles JO (1994) The changing microtubule arrangements in developing hair cells of the chick cochlea. Hear Res 81:100–108.

Tsai H, Hardisty RE, Rhodes C, Kiernan A, Roby P, Tymowska-Lalanne Z, Mburu P, Rastan S, Hunter AJ, Brown SDM, Steel KP (2001) The mouse *slalom* mutant demonstrates a role for Jagged1 in neuroepithelial patterning in the organ of Corti. Hum Mol Genet 10:507–512.

Uziel A, Gabrion J, Ohresser M, Legrand C (1981) Effects of hypothyroidism on the structural development of the organ of Corti in the rat. Acta Otolaryngol 92:469–480.

Verpy E, Leibovici M, Zwaenpoel I, Liu X-Z, Gal A, Salem N, Mansour A, Blanchard S, Kobayashi I, Keats BJB, Slim R, Petit C (2000) A defect in harmonin, a PDZ domain-containing protein expressed in the inner ear sensory hair cells, underlies Usher syndrome type 1c. Nat Genet 26:51–55.

Vollrath MA, Eatock RA (2003) Time course and extent of mechanotransducer adaptation in mouse utricular hair cells: comparison with frog saccular hair cells. J Neurophysiol 90:2676–2689.

von Bartheld CS, Patterson SL, Heuer JG, Wheeler EF, Bothwell M, Rubel EW (1991) Expression of nerve growth factor (NGF) receptors in the developing inner ear of chick and rat. Development 113:455–470.

Wallis D, Hamblen M, Zhou Y, Venken KJT, Schumacher A, Grimes HL, Zogghbi HY, Orkin SH, Bellen HJ (2003) The zinc finger transcription factor *Gfi1*, implicated in lymphomagenesis, is required for inner ear hair cell differentiation and survival. Development 130:221–232.

Walsh T, Walsh V, Vreugde S, Hertzano R, Shahin H, haika S, Lee MK, Kanaan M, King M-C, Avraham KB (2002) From flies' eyes to our ears: mutations in a human class III myosin cause progressive nonsyndromic hearing loss DFNB30. Proc Natl Acad Sci USA 99:7518–7523.

Weber T, Zimmerman U, Winter H, Mack A, Kopschall I, Rohbock K, Zenner H-P, Knipper M (2002) Thyroid hormone is a critical determinant for the regulation of the cochlear motor protein prestin. Proc Natl Acad Sci USA 99:2901–2906.

Weil D, El-Amraoui A, Masmoudi S, Mustapha M, Kikkawa Y, Laine S, Delmaghani S, Adata A, Nadifi S, Zina ZB, Hamel C, Gal A, Ayadi H, Yonekawa H, Petit C (2003) Usher syndrome type 1 G (USH1G) is caused by mutations in the gene encoding SANS, a protein that associates with the USH1C protein, harmonin. Hum Mol Genet 12:463–471.

Wells AL, Lin AW, Chen LQ, Safer D, Cain SM, Hasson T, Carragher BO, Milligan RA, Sweeney HL (1999) Myosin VI is an actin-based motor that moves backwards. Nature 401:505–508.

Wilson SM, Householder DB, Coppola V, Tessarollo L, Fritzsch B, Lee EC, Goss D, Carlson GA, Copeland NG, Jenkins NA (2001) Mutations in *Cdh23* cause nonsyndromic hearing loss in *waltzer* mice. Genomics 74:228–233.

Wu DK, Oh S-H (1996) Sensory organ generation in the chick inner ear. J Neurosci 16:6454–6462.

Wu YC, Ricci AJ, Fettiplace R (1999) Two components of transducer adaptation in auditory hair cells. J Neurophysiol 82:2171–2181.

Xiang M, Gan L, Li D, Chen XY, Zhou L, O'Malley Jnr BW, Klein W, Nathans J (1997) Essential role of POU-domain transcription factor Brn-3c in auditory and vestibular hair cell development. Proc Natl Acad Sci USA 94:9445–9450.

Xiang M, Gao W-Q, Hasson T, Shin JJ (1998) Requirement for Brn-3c in maturation and survival, but not in fate determination of inner ear hair cells. Development 125: 3935–3946.

Xu P-X, Adams J, Peters H, Brown MC, Heaney S, Maas R (1999) *Eya1*-deficient mice lack ears and kidneys and show abnormal apoptosis of organ primordia. Nat Genet 23:113–117.

Yin H, Pruyne D, Huffaker TC, Bretscher A (2000) Myosin V orientates the mitotic spindle in yeast. Nature 406:1013–1015.

Ylikoski J, Pirvola U, Eriksson U (1994) Cellular retinol-binding protein type I is prominently and differentially expressed in the sensory epithelium of the rat cochlea and vestibular organs. J Comp Neurol 349:596–602.

Zhang N, Martin GV, Kelley MW, Gridley T (2000) A mutation in the *Lunatic fringe* gene suppresses the effects of a *Jagged2* mutation on inner hair cell development in the cochlea. Current Biology 10:659–662.

Zheng J, Shen W, He DZZ, Long KB, Madison LD, Dallos P (2000) Prestin is the motor protein of cochlear outer hair cells. Nature 405:149–155.

Zheng JL, Gao W-Q (2000) Overexpression of *Math1* induces robust production of extra hair cells in postnatal rat inner ears. Nat Neurosci 3:580–586.

Zheng JL, Shou J, Guillemot F, Kageyama F, Gao W-Q (2000) *Hes1* is a negative regulator of inner ear hair cell differentiation. Development 127:4551–4560.

Zheng L, Sekerkova G, Vranich K, Tilney LG, Mugnaini E, Bartles JR (2000) The deaf jerker mouse has a mutation in the gene encoding the espin actin-bundling proteins of hair cell stereocilia and lacks espins. Cell 102:377–385.

Zheng W, Huang L, Wei Z-B, Silvius D, Tang B, Xu P-X (2003) The role of *Six1* in mammalian auditory system development. Development 130:3989–4000.

Zine A, Romand R (1996) Development of the auditory receptors of the rat: a SEM study. Brain Res 721:49–58.

Zine A, Hafidi A, Romand R (1995) Fimbrin expression in the developing rat cochlea. Hear Res 87:165–169.

Zine A, Van de Water TR, de Ribaupierre F (2000) Notch signalling regulates the pattern of auditory hair cell differentiation in mammals. Development 127:3373–3383.

Zine A, Aubert A, Qiu J, Therianos S, Guillemot F, Kageyama R, de Ribaupierre F (2001) *Hes1* and *Hes5* activities are required for the normal development of the hair cells in the mammalian inner ear. J Neurosci 21:4712–4720.

3
The Structure and Composition of the Stereociliary Bundle of Vertebrate Hair Cells

DAVID N. FURNESS AND CAROLE M. HACKNEY

1. Introduction

Mechanoelectrical transduction in the inner ear and lateral line organs of vertebrates depends on sensory hair cells. These usually have flask-shaped or cylindrical cell bodies (Fig. 3.1) and possess an apical bundle of cellular protrusions (Figs. 3.1 to 3.4) that look like tiny hairs, or cilia, under the light microscope. However, only one of the protrusions in each bundle, the kinocilium, is a true cilium (Fig. 3.3) containing an array of microtubules. The rest have either been called stereocilia or stereovilli (see, e.g., Neugebauer and Thurm 1984, 1985) but although the latter is a more accurate term because they more closely resemble microvilli than cilia, the former is in greater use and so is used here.

It is well accepted that deflection of the hair bundle toward the tallest row of stereocilia, the excitatory direction (Fig. 3.5), increases the rate of opening of relatively nonselective cationic channels in the hair cell apex; this results in the development of a depolarizing receptor potential (Hudspeth and Corey 1977; see also Fettiplace and Ricci, Chapter 4). Deflection in the direction of the shortest row leads to closure of these mechanoelectrical transducer (MET) channels. The stimulus direction is important, as it is most effective along the central axis and not effective at right angles to it (Shotwell et al. 1981). The process of mechanoelectrical transduction is extremely sensitive with displacements of the tip of the bundle of around 100 nm being sufficient to open all the MET channels. It is also so rapid that second-messenger systems are unlikely to be involved, suggesting that the channels must be mechanically gated (Corey and Hudspeth 1979, 1983). The motion of hair bundles may also be amplified by energy supplied by the hair cells themselves with hair bundles being capable of both force production and spontaneous oscillation (see Hudspeth 1997 and Fettiplace et al. 2001 for reviews). These phenomena may be manifestations of the adaptation mechanisms that have been developed by hair cells to keep the MET channels within their most sensitive operating range. Hair cell adaptation has a rapid component on a sub-millisecond time scale that begins almost immediately that the channels open followed by a slower component that can take tens of

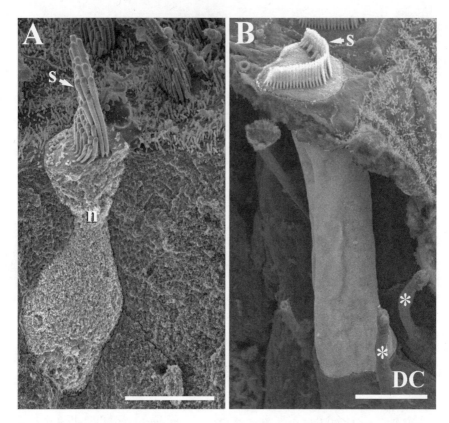

FIGURE 3.1. SEM showing the diverse morphology of hair cells. (**A**) A type I vestibular hair cell of the utricular macula of a guinea pig revealed by fracturing a critical point dried specimen. The cell is flask shaped with a narrow neck region (n) and the bundle of stereocilia (s) protruding from the apex. Many vertebrate hair cells resemble this type, although precise body and bundle shapes vary. The cell membrane has been removed during the fracturing process, revealing a granular interior. There are no spaces between supporting cells and the hair cell. (**B**) An outer hair cell in a fracture of the guinea pig organ of Corti revealing the cylindrical cell body and stereocilia (s). The cell body is enclosed by Deiters' cells (DC) only at the basal end. Processes (*), here broken, extend from the Deiters' cell bodies beneath the hair cell to the upper surface of the organ of Corti. The membrane of the hair cell is intact, giving it a relatively smooth appearance. Scale bars = 10 μm.

milliseconds to develop (Ricci and Fettiplace 1997; Wu et al. 1999). Mechanotransduction, amplification, and adaptation are likely to be affected by the precise morphological organization of the hair bundle and by the mechanical properties conferred by the cytoskeletal organization of the hair cell apex and the extracellular matrix associated with stereocilia. These processes are also calcium sensitive (see Fettiplace et al. 2001 for a review) and will therefore be critically influenced by the control of intracellular calcium levels by calcium pumps and buffers in the hair cell apex. Over the last few years, new data have emerged about the distribution and function of a number of proteins in hair cells. This review summarizes what is known about the morphology and protein composition of the hair cell apex with particular reference to findings from ultrastructural and immunocytochemical studies.

2. Immunocytochemical Techniques for Studying the Hair Cell Apex: Advantages and Disadvantages

A range of techniques has been applied to the analysis of the structure and composition of hair cells. These include phenotypic studies of mice and zebrafish with mutations or targeted deletions of genes important in human hearing which have revealed abnormalities in normal hair-bundle development and function (see, e.g., Nicolson et al. 1998 and reviews by Steel and Kros 2001; Bryant et al. 2002). These genetic and molecular approaches are being combined with light (LM), scanning (SEM), and transmission electron microscopy (TEM) and immunocytochemistry to determine the location and function of proteins directly responsible for particular processes in hair cells. In view of the importance of this latter method for determining the distributions of various hair cell proteins, it is perhaps first worth drawing attention to the pros and cons of some of the methods in common use.

2.1 Preembedding Labeling at the LM Level

In this technique, relatively lightly fixed material is prepared as whole mounts, isolated cells, or cryosections. The tissue is usually permeabilized with a detergent and then labeled with a primary antibody. This is then visualised using secondary antibodies conjugated to fluorochromes (Fig. 3.6) or to other intermediate molecules that can be used to generate a colored reaction product, for example immunoperoxidase, where chromogens such as diaminobenzidene provide the visible signal. This technique is relatively quick and easy to perform. However, false-positive stereociliary labeling is a hazard of this type of preparation. It is possible, for example, to produce stereociliary labeling in guineapig tissue by using rat, rabbit, or mouse serum in place of primary antibodies (see, e.g., Steyger et al. 1989). This labeling cannot readily be prevented by standard blocking steps and is thought to be due to the stereocilia being particularly prone to nonspecific binding of antibodies and other serum proteins, possibly because of glycoproteins or negative charges associated with their surfaces (see also Section 3.4).

2.2 Preembedding Labeling at the Electron Microscopic Level

Preembedding techniques for EM are similar to those used for LM but obviously require electron-dense markers such as immunoperoxidase/DAB or immunogold gold particles (Fig. 3.7A). The latter can also be used for SEM where backscattered electron detection can assist in the identification of gold particles (Fig. 3.7C, D). For TEM, labeled tissue is resin embedded with or without further fixation. However, this technique can produce the same potentially false-positive results described in Section 2.1. In addition, other drawbacks are that tissue preservation is often poor (because of detergent extraction) and when immunoperoxidase is used, the reaction with the chromogen also damages tissue structure (see, e.g., Danbolt et al. 1998). There are limitations with regard to antigen penetration; central regions of protein-rich structures such as the cuticular plate (Fig. 3.2; see also Section 4) or the core of the stereocilia may not label because antibodies are large molecules (immunoglobulin Gs) [IgGs] are approximately 16 nm long) and gold particles 5 to 30 nm in diameter are commonly used, both of which may be hindered from entering narrow spaces (Fig. 3.7A, B; see also Slepecky et al. 1990; Mahendrasingam et al. 1998). The use of smaller gold particles improves penetration but these are difficult to visualize unless enhanced by silver staining. These factors affect the distribution of labeling and

⎯⎯⎯⎯⎯⎯⎯⎯⎯⎯⎯⎯⎯⎯⎯⎯⎯⎯⎯⎯⎯⎯⎯⎯⎯▶

FIGURE 3.2. Detail of the outer hair cell apex. (**A**) TEM of a longitudinal section showing three rows of stereocilia (s) increasing in height from one side of the bundle to the other. The stereocilia are separated at the base but converge so that the tips of the shorter ones come into close apposition with the adjacent taller ones. Note the electron-dense patches at the tip of each stereocilium; the inset shows the density at the tip of the tallest stereocilium (*arrows*) at higher magnification. The stereociliary rootlets (r) extend into the cuticular plate (cp) and the underlying cytoplasm includes Golgi apparatus (*arrowhead*), mitochondria (*arrow*), and other organelles. A channel of cytoplasm (ch) extends through the cuticular plate and contains the basal body (bb) of the kinocilium. The kinocilium itself is absent in adult mammalian cochlear hair cells. Some cuticular plate material is seen lateral to the channel (*). Scale bar = 1 μm. Inset scale bar = 200 nm. (**B**) SEM of a fracture of an OHC along a plane that is slightly oblique compared with that shown in A showing the three-dimensional relationship of the hair bundle to the apical structures. The fracture was obtained by cryoprotecting a fixed segment of organ of Corti in a solution of 50% dimethyl sulfoxide, freezing the sample, and fracturing, followed by osmium maceration (see, e.g., Lim 1986). The three rows of stereocilia (s) are again visible, some having been broken where they cross the fracture plane to leave short stumps. The cuticular plate (cp) is relatively homogeneous compared with the granular cytoplasm, and the cytoplasmic channel (ch) and lateral cuticular plate material (*) are visible. Scale bar = 1 μm. Micrograph taken by the authors from material prepared by Dr. Y. Katori.

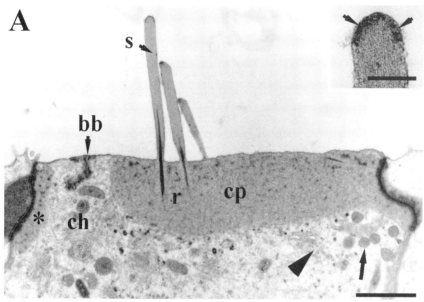

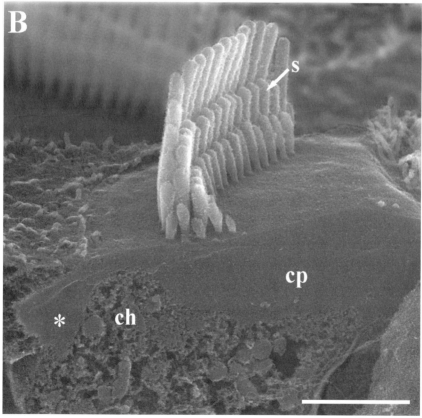

make estimation of relative amounts of antigen in different subcellular compartments subject to uncertainty.

2.3 Postembedding Immunogold Labeling at the Electron Microscopic Level

Many of the problems associated with preembedding labeling are reduced by using post-embedding methods (see, e.g., Fig. 3.7B, E–G). Here, labeling is performed after sectioning when the antigens are exposed on the surfaces of the section, ensuring equal accessibility. In TEM, the density of gold particles more accurately reflects the concentration of antigen. This method can therefore be used for comparing relative concentrations of one or more antigens in different compartments. It can also provide absolute quantitative data as long as the labeling is calibrated against known concentrations of the antigen prepared under similar circumstances (see, e.g., Gundersen et al. 1993; Hackney et al. 2003; and Section 5). The lateral resolution of this technique is good, on the order of 25 to 30 nm (Matsubara et al. 1996). However, the additional processing required to embed the samples prior to labeling them can diminish the antigenicity of the target proteins. Fixation with glutaraldehyde can itself cause loss of antigenicity by distortion of the fixed proteins, an effect that can be reduced by fixation with paraformaldehyde or by using antibodies that have been raised against antigens conjugated to carrier proteins using glutaraldehyde which then recognize glutaraldehyde-fixed antigens. Loss of antigenicity can also be reduced by low-temperature embedding or by embedding freeze-dried unfixed tissue (Slepecky and Ulfendahl 1992). However, the sensitivity of the postembedding technique may be lower than fluorescence methods; for example, Slepecky et al. (1990) found that phalloidin labeling showed actin to be present in regions where it was not detectable by postembedding labeling. Moreover, false-positive labeling of stereocilia cannot be completely ruled out by this method because the sites that make stereocilia "sticky" may remain attractive to antibodies even after processing. Thus it is important to perform a number of controls with this as with any immunocytochemical procedure.

2.4 Controls for Immunocytochemistry

In each of the above procedures, omission of the primary antibody will show whether the secondary antibody is binding nonspecifically to the tissue. However, this is not a specific control for the primary antibody used and is therefore of only limited value. A better control is to incubate the primary antibody with purified antigen in excess (preadsorption). Use of the preadsorbed mixture on the tissue should result in little or no specific labeling. This confirms that the epitopic site recognized by the antibody exists on the antigen and is present at the sites where labeling has been abolished. However, even if preadsorption has been completely successful, this control does not preclude the possibility that similar epitopic sites on different proteins might also be present in the tissue, as labeling of these epitopes would be reduced just as effectively as the presumed target because of elimination of the antibody. Preadsorption can also be used to remove unwanted cross-reactivity with other proteins if the pure form of those proteins is used as the absorbent. In addition, Western blotting can be used to check the molecular mass of the proteins being labeled is close to that expected, although different isoforms of the same protein may vary in this respect. Using monoclonal antibodies reduces the likelihood of an antibody directed at one protein recognizing another but has the drawback that such increased specificity reduces the sensitivity of the detection technique compared with using polyclonal antibodies where multiple epitopic sites on a given protein may be recognized (Harlow and Lane 1999).

Despite the problems associated with these various techniques, which mean results obtained from them should be interpreted carefully, their application has added significantly to our knowledge of the composition of the hair cell apex.

3. General Features of Hair Cells and Hair Cell Sensory Epithelia

All vertebrate hair cells studied so far appear to perform mechanoelectrical transduction in a similar way and share conserved morphological characteristics that relate to this basic function. However, there are also specializations of the different organs and the hair cells within them that are related to their differing physiological roles, including modifications to the hair bundles and the accessory structures to which they may be attached (Manley 2000).

3.1 Morphology and Ultrastructure of the Hair Bundle

Hair bundles are composed of a cluster of stereocilia arranged in rows that increase in height to form a staircase. The tips of stereocilia in different rows tend to converge so that many bundles have a conical appearance with a bevelled edge formed by the tips of the stereocilia as they change in height. The tips of stereocilia in each of the shorter rows approach a point on the side of the taller stereocilia in the adjacent taller row (Fig. 3.2). With the exception of some bundles in primitive vertebrates such as lampreys (Hoshino 1975), virtually all vertebrate hair cells show this basic staircase pattern but there is considerable variation in packing, number, height, and width of the stereocilia (Figs. 3.3, and 3.4). These differences in morphology may reflect the need to process different kinds of environmental stimuli and thus to respond to different ranges of frequency and stimulus intensity.

The stereocilia are arranged in a hexagonal pattern. In most cases, a single kinocilium is located next to the central stereocilia of the tallest row (Fig. 3.3A, D) and connected to them by extracellular material (Hillman 1969; Bagger-

Sjöbäck and Wersäll 1973; Neugebauer and Thurm 1985). In some hair cells, the stimulus is probably delivered partly or wholly via the kinocilium, which sometimes has a specialized bulb and which projects into the accessory membrane where it is connected by filaments (Lewis and Li 1975; Barber and Emerson 1979; Jacobs and Hudspeth 1990; Nagel et al. 1991; Tsuprun and Santi 1998). Mammalian and some avian cochlear hair cells, however, lose the kinocilium on maturation, leaving behind a basal body (Figs. 3.2A, 3.5, and 3.6B, C). The bundle has a central axis of symmetry that runs approximately through the position of the kinocilium which is parallel with the physiological excitatory–inhibitory axis.

Conical bundles are found in hair cell epithelia in different vertebrate classes (Fig. 3.3), although considerable variations in size and the number of rows of stereocilia occur even within a single organ. Thus, for example, six different types of hair bundle have been described in different regions of the bullfrog sacculus (Lewis and Li 1975). In the guinea pig, vestibular bundles have been classified according to differences in packing of stereocilia in adjacent rows into a "loose" type and a "tight" type; in "loose" bundles, one axis of the hexagonal array is aligned with the excitatory–inhibitory axis while in "tight" bundles, the hexagonal array is displaced by 30° from this axis (Bagger-Sjöbäck and Takumida 1988; see also Fig. 3.3B). However, in turtle and frog vestibular maculae, a continuum in these bundle geometries has been reported (Rowe and Peterson 2004). Differences in packing could affect the efficiency of activation of MET channels by off-axis stimulus directions and may therefore be relevant to the function of different hair cell types.

Hair bundles of mammalian cochlear and vestibular hair cells can have similar numbers of stereocilia but these are generally arranged in fewer rows in cochlear hair bundles (3–4; Furness and Hackney 1985) than in vestibular ones in the

FIGURE 3.3. SEM showing hair bundles from different vertebrate classes. (**A**) Teleost fish utricle (roach—*Rutilus rutilus*). Although the size of the bundles and packing of the stereocilia vary, the bundles share the same basic pattern, where the stereocilia form rows increasing in height in a stepwise manner. Note the kinocilium (k) to one side of the tallest row of stereocilia. Scale bar = 2 μm. Micrograph taken by the authors from material prepared by Dr. D-Ch. Neugebauer. (**B**) Mammalian utricle (guinea pig) again showing the variation in bundle size but the same basic pattern of stereociliary rows. Note that two forms of packing, "tight" (t) and "loose" (l) as defined by Bagger-Sjöbäck and Takumida (1988) can be seen in these hair bundles. Scale bar = 2 μm. (**C, D**) Reptilian cochlea (turtle—*Trachemys scripta elegans*). (C) View of the stereociliary rows looking along the direction in which the stepwise increments in height occur (i.e., along the excitatory–inhibitory axis) with the tallest row furthest away. View of another bundle in the opposite direction from C so that only the tallest stereocilia are clearly seen. Note the kinocilium (k) which here is similar in height to the tallest rows of stereocilia. Note also that the tallest stereocilia are thinner than the others in the same bundle. Scale bars = 1.5 μm.

same species (between 9 and 23; Bagger-Sjöbäck and Takumida 1988). The number of stereocilia per row is consequently greater and the bundle therefore becomes elongated in the plane orthogonal to the stimulus direction (Fig. 3.4). This arrangement may allow cochlear hair cells to respond more efficiently to higher frequencies than vestibular ones. In both inner and outer hair cells (IHCs and OHCs) (Fig. 3.4A, B), the bundle is arranged in a more or less pronounced W-shape. The kinociliary basal body lies just beyond the notch of the W in the excitatory direction, the loss of the kinocilium during development possibly being an adaptation to increase the sensitivity of the mammalian cochlear hair cells.

Variations in width of the stereocilia occur within and between hair bundles of the same organ. In the mouse utricle, the stereocilia of type II hair cells are finer than those of type I hair cells (Rüsch et al. 1998). In turtles, the shorter stereocilia are wider (approximately 500 nm) than the taller ones (approximately 250 nm) in the same bundle (Fig. 3.3C; Hackney et al. 1993). In OHCs in the rodent cochlea (Fig. 3.4), shorter stereocilia are sometimes narrower (approximately 150 nm) than taller ones (approximately 250 nm). The situation in rodent IHCs is more complex as the tallest stereocilia are narrower than the next row but are wider than the subsequent shorter rows (Lim 1986; Fig. 3.4B). In lizard hair cells, stereocilia approaching 1 µm in diameter have been noted (Tilney et al. 1980). Wider stereocilia may be stiffer than narrower ones because they contain more actin filaments, although variations in their cross-linking proteins and in the stiffness of rootlets may complicate the situation.

The heights of stereocilia also vary considerably. The shortest stereocilia in the guinea-pig cochlea are under 1 µm (see, e.g., Furness et al. 1997) but the tallest stereocilia in the free standing hair bundles in alligator lizard reach 38 µm (Mulroy and Williams 1987). It should be noted that the height of the kinocilium, when present, also varies in relation to the height of the tallest stereocilia within individual epithelia, for example, in the central regions of the turtle cochlea, the kinocilium is similar in height to the tallest stereocilia whereas the stereociliary bundles on the limbal edges may have kinocilia that are more than twice the height of the tallest stereocilia (Hackney et al. 1993).

FIGURE 3.4. SEM showing the morphology of mature mammalian cochlear hair bundles. The number of rows of stereocilia is reduced compared to hair cells from vestibular organs or the auditory organs of non-mammalian vertebrates. (**A**) A W-shaped OHC bundle from a guinea-pig cochlea. The stereocilia form three precisely organized rows in each of which they are nearly all of similar height. Scale bar = 2 µm. (**B**) In guinea-pig IHCs, the row structure is less precise. Scale bar = 2 µm. (**C**) An IHC bundle of a bat (*Pteronotus parnellii*) showing the presence of only two rows of stereocilia, the minimum number that would be required for mechanotransduction according to present models. Micrograph taken by the authors from material supplied by Dr. M. Vater. Scale bar = 2 µm.

3. The Structure and Composition of Vertebrate Hair Bundles 105

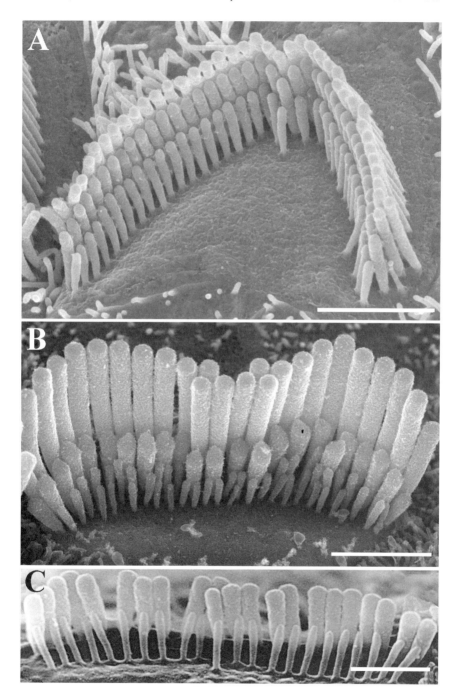

In auditory epithelia, systematic cochleotopic variations in bundle height occur with location, decreasing toward the high-frequency basal end of the cochlea or papilla. In turtle, the height decreases from about 8 µm to 4.5 µm (Hackney et al. 1993), in chick from 5.5 µm to 1.5 µm (Tilney and Tilney 1988), and in the chinchilla third OHC row, from 5.5 µm to 1.0 µm (Lim 1986). The number of stereocilia also changes; it increases toward the basal end. In the turtle basilar papilla, the number of stereocilia increases from about 60 to 100 (Hackney et al. 1993); in chick basilar papilla, it increases from 50 to 300 (Tilney and Tilney 1988); and in the chinchilla organ of Corti, in the third row OHCs, it increases from about 18 to 40 in the apical turns to about 80 in the base (Lim 1986). These changes in height and number of stereocilia are likely to contribute to the hair cell's mechanical properties; shorter bundles with more stereocilia are stiffer. In addition, the increased number of stereocilia is likely to result in an increased number of MET channels and may also be related to the fact that adaptation and active movements are faster in high-frequency hair bundles in some vertebrates (Fettiplace et al. 2001; see also Fettiplace and Ricci, Chapter 4).

The step size between stereocilia of different rows also varies and is likely to affect the sensitivity of the bundle to deflections. The excitation of hair cells is believed to depend on shearing between immediately adjacent shorter and taller stereocilia of neighboring rows. As the bundle is deflected, the stereocilia pivot about their ankles and this causes a shear to be developed so that the shorter stereocilium moves relative to the side of the taller stereocilium. The shearing is maximal at the tip of the shorter stereocilium. The ratio between this maximal shear distance and the size of the horizontal deflection of the tip of the tallest stereocilium is called the shear gain.

Kinematic modeling has revealed that the average shear gain in hair bundles of OHCs increases systematically from apical to basal locations in the guinea-pig cochlea (Fig. 3.5; Furness et al. 1997). Thus, higher frequency hair cells require less displacement of their tips to produce the same degree of shearing. However, the decreasing height of the tallest stereocilia offsets this, as the amount of shear developed remains relatively constant as a function of angular rotation. Interestingly, the shear gains for two substantially different bundle types are very similar; the mean gain of 38 individual stereociliary pairs from 21 guinea pig OHCs was approximately 0.15 (Furness et al. 1997) compared with the average gain for a bundle of 0.14 in bullfrog saccular hair cells (Jacobs and Hudspeth 1990). This means that for a 10-nm deflection of the hair bundle, a shearing movement of about 1.5 nm would occur between the tip of the shorter stereocilium and the side of the taller stereocilium. In the mouse utricle, a smaller average bundle shear gain has been found, approximately 0.047 (Holt et al. 1997), which appears to be consistent with a broader operating range in terms of bundle displacement for these cells.

Inhibitory deflections cause the shorter stereocilia to slide upwards relative to the taller stereocilia while excitatory deflections cause them to slide downwards. By mechanically linking the stereocilia together by means of a "gating spring" at the point of maximum shear, deformation produced by tension placed on the

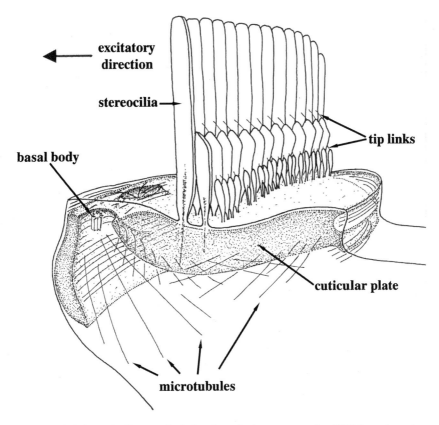

FIGURE 3.5. Schematic diagram depicting the apical structures of an IHC from the guinea pig cochlea and the complex supporting cytoskeleton of microtubules (*thin lines*). The cuticular plate is enclosed in a pericuticular basket of microtubules which includes rings around the apical and beneath the basal portions and a central underlying meshwork. Microtubules also extend into the channel of cytoplasm which penetrates the cuticular plate and contains the basal body (see also Fig. 3.2).

spring can deliver stimulus energy to MET channels located in this vicinity. An appropriate arrangement of the gating spring and its connection to the channels would detect the difference in shear direction, so that inhibitory deflections cause channels to close, and excitatory deflections cause them to open (see Fig. 4.3 in Fettiplace and Ricci, Chapter 4). This gating spring hypothesis can thus explain the asymmetry of the hair-bundle response (Corey and Hudspeth 1983) and has gained almost universal acceptance. More controversial, however, is the issue of the identity and location of the gating spring and the associated MET channels. It is most likely that this spring is represented by some structure in the bundle where maximum shear is developed (i.e., the tip of the shorter stereocilia), thereby maximizing the sensitivity of the bundle to deflections

(Thurm 1981). Attention has therefore been directed toward extracellular cross-links that connect adjacent stereocilia together and that may be specialized adaptations of the surface coat of the hair bundle.

3.2 The Surface Coat and the Cross-Links

Hair bundles have an extensive cell coat that has been demonstrated in chick vestibular hair cells by freeze-etching (Hirokawa and Tilney 1982) and in mammalian cochlear and vestibular hair cells by staining with cationic dyes such as ruthenium red (Slepecky and Chamberlain 1985a; Takumida et al. 1988) and Alcian blue (Santi and Anderson 1987). The cell coat of cochlear hair bundles appears to be more prominent than that of adjacent supporting cells. Staining with ruthenium red reveals a relatively thin (10-nm thick) surface coat on the hair bundle (Slepecky and Chamberlain 1985a; Santi and Anderson 1987). Long incubation in Alcian blue reveals a 90-nm thick inner coating that extends over the stereocilia and the apical cuticular plate and an outer coating that consists of filaments up to 1.5 μm long extending into the subtectorial space around the hair bundle (Santi and Anderson 1987).

In addition to this generalized cell coat, specific cross-links with well-defined spatial identities connect the adjacent stereocilia (Figs. 3.8 and 3.9). It has been suggested that these links could be artifactual condensations of the generalized cell coat (Hackney and Furness 1986). However, different categories of cross-link are immunologically distinct and have differential binding to, or susceptibility to, various agents, suggesting that they have specific identities and functions (Batanov et al. 2004; see also Section 3.4 and Goodyear et al., Chapter 2).

Several major categories of cross-links have been morphologically defined. Lateral links run horizontally (parallel to the apical surface) in probably all hair bundles and have been seen in, for example, fish (Neugebauer and Thurm 1985), reptiles (Bagger-Sjöbäck and Wersäll 1973; Czukas et al. 1987), birds (Goodyear and Richardson 1994), and mammals (Osborne et al. 1984; Pickles et al. 1984; Furness and Hackney 1985). These lateral links connect adjacent stereocilia both within and between rows in a hexagonal pattern. In the mammalian cochlea, they consist of filaments of variable width and length connecting the lateral membranes of the stereocilia (Fig. 3.8C–F). The central portions have a denser region (Fig. 3.8D, E; Furness and Hackney 1985) that stains specifically with tannic acid (Tsuprun and Santi 2002) and they tend to occupy discrete zones (Fig 3.8B, C). There are, for instance, clearly identifiable ankle links in fish (Fig. 3.8B; Neugebauer and Thurm 1985) and chicks (Goodyear and Richardson 1992). Fewer lateral links tend to be seen in adult hair cells in the mammalian cochlea, suggesting that they are reduced in high-frequency hair cells (cf. Figs. 3.8A, B and Fig. 3.8C). In adult guinea pigs, they are observed to varying extents along the shafts (Furness and Hackney 1985) and ankle links appear to be absent.

In addition to the lateral links, cross-links with various orientations can be found in the vicinity of the tips in many hair bundles (Fig. 3.9A, B). Among these is the tip link that connects the tip of each shorter stereocilium with its neighbor in the taller row behind (Figs. 3.8C, 3.9C–G; Osborne et al. 1984; Furness and Hackney 1985; Little and Neugebauer 1985). The discovery of tip links led to the suggestion that they represented the gating springs predicted by theories of mechanoelectrical transduction (Pickles et al. 1984). The tip link is a single 2-to-3-nm-wide filament that extends from the tip of the shorter stereocilium (Figs. 3.8C and 3.9C–G) and frequently bifurcates (Figs. 3.8C and 3.9D, E) to contact the membrane of the taller stereocilium at two adjacent points (Hackney and Furness 1995; Kachar et al. 2000). At its lower end, the tip link is anchored by submembrane linkers that occur in a narrow gap between the stereociliary membrane and an underlying dense plaque that caps the actin core (Osborne et al. 1988; Kachar et al. 2000). At its upper attachment point, there is a submembraneous electron-dense plaque (Fig. 3.9G) occupying the entire width of the gap between the outer actin filaments of the core and the membrane (Furness and Hackney 1985). In high-resolution studies of freeze-dried, deep-etched preparations, the lower part of the tip link appears to consist either of a pair of braided macromolecules or a single filament with double helical substructure. It may then, in either case, unwind further up to produce the two narrower anchoring strands (Kachar et al. 2000; see also Tsuprun and Santi 2000).

Tip links have been observed in all vertebrate classes and most hair cell sensory organs (fish: Little and Neugebauer 1985; amphibia: Assad et al. 1991; reptiles: Hackney et al. 1993; birds: Pickles et al. 1989). The electron-dense plaques at the upper and lower attachment sites are similarly common. These consistent features are thus likely to be important in the function of the tip link.

3.3 The Contact Region

Below the tip-link attachment points, the angulation of the stereocilia causes the shorter stereocilia to approach the taller ones in the row behind to a point where their membranes virtually touch (Fig. 3.10). Termed the contact region in mammals (Hackney et al. 1992), the membranes here frequently appear denser and to be connected by short pillar-like bridges (Figs. 3.9F, G and 3.10; Hackney and Furness 1995). These bridges may well correspond to the horizontal top connectors that have been described in other classes of vertebrate (Nagel et al. 1991; Goodyear and Richardson 1992). Treatment with cationic ferritin (Neugebauer and Thurm 1987) causes the stereocilia to fuse in this region, presumably by cancelling the negative charges keeping the membranes apart. Some form of morphological specialization is common in this contact region (see, e.g., Neugebauer and Thurm 1985; Hackney et al. 1993) but it varies in appearance, perhaps depending on the method and/or rapidity of fixation. In addition, calmodulin labeling has been noted to occur specifically at this point (Fig. 3.7G), suggesting that it has a particular function (Furness et al. 2002).

3.4 Composition of the Surface Coat and Links

The presence of carbohydrates in the coating over the membranes of the stereocilia and apical surface has been demonstrated by staining with ruthenium red and other cationic dyes (Santi and Anderson 1987; Tsuprun and Santi 2002). In addition, lectin binding histochemistry and immunocytochemistry have indicated more precisely the composition of the surface coat and shown that there are differences between hair bundles from different systems and during development (Warchol 2001).

Lectin binding in bullfrog has suggested the presence of residues of glucose, mannose (using concanavalin A), galactose (*Ricinus communis* agglutinin I), *N*-acetylglucosamine and *N*-acetylneuraminic acid (wheat germ agglutinin [WGA]) and *N*-acetylgalactosamine (*Vicia villosa* lectin) (Baird et al. 1993). However, there are differences between vestibular organs and regional differences within them. Similarly, some avian vestibular hair bundles but not auditory hair cells label for *N*-acetylgalactosamine using peanut agglutinin (Goodyear and Richardson 1994; Warchol 2001). WGA has also been demonstrated to bind to mammalian cochlear hair bundles (Gil-Loyzaga and Brownell 1988; Katori et al. 1996b) and chick vestibular hair bundles (Warchol 2001). This binding is associated with keratan sulfate proteoglycans (KSPGs). In fact, labeling with antibodies to KSPGs indicates they occur on the crown of the tallest stereocilia (Fig. 3.7C, D) and at tip and lateral link attachments in mammalian cochlear hair bundles (Katori et al. 1996a). Cuprolinic acid, which specifically binds to negatively charged sulfated residues, produces a similar labeling pattern (Tsuprun and Santi 2002). KSPGs may organize and determine the spacing of stereocilia, provide mechanical support, and maintain negative charges (Katori et al. 1996a). They are also present in the tectorial membrane (Munyer and Schulte 1994). The high molecular mass tectorin is a major component of tectorial membranes, and may be a KSPG (Killick and Richardson 1997). Antibodies to it label the stereociliary bundles of IHCs and OHCs.

Incubation of organ cultures of neonatal mouse cochlea with WGA (Richardson et al. 1988), conconavalin A, and neomycin (an aminoglycoside antibiotic) (Kössl et al. 1990) in vitro causes hair bundle stiffness to increase, indicating that the lectins affect the interaction between neighboring stereocilia, either by binding to cross-links or by changing the distribution of charge on the surface coat. Cationic ferritin particles bind all over the surface of the stereocilia and to all the different links, suggesting in general that negatively charged molecules coat all the structures of the hair bundle. The ferritin particles reduce the stereociliary spacing and, as indicated above, where they approach closely, the membranes of adjacent stereocilia fuse together (Neugebauer and Thurm 1987). Aminoglycoside antibiotics such as dihydrostreptomycin and gentamicin have also been shown to bind to the stereocilia (Tachibana et al. 1986) and cause stereociliary fusion (Furness and Hackney 1986). It is probable, therefore, that in vivo, stereociliary membranes are prevented from fusing by a combination of the repulsion of their negatively charged coating (Dolgobrodov et al. 2000) and the gel-like spacing properties of the KSPGs (Katori et al. 1996a). Conversely, they are kept from falling apart by the connections provided by the lateral cross-links. Thus there may be an internal balance of tension between these opposing forces that maintains bundle integrity. As discussed later, there are myosin isoforms and scaffolding and adhesion proteins present at the attachments of the lateral links (see Section 4), which suggests the possibility of an active involvement of the links in changing bundle properties. The surface coat and the distribution of the links may offer a way in which the mechanics of the bundle could be modified.

The composition of the lateral links and the tip links has yet to be fully determined, although both appear to have a coating of carbohydrate and a keratan sulfate component, implying that they have glycoprotein components. In chick auditory hair cells, ankle links are specifically labeled by an antibody to a 275-kDa protein, which also labels the whole of vestibular hair bundles (Richardson et al. 1990). An antigen of about 200 to 250 kDa associated with the tip links has also been detected in chick hair cells; this antigen is relatively insensitive to short (15 min) subtilisin treatment but disappears after prolonged (1 h) subtilisin exposure, although the main structure of the tip links does not (Goodyear and Richardson 2003). It remains to be determined, therefore, whether this antigen is on the tip link itself, or represents an epitope present in material surrounding the tip link. For example, it could be associated with material that binds the hair bundle to the overlying otolithic membrane, rather like the KSPGs at the tips of mammalian stereocilia (Katori et al. 1996a). In support of this idea, subtilisin treatment of 20 min is used to aid the removal of the tectorial membrane from the hair bundles in turtle papilla without preventing mechanotransduction (see Crawford et al. 1989), suggesting it weakens the attachment between the two without affecting the gating springs.

As indicated above, one approach to characterizing the composition of links is to use specific enzymes. In lightly fixed mammalian organ of Corti, elastase destroys tip and lateral links, whereas chondroitinase, hyaluronidase, keratanase, and collagenase have relatively little effect (Osborne and Comis 1990; Pickles et al. 1990; Meyer et al. 1998). In vestibular hair cells, elastase also affects the tip links and to a lesser extent the lateral links (Takumida et al. 1993). However, elastase is known to digest a range of extracellular matrix proteins, and antibodies to elastin do not label any components in the hair bundle, suggesting that elastin itself is probably not its target (Katori et al. 1996a). Treatment of isolated hair cells with the calcium chelator BAPTA also causes loss of tip links (Assad et al. 1991; Meyer et al. 1998) and lateral links in mammalian OHCs (Meyer et al. 1998), and the ankle links in chicks (Goodyear and Richardson 1999). These observations indicate that the links contain extracellular proteins with calcium-sensitive anchoring sites and that they vary in composition in different cells and locations.

It is possible that intercellular adhesion proteins of the cadherin superfamily may be components of the links. Protocadherin 15 appears in the stereocilia of developing mammalian hair cells and is then found all along their length in adult hair cells, suggesting that it may be part of the lateral links (Ahmed et al. 2003). Cadherin 23 also appears during the development of mammalian hair cells and was thought to disappear in adult ones (Boeda et al. 2002). However, recent studies in frog and zebrafish hair cells have suggested it is a component of the tip links as well as of the links that join the stereocilia to the kinocilium, and that it remains at the tips of adult mammalian stereocilia (Siemens et al. 2004; Sollner et al. 2004). A crucial role for cadherin 23 in mechanosensation is supported by the fact that mutations in the gene for this protein affect inner ear function in the zebrafish, human, and mouse inner ear (Nicolson et al. 1988; Holme and Steel 2004).

3.5 Possible Functions of the Cross-Links and the Identity of the Gating Spring

A great deal of effort has been made to localize MET channels and the gating springs as these are fundamental to the operation of these mechanosensory hair cells. As there may be no more than five functional transduction channels per stereocilium (see review by Hudspeth 1989), the amount of channel protein in most hair cell epithelia is small and the channels have not yet been purified or immunolocalized although various candidates are being investigated (see Fettiplace and Ricci, Chapter 4). Techniques for localizing them have included observing calcium influx into the stereocilia using calcium dyes (Denk et al. 1995; Lumpkin and Hudspeth 1995) and the application of blocking agents to different points on the hair bundle while monitoring channel activity (Jaramillo and Hudspeth 1991). These methods have provided strong evidence that the channels are located near the stereociliary tips but have insufficient resolution to indicate directly whether the channels are associated with any particular category of linkage in this region.

Various modeling studies have suggested that the horizontal links located along the stereociliary shaft would not provide an efficient way of operating the MET channels as they undergo minimal deformation during physiologically relevant deflections of the hair bundle (Geisler 1993; Pickles 1993a). Instead, the function of the lateral links may be to connect the stereocilia together, preventing them from becoming disarrayed and allowing them to be deflected as a unit, ensuring that the MET channels open simultaneously. Coupling between the stereocilia also reduces random responses caused by Brownian motion. The fact that the compositions of the lateral links and the associated glyocalyx on the stereocilia seem to show variation between different categories of hair bundle suggests that they might be associated with different mechanical properties, for example, responses to different frequency ranges as suggested by Goodyear and Richardson (1994).

The tip link has been, for some time, considered to be the main candidate for the gating spring. The evidence for and against this has been comprehensively reviewed on a number of occasions (see, e.g., Pickles and Corey 1992; Hackney and Furness 1995) so it will only briefly be summarized here. In favor of this idea, the tip link is in the vicinity of the channels and is in a good position to be stretched during excitatory deflections and relaxed during inhibitory ones (Pickles et al. 1984). In support of this view are data that suggest loss of the tip links produced by BAPTA treatment coincides with the loss of sensitivity in the transduction mechanism (Assad et al. 1991) and their subsequent regeneration after destruction in vitro coincides with reappearance of mechanosensitivity (Zhao et al. 1996). However, the tip link has been shown by high-resolution electron microscopy to have a double helical structure that is unlikely to be elastic, so rather than being the spring itself, it has been suggested to pull on elements at its ends that are the gating springs (Kachar et al. 2000). Thus tension generated by bundle deflection could open MET channels located at one or other end of the link.

Despite its position, it has yet to be demonstrated by a direct means that the channels are associated with the tip links, and instances have been documented in which the orientation and location of tip links are not ideal for efficient MET channel opening (Hackney et al. 1988). Hair bundles are also found that have numerous links at the tips of their stereocilia that are orientated in multiple off-axis directions (Figs. 3.8A, and 3.9A, B). In addition, experimental manipulations of tip-link integrity, such as the application of BAPTA, affect other structures (Goodyear and Richardson 1999) and the loss of the tip links may disturb the operation of gating springs located elsewhere. Also, Meyer et al. (1998) showed that if tip links were destroyed by either BAPTA or elastase treatment, the mechanotransducer mechanism became less sensitive but the channels remained in an open state blockable by dihydrostreptomycin. This suggests the tip link is not directly attached to the channel gate, as the channel would be expected to close in the absence of tension transmitted by the link. The loss of tip links may thus either be coincidental to the loss of mechanosensitivity or the tip link may be crucial but less directly involved in channel gating as postulated above. Distinguishing these possibilities will require knowledge of the precise ultrastructural localization of the MET channels and determination of the molecular arrangement of coupling between the gating spring and the channel itself.

In an attempt to localize the MET channels at an ultrastructural level, we sought to label them via their amiloride binding site as they had been shown to be blocked by amiloride (Jørgensen and Ohmori 1988), using polyclonal antibodies raised against whole epithelial sodium channels that also have an amiloride binding region (Tousson et al. 1989). Using the preembedding technique, we found labeling in the vicinity of the contact region (Hackney et al. 1992) that could be reduced by treatment with both amiloride and dihydrostreptomycin, another MET channel blocker (Furness et al. 1996). Controls suggested that the labeling was not simple trapping or nonspecific binding of the antibodies or gold label. In addition, a monoclonal antibody raised by an antiidiotypic approach against amiloride binding sites was found to functionally block transduction in a manner that suggested that they are an integral part of the MET channels (Schulte et al. 2002) but it remains to be confirmed by alternative techniques that the MET channels are located in the contact region rather than at the ends of the tip links.

Kinematic modeling of stereociliary motion has indicated that the shear in the contact region is greater than the relative elongation of the tip link for a given deflection of the bundle (Furness et al. 1997). Thus channels located in the contact region could detect bundle deflections efficiently. There are several ways in which such channels could be operated. First, the tip link could alternately pull on and relax the membrane at the tip of the shorter stereocilium between the tip-link attachment and the contact region. If the stereociliary membrane is anchored in the contact region, stretch-activated channels located between the tip link and the contact region could be operated. Second, the tip link may have no direct role in transduction but could simply be a connection maintaining tension in the contact region. In this case, the gating springs could be represented by the shorter cross bridges seen in the contact region. These may be arranged to detect shear in the plane of the membranes, for example, by rotating or deflecting an intermembrane molecular link that operates the channel gate. While these mechanisms are speculative, they are consistent with other channel gating mechanisms involving twisting of molecular components in ion channels or receptors (Unwin 1989, 2003).

The nature of the MET channels, their location, and gating mechanism thus remain major questions in research on hair cells. Other aspects of hair-bundle structure and in particular the internal cytoskeleton also play vital roles in bundle mechanics and physiology. In the next section, the major components of the hair-bundle cytoskeleton and their relationship to transduction are reviewed.

4. Cytoskeletal Composition of the Hair Cell Apex

The hair cell apex contains a range of structural and regulatory cytoskeletal proteins forming complex networks that have common features in all hair cells and display homologies with the terminal webs of other epithelial cells. The major component of these is actin (Fig. 3.6A), which will be described first.

FIGURE 3.6. Light microscope immunocytochemistry of components of the hair cell apex. All images are of guinea pig cochlea. (**A**) Immunofluorescence and confocal microscopy of actin in the hair-cell apex. Intense labeling of stereocilia (s) and weaker labeling of the cuticular plate (cp) of both IHC and OHCs can be seen. Note the cytoplasmic channel (*arrow in IHC*) through the cuticular plate is unlabeled (see also Fig. 3.2). The apices of supporting cells such as inner pillar cells (IPC) and Deiters cells (DC) also show some labeling. Scale bar = 10 μm. (**B, C**) Two focal planes of three outer hair cell apices immunolabeled for tubulin and viewed by epifluorescence microscopy. The location of the basal body (*arrow*) is indicated, and in one cell where the hair bundle is rotated (r), this location is displaced (*) compared with the two on either side. In the lower plane (C), microtubules can be seen directly beneath the plate forming a meshwork (m) and in a ring around the edge of the cell. The surrounding supporting cell apices are also labeled. Scale bar = 5 μm.

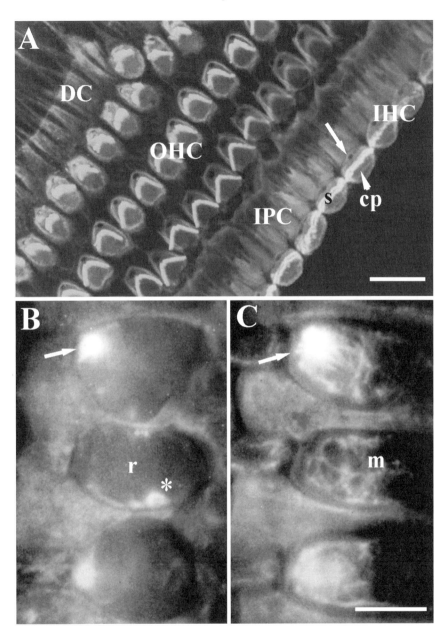

4.1 Actin Isoforms

G-actin is a globular monomeric protein of 43 kDa molecular mass that polymerizes in the presence of ATP to form 5- to 7-nm-diameter microfilaments (F-actin) of relatively high tensile strength that confer structural rigidity on cellular regions and protrusions. Actin is also involved in cell motility, for example, it is associated with myosin II in muscle to produce myofibril contraction and with non-muscle myosins ("unconventional myosins") for cell and organelle movement elsewhere (see, e.g., reviews by Mermall et al. 1998; Wu et al. 2000; Berg et al. 2001). Flock and Cheung (1977) first described 5-nm diameter microfilaments in stereocilia, their rootlets, and the cuticular plate, that could be decorated by the S-1 fragment of myosin (heavy meromyosin—the myosin head). The bound S-1 appeared as "arrowheads" in TEM pointing away from the stereociliary tip, the preferred end for addition of actin monomers during filament elongation. The presence of the actin was confirmed by immunofluorescence microscopy (Flock et al. 1982; Sobin and Flock 1983), while subsequent studies revealed that only non-muscle β- and γ-actin isoforms are present (Pickles 1993b; Slepecky and Savage 1994). Differences in the regional distributions of these two isoforms and the proteins commonly associated with them contribute to the formation of at least three distinct actin networks with different properties in the apex (Drenckhahn et al. 1991): (1) a paracrystalline bundle of actin filaments forming the core of the stereocilium; (2) a meshwork of randomly oriented actin filaments forming the matrix of the cuticular plate; and (3) a circumferential ring of actin filaments surrounding the cuticular plate.

4.1.1 Actin in the Stereocilia and Their Rootlets

The core of the stereocilium contains a bundle of actin filaments (Fig. 3.8E), composed of both β- and γ-actin (Hofer et al. 1997), the number of which varies with stereociliary width. In primitive vertebrate hair bundles from neuromasts of the lamprey, *Lampetra japonica*, the hair cells have a few very small stereocilia about 0.1 µm in diameter, similar in size to microvilli, containing approximately 30 to 40 actin filaments (data obtained from Fig. 4, Katori et al. 1994). In 0.8-µm wide alligator lizard stereocilia, there are about 3,000 filaments (Tilney et al. 1980) while in mammalian OHC stereocilia, which are approximately 0.2 to 0.25 µm in width (Lim 1986), there are approximately 300 (Engström and Engström 1978). The actin filaments lie in parallel in a paracrystalline array, with a center-to-center separation of approximately 10 nm in alligator-lizard (*Gerrhonotus multicarinatus*) hair cells (Tilney et al. 1980) and 8.3 nm in guinea-pig OHCs (Itoh 1982). The filaments are normally hexagonally packed, but in very thin cross sections of alligator-lizard (Tilney et al. 1980, 1983) and guinea-pig stereocilia (Fig. 3.8E), they appear to be arranged in festoons, which suggests that the packing is not completely symmetrical. It is possible that there are systematic changes in the packing density or arrangements of actin filaments that affect stereociliary stiffness, although this requires further investigation.

In nearly all hair cells, stereocilia have a narrower ankle (Fig. 3.2A), the majority of actin filaments terminating near the inward curving stereociliary membrane (see Fig. 4.3A in Fettiplace and Ricci, Chapter 4). However, several actin filaments (in alligator-lizard, between 18 and 30; Tilney et al. 1980) penetrate into the cuticular plate to form the rootlet (Fig. 3.2A). Within the rootlets, the filaments are wider, 6.5 nm in alligator lizard (Tilney et al. 1980) and 8 nm in chick (Tilney et al. 1983). In alligator lizard and turtle, they spread out in a cone as the rootlet goes deeper into the plate (Tilney et al. 1980; Hackney et al. 1993) but in guinea-pig cochlear hair cells, the rootlets are denser and tubular. The filaments in the latter are hexagonally packed (Itoh 1982), as in the stereocilium, but appear closer together with a smaller center-to-center spacing (8.0 nm).

The actin filaments of the stereocilium and rootlet all have the same polarity with the plus end toward the tip (Tilney et al. 1980; Slepecky and Chamberlain 1982a), suggesting they "grow" during development by the addition of monomers at the top of the stereocilia. Schneider et al. (2002), using fluorescently tagged β-actin, showed that monomers are continually added at the tips, renewing the β-actin in the stereocilium every 48 h. However, it has been proposed that the filaments in the rootlet can also grow from the cuticular plate end (Tilney and Tilney 1988) at different stages during development of the hair bundle (Tilney and DeRosier 1986). This latter suggestion is consistent with some data from the incorporation of radiolabeled amino acids into the hair bundle, which preferentially occurs at its base (Pickles et al. 1996).

The actin filaments have a helical substructure with a two-start helix turning once every 7.5 nm, and lie in register so that they are cross-linked to their neighbors by precisely spaced bridges once every 3.75 nm. This bridging can be visualized in TEM (Itoh 1982; Tilney et al. 1983) and suggests that the actin bundle is very stiff, as the bridges appear to remain attached to the same subunits even when the actin bundle is bent. The cross bridges may be composed of fimbrin (Tilney et al. 1989) and espin (Zheng et al. 2000), but although these proteins have been detected in the stereocilia at the LM level (Sobin and Flock 1983; Slepecky and Chamberlain 1985b; Zine et al. 1995; Lee and Cotanche 1996; Zheng et al. 2000), their localization to the cross bridges has yet to be confirmed by immunocytochemistry at the EM level (see Section 4.3.1). If cross bridges composed of two different proteins are present then differences in the length or angle of the bridges might give rise to the festoons described above. There are also linkers between the membrane and the peripheral actin filaments of the core along the length of the stereocilium (Hirokawa and Tilney 1982; Tilney et al. 1983; Furness and Hackney 1985) that may correspond to recently described myosin-associated proteins such as harmonin, vezatin, and whirlin (see Frolenkov et al. 2004 for a review and also Section 4.2).

As noted in Section 3.2, the stereocilia contain areas of electron-dense material associated with the actin core. These include the cap over the actin, which in the tallest stereocilia is located directly beneath the membrane and covers the actin bundle like a shower cap (see Fig. 3.2A), but in shorter stereocilia is separated from the membrane and may not completely cover the bundle (Fig. 3.9G). There is also a dense patch at the upper end of the tip link (see Section 3.2; Fig. 3.2A) and where the lateral links attach to the stereociliary membrane, the actin filaments just inside the attachment sites have a denser appearance (Fig. 3.8E, F; Furness and Hackney 1985). Dense material also surrounds the central actin filaments of the core, extending down into the cuticular plate with the rootlet filaments. Finally, toward the lower end of the actin bundle, a group of 10- to 12-nm-diameter electron-dense filaments (Fig. 3.8G) run parallel to the actin filaments at the periphery of the filament bundle (Furness and Hackney 1987). Proteins that may be associated with these electron-dense areas are discussed in subsequent sections.

4.1.2 Actin in the Cuticular Plate

The cuticular plate is composed predominantly of γ-actin (Hofer et al. 1997), which is interesting given that there are several mutations of the gene that encodes for this isoform that result in non-syndromic but progressive sensorineural hearing loss in humans (see review by Frolenkov et al. 2004). Given that γ-actin is found in many other cell types, this suggests that it has some specific function in the inner ear that may be related to unique interactions with hair-cell specific proteins or mechanisms in this region. The actin filaments in the center of the cuticular plate are randomly organized with no consistent polarity (Hirokawa and Tilney 1982; Slepecky and Chamberlain 1982a; Tilney et al. 1983) and are cross-linked together by fine, 3-nm whisker-like filaments (Hirokawa and Tilney 1982). The meshwork of actin filaments gives the plate a dense appearance (Fig. 3.2B) but an area immediately surrounding the rootlet appears devoid of filaments, giving the rootlets an electron-lucent halo (Kimura 1975; Fig. 3.2A) which, in alligator lizard, contains 3-nm filaments cross-linking the outermost filaments of the rootlet to the actin filaments of the matrix of the plate. Similar filaments are also present within the rootlet, giving it a wagon-wheel appearance in cross section (Tilney et al. 1980). In chick hair cells, as actin filaments from the cuticular plate approach the apical membrane, crows-foot-like projections link them to it (Hirokawa and Tilney 1982).

The cuticular plate of mammalian hair cells also contains dense strands and patches, often apparently arranged in layers (Fig. 3.2A) and a particular zone of parallel light and dark striations, called the Friedmann's body, has been noted commonly along the lower boundary of the plate in a number of hair cell types (Slepecky and Chamberlain 1982a,b).

Guinea-pig OHCs have a unique infracuticular network of actin (Angelborg and Engström 1973; Thorne et al. 1987; Carlisle et al. 1988) formed by a projection of the cuticular plate extending toward the nucleus. This protrusion extends deepest into the cell in apical OHCs, gradually diminishing toward the basal region of the cochlea (Thorne et al. 1987; Carlisle et al. 1988). This distribution pattern suggests that it may help stiffen or strengthen hair cells that have particularly long cell bodies such as those at the apex of the guinea-pig cochlea.

4.1.3 Actin in the Circumferential Ring

A circumferential ring of parallel actin filaments lies around the main cuticular plate, adjacent to the region of zonula adherens that holds the supporting cells and the hair cells together (see, e.g., Hirokawa and Tilney 1982; Tilney and Tilney 1988) and consists predominantly of γ-actin (Hofer et al. 1997). The ring contains filaments of opposite polarity lying alongside one another (Hirokawa and Tilney 1982) which could indicate that it is contractile given that this ring also contains myosin (see Section 4.2). If so, it could potentially compress or dilate the apical region of the cell, thereby changing the stiffness of the cuticular plate and the anchorage of the stereocilia.

4.1.4 How Does the Complex Organization of These Networks Arise?

It has been estimated that, at least within the chick cochlea, each hair-cell apex contains the same amount of actin (Tilney and Tilney 1988), despite the systematic differences in the height and number of stereocilia along the length of the cochlea (Tilney and Saunders 1983). Thus there must be very precise control of the organisation of actin filaments into the different networks during development. This is thought to be achieved by the regional sorting of the different isoforms of actin in conjunction with the different proteins that characteristically associate with them (Hofer et al. 1997). A model for the developmental control of hair bundles has been proposed that is based largely on data from analogous systems (Kollmar 1999); this suggests that the control of actin polymerization in the apex may be mediated through Rho GTPases, since one member of the Rho GTPase family (CDc42) is up-regulated during bundle regeneration after noise trauma in chick. This model is supported by the localization of other actin associated proteins in the apex (see Section 4.3.5). The distributions of these proteins together with their probable functional significance are now considered.

4.2 Myosin Isoforms and Their Distribution in the Hair Cell Apex

Myosin I and myosins III to XVIII are unconventional myosins found in non-muscle cells. The human genome contains some 40 genes coding 12 classes of myosins. Several of these relate to deafness (Berg et al. 2001) including myosin IIIa, myosin VI, myosin VIIa, and myosin XVa, where mutations in their genes affect the hair cell phenotype. All myosins consist of a head domain, which gives the molecule motor properties, and a tail domain which contains a variable number of IQ domains with calmodulin binding sites. Myosin II (muscle myosin) is a two-headed molecule that combines with other myosin II molecules to form the thick filaments recognizable in muscle cells by TEM. However, some unconventional myosins have only one head domain (see review by Mermall et al. 1998), and all myosins appear to have unique tail domains. Binding of the head domain to actin filaments and its modulation by ATP leads to myosin molecules moving along the actin filament away from the barbed end except for myosin VI, which appears to move in the opposite direction (see review by Wu et al. 2000). Movement of myosin II-coated beads has been demonstrated in vitro to occur along isolated actin cores of stereocilia (Shepherd et al. 1990) and the polarity of the actin suggests that myosins would move toward the tip, except potentially for myosin VI.

The earliest search for myosins in hair cells dates back to the discovery of active nonlinear processes involved in mammalian cochlear function. Since actin–myosin interactions produce motility using energy, it seemed likely that they would be involved in active feedback mechanism in the cochlea, for example, in the stereocilia because of their high actin content. Macartney et al. (1980) reported myosin in the cuticular plate and stereocilia using immunofluorescent microscopy of whole mounts with polyclonal antibodies to purified muscle myosin. More recently, it has been suggested that the slower of the two components of adaptation in hair cells could be based on myosin (Gillespie and Hudspeth 1993; see also review by Fettiplace et al. 2001) and several different myosins have now been described in the hair cell apex.

4.2.1 Myosin 1c

One of the more intensively studied isoforms is myosin 1β (related to brush border myosin), now called myosin 1c. The presence of this protein is further evidence of the similarity between stereocilia and microvilli, which contain brush border myosin. Myosin 1c has been immunolocalized in bullfrog hair cells by antibodies that do not cross-react with mammalian myosin 1c. In addition, two orthologues, *Myo1c* and *Myo1f*, have been identified by screening a mouse cochlear cDNA library (Crozet et al. 1997). Manipulation of *Myo1c* has been carried out in a mouse transgenic model in which physiological data suggest that myosin 1c plays a role in the slow form of adaptation in mammalian utricular hair cells (Holt et al. 2002).

Myosin 1c has been found in bullfrog stereocilia (Hasson et al. 1997; Metcalf 1998), particularly toward the tips where it is concentrated in the vicinity of both the lower and upper attachment site of the tip links (Garcia et al. 1998; Steyger et al. 1998). It has therefore been proposed to function in active tensioning of the tip link to mediate slow adaptation (Assad and Corey 1992); the direction of motion of the upper end of the tip link would be expected to be toward the top of the actin core. In rat and mouse, myosin 1c is distributed along the length of the stereocilia in cochlear hair cells but it is concentrated towards the tips in vestibular ones (Dumont et al. 2002). Labeling for this protein has also been noted along the periphery of the actin core and in a particular concentration in the ankle region of the stereocilia (Garcia et al. 1998).

Thus it could conceivably move the entire core relative to the plasma membrane or have some other transport function. In addition, labeling for myosin 1c can be denser in the cell body than in the stereocilia (Hasson et al. 1997), implying that it also has functions in hair cells unrelated to adaptation.

4.2.2 Myosin IIIa

Mutations in a human gene for myosin IIIa have been shown to cause nonsyndromic progressive hearing loss (Walsh et al. 2002). The ultrastructural location of this myosin has yet to be determined in hair cells but its *Drosophila* homologue binds to a PDZ (*p*ostsynaptic density, *d*isc large, *z*onula occludens) domain-containing scaffolding protein in the rhabdomeres to organize the phototransduction machinery into a signaling complex. It is therefore likely that it interacts with actin filaments in hair cells and may play a role in organizing the transduction mechanism there.

4.2.3 Myosin VI

Myosin VI has also been detected strongly in the cuticular plates and to a lesser extent, in the cytoplasm of hair cells of mammalian cochlea and vestibular system, and bullfrog saccule. Weak labeling has also been noted in bullfrog saccular stereocilia where it was particularly associated with the base of the hair bundle, although stronger labeling of stereocilia was detected in apparently newly forming hair bundles (Hasson et al. 1997). Defects in *Myo6* result in the *Snell's waltzer* mouse phenotype (Avraham et al. 1995), where cochlear hair bundles that initially appear normal become disorganized and the stereocilia fuse. It has therefore been suggested that myosin VI is necessary for proper development of the bundle and that it may be required to moor the apical plasma membrane to the base of the stereocilia or to anchor stereociliary rootlets (Self et al. 1999). In addition, its preferred direction of movement means that it could move down the actin filaments toward the cell body, making it a potential candidate for removal of molecular components from the stereocilia.

4.2.4 Myosin VIIa

Five of the relevant genes associated with Usher's syndrome type 1, in which patients suffer from sensorineural hearing loss, vestibular dysfunction, and visual impairment, have been identified (see Frolenkov et al. 2004 for a review). These genes encode PDZ-domain containing scaffold proteins called harmonin and SANS (scaffold protein containing ankyrin repeats and a sterile alpha motif domain); the intercellular adhesion and signaling molecules cadherin 23 and protocadherin 15; and the unconventional myosin, myosin VIIa (Boeda et al. 2002; Kikkawa et al. 2003; see also review by Frolenkov et al. 2004). Together these proteins may form adhesion complexes (Siemens et al. 2002) that form the basis of the stereociliary linkage systems. The myosin VIIa isoform has been localized throughout the cytoplasm and apical region of hair cells from both bullfrog saccule and guinea-pig vestibular system and cochlea. It occurs all along the stereocilia, in particular associated with regions where lateral links are concentrated (Hasson et al. 1997). Interestingly, vezatin, a protein that is known to link myosin VIIa to catenin–cadherin complexes, is also found at the attachment sites of the lateral links (Kussel-Andermann et al. 2000). This complex could correspond with the dense material noted at the edge of the actin core where the lateral links attach (Fig. 3.8E). Defects in the genes encoding myosin VIIa also cause disturbance of the hair bundles in mice, the bundles requiring more displacement than in the wild type to obtain a response, while showing stronger adaptation of the transducer currents (Kros et al. 2002). It has been suggested that myosin VIIa holds membrane-bound elements to the actin core. Thus it may be that vezatin and myosin VIIa together associate at the lateral link attachment sites to maintain the normal functional integrity of the hair bundle in response to displacements.

4.2.5 Myosin XVa

Another class of unconventional myosin implicated in hair cell function is myosin XVa. Defects in *Myo15* underlie the mouse *shaker 2* mutation (Probst et al. 1998) which is the homologue of the human *DFNB3* gene. The phenotype is shortened stereocilia and an additional long actin-containing process projecting from the basal end of the hair cell. Immunocytochemistry has shown that myosin XVa is present in the stereocilia and cuticular plate (Liang et al. 1999) and more recent studies have shown it to be located at the tip of every stereocilium where it appears just before the staircase is developed (Belyantseva et al. 2003). Furthermore, the longer stereocilia in hair bundles have more myosin XVa at their tips than do the shorter rows of stereocilia, suggesting that it adds components to the tips during growth (Rzadzinska et al. 2004). Another PDZ-domain containing protein, whirlin, has recently been discovered that causes deafness in humans (DFBN31) and whirler mice (Mburu et al. 2003). Whirler mice have short stereocilia like *shaker2* mice so it has been suggested that myosin XVa and *whirlin* may be part of a macromolecular complex involved in stereociliary elongation (Belyantseva et al. 2003).

In summary, myosin isoforms are clearly as important in hair-cell function as they are in many other cell types. The potential involvement of myosins in adaptation and development of the hair bundle are major areas of interest at present and the presence of several of them in the hair-cell body suggests that they may have a number of other as yet only poorly characterized roles in hair cells.

4.3 Other Actin-Associated Proteins in the Hair Cell Apex

4.3.1 Fimbrin and Espin

Fimbrin is a 65 to 70 kDa protein originally isolated from intestinal microvilli (Bretscher and Weber 1980). It has various isoforms (known as plastins in humans) that are actin-bundling proteins forming cross-links between adjacent actin filaments. Detergent treatment of the high-salt extracted hair cell cytoskeleton removes both fimbrin, as determined by immunoblots, and the cross bridges between actin filaments in the stereociliary core, as determined by TEM (Tilney et al. 1989) providing circumstantial evidence that it forms the bridges. Fimbrin contains calmodulin-like calcium binding domains (de Arruda et al. 1990) and one homologue, L-plastin, has been shown to be sensitive to calcium, its ability to cross-link actin becoming increasingly inhibited by higher calcium concentrations (Pacaud and Derancourt 1993). As noted above, fimbrin has been localized at the LM level in the stereocilia and cuticular plates of mammalian cochlear hair cells (Pack and Slepecky 1995), the stereocilia of mammalian vestibular hair cells (Sobin and Flock 1983), and in chick hair bundles (Lee and Cotanche 1996). A study using the reverse transcriptase polymerase chain reaction in chick utricles has also shown that message for fimbrin is reduced after aminoglycoside-induced hair cell loss in vitro, and then restored after 4 weeks, consistent with the recovery of the epithelium and regeneration of hair cells and/or their stereocilia (Stacey and McLean 2000). Although I-plastin is the isoform expressed in mature hair cells, T-plastin is transiently present in stereocilia during development (Daudet and Lebart 2002).

The major role of fimbrin is probably in the maintenance of the stiff actin core of the stereocilium, which is remarkably resistant to bending except for the rootlet region, where the core narrows and there are fewer actin filaments. Whether fimbrin can also alter the properties of the hair bundle in relation to stimulation remains to be determined. As discussed in Section 5, calcium and its regulatory proteins play important roles in the hair bundle and fimbrin is a potential target for calcium-dependent processes, and it is possible that the different isoforms have different sensitivities to Ca^{2+} (Lin et al. 1994). Fimbrin also has a higher binding affinity for β- than for γ-actin (Hofer et al. 1997) and the latter is found in higher concentrations in the periphery of adult guinea-pig stereocilia at least (Furness et al. in press). It will be interesting to see if fimbrin is also more concentrated around the periphery given that another actin bundling protein, espin, is also found in the stereocilia.

Espin is a 110-kDa actin-bundling protein first identified in Sertoli junctions (Bartles et al. 1996; Chen et al. 1999) and subsequently found as a 30-kDa splice variant of the same gene in microvilli (Bartles et al. 1998). Antibodies to a fusion protein containing almost half of the C-terminal sequence for the 110-kDa rat espin label the stereocilia strongly in both rat and mouse cochlear and vestibular hair cells (Zheng et al. 2000). *Espn* is found in the same region of mouse chromosome 4 as the recessive mutation that causes hair cell degeneration, deafness, and vestibular dysfunction in the *jerker* mouse. The mutant gene contains a frameshift mutation that results in a change in the sequence coding for the actin-binding domain. Although espin mRNAs are present at normal levels in the *jerker* mouse, the protein is not expressed in the stereocilia, thus probably giving rise to the *jerker* phenotype as a result of its failure to interact correctly in the formation of the actin core. Unlike fimbrin (plastins), espins bundle actin filaments efficiently even in the presence of Ca^{2+}, a property that has been suggested to be important to hair cells because they need to maintain their cross-linked actin structures even when intracellular Ca^{2+} levels rise as a result of influxes through the MET and other ion channels or via release from intracellular calcium stores (Sekerková et al. 2004).

4.3.2 α-Actinin

α-actinin is a 200-kDa homodimeric protein that enhances nucleation of actin (Wagner et al. 1999) and cross-links actin filaments dynamically (Sato et al. 1987), possibly in a calcium-dependent manner (Tang et al. 2001). α-actinin increases the viscosity of actin gels up to twofold compared with purified actin (Holmes et al. 1971) and also changes their response to a vibrating rod. Thus, as frequency increases, the elastic modulus of an actin/α-actinin gel increases, while for pure actin gels, it remains relatively constant (Wagner et al. 1999).

α-actinin has been detected by immunocytochemistry in the apex of mammalian cochlear hair cells (Drenckhahn et al. 1982). Its presence in the cuticular plate and the junctional regions associated with the supporting cells (Slepecky and Chamberlain 1985b; Slepecky and Ulfendahl 1992; Zine and Romand 1993) seems to be uncontroversial. However, it is less clear whether stereocilia contain α-actinin as there are both positive (Zine et al. 1995) and negative reports (Slepecky and Chamberlain 1985b). While a negative result cannot be considered definitive, false positives in stereocilia are also possible as discussed in Section 2.2, so this must be considered an unresolved point. Certainly at the EM level, using postembedding immunogold labeling, α-actinin was not found in stereocilia but was present throughout the cuticular plate, in the electron-dense patches in particular (Slepecky and Chamberlain 1985b).

Because of the influence of α-actinin on the stiffness of actin gels, the amount present in the cuticular plate will influence its rigidity. This may be one way in which stiffness properties of the apical region of the cell are optimized for the frequency response required of the hair bundle in different cochlear locations. It is possible that amounts of alpha-actinin in the apex of the hair cells vary systematically with the frequency of the stimulus that they receive, in which case a cochleotopic gradient might be expected.

4.3.3 Tropomyosin

Tropomyosin is involved in covering actin–myosin binding sites until displacement by troponin enables the actin–myosin interaction to occur. Immunocytochemical studies of tropomyosin in the hair cells (Slepecky and Chamberlain 1985b; Karkanevatos 2001) suggest that it occurs in rootlets and along the shaft of the stereocilia, where it may be involved in the interaction between the various myosin isoforms and the actin of the stereociliary core.

4.3.4 Kaptin and Spectrin

Kaptin (or 2E4) is an actin-associated protein of 45 kDa molecular mass that is found in growth cones (Bearer 1992), red blood cells, and stereocilia, particularly at the tips of the latter (Bearer and Abraham 1999). This protein has been implicated in human deafness because the gene maps close to the human deafness *DFN4* locus on chromosome 19 (Bearer et al. 2000). The function of kaptin is uncertain but from its position at the leading edge of lamellipodia, it has been proposed to be involved in the membrane-associated polymerization of actin filaments. Its presence at the barbed ends of actin filaments in stereocilia also suggests a similar, perhaps developmental role in stereociliary formation and maintenance. Its localization is close to or coincident with the position of dense material capping the actin core described earlier which also contains myosin XVa (Fig. 3.2A).

Spectrin is a 240-kDa structural protein that is associated with maintaining cell shape. It has been localized in the cuticular plate (Fig. 3.7B; Drenckhahn et al. 1991) and stereocilia of hair cells (Mahendrasingam et al. 1998). It has been suggested that spectrin cross-links the rootlets of the stereocilia (Nishida et al. 1993), and may connect the actin core to the membrane within the stereocilia.

4.3.5 Focal Adhesion Proteins

Several focal adhesion proteins may occur in the hair cell apex, for example, where the actin approaches the adjacent supporting cells. Vinculin appears to be present at the surface of fetal vestibular organs (Anniko et al. 1989). However, its ultrastructural distribution remains to be investigated. Keap1 is homologous to *Drosophila* kelch and interacts with myosin VIIa (Velichova et al. 2002). This protein could be associated with myosin VIIa in hair cells, along with vezatin as described earlier. Integrins also help to mediate surface interactions between cells and the extracellular matrix. Integrin $\alpha 8\beta 1$, fibronectin and the integrin-regulated focal adhesion kinase (FAK) colocalize to the apical hair cell surface during formation of stereocilia in the utricle and that defects in the integrin gene may cause malformation of the stereocilia (Littlewood-Evans and Müller 2000).

Proteins of the ezrin/radixin/moesin (ERM) family are actin-binding proteins often found associated with the microvilli of epithelial cells; they cross link actin filaments to the plasma membrane. These proteins discriminate between isoforms of actin, binding β- but not α-actin (Hirao et al. 1996), consistent with the presence of the former isoform in the hair cell apex. Both myosin VIIa and XVa have FERM binding domains (band 4.1, ezrin, radixin, moesin) which suggests their interaction with specific membrane proteins (Oliver et al. 1999; Anderson et al. 2000). In addition, radixin has been found in the stereocilia of zebrafish, bullfrogs, chicks, and mice (Fig. 3.8) and is concentrated in their ankle regions where it has been suggested to participate in the anchoring of the pointed end of the actin filaments to the membrane (Pataky et al. 2004).

Profilin is a low-molecular-weight protein (16 kDa) found in the cuticular plates of hair cells that binds primarily to actin monomers (Slepecky and Ulfendahl 1992). In other cells it is implicated in the organization of actin complexes along with actin-related proteins (see, e.g., Feldner and Brandt 2002). In

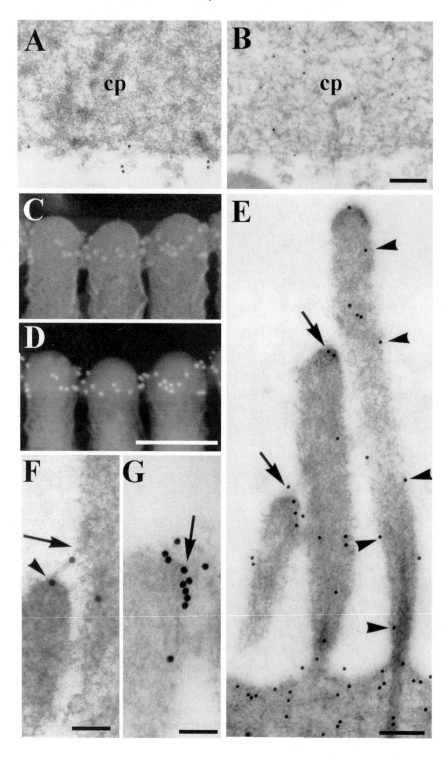

hair cells, its function is uncertain. However, one of its ligands is the human homologue of the *Drosophila* diaphanous protein, which is a target of Rho, a protein that regulates polymerization of actin during cytoskeletal development in the hair cell apex (see review by Kollmar 1999). The potential importance of this group of proteins is demonstrated by the fact that the diaphanous homologue is encoded by *DFNA1*, a gene responsible for an autosomal dominant, nonsyndromic sensorineural progressive hearing loss (Lynch et al. 1997). Diaphanous belongs to the formin family of proteins which accelerate actin nucleation while interacting with the barbed ends of the actin filaments. Thus, both profilin and diaphanous may play a role in the control of actin turnover in the stereocilia.

4.4 Tubulin and Microtubules in the Hair Cell Apex

The two major isoforms of tubulin, α-tubulin and β-tubulin, are each about 55 kDa molecular mass. They dimerize in the presence of guanosine triphosphate and then polymerize to form protofilament sheets that roll up to form microtubules. Microtubules vary in their stability and there are a range of microtubular-associated proteins (MAPs) that confer different properties on microtubular networks. Thus, microtubules are involved in development and maintenance of cell shape and structural support, in intracellular transport, and some in motility of cell extensions such as cilia (see review by Furness et al. 1990).

Microtubules occur in all hair cells in the kinocilium, when present, and its basal body, which is retained even when the kinocilium disappears during development, as in mammalian cochlear hair cells (Fig. 3.6B, C). Various isotypes of β-tubulin have been found in the vestibular and cochlear sensory epithelia, with different combinations being found in different types of hair cells and

FIGURE 3.7. Immunogold labeling of components of the hair-cell apex observed by electron microscopy. (**A, B**) Comparison of pre- and postembedding immunogold labeling for spectrin in the cuticular plate (cp) of inner hair cells. The location of immunoreactivity is revealed by gold particles (*black spots*). Note that only the edge of the plate is labeled following preembedding labeling (A) but the matrix is also labeled after postembedding (B). Scale bar = 100 nm. (**C, D**) SEM (C) and comparative backscattered (BSD) electron imaging (D) of the tips of stereocilia after immunogold labeling for keratan sulfate proteoglycans. In the BSD image, the gold particles appear white and are more clearly distinguished than with SEM. The images show a concentration of labeling crowning the tips of the tallest stereocilia of outer hair cells from the guinea-pig cochlea. Scale bar = 200 nm. (**E–G**) Postembedding labeling for calmodulin in outer hair cells. Note that labeling occurs at the periphery of the stereocilia (E, *arrowheads*) and around their tips (E, *arrows*). It is particularly concentrated at the upper (F, *arrow*) and lower (F, *arrowhead*) end of the tip link and in the contact region (G, *arrow*). Scale bars: (D) = 200 nm; (E, F) = 100 nm. (A) and (B) modified from Mahendrasingam et al. (1998); (E–G) modified from Furness et al. (2002).

supporting cells (Jensen-Smith et al. 2003; Perry et al. 2003), suggesting that they may have different functions although these are currently unknown. However, there are specific concentrations of microtubules in the apex of many hair cells of different types (see review by Furness et al. 1990). One of the most detailed descriptions of their organization has been provided for guinea-pig cochlear hair cells (Figs. 3.5, and 3.6B, C; Steyger et al. 1989; Furness et al. 1990). The apical microtubular networks form a basket around the cuticular plate and extend into cytoplasmic channels that run from top to bottom through it. Concentrations of microtubules are present in both IHC and OHCs, emanating from within the widest channel in the cuticular plate that contains the basal body, which may form a nucleating or organizing site for the microtubule networks. This channel lies below the position of the kinocilium in lower vertebrates, that is, on the abneural side of the hair cell (Figs. 3.5, and 3.6B). Other narrower channels in the periphery of the cuticular plate contain microtubules that emanate from them and contribute to a shallow network lying directly beneath the lower surface of the cuticular plate (Figs. 3.5, and 3.6C), and a circumferential ring that runs inside the actin ring. There is a second more apical ring more prominent in IHCs, usually incomplete, that encircles the cuticular plate just beneath the apical membrane and links the main channel and the smaller subsidiary channels. In the OHCs, microtubules usually extend down alongside the infracuticular network (Furness et al. 1990).

Studies in bullfrog hair cells have suggested that microtubules enter the cuticular plate and are associated with the actin filament matrix via a specialized

FIGURE 3.8. Morphology and ultrastructure of stereocilia and their cross links. (**A**) A fish (roach—*Rutilus rutilus*) hair bundle showing numerous filaments emanating from the tip and down the sides of each stereocilium, some of which connect adjacent stereocilia together. Scale bar = 500 nm. (**B**) The ankle links in a roach hair bundle (*arrows*). Scale bar = 750 nm. Micrographs in (A) and (B) taken by the authors of material prepared by Dr. D-Ch. Neugebauer. (**C**) SEM of guinea-pig cochlear stereocilia showing lateral links (l) in bands and tip links (t). Scale bar = 500 nm. (**D**) High-magnification SEM of lateral links between two guinea-pig cochlear stereocilia. Individual filaments emanate laterally from the shafts and enter a central stripe of material (*arrowhead*) that runs parallel to the sides of the stereocilia. Scale bar = 100 nm. (**E**) TEM of a cross section of guinea-pig stereocilia showing the packing of actin filaments within the core. Note the festoon pattern (*between arrows*) where rows of actin filaments appear to be curved and to meet adjacent rows at an angle (*at the central arrow*). Note also the lateral links with a medial density (*arrowhead*) corresponding to the central stripe visible in SEM (*see* D). The membrane and actin filaments adjacent to the origin of the links on the stereocilium shaft have associated electron densities. Scale bar = 100 nm. (**F**) TEM of a cross section of guinea-pig cochlear stereocilia showing lateral links connecting each neighbor. Scale bar = 100 nm. (E, F) taken by Dr. Y. Katori. (**G**) TEM of a guinea-pig cochlear stereocilium where thickened filaments occur around the periphery of the actin core. Scale bar = 50 nm.

3. The Structure and Composition of Vertebrate Hair Bundles   129

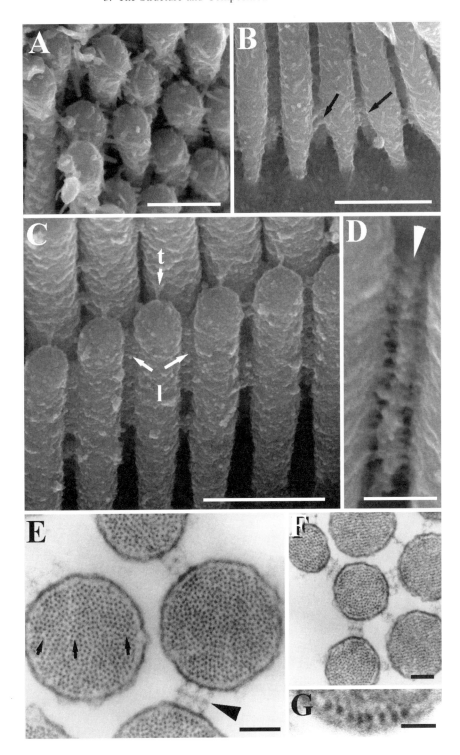

cap at their ends suggested to be composed of MAP-1A and MAP-1B (Jaeger et al. 1994). This capping material has not been noted in mammalian hair cells, where there is little evidence that microtubules penetrate the matrix of the plate (Furness et al. 1990). This may reflect variations in the packing of actin filaments in the cuticular plate. Moreover, neither MAP-1A nor MAP-1B was reported in immunocytochemical studies of guinea-pig organ of Corti, so this may indicate a further difference between mammalian and amphibian hair cells (Oshima et al. 1992; Pack and Slepecky 1995).

The microtubules in mammalian cochlear and vestibular hair cells label with an antibody to tyrosinated tubulin, implying that they are subject to frequent polyerization and depolymerization and probably involved in dynamic events (Ogata and Slepecky 1995; Pack and Slepecky 1995). Vesiculation, coated pits, and other signs of endocytosis are visible in the apical region of mammalian (Lim 1986; Furness and Hackney 1990; Furness et al. 1990), bullfrog (Kachar et al. 1997), and turtle hair cells (Hackney et al. 1993). Recent studies with the fluorescent membrane tracking dye FM1-43 have confirmed that there is considerable membrane trafficking in the apical region in living OHCs (Meyer et al. 2001) and IHCs (Griesinger et al. 2002). It is probable therefore that the microtubular networks in the apex not only provide structural support but are also involved in active movement of vesicles and membrane recycling in the hair cell apex.

5. Calcium Pumps, Buffers, and Binding Proteins

Cytoplasmic calcium regulates a number of fast events in sensory hair cells (see Fettiplace and Ricci, Chapter 4; Fuchs and Parsons, Chapter 6) including adaptation of MET channels at the apex (Assad et al. 1989; Crawford et al. 1989) and the release of neurotransmitter at the base (Parsons et al. 1994). For calcium ions to be effective as a signaling mechanism, their intracellular concentration has to be precisely controlled by calcium pumps, stores, and buffers.

5.1 Calcium Pumps

The calcium pump in hair cells appears to be a plasma membrane Ca^{2+}-ATPase (PMCA). Crouch and Schulte (1995) found that cochlear stereocilia immunolabeled with a monoclonal antibody to the conserved hinge region of PMCA and then showed that PMCA isoforms 1b, 2b, 3a, 3c, and 4b are present in the cochlea (Crouch and Schulte 1996). The fact that the presence of this calcium pump is essential to hair-cell function was confirmed by studies of the *deafwaddler* mouse mutant and mice with a targeted gene deletion of the gene for PMCA2 (Kozel et al. 1998; Street et al. 1998). In both cases, the mice become deaf and display vestibular defects. More recently, Dumont et al. (2001) reported that PMCA2a is the only PMCA isoform in the bundles of mammalian OHCs and vestibular hair cells, and in hair cells of the bullfrog sacculus, and that it is the predominant PMCA in the bundles of IHCs, while PMCA1b occurs in basolateral membranes. This suggests that specific PMCA isozymes with particular biochemical properties are targeted to specific membrane regions because of differing requirements for calcium extrusion by different subcellular compartments.

A significant fraction of the mechanotransduction current is carried by Ca^{2+} ions whose concentration rises rapidly in the tips of the stereocilia (Fig. 3.9) when the bundle is deflected in the excitatory direction (Denk et al. 1995; Ricci and Fettiplace 1997; Lumpkin and Hudspeth 1998). The influx of Ca^{2+} ions regulates the fast component of adaptation that is likely to be based on a direct interaction of the ions with the channel (Ricci et al. 1998). It is also implicated in regulating the rate of the slower component that is thought to be based on a myosin motor (Eatock et al. 1987; Assad et al. 1989). The rate of diffusion down the stereocilia and binding to both fixed and diffusible calcium-binding proteins will directly affect the spatial and temporal spread of the Ca^{2+} transient in the stereociliary tips and thus influence these processes. In addition, Ca^{2+} ions must also be extruded to remove them from the buffers so the latter can be reused. In the mammalian cochlea, the amount of PMCA appears to be greatest on the hair bundles of the OHCs (Crouch and Schulte 1995), suggesting these cells have to cope with a greater Ca^{2+} influx than the IHCs. In frog sacccular hair cells, Yamoah et al. (1998) estimated that the density of PMCA was approximately 2000 molecules per square micrometer of stereociliary membrane. They found that dialysis of the frog hair cells with ATPase inhibitors such as vanadate and carboxyeosin resulted in an increase in the Ca^{2+} concentration in the hair bundle, consistent with the regulatory role of this pump, and speculated that the activity of the PMCA molecules must locally increase the concentration of Ca^{2+} around hair bundles. Such a local extracellular increase could affect the MET channels by block affecting the fraction of current flowing at rest, the maximum current achieved on deflection and the time course of adaptation (Eatock et al. 1987; Ricci and Fettiplace 1997), thus affecting bundle twitches (Benser et al. 1996). Interestingly, one function proposed for the extracellular matrix around other cell types is that it may increase the concentration of divalent cations in the vicinity of the cell membrane and thus it is possible that these two systems, the Ca^{2+} pump and the extracellular matrix, may act together to alter the extracellular ionic environment close to the stereociliary membrane compared to that in the remaining endolymph.

5.2 Calcium Buffers

The effects of altering calcium buffering capacity on various calcium sensitive processes in hair cells have been investigated by including exogenous calcium buffers (e.g., BAPTA) in the recording pipette used to measure the hair cell's electrical responses. Comparison of the effects of applying different concentrations of exogenous buffers with results from perforated patch recordings, where soluble proteins are not eluted from the cytoplasm, have been used to determine

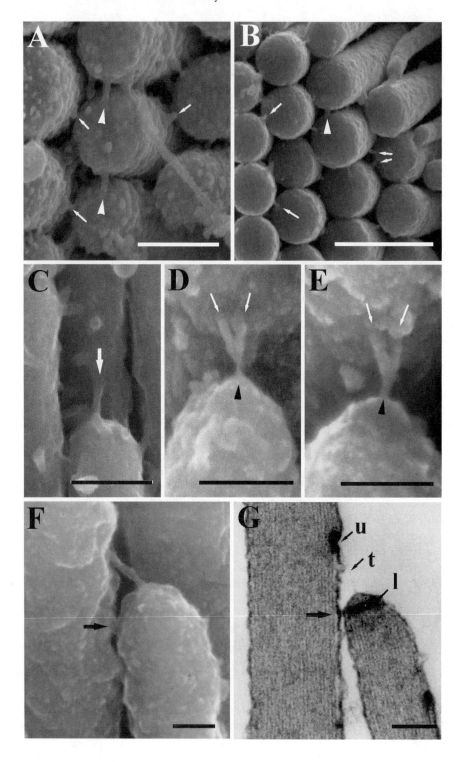

the concentration of the hair cell's native diffusible buffer (Roberts 1993; Tucker and Fettiplace 1996). The endogenous buffer in turtle auditory hair cells has similar concentrations in the hair bundle and cell body but increases with the cell's characteristic frequency (Ricci et al. 1998; 2000). Such variation probably reflects the extent of calcium loading via MET or Ca^{2+} channels, both of which increase in number as characteristic frequency increases. As indicated in the previous section, an important function of the endogenous buffer is to limit the spatial and temporal spread of the calcium signal associated with any of the calcium-dependent processes within the hair cell and thus prevent interference between the different signaling pathways and permit fast recovery and rapid responses (Roberts 1994). The question therefore arises, which proteins play this role?

A number of calcium-binding proteins have been found in hair cells (Slepecky and Ulfendahl 1993) and several of these have been proposed to act as calcium buffers. These include calbindin (Dechesne and Thomasset 1988; Oberholtzer et al. 1988; Rogers 1989), calretinin (Rogers 1989; Dechesne et al. 1991), and parvalbumin (Eybalin and Ripoll 1990; Sakaguchi et al. 1998; Heller et al. 2002). The parvalbumin family contains two sublineages, α and β (Goodman and Pechere 1977), the former being found mainly in muscle. Oncomodulin, a calcium binding protein discovered in rat hepatoma (MacManus 1979), was shown to resemble the latter and was then found in a range of tissues, including the mammalian cochlea (Thalmann et al. 1995) where it has been suggested to occur only in the OHCs (Sakaguchi et al. 1998; Yang et al. 2004).

Of these various proteins, wash-in of around 1 m*M* calretinin in frog vestibular hair cells has been found to have an effect similar to that of the native buffer (Edmonds et al. 2000) and biochemical analysis of the chick auditory

◀─────────────────

FIGURE 3.9. The ultrastructure of links at the tips of stereocilia. (**A**) SEM looking down onto the tips in a guinea-pig utricular hair bundle. Links emanate in several directions from the tip region including laterally between stereocilia of the same row (*white arrowheads*) and between rows in various orientations (*white arrows*). Scale bar = 200 nm. (**B**) Similar view of a turtle cochlear hair bundle. Links of various orientations are detectable between rows (*single white arrows*) and along the rows (*arrowhead*). Scale bar = 500 nm. (**C**) SEM showing a tip link from a roach utricular hair bundle. The link appears to be a single filament in this case. Scale bar = 100 nm. Micrograph taken by the authors from material prepared by Dr. D-Ch. Neugebauer. (**D, E**) Tip links from guinea pig (D) and turtle (E) cochlea showing that they have an almost identical appearance, with a relatively narrow single filament at the lower end (*black arrowhead*) bifurcating into two thicker upper strands (*white arrows*). Scale bars = 100 nm. (**F**) SEM of a side view of a tip link and contact region (arrow) between a shorter and taller stereocilium from a guinea-pig cochlea. Scale bar = 50 nm. (**G**) TEM of the tip link (T) and contact region (*arrow*) in guinea pig cochlea. Note the fine nature of the tip link and the dense patches at its upper end (u) and close to its lower end (l). The two membranes in the contact region show some thickening. Micrograph supplied by Dr. Y. Katori. Scale bar = 50 nm.

papilla gave a calbindin D-28k concentration of around 1 mM (Oberholtzer et al. 1988). Immunocytochemistry has also demonstrated a tonotopic gradient for calbindin in the mammalian (Pack and Slepecky 1995; Imamura and Adams 1996) and avian (Hiel et al. 2002) cochlea. However, the gradients are in opposite directions with the calbindin being most concentrated at the high frequency end in birds and least concentrated there in mammals, raising the question of how this protein is related to tonotopic variations in Ca^{2+}-dependent processes in these different classes of vertebrate. In addition, the concentration of parvalbumin-3, a calcium-binding protein that displays strong sequence similarity to parvalbumin-β (oncomodulin), in one morphological variety of frog saccular hair cells has been estimated to be as much as 3 mM by means of quantitative Western blot analysis. This could make it the major buffer in these cells (Heller et al. 2002).

Of the calcium buffers found so far in the inner ear, each has been reported to show a unique pattern of expression (Pack and Slepecky 1995; Baird et al. 1997; Edmonds et al. 2000) often restricted to one region of the hair cell epithelium, one type of cell or subcellular location. So far, there have been few ultrastructural studies of the subcellular distribution for any of these proteins. However, in a recent study, a postembedding immunohistochemical approach was used to investigate the distribution of calbindin, parvalbumin, and calretinin in turtle hair cells (Hackney et al. 2003). The likelihood of each of them acting as the major calcium buffer in this system was also investigated by determining their respective cytoplasmic concentrations. To calculate a protein concentration from the gold particle counts, immunogold labeling was performed on an ultrathin section of a gel containing a known amount of protein for comparison with a cochlear section processed at the same time in the same fluid drops. For each protein, the counts in the gel were matched to hair-cell measurements made at low-frequency and high-frequency positions in the cochlea. The concentration of calbindin-D28k was found to be 0.13 mM at the low-frequency end of the cochlea and 0.63 mM at the high-frequency end, with parvalbumin-β occurring at 0.25 mM all along the length.

Physiological measurements of the endogenous calcium buffering in the turtle cochlea indicate that 0.5 to 2 mM calbindin should be present if this is the main calcium buffer and that there should be a tonotopic gradient (Ricci et al. 1998, 2000). The findings by Hackney et al. (2003) show that calbindin and parvalbumin-β levels are nearly in this range but a gradient was found only for calbindin. However, it is possible that calbindin and parvalbumin contribute different components to calcium buffering in the hair-cell apex that change under different ionic conditions (Henzl et al. 2003) and thus both may be required for shaping the turtle hair cell's response to calcium influx. Parvalbumin-β is known to have two calcium binding sites while calbindin-D28K has four. The sites on calbindin-D28K are low affinity (K_D 0.2 to 0.5 μM) with fast binding kinetics (k_{on} = 1–8 × 10^7 $M^{-1} \cdot s^{-1}$) and a high selectivity site for Ca^{2+} over Mg^{2+} (Nagerl et al. 2000). Of the two sites in parvalbumin-β, one is a high-affinity site ($K_{Ca} > 10^8$ M^{-1}; $K_{Mg} > 10^4$ M^{-1}) while the other is a low-affinity site (K_{Ca} approximately 10^6 M^{-1}; $K_{Mg} < 10^3$ M^{-1}). The significant affinity for Mg^{2+} of the high-affinity site means that Mg^{2+} will compete with Ca^{2+} thus slowing its binding kinetics and lowering its calcium affinity (Pauls et al. 1996). Thus, in the presence of physiological millimolar concentrations of Mg^{2+}, both sites on parvalbumin-β will have similar low affinity for Ca^{2+} but one site will bind Ca^{2+} rapidly whereas the other will bind it slowly because it must first release Mg^{2+} (Pauls et al. 1996). As both calbindin-D28k and parvalbumin-β occur in turtle hair cells, they are likely to contribute both fast and slow components to calcium buffering. Both buffers are also found in rat hair cells but parvalbumin-β is found at levels of 1 to 2 mM in the OHCs and at micromolar levels (if at all) in IHCs while calbindin-D28K occurs at 0.1 to 0.2 mM in the OHCs and at micromolar levels in the IHCs. Calretinin and parvalbumin-α are also found at only micromolar levels in both types of cell (Hackney et al. 2005). These results suggest that the OHCs may have a greater calcium buffering requirement than the IHCs and that it is best met by the binding characteristics of parvalbumin-β.

In turtles, as in rats, calretinin was found at only micromolar levels, suggesting it does not play a major role as a buffer in this species either, although a gradient was observed for it in whole mounts using immunofluorescence. During development of the murine cochlea, calretinin first appears in the basal turn hair cells around E19 to P1 and then calretinin immunoreactivity occurs in a wave that moves apically with development, finally disappearing in the OHCs and remaining only in the IHCs by P22 (Dechesne et al. 1994). This pattern of expression suggests that calretinin may be predominantly involved in developmental processes in mammals at least. Clearly, its role needs to be investigated further by studying how much is present at different stages of development in various vertebrates.

In addition to the proteins mentioned above, other calcium binding proteins may contribute to calcium buffering (e.g., peptide 19; Imamura and Adams 1996) although some are fixed and not diffusible. How much of each is present and how they bind Ca^{2+} will also affect calcium handling in different types of hair cell and at different tonotopic locations in the cochlea.

5.3 Calmodulin

Calmodulin is a small, heat- and acid-stable calcium binding protein whose amino acid sequence has been highly conserved through evolution (Klee et al. 1980). It regulates a number of fundamental activities in eukaryotic cells (Means and Dedman 1980), acting as a calcium-activated facilitator of cellular metabolism and cell motility (Sobue et al. 1983). It possesses four calcium binding sites with equivalent affinities for calcium. Binding of calcium causes a change in the conformational state of the protein. This change is associated with calcium's ability to regulate multiple cell functions (Schulmann and Lou 1989), including cyclic nucleotide and glycogen metabolism, secretion, mitosis, and calcium transport (Means and Dedman 1980). In addition, a calcium-calmodulin–dependent activation of myosin light chain kinase regulates contraction in smooth muscle and non-muscle cells (Cheung 1980). Phosphorylation of the myosin light chain activates the actomysin ATPase, leading to hydrolysis of ATP and shortening. Calmodulin has been reported in the hair bundles in the bullfrog sacculus (Shepherd et al. 1989; Walker et al. 1993). Calmodulin has also been found in the adult mammalian cochlea in both IHCs and OHCs including the stereocilia (Slepecky and Ulfendahl 1993) and cuticular plate (Fig. 3.7E–G; Furness et al. 2002). In gerbil and guinea pig, Pack and Slepecky (1995) reported no difference in the intensity or labeling pattern for calmodulin between IHCs and OHCs. Similarly, Ogata and Slepecky (1998) reported equivalent levels of immunolabeling in type I and type II hair cells in the sensory epithelia of the gerbil utricle and crista ampullaris. However, Baird et al. (1997) reported that calmodulin is associated only with hair cells that can undergo adaptation in bullfrog otolith organs. This observation supports the findings of Walker and Hudspeth (1996), who found that calmodulin antagonists abolished adaptation to sustained mechanical stimuli and suggested that calmodulin binds directly to myosin 1c. More recently the interaction of myosin 1c with stereociliary receptors has been shown to be blocked by calmodulin (Cyr et al. 2002), leading to the suggestion that calcium-sensitive binding of calmodulin to myosin 1c may modulate the interaction of the adaptation motor with other components of the transduction apparatus. Postembedding immunogold labeling has shown calmodulin concentrations are highest at either end of the tip link and in the contact region between the stereocilia in hair cells (Fig. 3.7E, F; cf. Fig. 3.10; Furness et al. 2002), perhaps indicating that one or other

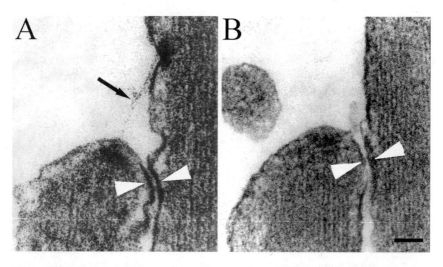

FIGURE 3.10. The ultrastructure of the contact region. (**A**) The relationship between the tip link (*arrow*) and the contact region (*arrowheads*) in a guinea-pig IHC. Note the increased electron density in the apposed membranes of the adjacent stereocilia and structures linking them in the contact region. (**B**) A comparable view of turtle stereocilia. The tip link is absent in this view. The contact region (*arrowheads*) is similar to that seen between guinea-pig stereocilia. Scale bar = 50 nm.

site might be involved in the slow component of adaptation. However, calmodulin is also known to interact with other molecules known to be abundant in the stereocilia such as PMCA and some ion channels proteins. Double labeling and further investigations of calmodulin interactions with stereociliary proteins will be required to determine its role in these two locations.

6. Functional Implications of the Apical Organization of Hair Cells

The foregoing description includes some examples of the likely roles of individual cytoskeletal and other apical proteins: a summary is provided in Table 3.1. However, the interaction between the various components of the apex make it desirable to consider the apex as a whole, in relation to the mechanosensory function of the hair cell, rather than as a series of independent components. Thus it is important to consider how the cytoskeleton will contribute in different ways to the bundles' response to displacement, how displacement is converted into gating of the MET channels, how the transduction process is modulated via regulatory proteins on both shorter and longer time scales, and finally how systematic variations in the cytoskeletal networks may contribute to differences in hair-cell responses in different organs and for different stimulus conditions.

A major facet of how the bundle works is dependent on the facts that the individual stereocilia are stiff and are coupled strongly together by cross-links. This stiffness, implied by the organization of the actin filaments and fimbrin/espin cross bridges, can be observed experimentally; the stereocilia do not bend when displaced, but pivot around their narrow ankles, and indeed, they snap at that point if displaced excessively (Flock et al. 1977). Functionally, stereociliary stiffness means that when a bundle is displaced, the stereocilia move together without individually bending, generating a point of maximum shear that occurs between adjacent ranks, where the tips of the shorter ones approach the sides of the taller ones. The energy produced by displacement is thus likely to be concentrated on channel gating structures in this region. MET channel opening occurs for very small positive deflections of the hair bundle (see, e.g., Russell et al. 1986). The tip links or structures in the contact region are ideally placed to operate them at the site of maximum shearing (Furness et al. 1997). If the stereocilia were not stiff, and instead began to bend, the energy of displacement would be dissipated throughout the bundle, thus making the process less efficient.

The presence of myosins that can interact with the actin core in the stereocilia gives the opportunity for the cells to actively modulate bundle movement and is one explanation for the slower component of adaptation (Holt et al. 2002). There are, in fact, at least two components, both of which are rapid events in terms of cellular responses but differing in that one is faster and located within 20 to 50 nm of the channel, while the other is slower and further away, approximately 150 to 200 nm. Both mechanisms are calcium dependent (Wu et

TABLE 3.1. Summary of the components of the hair cell apex discussed in this chapter.

Components	Location	Suggested function
Surface coat and cross-links		
Carbohydrate residues	Surface coat of stereocilia and cross-links	Stereociliary spacing
Keratan sulfate proteoglycans, tectorin	Surface coat of stereocilia and cross-links	Stereociliary spacing; links to accessory membranes
Tip-link antigen, cadherin 23	Stereociliary tips, kinociliary links	Mechanotransduction and adaptation, bundle integrity
Protocadherin 15, vezatin	Stereociliary shafts and ankle	Anchoring of links
Ankle-link antigen	Lateral links	Stereociliary spacing, bundle integrity
Cytoskeletal and associated proteins		
β- and γ-actin	Stereocilia and cuticular plate	Stereociliary stiffening and anchoring; substrate for motility and transport
Myosin IC	Stereociliary shaft and tip	Adaptation, transport (?)
Myosin IIIA	?	?
Myosin VI	Stereociliary ankles and cuticular plate	Bundle development, rootlet anchoring (?)
Myosin VIIa	Stereocilia	Link anchoring complex, adaptation (?)
Myosin XVa	Stereocilary tips	Stereociliary elongation
Fimbrin (T-plastin developmentally; I-plastin in adult)	Stereociliary core and cuticular plate	Ca^{2+}-sensitive (?) actin bundling
Espin	Stereociliary core	Ca^{2+}-insensitive actin bundling
α-actinin	Cuticular plate	Actin cross-linking
Tropomyosin	Stereociliary rootlets	Rootlet anchoring and motility
Kaptin (or 2E4)	Stereociliary tips	Actin polymerization
Spectrin	Cuticular plate and stereocilia	Actin-membrane linker
Profilin	Cuticular plate	Actin monomer binding
Diaphanous	?	Actin nucleation
β-Tubulin isotypes	Kinocilium and pericuticular meshwork	Structural support, transport
MAP-1A and MAP-1B	Microtubular network	Microtubule regulation
Other proteins		
Vinculin	Hair bundle (?)	Actin-membrane linker
Keap1	Lateral links	Link anchoring complex
Integrin α8β1, fibronectin, focal adhesion kinase	Hair cell apex	Bundle development
Radixin	Stereociliary ankles	Actin-membrane anchor
Whirlin	?	?
Harmonin b, SANS	Stereocilia	Link anchoring complex
Calcium pumps, buffers, and signaling proteins		
Plasma membrane Ca^{2+}-ATPase	Stereociliary membrane	Ca^{2+} extrusion
Calbindin D-28k, calretinin, parvalbumin-α and -β (oncomodulin)	Throughout hair cell	Diffusible Ca^{2+} buffers
Calmodulin	Stereociliary shaft and tips	Ca^{2+} signaling

The list is derived from studies of a range of species and hair cell epithelia and so not all molecules are necessarily found in all hair cells.

al. 1999). The bundle is therefore equipped with calcium buffers, pumps, and receptors that control calcium levels and use it for signaling. It is unlikely that the fast adaptation is based on actin–myosin interactions because it occurs too rapidly for them to be involved (see Fettiplace et al. 2001 for a fuller discussion). However, at least three of the different myosin isoforms, myosin 1c, VIIa, and XVa, may be present in the vicinity of the MET channels where calcium enters. Any of these could potentially be responsible for, or involved in, the slower components of adaptation. Although loss of myosin VIIa does not eliminate adaptation, it changes its properties and this may indicate a degree of involvement. Myosin 1c is, however, the currently favored candidate for an adaptation motor (Holt et al. 2002). The possibility of these myosin isoforms also acting on the attachment sites of the lateral links suggests another mechanism by which bundle motion can be modified. Changing the stiffness of the link attachments might affect the degree to which the stereocilia move together and thus the amount of shear developed. Tropomyosin in the stereociliary rootlets may also be involved in this process.

Other changes in the stiffness of the hair bundle may be mediated through the interaction between the stereociliary rootlets and the cuticular plate. Varying amounts of α-actinin could change the viscosity of the plate, affecting the rootlet movement, and provide another means by which the hair cell could adapt to different stimulus frequencies or strengths. Myosin, tropomyosin, and actin together may also allow a contractile response in the rootlet. Furthermore, myosin isoforms in the plate and the encircling actin ring could potentially provide a means for modifying the stiffness of the plate or producing a tilting action. The microtubular basket which seems to enclose and in places permeate the cuticular plate may also function in maintaining or altering its position within the apex and preventing mechanical deformation from disturbing the apical organization, as well as playing a role in membrane cycling.

In summary, the apical cytoskeleton appears to be designed for structural rigidity with pivot points that allow for appropriate hair bundle motion, and connections to prevent splaying of the bundle and ensuring its coherent movement. It also contains modulatory proteins that control stiffness by modifying the properties of the stereocilia, cuticular plate, and stereociliary anchors. These various events are regulated by calcium binding proteins, themselves controlled by calcium buffers and pumps that regulate calcium concentrations in the hair cell apex.

Focal adhesion proteins such as integrins are likely to be involved in attaching the hair bundle to its accessory structure, and offer possible mechanisms through which that interaction might be modulated. For example, changing the degree of coupling of the stereocilia to the tectorial membrane in OHCs is a potential way of modulating the degree of amplification that these cells could provide in the mammalian cochlea.

Several of the proteins, myosin VI, myosin VII, myosin XVa, profilin, the ERM proteins, and kaptin, have been suggested to be involved in shaping the cytoskeleton and the hair bundle during development, and potentially maintaining or repairing it, highlighting the fact that it is necessary to determine the role of a particular protein at different stages in the lifetime of a cell. There is therefore still much to be learned about the extent to which each of the different components that have been identified so far in the hair-cell apex are involved in its development, maintenance, or function.

Acknowledgments. Our research has been supported mainly by The Wellcome Trust and Defeating Deafness. Thanks are due to Y. Katori, D.-Ch. Neugebauer and M. Vater for providing us with material for EM.

References

Ahmed ZM Riazuddin S, Ahmad J, Bernstein SL, Guo Y, Sabar MF, Sieving P, Riazuddin S, Griffith AJ, Friedman TB, Belyantseva IA, Wilcox ER (2003) PCDH15 is expressed in the neurosensory epithelium of the eye and the ear and mutant alleles are responsible for both USH1F and DFNB23. Hum Mol Genet 12:3215–3223.

Anderson DW, Probst FJ, Belyantseva IA, Fridell RA, Beyer L, Martin DM, Wu D, Kachar B, Friedman TB, Raphael Y, Camper SA (2000) The motor and tail regions of myosin XV are critical for normal structure and function of auditory and vestibular hair cells. Hum Mol Genet 9:1729–1738.

Angelborg C, Engström H (1973) The normal organ of Corti. In: Møller A (ed), Basic Mechansisms in Hearing. London: Academic Press, pp. 125–183.

Anniko M, Thornell LE, Virtanen (1989) I. Actin-associated proteins and fibronectin in the fetal human inner ear. Am J Otolaryngol 10:99–109.

Assad JA, Corey DP (1992) An active motor model for adaptation by vertebrate hair cells. J Neurosci 12:3291–3309.

Assad JA, Hacohen N, Corey DP (1989) Voltage dependence of adaptation and active bundle movement in bullfrog saccular hair cells. Proc Natl Acad Sci USA 86:2918–2922.

Assad JA, Shepherd GM, Corey DP (1991) Tip-link integrity and mechanical transduction in vertebrate hair cells. Neuron 7:985–994.

Avraham KB, Hasson T, Steel KP, Kingsley DM, Russell LB, Mooseker MS, Copeland NG, Jenkins NA (1995) The mouse *Snell's waltzer* deafness gene encodes an unconventional myosin required for structural integrity of inner ear hair cells. Nat Genet 11:369–375.

Bagger-Sjöbäck D, Takumida M (1988) Geometrical array of the vestibular sensory hair bundle. Acta Otolaryngol 106:393–403.

Bagger-Sjöbäck D, Wersäll J (1973) The sensory hairs and tectorial membrane in the basilar papilla of the lizard, *Calotes versicolor*. J Neurocytol 2:329–350.

Baird RA, Schuff NR, Bancroft J (1993) Regional differences in lectin binding patterns of vestibular hair cells. Hear Res 65:151–163.

Baird RA, Steyger PS, Schuff NR (1997) Intracellular distributions and putative functions of calcium-binding proteins in the bullfrog vestibular otolith organs. Hear Res 103:85–100.

Barber VC, Emerson CJ (1979) Cupula-receptor cell relationships with evidence provided by SEM microdissection. Scan Electron Microsc 3:939–948.

Bartles JR, Wierda A, Zheng L (1996) Identification and characterization of espin, an actin-binding protein localized to the F-actin-rich junctional plaques of Sertoli cell ectoplasmic specializations. J Cell Sci 109:1229–1239.

Bartles JR, Zheng L, Li A, Wierda A, Chen B (1998) Small espin: a third actin-bundling protein and potential forked protein ortholog in brush border microvilli. J Cell Biol 143:107–119.

Batanov ME, Goodyear RJ, Richardson GP, Russell IJ (2004) The mechanical properties of sensory hair bundles: relative contributions of structures sensitive to calcium chelation and subtilisin treatment. J Physiol 559:3474–3479.

Bearer EL (1992) An actin-associated protein present in the microtubule organizing center and the growth cones of PC-12 cells. J Neurosci 12:750–761.

Bearer EL, Abraham MT (1999) 2E4 (kaptin): a novel actin-associated protein from human blood platelets found in lamellipodia and the tips of the stereocilia of the inner ear. Eur J Cell Biol 78:117–126.

Bearer EL, Chen AF, Chen AH, Li Z, Mark HF, Smith RJ, Jackson CL (2000) 2E4/Kaptin (KPTN)—a candidate gene for the hearing loss locus, DFNA4. Ann Hum Genet 64:189–196.

Belyantseva IA, Boger ET, Fredmann TB (2003) Myosin XVa localizes to the tips of inner ear sensory cell stereocilia and is essential for staircase formation of the hair bundle. Proc Natl Acad Sci 25:13958–13963.

Benser ME, Marquis RE, Hudspeth AJ (1996) Rapid, active hair bundle movements in hair cells from the bullfrog's sacculus. J Neurosci 16:5629–5643.

Berg JS, Powell BC, Cheney RE (2001) A millennial myosin census. Mol Biol Cell 12: 780–794.

Boeda B, El-Amraoui A, Bahloul A, Goodyear R, Daviet L, Blanchard S, Perfettini I, Fath KR, Shorte S, Reiners J, Houdusse A, Legrain P, Wolfrum U, Richardson G, Petit C (2002) Myosin VIIa, harmonin and cadherin 23, three Usher 1 gene products that cooperate to shape the sensory hair bundle. EMBO J 21:6689–6699.

Bretscher A, Weber K (1980) Fimbrin, a new microfilament-associated protein present in microvilli and other cell surface structures. J Cell Biol 86:335–340.

Bryant J, Goodyear RJ, Richardson GP (2002) Sensory organ development in the inner ear: molecular and cellular mechanisms. Br Med Bull 63:39–57.

Carlisle L, Zajic G, Altschuler RA, Schacht J, Thorne PR (1988) Species differences in the distribution of infracuticular F-actin in outer hair cells of the cochlea. Hear Res 33:201–205.

Chen B, Li A, Wang D, Wang M, Zheng L, Bartles JR (1999) Espin contains an additional actin-binding site in its N terminus and is a major actin-bundling protein of the Sertoli cell-spermatid ectoplasmic specialization junctional plaque. Mol Biol Cell 10:4327–4339.

Cheung WY (1980) Calmodulin plays a pivotal role in cellular regulation. Science 207: 19–27.

Corey DP, Hudspeth AJ (1979) Response latency of vertebrate hair cells. Biophys J 26: 499–506.

Corey DP, Hudspeth AJ (1983) Kinetics of the receptor current in bullfrog saccular hair cells. J Neurosci 3:962–976.

Crawford AC, Evans MG, Fettiplace R (1989) Activation and adaptation of transducer currents in turtle hair cells. J Physiol 419:405–434.

Crouch JJ, Schulte BA (1995) Expression of plasma membrane Ca-ATPase in the adult and developing gerbil cochlea. Hear Res 92:112–119.

Crouch JJ, Schulte BA (1996) Identification and cloning of site C splice variants of plasma membrane Ca-ATPase in the gerbil cochlea. Hear Res 101:55–61.

Crozet F, el Amraoui A, Blanchard S, Lenoir M, Ripoll C, Vago P, Hamel C, Fizames C, Levi-Acobas F, Depetris D, Mattei MG, Weil D, Pujol R, Petit C (1997) Cloning of the genes encoding two murine and human cochlear unconventional type I myosins. Genomics 40:332–341.

Cyr JL, Dumont RA, Gillespie PG (2002) Myosin-1c interacts with hair-cell receptors through its calmodulin-binding IQ domains. J Neurosci 22:2487–2495.

Czukas SR, Rosenquist TH, Mulroy MJ (1987) Connections between stereocilia in auditory hair cells of the alligator lizard. Hear Res 30:147–156.

Danbolt NC, Lehre KP, Dehnes Y, Chaudrhy FA, Levy LM (1998) Localization of transporters using transporter-specific antibodies. Meth Enzymol 296:388–407.

Daudet N, Lebart MC (2002) Transient expression of the t-isoform of plastins/fimbrin in the stereocilia of developing auditory hair cells. Cell Motil Cytoskel 53:326–336.

de Arruda, MV, Watson S, Lin C-S, Levitt J, Matsudaira, P (1990) Fimbrin is a homologue of the cytoplasmic phosphoprotein plastin and has domains homologous with calmodulin and actin gelation proteins. J Cell Biol 111:1069–1079.

Dechesne CJ, Thomasset M (1988) Calbindin (CaBP 28 kDa) appearance and distribution during development of the mouse inner ear. Brain Res 468:233–242.

Dechesne CJ, Winsky L, Kim HN, Goping G, Vu TD, Wenthold RJ, Jacobowitz DM (1991) Identification and ultrastructural localization of a calretinin-like calcium-binding protein (protein 10) in the guinea pig and rat inner ear. Brain Res 560:139–148.

Dechesne CJ, Rabejac D, Desmadryl G (1994) Development of calretinin immunoreactivity in the mouse inner ear. J Comp Neurol 346:517–529.

Denk W, Holt JR, Shepherd GM, Corey DP (1995) Calcium imaging of single stereocilia in hair cells: localization of transduction channels at both ends of tip links. Neuron 15:1311–1321.

Dolgobrodov SG, Lukashkin AN, Russell IJ (2000) Electrostatic interaction between stereocilia: I. Its role in supporting the structure of the hair bundle. Hear Res 150: 83–93.

Drenckhahn D, Kellner J, Mannherz HG, Groschel-Stewart U, Kendrick-Jones J, Scholey J (1982) Absence of myosin-like immunoreactivity in stereocilia of cochlear hair cells. Nature 300:531–532.

Drenckhahn D, Engel K, Hofer D, Merte C, Tilney L, Tilney M (1991) Three different actin filament assemblies occur in every hair cell: each contains a specific actin cross-linking protein. J Cell Biol 112:641–651.

Dumont RA, Lins U, Filoteo AG, Penniston JT, Kachar B, Gillespie PG (2001) Plasma membrane Ca^{2+}-ATPase isoform 2a is the PMCA of hair bundles. J Neurosci 21: 5066–5078.

Dumont RA, Yi-Dong Z, Holt JR, Bähler M, Gillespie PG (2002) Myosin I isozymes in neonatal rodent auditory and vestibular epithelia. J Assoc Res Otolaryngol 3:37–389.

Eatock RA, Corey DP, Hudspeth AJ (1987) Adaptation of mechanoelectrical transduction in hair cells of the bullfrog's sacculus. J Neurosci 7:2821–2836.

Edmonds B, Reyes R, Schwaller B, Roberts WM (2000) Calretinin modifies presynaptic calcium signaling in frog saccular hair cells. Nat Neurosci 3:786–790.

Engström H, Engström B (1978) Structure of the hairs on cochlear sensory cells. Hear Res 1:49–66.

Eybalin M, Ripoll C (1990) Immunolocalization of parvalbumin in two glutamatergic cell types of the guinea pig cochlea: inner hair cells and spiral ganglion neurons. C R Acad Sci III 310:639–644.

Feldner JC, Brandt BH (2002) Cancer cell motility—on the road from c-erbB-2 receptor steered signaling to actin reorganization. Exp Cell Res 272:93–108.

Fettiplace R, Ricci AJ, Hackney CM (2001) Clues to the cochlear amplifier from the turtle ear. Trends Neurosci 24:169–175.

Flock Å, Cheung HC (1977) Actin filaments in sensory hairs of inner ear receptor cells. J Cell Biol 75:339–343.

Flock Å, Flock B, Murray E (1977) Studies on the sensory receptor hairs of receptor cells in the inner ear. Acta Otolaryngol 83:85–91.

Flock Å, Bretscher A, Weber K (1982) Immunohistochemical localization of several cytoskeletal proteins in inner ear sensory and supporting cells. Hear Res 7:75–89.

Frolenkov GI, Belyantseva IA, Friedmann, Griffith AJ (2004) Genetic insights into the morphogenesis of inner ear hair cells. Nat Rev Genet 5:489–498.

Furness DN, Hackney CM (1985) Cross-links between stereocilia in the guinea pig cochlea. Hear Res 18:177–188.

Furness DN, Hackney CM (1986) Morphological changes to the stereociliary bundles in the guinea pig cochlea after kanamycin treatment. Br J Audiol 20:253–259.

Furness DN, Hackney, CM (1987) Cytoskeletal organization in the apex of cochlear hair cells. In: Syka JB, Masterton R (eds), Auditory Pathway. New York: Plenum, pp. 23–28.

Furness DN, Hackney CM (1990) Comparative ultrastructure of subsurface cisternae in inner and outer hair cells of the guinea pig cochlea. Eur Arch Otorhinolaryngol 247: 12–15.

Furness DN, Hackney CM, Steyger PS (1990) Organization of microtubules in cochlear hair cells. J Electron Microsc Tech 15:261–279.

Furness DN, Hackney CM, Benos DJ (1996) The binding site on cochlear stereocilia for antisera raised against renal Na^+ channels is blocked by amiloride and dihydrostreptomycin. Hear Res 93:136–146.

Furness DN, Zetes DE, Hackney CM, Steele CR (1997) Kinematic analysis of shear displacement as a means for operating mechanotransduction channels in the contact region between adjacent stereocilia of mammalian cochlear hair cells. Proc R Soc Lond B Biol Sci 264:45–51.

Furness DN, Karkanevatos A, West B, Hackney CM (2002) An immunogold investigation of the distribution of calmodulin in the apex of cochlear hair cells. Hear Res 173:10–20.

Furness DN, Katori Y, Mahendrasingam S, Hackney CM (2005) Differential distribution of β- and γ-actin in guinea-pig cochlear sensory and supporting calls. Hear Res (in press).

Garcia JA, Yee AG, Gillespie PG, Corey DP (1998) Localization of myosin-Ibeta near both ends of tip links in frog saccular hair cells. J Neurosci 18:8637–8647.

Geisler CD (1993) A model of stereociliary tip-link stretches. Hear Res 65:79–82.

Gillespie PG, Hudspeth AJ (1993) Adenine nucleoside diphosphates block adaptation of mechanoelectrical transduction in hair cells. Proc Natl Acad Sci USA 90:2710–2714.

Gil-Loyzaga P, Brownell WE (1988) Wheat germ agglutinin and *Helix pomatia* agglutinin lectin binding on cochlear hair cells. Hear Res 34:149–155.

Goodman M, Pechere JF (1977) The evolution of muscular parvalbumins investigated by the maximum parsimony method. J Mol Evol 9:131–158.

Goodyear R, Richardson G (1992) Distribution of the 275 kD hair cell antigen and cell surface specialisations on auditory and vestibular hair bundles in the chicken inner ear. J Comp Neurol 325:243–256.

Goodyear R, Richardson G (1994) Differential glycosylation of auditory and vestibular hair bundle proteins revealed by peanut agglutinin. J Comp Neurol 345:267–278.

Goodyear R, Richardson G (1999) The ankle-link antigen: an epitope sensitive to calcium chelation associated with the hair-cell surface and the calycal processes of photoreceptors. J Neurosci 19:3761–3772.

Goodyear R, Richardson G (2003) A novel antigen sensitive to calcium chelation that is associated with the tip links and kinocilial links of sensory hair bundles. J Neurosci 23:4878–4887.

Griesinger CB, Richards CD, Ashmore JF (2002) Fm1-43 reveals membrane recycling in adult inner hair cells of the mammalian cochlea. J Neurosci 22:3939–3952.

Gundersen V, Danbolt NC, Ottersen OP, Storm-Mathisen J (1993) Demonstration of glutamate/aspartate uptake activity in nerve endings by use of antibodies recognizing exogenous D-aspartate. Neuroscience 57:97–111.

Hackney CM, Furness DN (1986) Intercellular cross-linkages between the stereociliary bundles of adjacent hair cells in the guinea pig cochlea. Cell Tissue Res 245:685–688.

Hackney CM, Furness DN (1995) Mechanotransduction in vertebrate hair cells: structure and function of the stereociliary bundle. Am J Physiol 268:C1–13.

Hackney CM, Furness DN, Sayers DL (1988) Stereociliary cross-links between adjacent inner hair cells. Hear Res 34:207–212.

Hackney CM, Furness DN, Benos DJ, Woodley JF, Barratt J (1992) Putative immunolocalization of the mechanoelectrical transduction channels in mammalian cochlear hair cells. Proc R Soc Lond B Biol Sci 248:215–221.

Hackney CM, Fettiplace R, Furness DN (1993) The functional morphology of stereociliary bundles on turtle cochlear hair cells. Hear Res 69:163–175.

Hackney CM, Mahendrasingam S, Jones EMC, Fettiplace R (2003) The distribution of calcium buffering proteins in the turtle cochlea. J Neurosci 23:4577–4589.

Hackney CM, Mahendrasingam S, Penn A, Fettiplace R (2005) The concentrations of calcium buffering proteins in mammalian cochlear hair cells. J Neurosci 25:7867–7875.

Harlow E, Lane D (1999) Using Antibodies: A Laboratory Manual. Cold Spring Harbor, NY: Cold Spring Harbor Laboratory Press.

Hasson T, Gillespie PG, Garcia JA, MacDonald RB, Zhao Y, Yee AG, Mooseker MS, Corey DP (1997) Unconventional myosins in inner-ear sensory epithelia. J Cell Biol 137:1287–1307.

Heller S, Bell AM, Denis CS, Choe Y, Hudspeth AJ (2002) Parvalbumin 3 is an abundant Ca^{2+} buffer in hair cells. J Assoc Res Otolaryngol 3:488–498.

Henzl MT, Larson JD, Agah S (2003) Estimation of parvalbumin Ca^{2+} and Mg^{2+} binding constants by global least-squares analysis of isothermal titration calorimetry data. Anal Biochem 319:216–233.

Hiel H, Navaratnam DS, Oberholtzer JC, Fuchs PA (2002) Topological and developmental gradients of calbindin expression in the chick's inner ear. J Assoc Res Otolaryngol 3:1–15.

Hillman DE (1969) New ultrastructural findings regarding a vestibular ciliary apparatus and its possible functional significance. Brain Res 13:407–412.

Hirao M, Sato N, Kondo T, Yonemura S, Monden M, Sasaki T, Takai Y, Tsukita S, Tsukita S (1996) Regulation mechanism of ERM (ezrin/radixin/moesin)protein/plasma membrane association: possible involvement of phosphatidylinositol turnover and Rho-dependent signaling pathway. J Cell Biol 135:37–51.

Hirokawa N, Tilney LG (1982) Interactions between actin filaments and between actin filaments and membranes in quick-frozen and deeply etched hair cells of the chick ear. J Cell Biol 95:249–261.

Hofer D, Ness W, Drenckhahn D (1997) Sorting of actin isoforms in chicken auditory hair cells. J Cell Sci 110:765–770.

Holme R, Steel KP (2004) Progressive hearing loss and increased susceptibility to noise-induced hearing loss in mice carrying a *Cdh23* but not a *Myo7a* mutation. J Assoc Res Otolaryngol 5:66–79.

Holmes GR, Goll DE, Suzuki A (1971) Effect of α-actinin on actin viscosity. Biochim Biophys Acta 253:240–253.

Holt JR, Corey DP, Eatock RA (1997) Mechanoelectrical transduction and adaptation in hair cells of the mouse utricle, a low-frequency vestibular organ. J Neurosci 17:8739–8748.

Holt JR, Gillespie SK, Provance DW, Shah K, Shokat KM, Corey DP, Mercer JA, Gillespie PG (2002) A chemical-genetic strategy implicates myosin-1c in adaptation by hair cells. Cell 108:371–381.

Hoshino T (1975) An electron microscopic study of the otolithic maculae of the lamprey (*Entosphenus japonicus*). Acta Otolaryngol 80:43–53.

Hudspeth AJ (1989) How the ear's works work. Nature 341:397–404.

Hudspeth AJ (1997) How hearing happens. Neuron 19:947–950.

Hudspeth AJ, Corey DP (1977) Sensitivity, polarity, and conductance change in the response of vertebrate hair cells to controlled mechanical stimuli. Proc Natl Acad Sci USA 74:2407–2411.

Imamura S, Adams JC (1996) Immunolocalization of peptide 19 and other calcium-binding proteins in the guinea pig cochlea. Anat Embryol (Berl) 194:407–418.

Itoh M (1982) Preservation and visualization of actin-containing filaments in the apical zone of cochlear sensory cells. Hear Res 6:277–289.

Jacobs RA, Hudspeth AJ (1990) Ultrastructural correlates of mechanoelectrical transduction in hair cells of the bullfrog's internal ear. Cold Spring Harb Symp Quant Biol 55:547–561.

Jaeger RG, Fex J, Kachar B (1994) Structural basis for mechanical transduction in the frog vestibular sensory apparatus: II. The role of microtubules in the organization of the cuticular plate. Hear Res 77:207–215.

Jaramillo F, Hudspeth AJ (1991) Localization of the hair cell's transduction channels at the hair bundle's top by iontophoretic application of a channel blocker. Neuron 7:409–420.

Jensen-Smith HC, Eley J, Steyger PS, Luduena RF, Hallworth R (2003) Cell type-specific reduction of beta-tubulin isotypes synthesised in the developing gerbil organ of Corti. J Neurocytol 32:185–197.

Jørgensen F, Ohmori H (1988) Amiloride blocks the mechanoelectrical transduction channel of hair cells of the chick. J Physiol Lond 403:577–588.

Kachar B, Battaglia A, Fex J (1997) Compartmentalized vesicular traffic around the hair cell cuticular plate. Hear Res 107:102–112.

Kachar B, Parakkal M, Kurc M, Zhao Y, Gillespie PG (2000) High-resolution structure of hair-cell tip links. Proc Natl Acad Sci USA 97:13336–13341.

Karkanevatos A (2001) Ultrastructural localization of cytoskeletal proteins in guinea-pig cochlear hair cells. M. Phil Thesis, Keele University, UK.

Katori Y, Takasaka T, Ishikawa M, Tonosaki A (1994) Fine structure and lectin histochemistry of the apical surface of the free neuromast of *Lampetra japonica*. Cell Tissue Res 276:245–252.

Katori Y, Hackney CM, Furness DN (1996a) Immunoreactivity of sensory hair bundles of the guinea-pig cochlea to antibodies against elastin and keratan sulphate. Cell Tissue Res 284:473–479.

Katori Y, Tonosaki A, Takasaka T (1996b) WGA lectin binding sites of the apical surface of Corti epithelium: enhancement by back-scattered electron imaging in guinea-pig inner ear. J Electron Microsc (Tokyo) 45:207–212.

Kikkawa Y, Shitara H, Wakana S, Kohara Y, Takada T, Okamoto M, Taya C, Kamiya K, Yoshikawa Y, Tokano H, Kitamura K, Shimizu K, Wakabayashi Y, Shiroishi T, Kominami R, Yonekawa H (2003) Mutations in a new scaffold protein *Sans* cause deafness in Jackson shaker mice. Hum Mol Genet 12:453–461.

Killick R, Richardson GP (1997) Antibodies to the sulphated, high molecular mass mouse tectorin stain hair bundles and the olfactory mucus layer. Hear Res 103:131–141.

Kimura RS (1975) The ultrastructure of the organ of Corti. Int Rev Cytol 42:173–222.

Klee CB, Crouch TH, Richman PG (1980) Calmodulin. Annu Rev Biochem 49:489–515.

Kollmar R (1999) Who does the hair cell's 'do? Rho GTPases and hair-bundle morphogenesis. Curr Opin Neurobiol 9: 394–398.

Kössl M, Richardson GP, Russell IJ (1990) Stereocilia bundle stiffness: effects of neomycin sulphate, A23187 and concanavalin A. Hear Res 44:217–229.

Kozel PJ, Friedman RA, Erway LC, Yamoah EN, Liu LH, Riddle T, Duffy JJ, Doetschman T, Miller ML, Cardell EL, Shull GE (1998) Balance and hearing deficits in mice with a null mutation in the gene encoding plasma membrane Ca^{2+}-ATPase isoform 2. J Biol Chem 273:18693–18696.

Kros CJ, Marcotti W, van Netten SM, Self TJ, Libby RT, Brown SD, Richardson GP, Steel KP (2002) Reduced climbing and increased slipping adaptation in cochlear hair cells of mice with *Myo7a* mutations. Nat Neurosci 5:41–47.

Kussel-Andermann P, El-Amraoui A, Safieddine S, Nouaille S, Perfettini I, Lecuit M, Cossart P, Wolfrum U, Petit C (2000) Vezatin, a novel transmembrane protein, bridges myosin VIIA to the cadherin–catenins complex. EMBO J 19:6020–6029.

Lee KH, Cotanche DA (1996) Localization of the hair-cell-specific protein fimbrin during regeneration in the chicken cochlea. Audiol Neurootol 1:41–53.

Lewis ER, Li CW (1975) Hair cell types and distributions in the otolithic and auditory organs of the bullfrog. Brain Res 83:35–50.

Liang Y, Wang A, Belyantseva IA, Anderson DW, Probst FJ, Barber TD, Miller W, Touchman JW, Jin L, Sullivan SL, Sellers JR, Camper SA, Lloyd RV, Kachar B, Friedman TB, Fridell RA (1999) Characterization of the human and mouse unconventional myosin XV genes responsible for hereditary deafness DFNB3 and shaker 2. Genomics 61:243–258.

Lim DJ (1986) Functional structure of the organ of Corti: a review. Hear Res 22:117–146.

Lin C-S, Shen W, Chen ZP, Tu Y-H, Matsudaira P (1994) Identification of I-plastin, a human fimbrin isoform expressed in intestine and kidney. Mol Cell Biol 14:2457–2467.

Little KF, Neugebauer D-Ch (1985) Interconnections between the stereovilli of the fish inner ear. II. Systematic investigation of saccular hair bundles from *Rutilus rutilus* (Teleostei). Cell Tissue Res 242:427–432.

Littlewood-Evans A, Müller U (2000) Stereocilia defects in the sensory hair cells of the inner ear in mice deficient in integrin alpha8beta1. Nat Genet 24:424–428.

Lumpkin EA, Hudspeth AJ (1995) Detection of Ca^{2+} entry through mechanosensitive channels localizes the site of mechanoelectrical transduction in hair cells. Proc Natl Acad Sci USA 92:10297–10301.

Lumpkin EA, Hudspeth AJ (1998) Regulation of free Ca^{2+} concentration in hair-cell stereocilia. J Neurosci 18:6300–6318.

Lynch ED, Lee MK, Morrow JE, Welcsh PL, Leon PE, King MC (1997) Nonsyndromic deafness *DFNA1* associated with mutation of a human homolog of the *Drosophila* gene diaphanous. Science 278:1315–1318.

Macartney JC, Comis SD, Pickles JO (1980) Is myosin in the cochlea a basis for active motility? Nature 288:491–492.

MacManus JP (1979) Occurrence of a low-molecular-weight calcium-binding protein in neoplastic liver. Cancer Res 39:3000–3005.

Mahendrasingam S, Furness DN, Hackney CM (1998) Ultrastructural localisation of spectrin in sensory and supporting cells of guinea-pig organ of Corti. Hear Res 126: 151–160.

Manley GA (2000) Cochlear mechanisms from a phylogenetic viewpoint. Proc Natl Acad Sci USA 97:11736–11743.

Matsubara A, Laake JH, Davanger S, Usami S, Ottersen OP (1996) Organization of AMPA receptor subunits at a glutamate synapse: a quantitative immunogold analysis of hair cell synapses in the rat organ of Corti. J Neurosci 16:4457–4467.

Mburu P, Mustapha M, Varela A, Weil D, El-Amraoui A, Holme RH, Rump A, Hardisty RE, Blanchard S, Coimbra RS, Perfettini I, Parkinson N, Mallon AM, Glenister P, Rogers MJ, Paige AJ, Moir L, Clay J, Rosenthal A, Liu XZ, Blanco G, Steel KP, Petit C, Brown SD (2003) Defects in whirlin, a PDZ domain molecule involved in stereocilia elongation, cause deafness in the *whirler* mouse and families with *DFNB31*. Nat Genet 34:421–428.

Means AR, Dedman JR (1980) Calmodulin—an intracellular calcium receptor. Nature 285:73–77.

Mermall V, Post PL, Mooseker MS (1998) Unconventional myosins in cell movement, membrane traffic, and signal transduction. Science 279:527–533.

Metcalf AB (1998) Immunolocalization of myosin Ibeta in the hair cell's hair bundle. Cell Motil Cytoskel 39:159–165.

Meyer J, Furness DN, Zenner HP, Hackney CM, Gummer AW (1998) Evidence for opening of hair-cell transducer channels after tip-link loss. J Neurosci 18:6748–6756.

Meyer J, Mack AF, Gummer AW (2001) Pronounced infracuticular endocytosis in mammalian outer hair cells. Hear Res 161:10–22.

Mulroy MJ, Williams RS (1987) Auditory stereocilia in the alligator lizard. Hear Res 25:11–21.

Munyer PD, Schulte BA (1994) Immunohistochemical localization of keratan sulfate and chondroitin 4- and 6-sulfate proteoglycans in subregions of the tectorial and basilar membranes. Hear Res 79:83–93.

Nagel G, Neugebauer D-Ch, Schmidt B and Thurm U (1991) Structures transmitting stimulatory force to the sensory hairs of vestibular ampullae of fishes and frog. Cell Tissue Res 265:567–578.

Nagerl UV, Novo D, Mody I, Vergara JL (2000) Binding kinetics of calbindin-D28k determined by flash photolysis of caged Ca^{2+}. Biophys J 79:3009–3018.

Neugebauer D-Ch, Thurm U (1984) Chemical dissection of stereovilli from fish inner ear reveals differences from intestinal microvilli. J Neurocytol 13:797–808.

Neugebauer D-C, Thurm U (1985) Interconnections between the stereovilli of the fish inner ear. Cell Tissue Res 240:449–453.

Neugebauer D-C, Thurm U (1987) Surface charges of the membrane and cell adhesion substances determine the structural integrity of hair bundles from the inner ear of fish. Cell Tissue Res 249:199–207.

Nicolson T, Rusch A, Friedrich RW, Granato M, Ruppersberg JP, Nusslein-Volhard C (1998) Genetic analysis of vertebrate sensory hair cell mechanosensation: the zebrafish circler mutants. Neuron 20:271–283.

Nishida Y, Fujimoto T, Takagi A, Honjo I, Ogawa K (1993) Fodrin is a constituent of the cortical lattice in outer hair cells of the guinea pig cochlea: immunocytochemical evidence. Hear Res 65:274–280.

Oberholtzer JC, Buettger C, Summers MC, Matschinsky FM (1988) The 28-kDa calbindin-D is a major calcium-binding protein in the basilar papilla of the chick. Proc Natl Acad Sci USA 85:3387–3390.

Ogata Y, Slepecky NB (1995) Immunocytochemical comparison of posttranslationally modified forms of tubulin in the vestibular end-organs of the gerbil: tyrosinated, acetylated and polyglutamylated tubulin. Hear Res 86:125–131.

Ogata Y, Slepecky NB (1998) Immunocytochemical localization of calmodulin in the vestibular end-organs of the gerbil. J Vestib Res 8:209–216.

Oliver TN, Berg JS, Cheney RE (1999) Tails of unconventional myosins. Cell Mol Life Sci 56:243–257.

Osborne MP, Comis SD (1990) Action of elastase, collagenase and other enzymes upon linkages between stereocilia in the guinea-pig cochlea. Acta Otolaryngol 110:37–45.

Osborne MP, Comis SD, Pickles JO (1984) Morphology and cross-linkage of stereocilia in the guinea-pig labyrinth examined without the use of osmium as a fixative. Cell Tissue Res 237:43–48.

Osborne MP, Comis SD, Pickles JO (1988) Further observations on the fine structure of tip links between stereocilia of the guinea pig cochlea. Hear Res 35:99–108.

Oshima T, Okabe S, Hirokawa N (1992) Immunocytochemical localization of 205 kDa microtubule-associated protein (205 kDa MAP) in the guinea pig organ of Corti. Brain Res 590:53–65.

Pacaud M, Derancourt J (1993) Purification and further characterization of macrophage 70-kDa protein, a calcium-regulated, actin-binding protein identical to L-plastin. Biochemistry 32:3448–3455.

Pack AK, Slepecky NB (1995) Cytoskeletal and calcium-binding proteins in the mammalian organ of Corti: cell type-specific proteins displaying longitudinal and radial gradients. Hear Res 91:119–135.

Parsons TD, Lenzi D, Almers W, Roberts WM (1994) Calcium-triggered exocytosis and endocytosis in an isolated presynaptic cell: capacitance measurements in saccular hair cells. Neuron 13:875–883.

Pataky F, Pironkova R, Hudspeth AJ (2004) Radixin is a constituent of stereocilia in hair cells. Proc Natl Acad Sci USA 101:2601–2606.

Pauls TL, Cox JA, Berchtold MW (1996) The Ca^{2+}-binding proteins parvalbumin and oncomodulin and their genes: new structural and functional findings. Biochim Biophys Acta 1306:39–54.

Perry B, Jensen-Smith HC, Luduena RF, Hallworth R (2003) Selective expression of beta tubulin isotypes in gerbil vestibular sensory epithelia and neurons. J Assoc Res Otolaryngol 4:329–338.

Pickles JO (1993a) A model for the mechanics of the stereociliar bundle on acousticolateral hair cells. Hear Res 68:159–172.

Pickles JO (1993b) An analysis of actin isoforms expressed in hair-cell enriched fractions of the chick basilar papilla by the polymerase chain reaction technique. Hear Res 71: 225–229.

Pickles JO, Corey DP (1992) Mechanoelectrical transduction by hair cells. Trends Neurosci 15:254–259.

Pickles JO, Comis SD, Osborne MP (1984) Cross-links between stereocilia in the guinea pig organ of Corti, and their possible relation to sensory transduction. Hear Res 15: 103–112.

Pickles JO, Brix J, Comis SD, Gleich O, Köppl C, Manley GA, Osborne MP (1989) The organization of tip links and stereocilia on hair cells of bird and lizard basilar papillae. Hear Res 41:31–41.

Pickles JO, Brix J, Manley GA (1990) Influence of collagenase on tip links in hair cells of the chick basilar papilla. Hear Res 50:139–143.

Pickles JO, Billieux-Hawkins DA, Rouse GW (1996) The incorporation and turnover of radiolabelled amino acids in developing stereocilia of the chick cochlea. Hear Res 101:45–54.

Probst FJ, Fridell RA, Raphael Y, Saunders TL, Wang A, Liang Y, Morell RJ, Touchman JW, Lyons RH, Noben-Trauth K, Friedman TB, Camper SA (1998) Correction of deafness in shaker-2 mice by an unconventional myosin in a BAC transgene. Science 280:1444–1447.

Ricci AJ, Fettiplace R (1997) The effects of calcium buffering and cyclic AMP on mechanoelectrical transduction in turtle auditory hair cells. J Physiol 501:111–124.

Ricci AJ, Wu YC, Fettiplace R (1998) The endogenous calcium buffer and the time course of transducer adaptation in auditory hair cells. J Neurosci 18:8261–8277.

Ricci AJ, Gray-Keller M, Fettiplace R (2000) Tonotopic variations of calcium signalling in turtle auditory hair cells. J Physiol 524:423–436.

Richardson GP, Russell IJ, Wasserkort R, Hans M (1988) Aminoglycoside antibiotics and lectins cause irreversible increases in the stiffness of cochlear hair cell stereocilia In: Wilson JP, Kemp DT (eds), Cochlear Mechanisms. London: Plenum, pp. 57–65.

Richardson GP, Bartolami S, Russell IJ (1990) Identification of a 275-kD protein associated with the apical surfaces of sensory hair cells in the avian inner ear. J Cell Biol 110:1055–1066.

Roberts WM (1993) Spatial calcium buffering in saccular hair cells. Nature 363:74–76.

Roberts WM (1994) Localization of calcium signals by a mobile calcium buffer in frog saccular hair cells. J Neurosci 14:3246–3262.

Rogers JH (1989) Two calcium-binding proteins mark many chick sensory neurons. Neuroscience 31:697–709.

Rowe MH, Peterson EH (2004) Quantitative analysis of stereociliary arrays on vestibular hair cells. Hear Res 190:10–24.

Rüsch A, Lysakowski A, Eatock RA (1998) Postnatal development of type I and type II hair cells in the mouse utricle: acquisition of voltage-gated conductances and differentiated morphology. J Neurosci 18:7487–7501.

Russell IJ, Richardson GP, Cody AR (1986) Mechanosensitivity of mammalian auditory hair cells in vitro. Nature 321:517–519.

Rzadzinska AK, Schneider ME, Davies C, Riordan GP, Kachar B (2004) An actin molecular treadmill and myosins maintain stereocilia functional architecture and self-renewal. J Cell Biol 164:887–897.

Sakaguchi N, Henzl MT, Thalmann I, Thalmann R, Schulte BA (1998) Oncomodulin is expressed exclusively by outer hair cells in the organ of Corti. J Histochem Cytochem 46:29–40.

Santi PA, Anderson CB (1987) A newly identified surface coat on cochlear hair cells. Hear Res 27:47–65.

Sato M, Schwarz WH, Pollard TD (1987) Dependence of the mechanical properties of actin/alpha-actinin gels on deformation rate. Nature 325:828–830.

Schneider ME, Belyantseva IA, Azevedo RB, Kachar B (2002) Rapid renewal of auditory hair bundles. Nature 418:837–838.

Schulman H, Lou LL (1989) Multifunctional Ca^{2+}/calmodulin-dependent protein kinase: domain structure and regulation. Trends Biochem Sci 14:62–66.

Schulte CC, Meyer J, Furness DN, Hackney CM, Kleymann TR, Gummer AW (2002) Functional effects of a monoclonal antibody on mechanoelectrical transduction in outer hair cells. Hear Res 164:190–205.

Sekerková G, Zheng L, Loomis PA, Changyaleket B, Whitlon DS, Mugnaini E, Bartles JR (2004) Espins are multifunctional actin cytoskeletal regulatory proteins in the microvilli of chemosensory and mechanosensory cells. J Neurosci 24:5445–5456.

Self T, Sobe T, Copeland NG, Jenkins NA, Avraham KB, Steel KP (1999) Role of myosin VI in the differentiation of cochlear hair cells. Dev Biol 214:331–341.

Shepherd GM, Barres BA, Corey DP (1989) "Bundle blot" purification and initial protein characterization of hair cell stereocilia. Proc Natl Acad Sci USA 86:4973–4977.

Shepherd GM, Corey DP, Block SM (1990) Actin cores of hair-cell stereocilia support myosin motility. Proc Natl Acad Sci USA 87:8627–8631.

Shotwell SL, Jacobs R, Hudspeth AJ (1981) Directional sensitivity of individual vertebrate hair cells to controlled deflection of their hair bundles. Ann NY Acad Sci 374:1–10.

Siemens J, Kazmierczak P, Reynolds A, Sticker M, Littlewood-Evans A, Muller U (2002) The Usher syndrome proteins cadherin 23 and harmonin form a complex by means of PDZ-domain interactions. Proc Natl Acad Sci USA 99:14946–14951.

Siemens J, Lillo C, Dumont RA, Reynolds A, Williams DS, Gillespie PG, Müller U (2004) Cadherin 23 is a component of the tip link in hair-cell stereocilia. Nature 428:950–954.

Slepecky NB, Chamberlain SC (1982a) Distribution and polarity of actin in the sensory hair cells of the chinchilla cochlea. Cell Tissue Res 224:15–24.

Slepecky NB, Chamberlain SC (1982b) Actin in cochlear hair cells-implications for stereocilia movement. Arch Otorhinolaryngol 234:131–134.

Slepecky NB, Chamberlain SC (1985a) The cell coat of inner ear sensory and supporting cells as demonstrated by ruthenium red. Hear Res 17:281–288.

Slepecky NB, Chamberlain SC (1985b) Immunoelectron microscopic and immunofluorescent localization of cytoskeletal and muscle-like contractile proteins in inner ear sensory hair cells. Hear Res 20:245–260.

Slepecky NB, Savage JE (1994) Expression of actin isoforms in the guinea pig organ of Corti: muscle isoforms are not detected. Hear Res 73:16–26.

Slepecky NB, Ulfendahl M (1992) Actin-binding and microtubule-associated proteins in the organ of Corti. Hear Res 57:201–215.

Slepecky NB, Ulfendahl M (1993) Evidence for calcium-binding proteins and calcium-dependent regulatory proteins in sensory cells of the organ of Corti. Hear Res 70:73–84

Slepecky NB, Hozza MJ, Cefaratti L (1990) Intracellular distribution of actin in cells of the organ of Corti: a structural basis for cell shape and motility. J Electron Microsc Tech 15:280–292.

Sobin A, Flock Å (1983) Immunohistochemical identification and localization of actin and fimbrin in vestibular hair cells in the normal guinea pig and in a strain of the waltzing guinea pig. Acta Otolaryngol 96:407–412.

Sobue K, Kanda K, Adachi J, Kakiuchi S (1983) Calmodulin-binding proteins that interact with actin filaments in a Ca^{2+}-dependent flip-flop manner: survey in brain and secretory tissues. Proc Natl Acad Sci USA 80:6868–6871.

Sollner C, Rauch G-J, Siemens J (2004) Mutations in cadherin 23 affect tip links in zebrafish sensory hair cells. Nature 428:955–959.

Stacey DJ, McLean WG (2000) Cytoskeletal protein mRNA expression in the chick utricle after treatment in vitro with aminoglycoside antibiotics: effects of insulin, iron chelators and cyclic nucleotides. Brain Res 871:319–332.

Steel KP, Kros CJ (2001) A genetic approach to understanding auditory function. Nat Genet 27:143–149.

Steyger PS, Furness DN, Hackney CM, Richardson GP (1989) Tubulin and microtubules in cochlear hair cells: comparative immunocytochemistry and ultrastructure. Hear Res 42:1–16.

Steyger PS, Gillespie PG, Baird RA (1998) Myosin Ibeta is located at tip link anchors in vestibular hair bundles. J Neurosci 18:4603–4615.

Street VA, McKee-Johnson JW, Fonseca RC, Tempel BL, Noben-Trauth K (1998) Mutations in a plasma membrane Ca^{2+}-ATPase gene cause deafness in *deafwaddler* mice. Nat Genet 19:390–394.

Tachibana M, Morioka H, Machino M, Amagai T, Mizukoshi O (1986) Aminoglycoside binding sites in the cochlea as revealed by neomycin-gold labelling. Histochemistry 85:301–304.

Takumida M, Wersäll J, Bägger-Sjöbäck D (1988) Stereociliary glycocalyx and interconnections in the guinea pig vestibular organs. Acta Otolaryngol 106:130–139.

Takumida M, Harada Y, Kanemia Y (1993) Influence of elastase and hyaluronidase on the ciliary interconnecting systems in frog vestibular sensory cells. ORL J Otorhinolaryngol Relat Spec 55:77–83.

Tang J, Taylor DW, Taylor KA (2001) The three-dimensional structure of alpha-actinin obtained by cryoelectron microscopy suggests a model for $Ca(^{2+})$-dependent actin binding. J Mol Biol 310:845–858.

Thalmann I, Shibasaki O, Comegys TH, Henzl MT, Senarita M, Thalmann R (1995) Detection of a beta-parvalbumin isoform in the mammalian inner ear. Biochem Biophys Res Commun 215:142–147.

Thorne PR, Carlisle L, Zajic G, Schacht J, Altschuler RA (1987) Differences in the distribution of F-actin in outer hair cells along the organ of Corti. Hear Res 30:253–265.

Thurm U (1981) Mechano-electric transduction. Biophys Struct Mech 7:245–246.

Tilney LG, DeRosier DJ (1986) Actin filaments, stereocilia, and hair cells of the bird cochlea. IV. How the actin filaments become organized in developing stereocilia and in the cuticular plate. Dev Biol 116:119–129.

Tilney LG, Saunders JC (1983) Actin filaments, stereocilia, and hair cells of the bird cochlea. I. Length, number, width, and distribution of stereocilia of each hair cell are related to the position of the hair cell on the cochlea. J Cell Biol 96:807–821.

Tilney LG, Tilney MS (1988) The actin filament content of hair cells of the bird cochlea is nearly constant even though the length, width, and number of stereocilia vary depending on the hair cell location. J Cell Biol 107:2563–2574.

Tilney LG, Derosier DJ, Mulroy MJ (1980) The organization of actin filaments in the stereocilia of cochlear hair cells. J Cell Biol 86:244–259.

Tilney LG, Egelman EH, DeRosier DJ, Saunder JC (1983) Actin filaments, stereocilia, and hair cells of the bird cochlea. II. Packing of actin filaments in the stereocilia and in the cuticular plate and what happens to the organization when the stereocilia are bent. J Cell Biol 96:822–834.

Tilney MS, Tilney LG, Stephens RE, Merte C, Drenckhahn D, Cotanche DA, Bretscher A (1989) Preliminary biochemical characterization of the stereocilia and cuticular plate of hair cells of the chick cochlea. J Cell Biol 109:1711–1723.

Tousson A, Alley CD, Sorscher EJ, Brinkley BR, Benos DJ (1989) Immunochemical localization of amiloride-sensitive sodium channels in sodium-transporting epithelia. J Cell Sci 93:349–362.

Tsuprun V, Santi P (1998) Structure of outer hair cell stereocilia links in the chinchilla. J Neurocytol 27:517–528.

Tsuprun V, Santi P (2000) Helical structure of hair cell stereocilia tip links in the chinchilla cochlea. J Assoc Res Otolaryngol 1:224–231.

Tsuprun V, Santi P (2002) Structure of outer hair cell stereocilia side and attachment links in the chinchilla cochlea. J Histochem Cytochem 50:493–502.

Tucker TR, Fettiplace R (1996) Monitoring calcium in turtle hair cells with a calcium-activated potassium channel. J Physiol 494:613–626.

Unwin N (1989) The structure of ion channels in the membrane of excitable cells. Neuron 3:665–676.

Unwin N (2003) Structure and action of the nicotinic acetylcholine receptor explored by electron microscopy. FEBS Lett 555:91–95.

Velichkova M, Guttman J, Warren C, Eng L, Kline K, Vogl AW, Hasson T (2002) A human homologue of *Drosophila* kelch associates with myosin-VIIa in specialized adhesion junctions. Cell Motil Cytoskel 51:147–164.

Wagner O, Zinke J, Dancker P, Grill W, Bereiter-Hahn J (1999) Viscoelastic properties of f-actin, microtubules, f-actin/alpha-actinin, and f-actin/hexokinase ddetermined in microliter volumes with a novel non-destructive method. Biophys J 76:2784–2796.

Walker RG, Hudspeth AJ (1996) Calmodulin controls adaptation of mechanoelectrical transduction by hair cells of the bullfrog's sacculus. Proc Natl Acad Sci USA 93:2203–2207.

Walker RG, Hudspeth AJ, Gillespie PG (1993) Calmodulin and calmodulin-binding proteins in hair bundles. Proc Natl Acad Sci USA 90:2807–2811.

Walsh T, Walsh V, Vreugde S, Hertzano R, Shahin H, Haika S, Lee MK, Kanaan M, King M-C, Avraham KB (2002) From flies' eyes to our ears: mutations in a human class III myosin cause progressive on-syndromic hearing loss DFNB30. Proc Natl Acad Sci USA 99:7518–7523.

Warchol ME (2001) Lectin from *Griffonia simplicifolia* identifies an immature-appearing subpopulation of sensory hair cells in the avian utricle. J Neurocytol 30:253–264.

Wu X, Jung G, Hammer JA 3rd (2000) Functions of unconventional myosins. Curr Opin Cell Biol 12:42–51.

Wu YC, Ricci AJ, Fettiplace R (1999) Two components of transducer adaptation in auditory hair cells. J Neurophysiol 82:2171–2181.

Yamoah EN, Lumpkin EA, Dumont RA, Smith PJ, Hudspeth AJ, Gillespie PG (1998) Plasma membrane Ca^{2+}-ATPase extrudes Ca^{2+} from hair cell stereocilia. J Neurosci 18:610–624.

Yang D, Thalmann I, Thalmann R, Simmons DD (2004) Expression of alpha and beta parvalbumin is differentially regulated in the rat organ of Corti during development. J Neurobiol 58:479–492.

Zhao Y, Yamoah EN, Gillespie PG (1996) Regeneration of broken tip links and restoration of mechanical transduction in hair cells. Proc Natl Acad Sci USA 93:15469–15474.

Zheng L, Sekerkova G, Vranich K, Tilney LG, Mugnaini E, Bartles JR (2000) The deaf *jerker* mouse has a mutation in the gene encoding the espin actin-bundling proteins of hair cell stereocilia and lacks espins. Cell 102:377–385.

Zine EA, Romand R (1993) Expression of alpha-actinin in the stereocilia of hair cells of the inner ear: immunohistochemical localization. NeuroReport 4:1350–1352.

Zine A, Hafidi A, Romand R (1995) Fimbrin expression in the developing rat cochlea. Hear Res 87:165–169.

4
Mechanoelectrical Transduction in Auditory Hair Cells

ROBERT FETTIPLACE AND ANTHONY J. RICCI

1. Introduction

The detection of a sound stimulus and its conversion into an electrical signal by hair cells of the vertebrate inner ear is a special case of mechanoreception, which is arguably a most ancient and ubiquitous sensation. The mechanical sensitivity of the cell membrane underlies such fundamental processes as cell volume regulation and detection of osmotic stress, touch sensation that pervades the animal and plant kingdoms, and the signaling of muscle stretch and joint position indispensable for coordinated motion in higher animals. But compared with these disparate forms of mechanoreception, transduction in the inner ear is exceptional both in terms of its speed and sensitivity. Auditory hair cells can detect vibrations of atomic dimension on the order of 0.2 nm, and yet respond in mammals like bats up to 100,000 times a second, thus evading the frequency limits imposed on other nerve cells. How they achieve this feat is still not fully understood, but it probably requires local amplification of the force stimulus that is directed at a mechanically gated ion channel with ultrafast activation kinetics. The possibility that all forms of mechanoreception share a common molecular basis has not been substantiated, and over the past decade quite different types of ion channel believed to serve as mechanoreceptors have been cloned from *Escherichia coli* (Sukahrev et al. 1993), *Caenorhabditis elegans* (Huang and Chalfie 1994), and *Drosophila* (Walker et al. 2000). In both the worm and insect, channel identification was aided by the availability of mechanically insensitive mutants. A similar approach has recently indicated that a channel of the transient receptor potential (TRP) family orthologous to a *Drosophila* mechanoreceptor may underlie hair cell transduction in zebra fish, *Danio rerio* (Sidi et al. 2003). Although it is still unclear whether TRP channels mediate transduction in the inner ears of other vertebrates, especially mammals, there are many observations about the in situ properties of the channel that provide clues to its molecular nature and classification. This chapter summarizes such properties, including the conductance, ionic permeation, and blockade of single channels, and their control by Ca^{2+}. The ability of the mechanotransducer channel to generate force during gating, which has attracted attention as a potential stimulus amplification mechanism (Hudspeth 1997; Fettiplace et al. 2001; Manley 2001), is also addressed. A recurring theme in discussing auditory hair cells is the variation in cellular properties with the sound frequency to which the cell is tuned. Here tonotopic gradients in both single-channel and macroscopic aspects of mechanotransduction are described.

2. Overview of Hair Cell Transduction

2.1 Hair Bundles and Tip Links

Hair cells of the inner ear are tightly anchored to nonsensory supporting cells in an epithelial sheet with their mechanically sensitive appendage, the hair bundle, projecting from the surface of the epithelium. Forces transmitted to the sensory epithelium result in submicron deflections of the hair bundles that open mechanoelectrical transducer (MET) channels, causing influx of small cations and depolarization of the hair cell. The hair bundle is composed of a few tens to a few hundred modified microvilli, stereocilia, stacked in rows of increasing height. Abutting the tallest stereociliary row in many hair bundles (but not those in the adult mammalian cochlea) is a kinocilium of uncertain function. The number and maximum height of the stereocilia vary with hair cell location; for example, in a wide array of species, auditory hair cells tuned to higher frequencies have hair bundles that are shorter and often contain more stereocilia (Mulroy 1974; Tilney and Saunders 1983; Lim 1986; Hackney et al. 1993). The stereocilia are interconnected by various types of extracellular filamentous links (Bagger-Sjöback and Wersäll 1973; Little and Neugebauer 1985; Goodyear and Richardson 1999). Lateral links connect adjacent stereocilia along their entire length (Flock et al. 1977), while tip links extend from the vertex of each stereocilium to the side of its taller neighbor (Pickles et al. 1984; Furness and Hackney 1985; Kachar et al. 2000). Together the links constitute an extracellular matrix coupling the stereocilia, so that when force is delivered at the tip of the bundle, all stereocilia rotate in unison (Flock and Strelioff 1984; see Fig. 3 of Crawford et al. 1989).

The standard model for hair cell transduction is that deflection of the stereocilia exerts tension on the tip links, which transmits force directly to the MET channels (Pickles et al. 1984; Hudspeth 1985). Since the tip links run approximately parallel to the axis of symmetry of the bundle, they are properly oriented to apply force during rotations of the bundle toward its taller edge, and to unload with deflections toward its shorter edge. This explains the functional polarization of transduction (Pickles et al. 1984) whereby only those movements of the bundle along its axis of symmetry, toward or away from the tallest stereociliary rank, are detected (Shotwell et al. 1981). The clearest evidence for the role of tip links is that lowering extracellular Ca^{2+} (usually with a calcium chelator such as BAPTA) irreversibly abolishes transduction (Crawford et al. 1991) and destroys these links (Assad et al. 1991). Furthermore, subsequent regeneration of the tip links after exposure to BAPTA is correlated with restoration of mechanotransduction (Zhao et al. 1996). However, low-calcium BAPTA treatment also destroys the lateral links as well as the tip links and causes substantial rotation of the bundle toward its taller edge (Hackney and Furness 1995). Even assuming the tip links are involved in force transmission, their exact relationship to the channels is still debatable, and whether they have a direct molecular connection to the channels or merely insert into the plasma membrane around the channels is unknown.

Various physiological methods have been used to localize the MET channels in the hair bundle. Evidence derived from mapping the flow of transducer current during excitatory stimuli (Hudspeth 1982), or from blocking transduction by local iontophoresis of dihydrostreptomycin (Jaramillo and Hudspeth 1991), has placed the channels at or near the top of the bundle. Finer spatial resolution was achieved by exploiting the high Ca^{2+} permeability of the channels and by using confocal microscopy to trace the site at which Ca^{2+} entered the stereocilium (Denk et al. 1995; Lumpkin and Hudspeth 1995). These experiments, demonstrating the first appearance of Ca^{2+} fluorescence at the tips of the stereocilia, are remarkable given the small stereociliary diameter (0.1 to 0.4 µm), but light microscopic observations have inadequate resolution to demonstrate that the links are directly connected to the channels. Based on the presence of Ca^{2+} signals in both the shortest and tallest stereociliary rows, Denk et al. (1995) argued that channels occur at both ends of the tip link. However, such an arrangement implies a negative cooperativity between pairs of channels at the two ends of the link (the opening of one of the pair relieving force on the other) for which there is no experimental support (Martin et al. 2000). It is worth noting that many cartoons of the transduction apparatus locate the MET channel solely at the upper end of the tip link and thus free to slip down the side of the stereocilium during adaptation. It is argued that the myosin motor underlying slow adaptation is located in the osmiophilic plaque at the upper end of the tip link in order to tension the link, and that the MET channel must be in close proximity to enable Ca^{2+} influx to control the motor (see Section 4.1). There is no firm evidence for this arrangement.

An alternative site for the MET channel is in the contact region just below the stereociliary tips where short lateral connections can be seen between the membranes of neighboring stereocilia (Hackney et al. 1992). Placement there is supported by immunogold labeling with an antibody raised against the binding site for amiloride that blocks the transducer channels (Jørgensen and Ohmori 1988; Rüsch et al. 1994). One interpretation of these results is that the tip links are not directly connected to the channels, an idea endorsed by the surprising observation that the MET channels can still open after tip link destruction with either BAPTA or elastase (Meyer et al. 1998). This latter work was done on mammalian outer hair cells, and could not be repeated in similar experiments on turtle hair cells (Ricci et al. 2003). Whether restriction of the channel to the contact region (Hackney et al. 1992) is correct or not, the work emphasizes the subcellular resolution afforded by electron microscopy that is essential for precisely localizing the channel. The full benefit of this technique could be realized with the availability of a high-affinity channel-blocking agent whose binding could be revealed in electron micrographs.

2.2 Mechanoelectrical Transducer Currents

The properties of transduction, its high sensitivity and fast kinetics, have been studied using in vitro preparations, in which the electrical response of the cells can be monitored while directly manipulating the hair bundle. Original evidence came from hair cells of animals such as frogs (Hudspeth and Corey 1977; Holton and Hudspeth 1986), chickens (Ohmori 1985), and turtles (Crawford and Fettiplace 1985), but the conclusions also extend to mammalian hair cells (Kros et al. 1992; Géléoc et al. 1997; Holt et al. 1997; Vollrath and Eatock 2003; Kennedy et al. 2003). Rotation of the bundle toward its taller edge opens mechanosensitive ion channels, which at the resting potential, −50 to −70 mV, rapidly activates an inward transducer current that peaks and then declines to a steady level in multiple processes of adaptation (Fig. 4.1). The mechanisms underlying

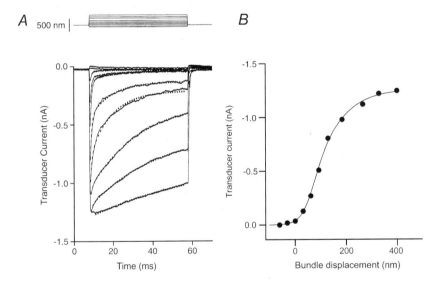

FIGURE 4.1. (**A**) Family of mechanotransducer currents recorded from a voltage-clamped turtle hair cell in response to step displacements of the hair bundle. Positive deflections toward the tallest stereociliary row evoke an inward transducer current at −84 mV holding potential that rapidly reaches a peak and then declines in multiple phases of adaptation. The decline in the current has been fitted with the sum of two exponential processes (*dashed lines*) representing fast and slow adaptation. Fast adaptation, most prominent for small stimuli, has a time constant of 0.7 ms; slow adaptation appears with larger stimuli and has a time constant of 10 to 70 ms. Extracellular Ca^{2+}, 2.8 mM. (**B**) Relationship between the peak current, I, and bundle displacement, X, measured from the records in (**A**). Smooth curve is Eq. (4.2): $I = I_{max}/[1 + \{\exp(z_0 \cdot (X_0 - X))\}\{1 + \exp(z_1 \cdot (X_1 - X))\}]$, where $I_{max} = 1.26$ nA, $z_0 = 12.5$ µm^{-1}, $z_1 = 31$ µm^{-1}, and $X_0 = X_1 = 0.177$ µm.

adaptation are discussed in Section 4.1. In the intact ear, where the bundles are bathed in endolymph, the inward current is carried mainly by K^+, with a contribution (up to 10%) from Ca^{2+} (Ricci and Fettiplace 1998). Rotation of the bundle toward its shorter edge closes that small fraction of channels open at rest, which varies from 2% to 30% depending on conditions (e.g., concentration of extracellular Ca^{2+} or cytoplasmic calcium buffer; Ricci et al. 1998). Even at room temperature, the current develops with a sub-millisecond time course signifying fast gating kinetics for the channel. The minimal delay (< 20 μs) between the bundle deflection and channel opening argues that external force directly gates the MET channels without any of the biochemical amplification seen in other sensory receptors (Corey and Hudspeth 1983).

As might be expected if the mechanical stimulus directly affects the channel's opening rate, fast bundle deflections (rise time of 50 μs or less) elicit a transducer current with an onset time constant that depends on stimulus amplitude (Corey and Hudspeth 1983; Crawford et al. 1989). Information about channel kinetics was originally derived from measuring extracellular "microphonic" currents summed across the voltage-clamped frog saccular epithelium (Corey and Hudspeth 1983), but subsequent experiments on individual voltage-clamped turtle hair cells gave similar results (Crawford et al. 1989; Fettiplace et al. 2003).

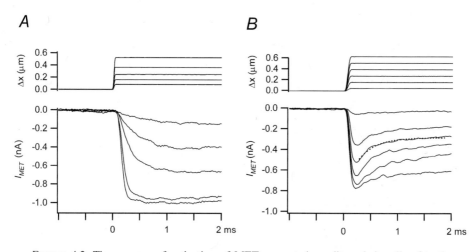

FIGURE 4.2. Time course of activation of MET currents in auditory hair cells of turtle (A) and rat (B). Rapid deflection (Δx) of the hair bundle with a piezoelectric stimulator connected to a rigid glass fiber evokes fast-onset MET currents. In the turtle (A), the activation time constant decreases (from approximately 400 μs to 50 μs) with increasing stimulus amplitude. In the rat outer hair cell, the time course of current activation is faster than in the turtle and is limited by the stimulus onset, time constant approximately 50 μs. The mammalian responses also show adaptation at a faster rate than in the turtle, *dotted line* denoting fit with a time constant of 300 μs. In both experiments, the fluid bathing the hair bundle contained 0.05 mM Ca^{2+}; holding potential, -80 mV at room temperature (approximately 22°C).

The activation time constant measured in turtle hair cells decreased with the size of bundle displacement (Fig. 4.2) from approximately 400 μs for small stimuli to less than 50 μs for saturating ones at room temperature (approximately 22°C). These time constants also depended on the local Ca^{2+} concentration and were lengthened by lowering Ca^{2+} (Fettiplace et al. 2003). Values from single cell measurements were twofold slower than those obtained from microphonics when adjusted to the same conditions. Even so, it is unclear that either is sufficiently fast to account for the speed of transduction in mammalian outer hair cells, where cochlear microphonics indicate that channel gating occurs on a cycle-by-cycle basis at frequencies of 50 kHz or more (Pollak et al. 1972). Thus extrapolating existing kinetic measurements to 37°C using a Q_{10} of 2.0 (Corey and Hudspeth 1983; Crawford et al. 1989) yields an activation time constant of 60 to 100 μs for small stimuli and 10 to 20 μs for large stimuli. To generate a corner frequency of 50 kHz, the required activation time constant is $(2\pi \cdot 50)^{-1}$ ms or 3 μs, substantially briefer than inferred from kinetic measurements on frog or turtle hair cells. It is conceivable that there are variations in kinetics originating from small differences in channel structure or bundle mechanics in different species, and that mammalian cochlear hair cells are optimized to achieve the rapid transduction required for high-frequency hearing (Fig. 4.2; Kennedy et al. 2003).

2.3 Kinetic Schemes for the Mechanoelectrical Transducer Channel

A basic assumption of models for hair cell transduction is that force is transmitted to the channel via elastic elements, the "gating springs," which are stretched when the bundle is displaced toward its taller edge (Howard and Hudspeth 1988). The gating springs are usually identified with the tip links. Their location would admirably suit them for this role, because the tension in them would rise with positive deflections and fall with negative deflections. However, there is no evidence that the tip links are significantly extensible: high-resolution electron microscopy reveals a coiled double-filamentous structure, suggesting a protein structure too stiff for them to act as gating springs (Kachar et al. 2000). The proposal that cadherin-23 is a major component of the tip links (Siemens et al. 2004; Söllner et al. 2004) accords with the notion that they are not very elastic. An alternative site for the series compliance of the gating springs is in the channel or its intracellular attachments to the cytoskeleton (Fig. 4.3).

In the simplest form of the model, the MET channel occupies two states, closed and open (C ↔ O), with the transition rates modulated by stimulus energy. The open probability of the channel (p_O) is then a sigmoidal function of bundle displacement (X):

$$p_o(X) = [1 + \exp(-z \cdot (X - X_0) / k_B T)]^{-1} \quad (4.1)$$

where k_B is the Boltzmann constant, T is absolute temperature, X_0 is the displacement for half-activation, and z is the single channel gating force applied at

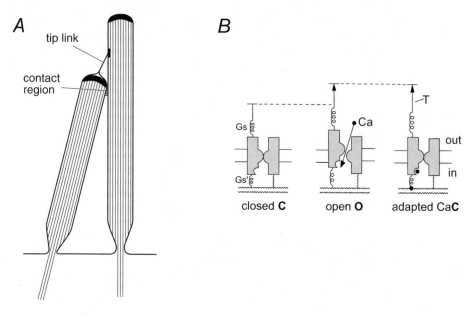

FIGURE 4.3. (**A**) Each stereocilium is packed with actin filaments that insert into the osmiophilic region at its apex which may be the location of the MET channels as well as several myosin isoforms. The tip link stretches from the apex of one stereocilium to the side wall of the next, inserting into a plaque that has been proposed to contain myosin IC. It is suggested that the tip link is tensioned by myosin IC climbing the actin backbone. Also shown is the contact region which has been suggested as an alternative MET channel location. (**B**) Schematic of MET channel gating. Hair bundle displacement tensions the tip link (T), which acts via gating springs that deliver force to the channel to increase its probability of being open. The sites of the gating springs are unknown, but they are depicted as connecting to the channel both externally (Gs) and internally (Gs'). Ca^{2+} entering the stereocilium through the open channel may promote channel closure and evoke fast adaptation either by binding at the inner face of the channel or by altering the stiffness of the internal gating spring Gs'. Ca^{2+} may also produce slow adaptation by causing slippage of the myosin IC motor to slackening the tip link.

the top of the bundle. Equation (4.1) has a symmetrical form that often inadequately describes experimental measurements (Fig. 4.1) which can display a sharp corner at low p_O and a more gradual asymptote to saturation at high p_O. The asymmetry is accommodated by postulating a three-state channel, two closed states and one open state ($C_1 \leftrightarrow C_2 \leftrightarrow O$). The open probability of the channel (p_O) is then given by:

$$P_O(X) = [1 + \{\exp(-z_0 \cdot (X - X_0) / k_B T)\} \cdot \{1 + \exp(-z_1 \cdot (X - X_1) / k_B T)\}]^{-1} \quad (4.2)$$

where z_0, X_0, z_1 and X_1 are constants the meanings of which depend on the specific model (Markin and Hudspeth 1995; van Netten and Kros 2000). Such a three-state channel scheme has been used to fit hair cell current–displacement (*I–X*) relationships in various preparations (Corey and Hudspeth 1983; Holton and Hudspeth 1986; Crawford et al. 1989; Kros et al. 1992; Géléoc et al. 1997). To accommodate the Ca^{2+} dependence of the kinetics, more complex gating schemes in which both closed and open states can also bind Ca^{2+} ions may be required.

2.4 Sensitivity and Working Range of Transduction

The operating range of the MET channels can be specified in terms of the width of the *I–X* relationship, the differences in bundle position from where the transducer current starts activating to where it is fully saturated. Because of the curvature of the relationship, the width can be quantified by plotting the current scaled to its maximum value (I/I_{max}) against bundle deflection, determining the tangent line of greatest slope and calculating its inverse value. The operating range varies from approximately 50 to 250 nm for hair cells of the turtle cochlea (Crawford and Fettiplace 1985; Crawford et al. 1989), mouse cochlea (Géléoc et al. 1997), and frog saccule (Howard and Hudspeth 1988; Assad et al. 1989). These values represent motion at the tip of the hair bundle and to obtain the displacements imposed on the channel they must be scaled by a geometrical factor (γ) that depends on bundle geometry (Howard and Hudspeth 1988; Geisler 1993). This factor is given by $\gamma \approx s/h$ where s is the separation between the rootlets of successive stereocilia along the excitatory axis of the hair bundle and h is the average stereociliary height. For turtle auditory hair cells, with $s \approx 0.4$ μm and an average h of 4 μm, $\gamma \approx 0.1$. Since γ is inversely proportional to stereociliary height, a change in bundle dimensions provides a means of altering the sensitivity of transduction, matching it to the needs of the hair cell. The hair bundle can thus be regarded as an accessory structure for converting displacements occurring during inner ear stimulation to the molecular scale appropriate for deforming the transducer channel protein. In the mammalian cochlea a fivefold decrease in bundle height from the the low- to the the high-frequency end of the cochlea will produce an equivalent increase in γ, and hence sensitivity.

Because the *I–X* relationship is asymmetric, it has the steepest slope, and hence maximal sensitivity, around the bundle's resting position near the foot of the curve. The largest transducer current sensitivities of approximately 10 pA · nm^{-1} have been measured from auditory hair cells in turtle (Fig. 4.1; Ricci and Fettiplace 1997) and mouse (Géléoc et al. 1997; van Netten et al. 2003). In both preparations, the stimuli were delivered to bundles of comparable height (6 and 4 μm, respectively) and at about the same holding potential, −80 mV. Since hair cells signal by changes in membrane potential, a more pertinent measure is the linear voltage sensitivity, which, for a cell with 100 MΩ input resistance, is 1.0 mV · nm^{-1}. A significant receptor potential can therefore be generated by displacements less than 1 nm. At auditory threshold, turtle auditory hair cells can transmit information about bundle displacements of approximately 0.2 nm corresponding to 0.2 mV of receptor potential (Crawford and Fettiplace 1980, 1983). However, reliable signaling may be limited by the signal-to-noise ratio. Sources of noise include the stochastic chattering of the MET channels, with large unitary current of 10 pA (see Section 3.2), Brownian motion in the gating springs and fluctuations in the adaptation motor. Each of these is equivalent to a noise stimulus to the hair bundle of several nanometers (Frank et al. 2002; van Netten et al. 2003). Such values imply a signal-to-noise ratio less than 1 during hair cell transduction at threshold. One mechanism for extracting the signal is for the hair cell to respond only over a limited frequency range, band-limiting the noise and hence improving the signal-to-noise ratio. However, the tuning process must be an active rather than a passive one, since a passive filter would compress all the noise power into the pass-band and amplify both signal and noise alike (Bialek 1987; Block 1992).

3. The Mechanoelectrical Transducer Channel

3.1 Ionic Selectivity and Block

The MET channel is a nonspecific cation conductance, showing limited voltage dependence or rectification (Ohmori 1985; Crawford et al. 1989, 1991). The channel discriminates little between monovalent cations but is highly permeable to divalent cations, especially Ca^{2+} (Corey and Hudspeth 1979; Ohmori 1985). The permeability ratios for monovalent cations determined from measuring reversal potentials for the transducer current are: Li^+, 1.14; Na^+, 1 ; K^+, 0.96 ; Rb^+, 0.92 ; Cs^+, 0.82 (Ohmori 1985). The permeability ratios for divalent cations relative to Na^+ are: Ca^{2+}, 3.8; Sr^{2+}, 2.3 ; Ba^{2+}, 2.2; Mg^{2+}, 2.0 ; Na^+, 1, and indicate only a modest preference for Ca^{2+} over monovalent cations (Ohmori 1985). However these measurements were also based on interpreting reversal potentials using the constant field equation, which assumes no interaction between different ionic species in the pore. This is unlikely to be the case given the blocking action of external Ca^{2+} on the flux of monovalent ions (Crawford et al. 1991; Ricci and Fettiplace 1998) and the anomalous mole fraction effects seen with ion mixtures (Lumpkin et al. 1997). A more pertinent measure of the channel's selectivity is obtained from the fraction of current carried by the different ions. This indicates an effective permeability ratio for Ca^{2+} over monovalent ions of more than 100: 1 (Jørgensen and Kroese 1995; Ricci and Fettiplace 1998). Thus, in saline with 2.8 mM Ca^{2+} and 150 mM Na^+, more than half the current will be carried by Ca^{2+} (Ricci and Fettiplace 1998), and even in an endolymph-like solution containing only 50 μM Ca^{2+}, Ca^{2+} carries at least a tenth of the current. The Ca^{2+} content of cochlear endolymph has been estimated in different preparations as between 20 and 65 μM (Bosher and Warren 1978; Salt et al. 1989; Crawford et al. 1991).

The high flux of Ca^{2+} through the channel when exposed to endolymph is important physiologically because of the ion's role in regulating transducer adaptation. Furthermore, though the channel is equally permeable to Na^+ and K^+, substitution of K^+ for Na^+ as the chief monovalent ion increases the fractional current carried by Ca^{2+} and augments its intracellular effect in adaptation (Ricci and Fettiplace 1998). Thus endolymph composition is specialized to allow a compromise between the largest monovalent current and the largest Ca^{2+} influx. In the mammalian cochlea, both monovalent and divalent currents are further augmented in vivo by the +80 mV endocochlear potential that increases the driving force through the MET channel.

A variety of larger organic cations can traverse the channel (Corey and Hudspeth 1979) with permeability ratios relative to Na^+ of choline, 0.27; tetramethylammonium (TMA), 0.16; tetraethylammonium (TEA), 0.14 (Ohmori 1985). These permeability values suggest that the channel has a minimum pore size of approximately 0.7 nm, similar to the diameter of the largest permeant ion, TEA. More recently, the MET channel pore diameter was estimated using methods first developed while investigating pore dimensions of the nicotinic acetylcholine channel (Dwyer et al. 1980). A series of alkyl ammonium compounds that increase in molecular dimensions were substituted for Na^+ as the monovalent ion and the corresponding change in MET current measured (Farris et al. 2004). The plot of current ratio against the size of the substituted ammonium indicated that the pore was approximately 1.26 nm, considerably larger than that of other nonselective cation channels including the nicotinic acetylcholine receptor channel (Dwyer et al. 1980) and olfactory cyclic nucleotide gated channels (Balasubramanian et al. 1995). However, the hair cell MET channel is much smaller than the 4 nm measured for the bacterial mechanically gated (MscL) channel (Cruickshank et al. 1997).

The cationic styryl dye FM1-43 (MW = 451) can also pass through the channel and its fluorescence on binding to intracellular membranes provides a sensitive monitor of rapid intracellular accumulation (Gale et al. 2001; Meyers et al. 2003). The ability of such a large molecule to pass through the channel has been explained by its elongated structure with cross-section, 0.78 nm × 0.5 nm, similar to TEA. FM1-43 acts as a permeant blocker of the MET channel, as do Ca^{2+} and Mg^{2+}, which block the channel in millimolar concentrations (Crawford et al. 1991; Ricci and Fettiplace 1998). The position of the Ca^{2+} binding site responsible for the blocking effect was estimated by fitting a single-site binding model to the current–voltage relationships generated in the presence and absence of the blocker. A distance of approximately 0.5 through the membrane electric field was obtained (Kros et al. 1992; Gale et al. 2001; Farris et al. 2004). Similar distances were found for FM1-43 dye and a variety of other antagonists, which implies an interaction between the Ca^{2+} binding site and these compounds (Gale et al. 2001; Farris et al. 2004).

Familiar blocking agents of hair cell transduction, such as the aminoglycoside antibiotic dihydrostreptomycin (DHS), are also polycations. These antagonists share a number of properties that indicate they may compete with Ca^{2+} for a binding site in external mouth of the pore. They are effective only from the extracellular surface and their block is voltage dependent and diminished by depolarization that opposes the entry of the cation into the pore (Kroese et al. 1989). Furthermore the block is subject to competition from extracellular Ca^{2+}:

raising the Ca^{2+} concentration alleviates channel block with FM1-43 (Gale et al. 2001) and increases the half-blocking concentration for DHS (Table 4.1; Kroese et al. 1989; Ricci 2002). An alternative explanation for the calcium and voltage effects on DHS block is that these manipulations also alter the probability of opening of the MET channel at rest, thereby indirectly affecting the half-blocking concentration for open channel blockers (Ricci 2002). On this hypothesis, an increased probability of opening will allow more time for the blocker to enter the channel and therefore increase its potency. A similar explanation may account for the dependence of DHS block on hair cell location within the turtle cochlea with a half-blocking concentration of 14 μM at the low-frequency end and 75 μM at the high-frequency end of the papilla respectively (Ricci 2002). This difference parallels a two- to threefold difference in resting open probability between hair cells at the low-frequency and high-frequency locations (Ricci et al. 1998). A related phenomenon is the decrease in aminoglycoside-induced toxicity in mice where myosin VIIA was knocked out (Richardson et al. 1997). In these mice, the activation curve for the MET channel is profoundly shifted to the right, reducing the probability of opening at rest to near zero and thus limiting the effectiveness of a permeant blocker (Kros et al. 2002).

Based on the observation that FM1-43 not only blocks the channel but also (probably in small amounts) permeates it (Nishikawa and Sasaki 1996), similar behavior may apply to the aminoglycoside antibiotics, which may gain access to the hair cell cytoplasm via the MET channel. There they have a long-term deleterious effect on energy supply from the mitochondria (Cortopassi and

TABLE 4.1. Some blocking agents of the hair cell's mechanotransducer channel.

Blocking agent	K_I (μM)	n_H	Reference
Ca^{2+}	1000	1	Ricci and Fettiplace 1998
Gd^{3+}	10	1.1	Kimitsuki et al. 1996
La^{3+}	3.8	1.4	Farris et al. 2004
Tetracaine	579	2.7	Farris et al. 2004
Diltiazem	228	1.8	Farris et al. 2004
D600	110	1	Baumann and Roth 1986
			Farris et al. 2004
Amiloride	50		Jørgensen and Ohmori 1988
Amiloride	53	1.7	Rüsch et al. 1994
Amiloride	25	2.2	Ricci 2002
Dihydrostreptomycin	44 (Ca = 5 mM)		Kroese et al. 1989
Dihydrostreptomycin	23 (Ca = 2.5 mM)	0.9	Kimitsuki and Ohmori 1993
Dihydrostreptomycin	8 (Ca = 0.25 mM)	1	Kroese et al. 1989
Benzamil	6	1.6	Rüsch et al. 1994
Ruthenium Red	3.6	1.4	Farris et al. 2004
Curare	2.3	1.0	Glowatzki et al. 1997
	6.3	2.1	Farris et al. 2004
FM1-43	2.4	(1.2)	Gale et al. 2001

K_I is the half-blocking concentration and n_H is the Hill coefficient.

Hutchin 1994). Interestingly, hair cell destruction by aminoglycoside antibiotics proceeds more rapidly at the high-frequency end (Forge and Schacht 2000), which is similar to the cochlear gradient in intracellular accumulation of FM1-43 (Gale et al. 2001). One explanation for this effect is that cells tuned to higher frequencies have more MET channels (Ricci and Fettiplace 1997) or larger single channel conductances (Ricci et al. 2003). This may also account for the variation in DHS block with cochlear location.

3.2 Single-Channel Conductance

A basic property of the transducer channel that may provide a clue to its molecular identity is single-channel conductance but there have been few systematic studies of this attribute. It has not been possible to measure by the usual method of recording channels in cell-attached or detached membrane patches most likely because of the small diameter of the stereocilia. An alternative approach is to inactivate the majority of channels in a hair bundle and to monitor the one or two channels remaining by whole-cell recording (Crawford et al. 1991; Ricci et al. 2003). The number of channels can be reduced by brief exposure to submicromolar concentrations of Ca^{2+} (buffered with EGTA or BAPTA) which probably severs most of the tip links (Assad et al. 1991; Hackney and Furness 1995). When applied to turtle hair cells this method gave single channels with surprisingly large unitary conductances of 80 to 160 pS (mean = 118 pS) in high 2.8 mM Ca^{2+} (Crawford et al. 1991; Ricci et al. 2003). Similar values have also been obtained by whole-cell recordings in other species: 50 to 100 pS in hair cells of chick (Ohmori 1985) and 112 pS in mouse (Géléoc et al. 1997). Without observing single channel events, a unitary conductance of 87 pS was inferred in frog hair cells from measurement of the macroscopic current and the number of active stereocilia (Denk et al. 1995). This inference assumed that each active stereocilium possessed two channels.

The technique of inactivating most of the channels in the hair bundle by exposure to submicromolar Ca^{2+} allowed a systematic study of the channel remaining (Ricci et al. 2003). The single channel behaved similarly to the macroscopic transduction current, suggesting it was not affected by the isolation procedure. Thus it was gated by submicron displacements of the hair bundle, displayed fast activation and adaptation, and was blocked by 0.2 mM DHS (Ricci et al. 2003). Furthermore, changing external Ca^{2+} had two actions on channel properties mirroring its effects on the macroscopic current (Fig. 4.4). Reducing Ca^{2+} slowed the time course of channel activation and adaptation and also increased the channel amplitude. When the Ca^{2+} concentration was by lowered from 2.8 mM to 50 µM, the range of single-channel conductances nearly doubled to 150 to 300 pS (Fig. 4.4). Since the half-blocking concentration by Ca^{2+} is approximately 1 mM (Table 4.1; Ricci and Fettiplace 1998) measurements in 50 µM Ca^{2+} should reflect the maximum conductance of the channel in its unblocked state.

In the vertebrate cochlea, hair cells are arranged tonotopically such that their

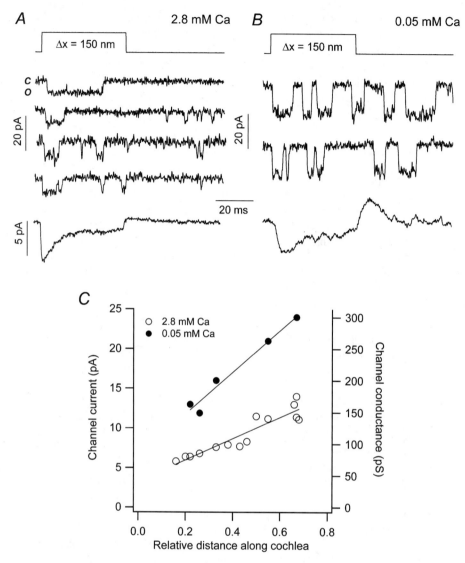

FIGURE 4.4. Single MET channels in turtle hair cells. (**A**) Four single-channel responses for hair bundle deflections (Δx) of 150 nm in 2.8 mM external Ca^{2+}. *Bottom* trace is the ensemble average of 140 individual responses. ***C*** and ***O*** denote the closed and open levels of the channel. (**B**) Two single-channel responses in the same cell in 0.05 mM external Ca^{2+}; *bottom* trace shows the ensemble average of 250 channel responses. Reducing extracellular Ca^{2+} had two effects increasing the mean single-channel amplitude from 7 pA to 15 pA and slowing the time course of channel activation and adaptation. (**C**) Plot of the single-channel current and conductance for hair cells at different fractional distances along the turtle cochlea from the low-frequency end. The holding potential was −80 mV.

characteristic frequency (CF), the sound frequency to which they are most sensitive, changes systematically along the cochlear length. The variation in single-channel conductance, either in high or low Ca^{2+}, was correlated with the CF of the hair cell: cells with higher CFs possessed channels with larger conductance (Fig. 4.4; Ricci et al. 2003). The change in channel conductance with CF in the turtle cochlea is unlikely to result from differences in Ca^{2+} homeostasis within the stereocilia because the relative increase in conductance was virtually identical in both high and low extracellular Ca^{2+} (Ricci et al. 2003). Changes in MET channel conductance are most simply explained if hair cells in different regions of the cochlea contain channels with unique structure or subunit composition. Further support for this notion comes from the finding that MET channels in hair cells with higher CFs also differ functionally in having faster activation kinetics (Ricci 2002; Fettiplace et al. 2003). The tonotopic distribution of MET channel properties is unexpected, but it parallels gradients in the properties of the Ca^{2+} activated K^+ channel that underlie changes in the electrical resonant frequency in the turtle cochlea (Art et al. 1995). Such variation may reflect differential expression of alternatively spliced isoforms of the Ca^{2+} activated K^+ channel α-subunit as well as a cochlear gradient in an accessory β-subunit (Jones et al. 1999; Ramanathan et al. 1999). The consequences of differences in single-channel conductance for tonotopic variation in mechanotransduction are further examined in Section 5.1.

3.3 Molecular Identity of the Mechanoelectrical Transducer Channel

The paucity of starting material, several thousand hair cells per animal and only a few hundred MET channels per cell, has hampered isolation of the channel by either biochemical purification or molecular cloning. Nevertheless, there are several distinctive features of the channel gleaned from measurements on intact hair cells that may aid in its molecular classification. These include a high selectivity for Ca^{2+} over other cations, a broad spectrum of blocking agents, large unitary conductance, and regulation by intracellular Ca^{2+}. These attributes eliminate several channel contenders (Strassmaier and Gillespie 2002). One is the epithelial Na^+ channel (ENaC) that has subunits orthologous to the MEC-4 and MEC-10 proteins that form a mechanoreceptor channel in touch neurons of the nematode worm *C. elegans* (Huang and Chalfie 1994; Goodman et al. 2002). Like the hair cell MET channel, ENaC is blocked by amiloride. However, known forms of ENaC are Na^+-selective channels with low Ca^{2+} permeability and small unitary conductance (13 to 40 pS; Ismailov et al. 1996). Moreover, the characteristics of their block by amiloride block are substantially different, with a 100-fold greater affinity for the drug (half-blocking concentration = approximately 0.5 μ*M*) and a Hill coefficient of 1 (Sariban-Sohraby and Benos 1986). Although subunits of the ENaC channel do occur in the cochlea, they have been localized to the epithelial cells lining the scala media, and may be involved in regulating the Na^+ content of the cochlear fluids (Grunder et al. 2001).

In its ionic selectivity and pharmacological profile, the MET channel shows some similarity to the cyclic-nucleotide-gated (CNG) channel that underlies transduction in the visual and olfactory systems. Both are nonselective cation channels with a high permeability to Ca^{2+}, show little voltage dependence, and are also blocked by approximately 1 mM Ca^{2+} at the external face, with the divalent binding site near the center of the membrane electric field (Matulef and Zagotta 2003; Farris et al. 2004). Many blocking agents for the CNG channel, including amiloride, diltiazem, D600, and tetracaine (Frings et al. 1992; Fodor et al. 1997), are effective antagonists of the MET channel with similar half-blocking concentrations (Table 4.1). However, some CNG channel blockers, such as pseudecatoxin and LY83583, are ineffective (Farris et al. 2004). Although CNG transcripts have been found in hair cells (Drescher et al. 2002), the case for a CNG channel as the hair cell mechanotransducer is not conclusive. If the MET channel is a member of the CNG channel family, it is unlike known channels that have small single–channel conductance (35 pS; Bönigk et al. 1999) and lower permeability to organic cations such as TEA (Balasubramanian et al. 1995).

The most likely channel candidate based on present evidence belongs to the TRP superfamily, some members of which possess channel properties resembling the hair cell MET channel and are linked with sensory transduction (Minke and Cook 2002). All are nonselective cation channels with a high permeability for Ca^{2+} over Na^+ (P_{Ca}/P_{Na}) and a large unitary conductance. For example, in the TRPV subfamily, TRPV3, a temperature receptor, has a P_{Ca}/P_{Na} of 12:1 and maximum channel conductance of 172 pS (Xu et al. 2002) and TRPV5, an epithelial Ca^{2+} transporter, has a P_{Ca}/P_{Na} of 100:1 and a maximum channel conductance of 77 pS (Nilius et al. 2000; Vennekens et al. 2000). Values for single-channel conductance were obtained in Ca^{2+}-free media and were reduced in the presence of Ca^{2+}. MET channels and TRP channel also show a common set of antagonists including Gd^{3+} (Kimitsuki et al. 1996), La^{3+} (Ohmori 1985; Farris et al. 2004), and ruthenium red (Farris et al. 2004; Table 4.1). There is thus substantial evidence suggesting the MET channel has strong similarity with the TRP class of ion channels.

Genetic screens of *C. elegans* mutants lacking a touch response, besides uncovering the MEC channels, also generated a separate channel, OSM–9, from animals with defects in osmotic avoidance and nose touch (Colbert et al. 1997). OSM-9 was the first TRP channel implicated in mechanosensitivity, but two relatives were subsequently cloned from *Drosophila melanogaster*. One of these is NOMPC, the mutation of which largely abolishes receptor potentials in the touch-sensitive bristle organs (Walker et al. 2000) but only mildly affects hearing in *Drosophila* (Eberl et al. 2000). The other is Nanchung (NAN), which is localized to the ciliary neurons in Johnston's organ and whose mutation deafens the insects (Kim et al. 2003). Both NOMPC and NAN have TRP-like structures with six transmembrane domains, a pore region between S5 and S6, intracellular

N- and C-termini, and multiple ankyrin repeats at the N-terminus (29 in NOMPC and 5 in NAN). A vertebrate NOMPC with 62% similarity at the amino acid level to the *Drosophila* version has now been identified in hair cells of the zebra fish, *Danio rerio* (Sidi et al. 2003). Removal of NOMPC gene function by injection of morpholino antisense oligonucleotides into larvae caused initial deafness and imbalance which later reversed as the morpholino was diluted by the endogenous transcript, thus confirming its role in hair cell transduction. However, there is as yet no evidence for the occurrence of NOMPC in hair cells of other vertebrates including mammals. Another TRP channel, TRPA1, was recently identified in both mouse and zebrafish hair cells and proposed as a component of the mechanotransduction channel in these animals (Corey et al. 2004). Strong evidence in support of this suggestion was that mechanotransduction in mouse hair cells was inhibited by transfection with small interfering RNAs that reduced TRPA1 protein expression. This leaves open the question of how TRPA1 and NOMPC might interact to mediate transduction in zebrafish hair cells.

4. Adaptation

4.1 Multiple Mechanisms of Adaptation

The MET channels of auditory hair cells encode hair bundle displacements of no more than a few hundred nanometers, which is similar to the diameter of the constituent stereocilia. Therefore to preserve high sensitivity in the face of larger fluctuations in bundle position, the MET channels are subject to multiple mechanisms of adaptation (reviewed in Eatock 2000; Fettiplace and Ricci 2003). Adaptation is manifested as a decline in the transducer current during a prolonged bundle displacement, reflecting a shift of the *I–X* relationship along the displacement axis in the direction of the adapting stimulus (Eatock et al. 1987; Crawford et al. 1989; Assad and Corey 1992). The kinetics of adaptation, which may be important for assigning a mechanism, can be determined from the time course of decay of current to the maintained stimulus (Crawford et al. 1989) or of the shift in the *I–X* relationship (Assad et al. 1989). For small adapting steps in the linear range, the two methods should yield the same time constant. Recordings from hair cells in frog saccule and turtle cochlea have suggested two distinct processes with different kinetics. In turtle auditory hair cells, adaptation is predominantly fast with a time constant (τ_A) of 0.3 to 5 ms (Crawford et al. 1989; Ricci and Fettiplace 1997), whereas adaptation in frog saccular cells is usually slower with τ_A reported to be in the range 10 to 100 ms (Assad et al. 1989; Assad and Corey 1992). Nevertheless, slow adaptation can be observed as a secondary component in turtle, especially with larger stimuli (Crawford et al. 1989; Wu et al. 1999), and a fast adaptive component has been seen in frog (Howard and Hudspeth 1987; Vollrath and Eatock 2003). For example, adaptation in turtle MET currents in Figure 4.1 show a fast component with τ_A of

0.7 ms prominent at low stimulus levels and a slower component with τ_A of more than 10 ms at high levels. The ability to see two phases of adaptation may to some extent depend on the mode and speed of stimulation (Wu et al. 1999; Vollrath and Eatock 2003). It appears therefore that multiple adaptation mechanisms exist in both auditory and vestibular hair cells. The different balance between the fast and slow components in turtles and frogs may reflect the fact that turtle hair cells are auditory and respond to higher frequencies of vibration than frog vestibular hair cells. Despite differences in kinetics and underlying mechanism, both fast and slow adaptations are regulated by Ca^{2+} influx through the MET channels (Assad et al. 1989; Crawford et al. 1989; Ricci and Fettiplace 1997).

Fast adaptation, because it can occur on a sub-millisecond time scale, probably requires a direct interaction of Ca^{2+} with the MET channel to modulate its probability of opening (Crawford et al. 1989; Jaramillo et al. 1990; Ricci et al. 1998). The distance that Ca^{2+} diffuses must be short, 15 to 35 nm from the mouth of the channel (Ricci et al. 1998), and its resulting action must occur in well under a millisecond (Ricci and Fettiplace 1998). Furthermore, Ca^{2+} can alter the time constant of channel activation as well as adaptation, which argues that it is closely associated with the channel gating mechanism (Fettiplace et al. 2003). By contrast, slow adaptation is regarded as an input control in which the tension in the tip link is adjusted by moving its upper attachment point along the side of the stereocilium (Howard and Hudspeth 1987; Assad and Corey 1992). In a specific molecular model, the upper end of the tip link is proposed to connect through the membrane to an array of unconventional myosins that are capable of ratcheting along the actin backbone of the stereocilium (Gillespie and Corey 1997; Gillespie and Cyr 2004). Ca^{2+} influx through an adjacent MET channel is postulated to detach the myosin from the actin core of the stereocilium, allowing the link's attachment to slip. This would slacken the tip link and reduce the force on channels, causing them to close. There is now evidence implicating myosin 1c in slow adaptation (Holt et al. 2002; Gillespie and Cyr 2004). Myosin-1c has an ATPase cycle time, measured in vitro as approximately 6 s^{-1} (Ostap and Pollard 1996; Howard 2001), which is clearly too slow to explain fast sub-millisecond adaptation. Furthermore, fast adaptation in turtle hair cells is insensitive to inhibitors of myosin-based motors, such as butanedione monoxime, and is therefore unlikely to directly involve the myosin motor cycle (Wu et al. 1999).

4.2 Calcium Control of Adaptation

Although fast and slow adaptation may operate by different mechanisms, both are regulated by Ca^{2+} that enters the stereocilia through the MET channels (Assad et al. 1989; Crawford et al. 1989; Hacohen et al. 1989; Ricci and Fettiplace 1997). Several types of experimental manipulation argue for a regulatory role of intracellular calcium. Suppressing Ca^{2+} influx either by lowering its extracellular concentration or by depolarizing to near the Ca^{2+} equilibrium potential slows or abolishes adaptation. Such manipulations also shift the I–X curve to the left, increasing the fraction of MET current activated at the bundle's resting position. Furthermore, the rate of fast adaptation in the turtle is directly proportional to the amount of Ca^{2+} entering the stereocilia through the MET channels; this was assayed by the fluorescence change in hair bundles loaded with a calcium-sensitive dye (Ricci and Fettiplace 1998). A rise in cytoplasmic Ca^{2+} by release from an intracellular caged source increases the rate of adaptation (Kimitsuki and Ohmori 1992). Finally, adaptation is susceptible to the nature and concentration of the cytoplasmic calcium buffer. Raising the concentration of BAPTA in the patch electrode solution had effects similar to lowering external Ca^{2+}, decreasing the speed and extent of adaptation and shifting the I–X relationship to more negative displacements (Ricci and Fettiplace 1997; Ricci et al. 1998).

A drawback of an MET channel control mechanism that relies on Ca^{2+} influx is that the endolymph bathing the hair bundles in vivo has a low Ca^{2+} concentration, especially in the auditory division of the inner ear: 20 to 30 μM in mammals (Bosher and Warren 1978; Ikeda et al. 1987; Salt et al. 1989) and 65 μM in turtles (Crawford et al. 1991). Nevertheless, recordings with perforated-patch electrodes, where the endogenous calcium buffer was retained in the cytoplasm, showed that adaptation in turtle auditory hair cells persisted even in physiological levels of extracellular Ca^{2+} (70 μM; Ricci et al. 1998). Moreover, such recordings revealed that when in vivo like extracellular Ca^{2+} and intracellular calcium buffering were used, the usual adaptive decline in current to a displacement step could sometimes become resonant and generate damped oscillations (Fig. 4.5). These oscillations occur at a frequency near the hair cell's CF, implying that fast adaptation may amplify and tune the MET current. Oscillation frequencies in different cells ranged between 58 and 230 Hz, which represents a significant portion of the auditory range of the turtle (30 to 600 Hz;

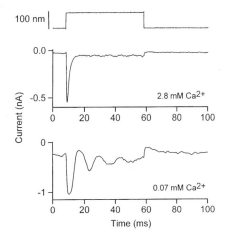

FIGURE 4.5. Tuning in the MET current of a turtle hair cell. When the hair bundle was bathed in 2.8 mM Ca^{2+} the MET current for a step deflection of the bundle showed an exponential decline, reflecting adaptation. When the Ca^{2+} was reduced to 0.07 mM, a concentration similar to that in turtle endolymph, the current in response to the same bundle displacement developed damped oscillations at 77 Hz. The calcium buffer in the patch pipette solution was 1 mM EGTA, similar to that expected for the endogenous cytoplasmic calcium buffer.

Crawford and Fettiplace 1980). Under some conditions, the damping was diminished to the point where the transducer current oscillated continuously with large amplitude in a limit-cycle mode (Fig. 4.6), which may be accompanied by spontaneous oscillatory motion of the hair bundle (Crawford and Fettiplace 1985; Howard and Hudspeth 1987; Martin and Hudspeth 1999; Martin et al. 2003). The importance of this mechanism in vivo is unknown but it is significant that the resonance is most prominent in isolated preparations at reduced Ca^{2+} concentrations similar to those in endolymph (Ricci et al. 1998; Martin and Hudspeth 1999).

Besides the two calcium-driven mechanisms discussed so far, other signals, such as cyclic adenosine monophosphate (cAMP), may control the operating position of the MET channel on a slower time scale. Perfusion of 8-Br-cAMP shifted the I–X relationship along the displacement axis in the positive direction with no effect on fast adaptation (Ricci and Fettiplace 1997; Géléoc and Corey 2001). A similar shift was produced by the phosphodiesterase inhibitor 3-isobutyl-1-methylxanthine (IBMX), which elevates cAMP by preventing its breakdown. Application of protein kinase A inhibitors H89 or RpcAMP had the opposite effect in some cells (Géléoc and Corey 2001), suggesting that cAMP acts through a protein kinase A that is constitutively active. Possible

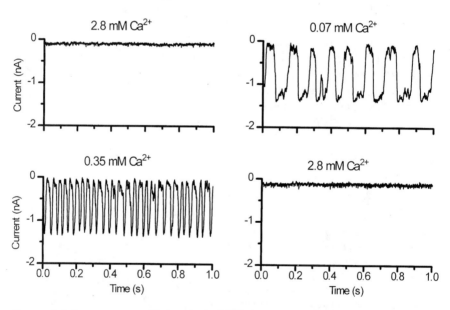

FIGURE 4.6. Spontaneous oscillations in the MET current of a turtle hair cell. The resting current in the absence of hair bundle stimulation was quiescent in 2.8 mM Ca^{2+}, but when the bundle was bathed in 0.35 mM Ca^{2+} it became oscillatory at 26 Hz and in 0.07 mM Ca^{2+} the oscillation frequency dropped to 8 Hz. The oscillations in the current disappeared on returning to saline with 2.8 mM Ca^{2+}.

targets for phosphorylation include the transducer channel or the myosin motor. However, the large (0.5 μm) shifts in the *I–X* relation evoked by cAMP, and the lack of effect on the fast adaptation are more consistent with an action on the motor. Protein kinase A may in turn be activated by an ambient cAMP level produced by ongoing activity of a calcium-calmodulin–activated type I adenyl cyclase found in hair cells (Drescher et al. 1997). Stereociliary Ca^{2+} would thus also influence the activity of this cAMP pathway to regulate the set position of the transducer channels.

4.3 Functions of Adaptation

The usual role of adaptation in sensory transduction is to preserve sensitivity for small changes in stimulus occurring on a larger background. In the ear there are adaptive processes peripheral to the hair cells that prevent large static displacement being imposed on the cells, shielding them from overstimulation and damage. For example, the helicotrema acts like a high-pass filter for sound frequencies below 100 Hz (Franke and Dancer 1982; Cheatham and Dallos 2001). More precise control may be exerted at the hair cell level to maintain the MET channels near their maximal sensitivity on the steepest slope of the *I–X* relationship. But why have multiple processes of adaptation? One explanation is that the slower myosin-based mechanism has a wider dynamic range to maintain tension in the tip links and orient the bundle to where the fast channel mechanism can produce finer control at frequencies used by the auditory system. In support of this notion, large (0.5 to 1 μm) shifts in the *I–X* relationship with no effect on the fast adaptation time constant were seen after treatment with phosphate analogs such as vanadate, which would block a myosin-based motor (Yamoah and Gillespie 1996; Wu et al. 1999). In contrast the compensatory range of fast adaptation may be no more than 0.1 μm, comparable to the channel's dynamic range. Fast adaptation, with its small limits of adjustment and its rapid kinetics, may have a more subtle function.

A possible clue to the role of fast adaptation comes from the turtle auditory papilla, where the fast adaptation time constant (τ_{Af}) varies inversely with hair cell characteristic frequency or CF (Ricci and Fettiplace 1997; Ricci et al. 1998). Fast adaptation acts as a first-order high-pass filter whose corner frequency ($2\pi \cdot \tau_{Af})^{-1}$ is approximately two thirds of the CF. This implies that fast adaptation may contribute to hair cell frequency selectivity, a notion reinforced by the observation that under physiological conditions fast adaptation can display under-damped resonance at frequencies in the turtle's auditory range (Ricci et al. 1998). In the mammalian cochlea, where the CFs are much higher than those in the turtle, τ_{Af} is correspondingly smaller with a value of approximately 100 μs (Kennedy et al. 2003).

In the turtle cochlea the activation time constant of the MET channel also changes with CF (Fettiplace et al. 2003); the kinetics of activation and fast adaptation therefore bestow on transduction a variable band-pass filter matched to the CF. To a first approximation, this filter consists of a first-order low-pass filter, attributable to the principal time constant of activation, and a first-order high-pass filter produced by fast adaptation. It is important to note that the tuning of the MET current in turtle hair cells does not require simultaneous active motion of the hair bundle (see Section 5.2) because it occurs even when the bundle is displacement clamped with a rigid stimulator. However, this transduction filter is unlikely to be the major source of hair cell frequency selectivity in the turtle, which instead stems from a sharply tuned electrical resonance (Crawford and Fettiplace 1981). Electrical tuning is produced by combined action of a voltage-dependent Ca^{2+} channels and a Ca^{2+}-activated K^+ channels in the hair cell basolateral membrane (Art and Fettiplace 1987). The transduction filter may provide a mechanism for actively restricting the input bandwidth to improve the signal-to-noise ratio of transduction within the frequency range encoded by the hair cell. This must be done on a cycle-by-cycle basis and to be optimal should therefore vary with CF. It has been previously argued (Bialek 1987; Block 1992) that using an active filter to restrict the stimulus bandwidth is a way of extending the physical detection limits of sensory transduction in the face of intrinsic thermal noise.

4.4 An Unconventional Myosin as the Motor for Slow Adaptation

Since the original proposal that a myosin motor caused adaptation (Howard and Hudspeth 1987), six myosin isoforms have been found in the inner ear and have been localized to hair cells: myosin 1c, IIA, IIIA, VI, VIIA, and XV (Gillespie et al. 1993; Hasson et al. 1997; Walsh et al. 2002; Belyantseva et al. 2003; Mhatre et al. 2004; Rzadzinska et al. 2004; see Furness and Hackney, Chapter 3). Remarkably all appear vital for sensory function because mutations in their genes cause progressive deafness and vestibular dysfunction (Petit et al. 2001; Walsh et al. 2002; Donaudy et al. 2003). Several of these myosins may be crucial for development and maintenance of the hair bundle structure and may not therefore be directly involved in transducer function. However, there is evidence for the adaptation motor being myosin 1c (Gillespie et al. 2002; Gillespie and Cyr 2004) or myosin VIIA (Kros et al. 2002). Myosin 1c occurs throughout the frog hair cell soma and the hair bundle (Metcalf 1998) but some work shows it to be concentrated in the hair bundle at the two ends of the tip link (Garcia et al. 1998; Steyger et al. 1998): in the osmiophilic plaque marking its upper attachment point and at the tip of the stereocilium. Direct support for the role of myosin 1c in slow adaptation has come from introducing a point mutation in its ATP-binding site to confer susceptibility to inhibition by certain ADP analogues (Gillespie et al. 1999). Expression of the mutant myosin 1c in mouse utricular hair cells rendered slow adaptation sensitive to block by the ADP analogues introduced through the recording pipette (Holt et al. 2002). The mutation did not alter slow adaptation in the absence of the inhibitor, nor was fast adaptation affected even in the presence of the inhibitor, which accords with other evidence for the lack of involvement of a myosin in fast adaption (Wu et al. 1999). Block of adaptation was not accompanied by a change in the resting open probability or the position of the I–X relationship, an unexpected observation if myosin 1c does indeed underlie the optimal positioning of the bundle by slow adaptation; furthermore, the result differs from that seen with mutation of myosin VIIA, where the I–X relationship is shifted positive.

Myosin VIIA is distributed along the entire length of the stereocilia and there is no evidence for its accumulation at the stereociliary tips near the transduction apparatus in contrast to myosin 1c and myosin XV (Rzadzinska et al. 2004). Nevertheless, mutation of myosin VIIA in *shaker1* mice caused a substantial positive shift of the I–X relationship along the displacement axis so that the MET channels were no longer poised to open at the bundle's resting position (Kros et al. 2002). However, the effects of the mutation on the response to a maintained hair bundle displacement were inconsistent because adaptation was barely evident in the wild type but it became more conspicuous in the mutant. It is possible that adaptation in the control, because of its speed, was underestimated owing to the slowness of bundle deflection by the water jet stimulator, but that adaptation in the mutant was sufficiently sluggish to be monitored faithfully. Other important consequences of myosin VIIA mutation included progressive disorganization of the hair bundle and reduction in its stiffness which may have contributed to the effects on the MET currents (Kros et al. 2002). Although both myosin 1c and VIIA may be important in optimizing transduction, their relative roles in setting the tension in the tip links are not well understood. Indeed it is not even known whether they operate together in the same cell because myosin-1c has been studied exclusively in vestibular hair cells whereas work on myosin VIIA has been confined to auditory hair cells.

A significant area of uncertainty in understanding myosin-based adaptation lies in the regulation by intracellular Ca^{2+} (Gillespie and Cyr 2004). Increased tension in the tip link opens MET channels, thus elevating stereociliary calcium. According to the current hypothesis, this must promote detachment of the myosin head from actin and its slippage down the stereocilium. Channel closure by a slackening of the tip link and the ensuing drop in Ca^{2+} causes the myosin to ascend the stereocilium. At first sight these effects seem opposite to what is expected for Ca^{2+} stimulation of myosin motility. Ca^{2+} is known to regulate the enzymatic and mechanical activities of myosin I by interaction with calmodulin (Zhu et al. 1998; Perrault-Micale et al. 2000). At low concentrations Ca^{2+} binds to calmodulins that are constitutively attached to myosin I, but high concentrations of Ca^{2+} cause dissociation of the calmodulins. The involvement of calmodulin in hair cell adaptation is supported by the observation it is blocked by calmodulin antagonists (Walker and Hudspeth 1996). When the MET channel is closed, the resting Ca^{2+} must be adequate to stimulate myosin to climb the stereocilium. During positive bundle deflection, an increase in Ca^{2+} with MET channel opening promotes detachment of myosin from the actin filament. This may occur either by dissociation of the calmodulin or by the recently reported mechanism of shortening the lifetime of attachment of the myosin (Batters et al. 2004).

5. Calcium in the Hair Bundle

5.1 Calcium Summation and the Time Constant of Fast Adaptation

There are various conditions under which the time constant of fast adaptation, τ_{Af}, and the amplitude of the MET current are inversely correlated. For example, hair cells tuned to higher frequencies have on average larger MET currents and faster adaptation than those tuned to lower frequencies (Fig. 4.7; Ricci and Fettiplace 1997). Even at one cochlear location, different hair cells show a range of current sizes with a corresponding variation in τ_{Af} (Ricci and Fettiplace 1997; Kennedy et al. 2003). Reducing the MET current in a single hair cell by partially blocking the channels with streptomycin also slows adaptation (Ricci 2002), often to the point where it disappears (Kimitsuki and Ohmori 1993). These collected observations could be explained if the differences in τ_{Af} arise at least partly from a variation in the Ca^{2+} influx, which in turn depends on the magnitude of the MET current. The adaptation rate $(\tau_{Af})^{-1}$ has been shown to

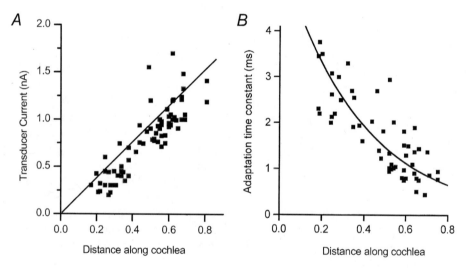

FIGURE 4.7. Variation of maximum MET current and time constant of fast adaptation with hair cell location in the turtle cochlea. (**A**) Maximum MET current (I_{max}), measured at -80 mV holding potential, is plotted against the relative distance of the hair cell along the cochlea for 171 cells. The straight line was calculated from the single-channel conductance data (Fig. 4.4C) assuming one channel/stereocilium with the number of stereocilia per bundle increasing from 60 at the low-frequency end to 90 at the high-frequency end (Hackney et al. 1993). (**B**) Time constant of fast adaptation (τ_{Af}) plotted against relative distance of the hair cell along the cochlea for 83 hair cells. In both (A) and (B), the distance of the cell from the apical (low-frequency) end of the cochlea was normalized to the total length of the cochlea (approximately 830 μ). All measurements were made in saline containing 2.8 mM Ca^{2+}. Smooth curve in (B) is a fit given by: $\tau_{Af} = 5.7 \cdot \exp(-d/0.4)$, where d = distance of the hair cell along the cochlea.

increase in proportion to the amount of Ca^{2+} entering the hair bundle (Ricci and Fettiplace 1998). A larger Ca^{2+} influx will produce a higher stereociliary Ca^{2+} concentration, which will accelerate adaptation. This sequence of events assumes that the intracellular Ca^{2+} concentration is raised by the summation of Ca^{2+} originating from multiple MET channels (Ricci and Fettiplace 1997; Ricci 2002). However, summation must occur within each stereocilium because interaction between channels located at the tips of neighboring stereocilia will not occur sufficiently rapidly to affect adaptation. For example, the time required for Ca^{2+} to diffuse along a single 4-μm stereocilium is 20 milliseconds assuming a Ca^{2+} diffusion coefficient of 400 $\mu m^2 \cdot s^{-1}$. The Ca^{2+} concentration will also be attenuated along the stereocilium, thus diminishing its signaling capacity, as a consequence of binding to cytoplasmic proteins and extrusion by the Ca^{2+} pump in the stereociliary membrane (Yamoah et al. 1998).

It has been generally acknowledged that there are only a few MET channels (fewer than five) per stereocilium (Ohmori 1985; Holton and Hudspeth 1986; Howard and Hudspeth 1987; Denk et al. 1995; Ricci and Fettiplace 1997), but are these sufficient to produce a large enough range of values of τ_{Af}? The number of channels per stereocilium has been estimated in turtle auditory hair cell from the single-channel conductance and the maximum size of the macroscopic transduction current, both of which increase with CF. If the MET channels are uniformly distributed across the bundle, each stereocilium possesses between one and two channels regardless of hair cell CF (Ricci et al. 2003). A combined increase with CF in the stereociliary complement (Hackney et al. 1993) and single-channel conductance (Fig. 4.4) produces a several-fold increase in the peak amplitude of the macroscopic MET current from the low- to high-frequency ends of the turtle cochlea (Fig 4.7). Therefore Ca^{2+} summation in each stereocilium occurs between at most two channels, which cannot account for the range of time constants. Instead the tonotopic gradient in τ_{Af} must be largely explicable by variations in single-channel properties. Adaptation was evident in the ensemble averages of single-channel activity (Fig. 4.4), arguing that there is sufficient Ca^{2+} entering through a single channel to elicit adaptation. The fast adaptation time constant extractable from single-channel records showed a similar range to that seen in macroscopic (multichannel) currents and also varied similarly with CF (Ricci et al. 2003). Thus the increase in the rate of fast adaptation with hair cell CF may be largely attributable to an increase in channel size: doubling the channel conductance doubles the amount of Ca^{2+} entering and thus halves the adaptation time constant.

Cochlear gradients in other properties of the MET channel could also contribute to the tonotopic variation in τ_{Af}. Spectral analysis of MET current noise has suggested a difference in channel kinetics between high- and low-frequency hair cells (Ricci 2002). This difference is endorsed by measurements of activation kinetics obtained by fitting the onset of the MET current in response to rapid bundle deflections (Fettiplace et al. 2003). Speeding up the activation and deactivation kinetics of the channel might produce additional acceleration of adaptation rate over that realizable by changes in single-channel conductance. Altering the calcium sensitivity of the MET channel (as occurs with the Ca^{2+}-

activated K$^+$ channel; Jones et al. 1999) is another possible source of variability. However, there is no evidence that the tonotopic gradient in channel conductance is accompanied by a concomitant augmentation of Ca^{2+} permeability, which would theoretically speed up adaptation (Ricci 2002).

If there are at most two channels per stereocilium, it is less easy to understand why τ_{Af} is inversely correlated with the maximum MET current at a one cochlear location. In both turtle and rat, a fivefold reduction in the amplitude of the MET current is associated with an almost equivalent slowing of the adaptation time constant (Ricci and Fettiplace 1997; Kennedy et al. 2003). MET currents smaller than normal in an excised preparation are often attributed to mechanical trauma to the bundle causing destruction of tip links. Loss of one of the two channels per stereocilium could at most double τ_{Af} but much larger variations in time constant are seen. This suggests that the residual channels in a damaged bundle have a reduced conductance compared to that in healthy intact bundles. It is possible that one consequence of hair cell injury is raised intracellular Ca^{2+} which might lead, directly or indirectly, to a modification or block of the MET channels, thus accounting for the correlation between maximum current and τ_{Af}. It is important therefore to understand the factors that contribute to hair cell calcium balance.

5.2 Calcium Homeostasis in the Stereocilia

The free Ca^{2+} in the stereocilia is a crucial parameter in regulating mechanotransduction. By affecting adaptation, it sets the position of the bundle and the open probability at rest; it also speeds up the kinetics of both activation and adaptation of the MET channels (Ricci et al. 1998; Fettiplace et al. 2003); it may control phosphorylation of stereociliary proteins by direct or indirect (via cAMP) activation of protein kinases; and it may influence tip link regeneration (Zhao et al. 1996). Under conditions where the MET current is oscillatory (Figs. 4.5 and 4.6) free Ca^{2+} will determine the resonant frequency. Stereociliary Ca^{2+} concentration is determined mainly by the balance between influx through the MET channels and extrusion via plasma membrane Ca^{2+}-ATPase pumps (Tucker and Fettiplace 1995; Yamoah et al. 1998). Other factors that modify the magnitude and time course of calcium transients include buffering by proteins such as calbindin-D28k and parvalbumin-β (Hiel et al. 2001; Heller et al. 2002; Hackney et al. 2003) that may be present at near millimolar concentrations.

Stereociliary Ca^{2+} concentration may reach several hundred micromolar around the MET channel during its opening (Ricci et al. 1998; Fig. 4.9) which places a burden on the Ca^{2+} pump to aid with its removal. Hair bundles of both frogs and rats express exclusively the PMCA2a isoform of the Ca^{2+} pump, which differs from the PMCA1 isoform localized to the basolateral membrane (Dumont et al. 2001). Mutation of the PMCA2 in the *deaf waddler* mouse causes deafness and balance defect which attests to the importance of the extrusion by the hair bundle pump (Street et al. 1998; Kozel et al. 1998). Furthermore, *PMCA2*$^{+/-}$

heterozygotes show an increased susceptibility to noise-induced hearing loss (Kozel et al. 2002). Thus Ca^{2+} loading may be an important factor that influences transducer performance. Because of adaptation, the MET channels remain closed for a significant period of time after a single large stimulus and the resting open probability returns to its initial level with a time course that parallels the recovery of sensitivity (Fig. 4.8). These slow recoveries, which can take several

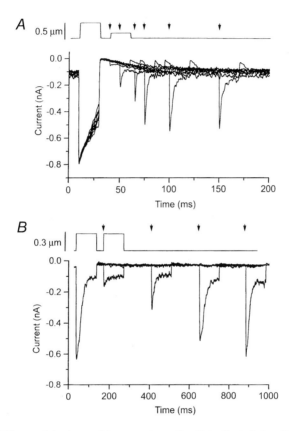

FIGURE 4.8. Effects of the interval between two stimuli to the hair bundle on the MET current. (**A**) A large conditioning stimulus, 20 ms duration, was followed by a smaller test stimulus delivered with a variable delay. The *arrows* indicate the time of onset of the test stimuli, the responses to which are superimposed. (**B**) An initial conditioning stimulus, 100 ms duration, was followed by an identical test stimulus delivered with a variable delay. Interpulse intervals of more than 0.8 s are needed for the recovery from the effects of the first pulse. Note that in both (A) and (B), with short interpulse intervals, the peak current and the time course and extent of adaptation for the second stimulus are reduced. The time course of recovery from the initial stimulus is slower in (B) owing to the larger Ca^{2+} load. Both recordings were in 2.8 mM external Ca^{2+}, 1 mM intracellular BAPTA.

hundred milliseconds, are probably limited by the clearance of Ca^{2+} from the stereocilia following its accumulation during the initial stimulus rather than by the kinetics of adaptation. In accord with this hypothesis, when the conditioning step is lengthened so increasing the Ca^{2+} load, the recovery time is markedly prolonged to close to 1 second (Fig. 4.8B).

The importance of the Ca^{2+}-ATPase in this recovery process can be appreciated from models that explore the calcium homeostasis in the stereocilia. Figure 4.9 shows calculations of the open probability of the transducer channels during a repetitive stimulus using the model of mechanotransduction described in Wu et al. (1999). The model also displays the local concentrations of free

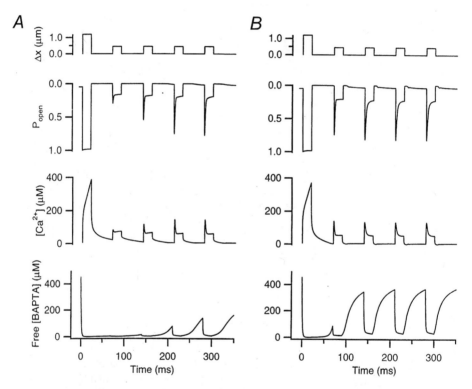

FIGURE 4.9. Effects of Ca^{2+} extrusion rate on the time course of recovery of theoretical responses to a maximal stimulus. (**A**) A conditioning 1-μm displacement step, 20 ms duration, was followed by a series of 0.4 μm test stimuli at 70-ms repetition intervals. The traces give the hair bundle displacement, open probability of MET channels (P_{open}), and the concentrations of Ca^{2+} and BAPTA measured 20 nm from the channel. Density of Ca^{2+}-ATPase pumps, 2000 μm^{-2}. (**B**) Same calculations as in (A) except that the Ca^{2+}-ATPase pump density was increased to 20,000 μm^{-2}. Note that in (A), although the responses have apparently recovered, the BAPTA remains depleted near the channel. 2.8 mM external Ca^{2+}. Total intracellular calcium buffer was 1 mM BAPTA. For details of computation see Wu et al. (1999).

Ca^{2+} and the calcium buffer (BAPTA) and incorporates the fast calcium-controlled adaptation mechanism. After the first large stimulus, the Ca^{2+} concentration close to the channel approached 0.4 mM, and took more than 100 ms to return to its resting level. Subsequent test stimuli during the recovery evoked smaller transducer responses and Ca^{2+} excursions. It is important to note that the calcium buffer remained depleted throughout the entire period even when the response had reached a steady state. The calcium buffer acts as a large sink and although a quasi-steady state was attained, the stereocilia were still effectively Ca^{2+}-loaded. Regeneration of free buffer can occur only by extruding Ca^{2+} from the cell. Recovery of the responses and restoration of the free buffer was accelerated by a 10-fold increase in the density of Ca^{2+}-ATPase pumps from a standard value of 2000/µm^2 (Ricci and Fettiplace 1998; Yamoah et al. 1998). The effect of pump density was evaluated for a single location in the cochlea corresponding to a particular Ca^{2+} flux per channel. If the MET channel conductance and with it the Ca^{2+} influx per stereocilium increases with CF, an accompanying increase in the number of Ca^{2+}-ATPase pumps per stereocilium might be expected to handle the larger Ca^{2+} load.

These simulations attest to the importance of the long-term calcium balance in the stereocilia for optimizing the sensitivity of the transduction process. Failure to extrude the Ca^{2+} accumulated in the hair bundle during intense or frequent stimulation may be an important factor in accounting for the threshold shifts and hearing loss known to result from acoustic over stimulation (Saunders et al. 1991; Fridberger et al. 1998).

6. Mechanical Properties of the Hair Bundle

6.1 Passive and Active Components of Hair Bundle Compliance

Hair cells are stimulated in vivo by a force produced by lateral motion of the overlying gelatinous membrane, the tectorial or otolithic membranes, which causes the hair bundle to rock about its insertion into the top of the cell. The displacement caused by a given force stimulus depends on the compliance of the hair bundle. This compliance has been measured in isolated preparations by delivering calibrated force stimuli, with a glass fiber more flexible than the bundle (Strelioff and Flock 1984; Crawford and Fettiplace 1985; Howard and Ashmore 1986; Howard and Hudspeth 1987), or a water jet (Szymko et al. 1992; Géléoc et al. 1997). The forces delivered with the water jet were estimated from the viscous force on an idealized geometrical shape resembling the bundle, and are therefore less well defined than using the flexible fiber. The ensuing motion of the bundle was usually inferred from the change in photocurrent as the shadow of the bundle or the attached fiber traversed a photodiode array (Crawford and Fettiplace 1985; Denk and Webb 1992). Such measurements have shown that the bundle compliance is nonlinear (Fig. 4.10) and has both passive

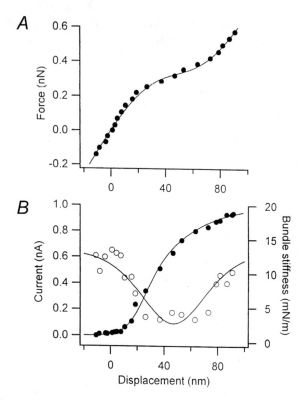

FIGURE 4.10. Nonlinear compliance of the stereociliary bundle of a turtle hair cell. (**A**) The force–displacement relationship (*top*) and the current–displacement relationship (*bottom, filled circles*) were derived from measuring the displacement of the hair bundle and the MET current evoked by a force step of the hair bundle delivered with a flexible glass fiber. Note that the force–displacement relationship is nonlinear with a reduction in slope stiffness over the range where the MET channels are gated (Eq. [4.3]). (**B**) Differentiating the force–displacement results gives the hair bundle slope stiffness which is plotted against displacement (*bottom, open circles*).

and active components (Howard and Hudspeth 1988; Russell et al. 1992; van Netten and Kros 2000; Ricci et al. 2002). The passive component is attributable to the flexibility of the stereociliary ankles, the interciliary connections, and the gating springs. The active component, termed the "gating compliance" (Howard and Hudspeth 1988), is linked to the opening and closing of the MET channels. The contributions of the components are described by the gating-spring model (Howard and Hudspeth 1988; Markin and Hudspeth 1995; van Netten and Kros 2000) which predicts a relationship between the force, F_B, applied to the bundle and X, the bundle displacement given by:

$$F_B = X \cdot K_s - Nz \cdot p_o + F_o \qquad (4.3)$$

where K_s is the passive linear stiffness, p_O is the probability of opening of the N transducer channels, and z is the gating force per channel. F_0 is a constant to make F_B zero at the bundle's resting position and is presumably generated by the motors tensioning the tip links.

The passive stiffness, K_s in Eq. (4.3), is contributed by the parallel combination of the stereociliary pivots, interciliarly links and gating springs (Pickles 1993). In most preparations, K_s is 0.5 to 1 mN · m^{-1} (Crawford and Fettiplace 1985; Howard and Ashmore 1986; Howard and Hudspeth 1987; Szymko et al. 1992; Russell et al. 1992) measured at the tip of 6 μm-tall hair bundles. The passive stiffness varies directly with the number of stereocilia and inversely with the square of their maximum height (Crawford and Fettiplace 1985; Howard and Ashmore 1986). Thus smaller hair bundles, such as those in the basal turn of the mammalian cochlea, will have a larger stiffness than those in the apical turn (Strelioff and Flock 1984). Since K_s includes a contribution from the gating springs connected to the active channels, it is larger in cells with more functional channels. Recent measurements on hair cells with large MET currents that may approximate those in vivo gave stiffnesses of 3 to 5 mN · m^{-1} in mouse (Géléoc et al. 1997) and turtle (Ricci et al. 2002). Destruction of the tip links, and hence loss of the gating springs, by treatment with low-calcium BAPTA, reduced frog hair bundle stiffness by about 30% (Jaramillo and Hudspeth 1993). The relative bundle stiffness contributed by the gating springs represents the fraction of work done in deflecting the bundle that is funneled into the gating springs to open the channels. In cells with increased bundle stiffness attributable to more intact gating springs and channels, this fraction is likely to be even higher, implying that the stimulus energy is very efficiently coupled to the MET channel.

The negative term in Eq. (4.3) embodies an active component in the bundle stiffness. The effect of channel gating is to generate a force in the same direction as the imposed displacement, thus effectively lowering bundle stiffness. The stiffness of the hair bundle is therefore nonlinear, and it decreases as the channels open, reaches a minimum when the probability of opening is approximately 0.5, and then increases again at larger open probabilities (Howard and Hudspeth 1988; Fig. 4.10). The relative magnitudes of the passive and active components determine the extent of the nonlinearity. For some hair bundles, the passive compliance is much larger than the active one and so the bundle approximates a simple spring (van Netten and Kros 2000). For others, the stiffness is dominated by the active component to the extent that over the central range of channel gating, the stiffness can become negative. This property had been linked to spontaneous oscillations of the bundle (Martin et al. 2000). The more usual behavior lies somewhere between these extremes with the active component supplying about half the total stiffness of the bundle (Howard and Hudspeth 1988; Russell et al. 1992; Ricci et al. 2002). Whenever there is a substantial active component, factors that affect the probability of opening of the MET channels, such as adaptation, will alter compliance of the hair bundle and may cause it to move. This is termed "active movement" because it is powered by energy supplied from the hair cell. Adaptation is driven by the influx of Ca^{2+}

and its interaction with the MET channels, and a source of energy for the active bundle motion derives from the large (up to 1000-fold) gradient in Ca^{2+} concentration across the stereociliary membrane.

6.2 Active Hair Bundle Movements

There are three manifestations of active hair bundle movements: (1) spontaneous periodic motion of free-standing bundles; (2) nonlinear responses of the bundle to force stimuli delivered with a compliant probe; (3) bundle displacements to voltage steps which depolarize the hair cell positive to 0 mV. Spontaneous oscillations occur with amplitudes up to 50 nm, greater than those expected for Brownian motion of the bundle, and at frequencies of 5 to 50 Hz (Crawford and Fettiplace 1985; Denk and Webb 1992; Martin and Hudspeth 1999; Martin et al. 2003). Their under-damped nature and low frequency make it unlikely that they arise from a passive bundle resonance (Crawford and Fettiplace 1985) and they provide the most convincing evidence for an active process (Martin and Hudspeth 1999). Like the oscillatory transducer currents (Figs. 4.5 and 4.6), they are exaggerated when the hair bundles are bathed in artificial endolymph with low Ca^{2+} concentrations similar to those in vivo (Martin and Hudspeth 1999). Active bundle movements can also be elicited by mechanical or electrical stimuli. The connection between these two manipulations is the intracellular Ca^{2+} concentration, which is reduced by displacements of the hair bundle toward its shorter edge or by closing off Ca^{2+} influx through the MET channels by depolarization to the Ca^{2+} equilibrium potential. The evoked movements can be classified according to their polarity and kinetics, which match those of fast and slow adaptation. For an increase in stereociliary calcium, the fast movement is in the negative direction (toward the short edge of the bundle) and therefore opposes applied deflection of the bundle, whereas the slow movement is in the positive direction and reinforces the applied deflection. The fast active response (Fig. 4.11) is complete in a few milliseconds (Crawford and Fettiplace 1985; Benser et al. 1996; Ricci et al. 2000), but the slow response can extend from 10 to 100 ms (Howard and Hudspeth 1987; Assad and Corey 1992).

The slow movement is probably a consequence of the operation of the myosin-based motor proposed to regulate tip link tension and slow adaptation (Assad and Corey 1992). Thus during a positive force stimulus to the bundle, the increase in tip link tension initially opens the channel, but slipping of the upper attachment point of the link increases the compliance, causing further displacement of the bundle. The fast movement, a recoil that opposes the stimulus, is synchronous with the channel closure attributed to fast adaptation (Ricci et al. 2000). As with fast adaptation, the time constant of the fast recoil in turtle auditory hair cells varies with cochlear location, and hence CF of the cell (Fig. 4.11). A plausible mechanism is a shift along the displacement axis of the active component of the bundle compliance as the channel adapts (Fig. 4.12; Ricci et al. 2000). The fast bundle movements, similar to fast adaptation, are slowed by reducing extracellular Ca^{2+}. Counterparts of these active movements are also

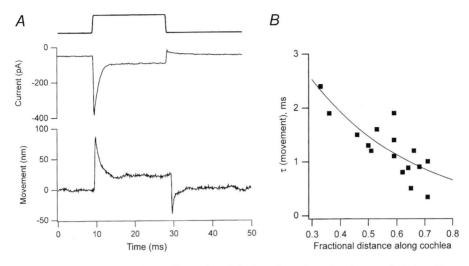

FIGURE 4.11. Active hair bundle motion linked to fast adaptation in a turtle hair cell. (**A**) Hair bundle was stimulated with a 40 pN force step (*top*) delivered with a flexible glass fiber that produced an MET current (*middle*) with time course resembling the bundle displacement (*bottom*). The bundle displacement showed a decline (or recoil) that could be fitted with a single time constant τ(movement) that was virtually identical to the time constant of fast adaptation. (**B**) Time constant of decay of displacement transient, τ(movement), for 16 cells is plotted against fractional distance of hair cell from the low-frequency end (d). Smooth line calculated from tonotopic variation in time constant of fast adaptation with cochlear location (Fig. 4.7B): τ(movement) = $5.7 \exp(-d/0.37)$.

seen during depolarizing voltage steps, which can produce a slow deflection of the hair bundle toward its shorter edge (Assad et al. 1989) or a fast displacement toward its taller edge (Ricci et al. 2000). The relative prominence of these two, as with adaptation, differs in frog vestibular hair cells and turtle auditory hair cells but both polarities of movement may occur, depending on the conditions, in the same cell. Thus in turtle hair cells, depolarization normally evokes a fast positive motion of the bundle, but the response is slowed and reverses polarity when a sustained positive bias is imposed on the bundle (Ricci et al. 2002).

6.3 Role of Active Bundle Movements and Cochlear Amplification

The significance of the active bundle movements in transduction is not fully understood. They could simply be an epiphenomenon, resulting from instabilities in an adaptation mechanism the primary role of which is to adjust the mechanical input and thereby set the working point of a mechanically sensitive channel. Alternatively they might sum with the forces of the external stimulus providing amplification to enhance the signal-to-noise ratio of transduction es-

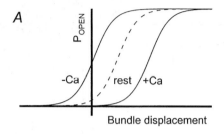

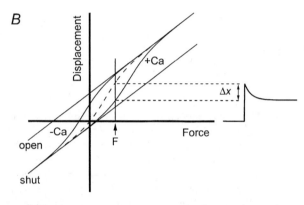

FIGURE 4.12. A model for the effects of intracellular Ca^{2+} on hair bundle mechanics. (**A**) Probability of opening of MET channels with bundle displacement in different intracellular Ca^{2+} concentrations. The relationship under resting conditions is indicated by the *dashed curve*. The curve is shifted to the right with an increase in intracellular calcium (+Ca) as occurs during adaptation to positive steps; the curve is shifted to the left with a decrease in intracellular calcium (−Ca) as occurs during depolarization or reduction in external Ca^{2+}. (**B**) Displacement–force relationship of the hair bundle depends on the state of the channels. If all the channels are fixed closed or open, the plot is linear with a slope reflecting the compliance of the stereociliary pivots, ciliary links, and gating springs (Eq. [4.3]). However, the channels are free to open with displacement producing the *dashed curve* under resting conditions. Because of the effects in (A), an increase (+Ca) or decrease (−Ca) in Ca^{2+} shifts the inflection in the displacement–force relationship. Thus if the bundle is subject to a force step, F, the bundle initially moves along the dashed curve, but as Ca^{2+} enters and binds to the channel (+Ca), there is a reduction in the displacement (Δx) reflecting channel closure.

pecially near threshold (Hudspeth 1997; Fettiplace et al. 2001). Amplification has been demonstrated by active motion of individual hair bundles in isolated epithelia (Crawford and Fettiplace 1985; Martin and Hudspeth 1999), but in the intact inner ear, the hair bundles are restrained by attachment to a tectorial (or otolithic) membrane. For active bundle movement to fulfill a physiological role, the hair bundles must contribute significantly to cochlear mechanics, and the size and speed of the forces produced by the active process must be comparable to those of the external physiological stimulus. In the frog saccule, the stiffness of the otolithic membrane is determined largely by the stiffness of the hair bundles to which it is coupled and therefore much of the stimulus energy goes into deflecting the hair bundle (Benser et al. 1993). Consequently, stiffness changes or active motion of the bundle must affect the mechanics of the entire organ. No equivalent measurements have been performed for the cochlea, though available values suggest that the basilar membrane is stiffer than the hair bundles (Kolston 1999; Dallos 2003).

Evidence of a mechanical output from cochlear hair cells is available in the form of otoacoustic emissions. Such emissions, where the external ear radiates sound energy (Kemp 1978), have been recorded in sub-mammalian vertebrates (frogs, lizards and birds; reviewed in Köppl 1995; Manley 2001) as well as mammals (Probst et al. 1991). Otoacoustic emissions can be measured under conditions similar to those where active hair bundle motion is found: they can occur spontaneously, or they can be evoked by acoustic or electrical stimulation. The combined results of electrical and acoustic stimulation have provided strong evidence for the active involvement of the hair cells in generating otoacoustic emissions in mammals (Mountain and Hubbard 1989; Ren and Nuttall 1996; Yates and Kirk 1998). The interaction between electrical and acoustic stimuli in the bobtail lizard, which has rows of hair cells with oppositely oriented hair bundles, argues that emissions in that animal originate from motion of the bundle (Manley et al. 2001).

The narrow-band nature of spontaneous otoacoustic emissions is reminiscent of the spontaneous periodic motion of hair bundles, but there is a significant discrepancy in the frequency ranges of the two phenomena. Spontaneous otoacoustic emissions occur at frequencies within the audible range of the animal, from a minimum of approximately 600 Hz in frogs (van Dijk et al. 1996) to approximately 60 kHz in certain mammals (Kössl 1994), but the highest frequency reported for spontaneous bundle motion is less than 100 Hz. In the mammalian cochlea, it is not even established that the otoacoustic emissions are the product of a local hair cell oscillator, rather than the global properties of the cochlear duct (Shera 2003). A causal link between narrow-band otoacoustic emissions and hair bundle oscillations would be greatly strengthened by the demonstration of spontaneous bundle motion in the kilohertz range of a mammalian preparation. Amplification at such high frequencies, if present, is most likely to involve the fast active process mediated by Ca^{2+} binding directly to the transducer channels. This mechanism can provide both under-damped resonance and amplification, and generate a range of resonant frequencies depending on the feedback parameters, such as the Ca^{2+} influx and the number of MET channels. In the turtle, damped oscillations of the transducer current were observed at frequencies between 58 and 230 Hz (Ricci et al. 1998), and damped oscillations in bundle motion were seen at 31 to 171 Hz (Crawford and Fettiplace 1985). Both are within the auditory range of the animal (20 to 600 Hz). Damped oscillations of a hair bundle in the chick cochlea have been reported at a frequency of 235 Hz (Hudspeth et al. 2000), also within the auditory range of that animal. Spontaneous oscillations of hair bundles could be entrained up to 300 Hz or more by delivering sinusoidal electrical currents across the frog saccular epithelium (Bosovic and Hudspeth 2003). This response was linked to hair cell transduction since it disappeared on blocking the MET channels with gentamycin.

Despite theoretical arguments, there is no direct evidence that an active mechanical process in the hair bundle is employed to tune hair cell responses in any auditory end organ. In many sub-mammalian vertebrates, including the turtle, sufficient frequency selectivity is available from an electrical tuning mechanism (Crawford and Fettiplace 1981; Fettiplace and Fuchs 1999). By contrast, in the mammalian cochlea augmentation of the intrinsic mechanical tuning of the basilar membrane has been largely attributed to the contractile behavior of the outer hair cells (Brownell et al. 1985: Ashmore 1987), which is mediated by prestin (Zheng et al. 2000). The best case in support of a role for hair bundle amplification has been made for the abneurally placed short hair cells of the avian cochlea, which by their position and sparse afferent innervation resemble outer hair cells of the mammalian cochlea. It has been proposed that their function, similar to that of the outer hair cells, is to amplify the vibrations of the tectorial membrane and thereby modify the input to the tall (inner) hair cells that contact the majority of the afferent nerve fibers (Köppl et al. 2000). The attachment of the short hair cells to adjacent supporting cells effectively precludes contractions of their cell body as envisioned for outer hair cells, and force generation by the hair bundle is the more attractive proposition. Although tall avian hair cells possess electrical tuning (Fuchs et al. 1988), this mechanism may not operate up to the highest frequencies of 5 to 10 kHz of bird hearing (Fettiplace and Fuchs 1999). Modeling of the amplification produced by Ca^{2+} binding to the transducer channel has suggested that with the distribution of avian hair bundle morphologies resonant frequencies from 0.05 to 5 kHz might be achievable (Choe et al. 1998). In the model the resonant frequency is partly dictated by the number of stereocilia per bundle, which increases from 50 to 300 from the low-frequency to high-frequency end of the chick cochlea (Tilney and Saunders 1983).

7. Mechanotransduction in the Mammalian Cochlea

7.1 Properties of Mammalian Mechanotransducer Channels

The current view of hair cell transduction is largely formulated from experiments on sub-mammalian vertebrates where it has been possible to combine recordings of electrical activity in individual cells with direct manipulation of the hair bundle. Understanding auditory transduction in the mammalian cochlea is the ultimate goal of much hair cell research, however, and it is essential to examine the generality of the conclusions because of their relevance to constructing cochlear models that incorporate realistic hair cell properties (Nobili and Mammano

1993; Kolston 1999; Dallos 2003). The introduction of an isolated neonatal mouse cochlea (Russell and Richardson 1987) allowed repetition of many of the experiments already performed on sub-mammalian vertebrates. These demonstrated a high sensitivity for hair bundle displacements toward the tallest row of stereocilia, and a nonselective cationic transducer conductance up to 9 nS, as large as any previously measured (Russell and Richardson 1987; Kros et al. 1992; Géléoc et al. 1997). The transducer conductance is composed of single channels with unitary amplitude of 112 pS (Géléoc et al. 1997), and is blocked by aminoglycoside antibiotics (Kros et al. 1992) and amiloride (Rüsch et al. 1994). Recordings in mouse hair cells also showed evidence of the fast adaptation mechanisms seen in lower vertebrates (Kros et al. 1992). Experiments using a rapid piezoelectric stimulator to deflect the hair bundles of rat outer hair cells have shown sub-millisecond adaptation kinetics even at room temperature (Fig. 4.13; Kennedy et al. 2003). As in the turtle, this fast adaptation was Ca^{2+} dependent. Correcting τ_{Af} to in vivo conditions of endolymph composition and higher body temperature predicted adaptation time constants of approximately 50 μs, which fits with the high CF of the hair cells studied (Kennedy et al.

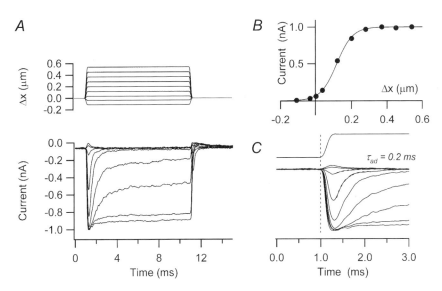

FIGURE 4.13. Hair cell transduction in the mammalian cochlea. (**A**) Average MET currents recorded from an outer hair cell from a P7 rat in response to hair bundle deflections (Δx) evoked with a fast piezoelectric actuator. Note the predominance of fast adaptation, but otherwise the currents resemble those seen in turtle hair cells. (**B**) Current–displacement relationship, which has been fitted with Eq. (4.1): $I = I_{max}/[1 + \{\exp(a \cdot (X_0 - X))\}]$, where $I_{max} = 1$ nA, $a = 22.2$ μm^{-1}, and $X_0 = 0.12$ μm. (**C**) Current onsets on a faster time scale, showing that current activation is sufficiently fast to be limited by the stimulus rise time, and that fast adaptation occurs with a time constant of 0.2 ms. Holding potential −84 mV, temperature 22°C, and extracellular Ca^{2+} 2.8 mM.

2003). These results indicate that the MET channels in mammalian auditory hair cells display the same characteristics as those in sub-mammalian hair cells, although their kinetics may be faster for processing the higher sound frequencies.

A drawback of the mouse experiments is that they employ neonatal preparations up to about 10 days of age, prior to the onset of hearing which occurs between postnatal days 12 and 14. Extending the age at which a rodent preparation can be isolated and still transduce has proved difficult, possibly because of alterations in cochlear anatomy or increased mechanical sensitivity of the hair cells at maturation. Some of the vulnerability may reflect changes in expression of a number of proteins that occur at this point in development. For example, in outer hair cells, both the KCNQ4 potassium channel (Kharkovets et al. 2000; Marcotti and Kros 1999) and the motor protein, prestin (Belyantseva et al. 2000), are up-regulated at a time coinciding with the emergence of electomotility. It is important to know whether there are concurrent changes in the number or properties (e.g., kinetics, adaptation, subunit composition) of the MET channels. Recent observations in neonatal rat outer hair cells failed to find any major change in the magnitude of the MET current or in the time constant of adaptation as the cochlea matured through the onset of hearing (Kennedy et al. 2003). More extensive measurements in adults are still needed to verify this conclusion, however. Another point of comparison with the turtle is whether there is an increase in the number of channels per hair cell toward the high-frequency end of the mammalian cochlea. A gradient in hair cell transduction has been assumed in some cochlear models (Nobili and Mammano 1996). Recordings from outer hair cells in a novel gerbil hemicochlea preparation during vibration of the intact basilar membrane have found large MET currents whose amplitude and sensitivity increase with more basal high-frequency locations (He et al. 2004). These changes may stem from increases in the number of stereocilia per bundle and size and number of MET channels, as in the turtle. Preliminary measurements at multiple locations in the rat cochlea have also confirmed that more basal outer hair cells have larger currents and correspondingly faster adaptation time constants (Ricci et al. 2005).

It is known that there are positional gradients of K^+ currents in outer hair cells (Mammano and Ashmore 1996) with $I_{K,n}$, the current that flows through ion channels containing the KCNQ4 subunit (Marcotti and Kros 1999), increasing toward the basal turn of the guinea pig cochlea. One hypothesis for the function of $I_{K,n}$ is that it is an exit route for K^+ ions that have entered the outer hair cells through the MET channels. It may therefore be part of the pathway for recycling K^+ back to the endolymphatic space (Kikuchi et al. 2000). The tonotopic gradient in $I_{K,n}$ would be consistent with an increased K^+ loading in the high-frequency cells resulting from a larger MET current.

7.2 Hair Bundle Mechanics and Active Bundle Movements

The properties of mammalian hair bundles are important for the part they might play in cochlear mechanics, contributing to the passive stiffness of the cochlear partition and generating force as part of the cochlear amplifier. Several measurements on rodent inner and outer hair cells have given passive stiffness values of 1 to 10 mN · m^{-1} at the tip of the bundle (Strelioff and Flock 1984; Russell et al. 1992; Géléoc et al. 1997; Langer et al. 2001). The mechanics of mammalian hair bundles, like those of sub-mammalian vertebrates, show signs of a nonlinear stiffness attributable to the gating of the MET channels (Russell et al. 1992; van Netten and Kros 2000). However, there is no evidence so far that such nonlinearity is accompanied by fast active movements of the hair bundle such as those seen in turtle and bullfrog. Even if they did occur, it is not clear that compared to the somatic contractility of outer hair cells they are an important component of the mechanical feed back that underlies the cochlear amplifier.

The relative benefits of mechanical amplification by active motion of the hair bundle ("bundle amplifier") compared to the cell body ("somatic amplifier") have been recently summarized (Fettiplace and Fuchs 1999; Robles and Ruggero 2001). Three points of comparison are the amount of force that each develops, the speed of the movements, and their potential for frequency selectivity. The maximum force produced by the somatic amplifier is at least 10 times larger than that produced by the bundle amplifier, assuming that the cell's complement of MET channels is similar to turtle auditory hair cells (Fettiplace et al. 2001). However, recent measurements in the semi-intact gerbil cochlea have indicated that the MET currents are significantly larger than previously thought (He et al. 2004). The force produced by the bundle amplifier will increase with the number of MET channels, which may be more numerous in adult mammalian outer hair cells than in sub-mammalian vertebrates. A recurring criticism of a voltage-controlled somatic amplifier is the low-pass filtering by the membrane time constant, which will attenuate the receptor potential above 1 kHz (Santos-Sacchi 1992; Preyer et al. 1996). The decline in sensitivity imposed by the membrane time constant may be partially offset by the increase in amplitude of the MET current with CF of mammalian outer hair cells. The bundle amplifier, in contrast, is driven by the MET current, which is expected to be wideband and limited only by the kinetics of the transducer channels. A third factor is the frequency selectivity of the amplifier. For cochlear models to produce realistically sharp tuning, the active mechanical feedback from the outer hair cells must occur in a frequency-selective manner to ensure it supplies force at the appropriate phase of basilar membrane vibration (Neely and Kim 1983; Geisler and Sang 1995; Nobili and Mammano 1996). The required tuning has been postulated as a mechanical resonance in the tectorial membrane but it could equally reside in the outer hair cells. There is no evidence that the "somatic amplifier" is frequency selective but the bundle amplifier, if it operates as in the turtle cochlea, will possess frequency selectivity and the appropriate phase delay imparted by the variation in the speed of fast adaptation with hair cell location (Ricci et al. 2000). As discussed in Section 5, this variation reflects differences in Ca^{2+} influx through MET channels of different unitary conductance.

7.3 Conclusions

Appreciation of the complexities of mechanoelectrical transduction in hair cells has grown steadily since the first success in manipulating hair bundles and measuring receptor potentials in isolated epithelia (Flock et al. 1977; Hudspeth and Corey 1977). Nevertheless, nearly 30 years on there are still important questions remaining about the molecular and functional attributes of mechanotransduction in hair cells. A major goal of future research is to identify the components of the transduction apparatus, including the MET channel itself and the ancillary proteins that bridge each to the actin core of the stereocilium. Together these may comprise a large submembranous complex (conceivably the electron-dense region at the tips of the stereocilia) as has been proposed for other transduction channels (Huber 2001). Development of specific antibodies for the MET channel for use in electron microscopy will confirm whether they are localized at one or both ends of the tip link. It will be important to document more fully the properties of the MET channels in mammalian cochlear hair cells. Are they identical to those channels that have been characterized in lower vertebrates? Are their properties, size and kinetics, tailored to the frequencies encoded by the hair cell? Are they subject to the same control by Ca^{2+} binding that can create frequency tuning of the transducer current and hair bundle motion? The narrow stimulus limits of transduction, corresponding to a total excursion of approximately 100 nm at the tip of the hair bundle, demand tight regulation of the mechanical input to ensure that the MET channels operate in their optimal range. This regulation is provided by multiple components of adaptation involving one or more unconventional myosins. More information is needed about whether the same myosin isoforms are used in vestibular and auditory hair cells, how these are regulated by Ca^{2+}, and how they connect to the channel. In the mammalian cochlea, the somatic contractions of the outer hair cell may also perform an adaptive function by adjusting the position of the reticular lamina to appropriately orient the hair bundles. Such adaptation would be fast (Frank et al. 1999), have a broad dynamic range, and may facilitate active high-frequency mechanical resonance of the hair bundle. The relative contribution of active bundle motion and outer hair cell contractility to amplification and frequency selectivity in the mammalian cochlea is still an open question. Two recent papers (Chan and Hudspeth 2005; Kennedy et al. 2005) provide evidence for a mechanical output attributable to MET channel gating in mammalian cochlear hair cells, but the significance and mechanism remain to be established.

Acknowledgments. Work in the authors' laboratories was supported by National Institutes on Deafness and other Communicative Disorders Grants RO1 DC 01362 to R. Fettiplace and RO1 DC 03896 to A.J. Ricci. We wish to thank our colleagues Andrew Crawford and Helen Kennedy, who participated in some of the experiments illustrated, and Andrew Crawford and Carole Hackney for valuable comments and suggestions.

References

Art JJ, Fettiplace R (1987) Variation of membrane properties in hair cells isolated from the turtle cochlea. J Physiol 385:207–242.
Art JJ, Wu YC, Fettiplace R (1995) The calcium-activated potassium channels of turtle hair cells. J Gen Physiol 105:49–72.
Ashmore JF (1987) A fast motile response in guinea-pig outer hair cells: the cellular basis of the cochlear amplifier. J Physiol 388:323–347.
Assad JA, Corey DP (1992) An active motor model for adaptation by vertebrate hair cells. J Neurosci 12:3291–3309.
Assad JA, Hacohen N, Corey DP (1989) Voltage dependence of adaptation and active bundle movement in bullfrog saccular hair cells. Proc Natl Acad Sci USA 86:2918–2922.
Assad JA, Shepherd GM, Corey DP (1991) Tip-link integrity and mechanical transduction in vertebrate hair cells. Neuron 7:985–994.
Bagger-Sjöback D, Wersäll J (1973) The sensory hairs and tectorial membrane of the basilar papilla in the lizard *Calotes versicolor*. J Neurocytol 2:329–350.
Balasubramanian S, Lynch JW, Barry PH (1995) The permeation of organic cations through cAMP-gated channels in mammalian olfactory receptor neurons. J Membr Biol 146:177–191.
Batters C, Arthur CP, Lin A, Porter J, Geeves MA, Milligan RA, Molly JE, Coluccio LM (2004) Myo 1c is designed for the adaptation responsxe in the inner ear. EMBO J 23:1433–1440.
Baumann M, Roth A (1986) The Ca^{++} permeability of the apical membrane in neuromast hair cells. J Comp Physiol [A] 158:681–688.
Belyantseva IA, Adler HJ, Curi R, Frolenkov GI, Kachar B (2000) Expression and localization of prestin and the sugar transporter GLUT-5 during development of electromotility in cochlear outer hair cells. J Neurosci 20:RC116.
Belyantseva IA, Boger ET, Friedman TB (2003) Myosin XVa localizes to the tips of the inner ear stereocilia and is essential for staircase formation of the hair bundle. Proc Natl Acad Sci USA 100:13958–13963.
Benser ME, Issa NP, Hudspeth AJ (1993) Hair-bundle stiffness dominates the elastic reactance to otolithic- membrane shear. Hear Res 68:243–252.
Benser ME, Marquis RE, Hudspeth AJ (1996) Rapid, active hair bundle movements in hair cells from the bullfrog's sacculus. J Neurosci 16:5629–5643.
Bialek W (1987) Physical limits to sensation and perception. Annu Rev Biophys Biophys Chem 16:455–478.
Block SM (1992) Biophysical principles of sensory transduction. In: Corey DP, Roper SD (eds), Sensory Transduction. New York: Rockefeller University Press, pp. 1–17.
Bönigk W, Bradley J, Müller F, Sesti F, Boekhoff I, Ronnett V, Kaupp UB, Frings S (1999) The native rat olfactory nucleotide-gated channel is composed of three distinct subunits. J Neurosci 19:5332–5347.
Bosher SK, Warren RL (1978) Very low calcium content of cochlear endolymph, an extracellular fluid. Nature 273:377–378.

Bosovic D, Hudspeth AJ (2003) Hair-bundle movements elicited by transepthelial electrical stimulation of hair cells in the sacculus of the bullfrog. Proc Natl Acad Sci USA 100:958–963.

Brownell WE, Bader CR, Bertrand D, de Ribaupierre Y (1985) Evoked mechanical responses of isolated cochlear outer hair cells. Science 227:194–196.

Chan DK, Hudspeth AJ (2005) Ca^{2+} current-driven nonlinear amplification by the mammalian cochlea in vitro. Nat Neurosci 8:149–155.

Cheatham MA, Dallos P (2001) Inner hair cell response patterns: implications for low-frequency hearing. J Acoust Soc Am 110:2034–2044.

Choe Y, Magnasco MO, Hudspeth AJ (1998) A model for amplification of hair-bundle motion by cyclical binding of Ca^{2+} to mechanoelectrical-transduction channels. Proc Natl Acad Sci USA 95:15321–15326.

Colbert HA, Smith TL, Bargmann CI (1997) OSM-9, a novel protein with structural similarity to channels, is required for olfaction, mechanosensation, and olfactory adaptation in *Caenorhabditis elegans*. J Neurosci 17:8259–8269.

Corey DP, Hudspeth AJ (1979) Ionic basis of the receptor potential in a vertebrate hair cell. Nature 281:675–677.

Corey DP, Hudspeth AJ (1983) Kinetics of the receptor current in bullfrog saccular hair cells. J Neurosci 3:962–976.

Corey DP, Garcia-Anoveros J, Holt JR, Kwan KY, Lin SY, Vollrath MA, Amalfitano A, Cheung EL, Derfler BH, Duggan A, Géléoc GS, Gray PA, Hoffman MP, Rehm HL, Tamasauskas D, Zhang DS (2004) TRPA1 is a candidate for the mechanosensitive transduction channel of vertebrate hair cells. Nature 9:723–730.

Cortopassi G, Hutchin T (1994) A molecular and cellular hypothesis for aminoglycoside-induced deafness. Hear Res 78:27–30.

Crawford AC, Fettiplace R (1980) The frequency selectivity of auditory nerve fibres and hair cells in the cochlea of the turtle. J Physiol 306:79–125.

Crawford AC, Fettiplace R (1981) An electrical tuning mechanism in turtle cochlear hair cells. J Physiol 312:377–412.

Crawford AC, Fettiplace R (1983) Auditory nerve responses to imposed displacements of the turtle basilar membrane. Hear Res 12:199–208.

Crawford AC, Fettiplace R (1985) The mechanical properties of ciliary bundles of turtle cochlear hair cells. J Physiol 364:359–379.

Crawford AC, Evans MG, Fettiplace R (1989) Activation and adaptation of transducer currents in turtle hair cells. J Physiol 419:405–434.

Crawford AC, Evans MG, Fettiplace R (1991) The actions of calcium on the mechano-electrical transducer current of turtle hair cells. J Physiol 434:369–398.

Cruickshank CC, Minchin RF, Le Dain AC, Martinac B (1997) Estimation of the pore size of the large-conductance mechanosensitive ion channel of *Escherichia coli*. Biophys J 73:1925–1931.

Dallos P (2003) Some pending problems in cochlear mechanics. In: Gummer AW (ed), Biophysics of the Cochlea: From Molecules to Models. Singapore: World Scientific, pp. 97–109.

Denk W, Webb WW (1992) Forward and reverse transduction at the limit of sensitivity studied by correlating electrical and mechanical fluctuations in frog saccular hair cells. Hear Res 60:89–102.

Denk W, Holt JR, Shepherd GM, Corey DP (1995) Calcium imaging of single stereocilia in hair cells: localization of transduction channels at both ends of tip links. Neuron 15:1311–1321.

Donaudy F, Ferrara A, Esposito L, Hertzano R, Ben-David O, Bell RE, Melchionado S, Zelante L, Avraham K, Gasparini P (2003) Multiple mutations of *MYO1A*, a cochlear-expressed gene, in sensorineural hearing loss. Am J Hum Genet 72:1571–1577.

Drescher MJ, Khan KM, Beisel KW, Karadaghy AA, Hatfield JS, Kim SY, Drescher AJ, Lasak JM, Barretto RL, Shakir AH, Drescher DG (1997) Expression of adenylyl cyclase type I in cochlear inner hair cells. Mol Brain Res 45:325–330.

Drescher MJ, Barretto RL, Chaturvedi D, Beisel KW, Hatfield JS, Khan KM, Drescher DG (2002) Expression of subunits for the cAMP-sensitive 'olfactory' cyclic nucleotide-gated ion channel in the cochlea: implications for signal transduction. Brain Res Mol Brain Res 98:1–14.

Dumont RA, Lins U, Filoteo AG, Penniston JT, Kachar B, Gillespie PG (2001) Plasma membrane Ca^{2+}-ATPase isoform 2a is the PMCA of hair bundles. J Neurosci 21: 5066–5078.

Dwyer TM, Adams DJ, Hille B (1980) The permeability of the endplate channel to organic cations in frog muscle. J Gen Physiol 75:469–492.

Eatock RA (2000) Adaptation in hair cells. Annv Rev Neurosci 23:285–314.

Eatock RA, Corey DP, Hudspeth AJ (1987) Adaptation of mechanoelectrical transduction in hair cells of the bullfrog's sacculus. J Neurosci 7:2821–2836.

Eberl DF, Hardy RW, Kernan MJ (2000) Genetically similar transduction mechanisms for touch and hearing in *Drosophila*. J Neurosci 20:5981–5988.

Farris HE, LeBlanc CL, Goswami J, Ricci AJ (2004) Probing the pore of the auditory hair cell mechanotransducer channel in turtle. J Physiol (epub June).

Fettiplace R, Fuchs PA (1999) Mechanisms of hair cell tuning. Annv Rev Physiol 61: 809–34.

Fettiplace R, Ricci AJ (2003) Adaptation in auditory hair cells. Curr Opin Neurobiol 13:446–451.

Fettiplace R, Ricci AJ, Hackney CM (2001) Clues to the cochlear amplifier from the turtle ear. Trends Neurosci 24:169–175.

Fettiplace R, Crawford AC, Ricci AJ (2003) The effects of calcium on mechanotransducer channel kinetics in auditory hair cells. In: Gummer AW (ed), Biophysics of the Cochlea: From Molecules to Models. Singapore: World Scientific, pp. 65–72.

Flock A, Flock B, Murray E (1977) Studies on the sensory hairs of receptor cells in the inner ear. Acta Otolaryngol 83:85–91.

Flock AF, Strelioff D (1984) Studies on hair cells in isolated coils from the guinea pig cochlea. Hear Res 15:11–18.

Forge A, Schacht J (2000) Aminoglycoside antibiotics. Audiol Neurootol 5: 3–32.

Frank G, Hemmert W, Gummer AW (1999) Limiting dynamics of high-frequency electromechanical transduction of outer hair cells. Proc Natl Acad Sci USA 96:4420–4425.

Frank JE, Markin V, Jaramillo F (2002) Characterization of adaptation motors in saccular hair cell by fluctuation analysis. Biophys J 83:3188–3201.

Franke R, Dancer A (1982) Cochlear mechanisms at low frequencies in the guinea pig. Arch Otorhinolaryngol 234:213–218.

Fridberger A, Flock A, Ulfendahl M, Flock B (1998) Acoustic overstimulation increases outer hair cell Ca^{2+} concentrations and causes dynamic contractions of the hearing organ. Proc Natl Acad Sci USA 95:7127–7132.

Frings S, Lynch JW, Lindemann B (1992) Properties of cyclic nucleotide-gated channels mediating olfactory transduction. Activation, selectivity and blockage. J Gen Physiol 100:45–67.

Fodor AA, Gordon SE, Zagotta WN (1997) Mechanism of tetracaine block of cyclic nucleotide-gated channels. J Gen Physiol 109:3–14.

Fuchs PA, Nagai T, Evans MG (1988) Electrical tuning in hair cells isolated from the chick cochlea. J Neurosci 8:2460–2467.

Furness DN, Hackney CM (1985) Cross-links between stereocilia in the guinea pig cochlea. Hear Res 18:177–188.

Gale JE, Marcotti W, Kennedy HJ, Kros CJ, Richardson GP (2001) FM1-43 dye behaves as a permeant blocker of the hair-cell mechanotransducer channel. J Neurosci 21: 7013–7025.

Garcia JA, Yee AG, Gillespie PG, Corey DP (1998) Localization of myosin-Ibeta near both ends of tip links in frog saccular hair cells. J Neurosci 18:8637–8647.

Geisler CD (1993) A model of stereociliary tip-link stretches. Hear Res 65:79–82.

Geisler CD, Sang C (1995) A cochlear model using feed-forward outer-hair-cell forces. Hear Res 86:132–146.

Géléoc G, Corey DP (2001) Modulation of mechanoelectrical transduction by protein kinase A in utricular hair cells of neonatal mice. Assoc Res Otolaryngol Abst 24:242.

Géléoc GS, Lennan GW, Richardson GP, Kros CJ (1997) A quantitative comparison of mechanoelectrical transduction in vestibular and auditory hair cells of neonatal mice. Proc R Soc Lond B 264:611–621.

Gillespie PG, Corey DP (1997) Myosin and adaptation by hair cells. Neuron 19:955–958.

Gillespie PG, Cyr JL (2004) Myosin-1c, the hair cell's adaptation motor. Ann Rev Physiol 66:521–545.

Gillespie PG, Wagner MC, Hudspeth AJ (1993) Identification of a 120 kd hair-bundle myosin located near stereociliary tips. Neuron 11:581–594.

Gillespie PG, Gillespie SK, Mercer JA, Shah K, Shokat KM (1999) Engineering of the myosin-1beta nucleotide-binding pocket to create selective sensitivity to $N(6)$-modified ADP analogs. J Biol Chem 274:31373–31381.

Glowatzki E, Ruppersberg JP, Zenner HP, Rüsch A (1997) Mechanically and ATP-induced currents of mouse outer hair cells are independent and differentially blocked by d-tubocurarine. Neuropharmacology 36:1269–1275.

Goodman MB, Ernstrom GG, Chelur DS, O'Hagan R, Yao CA, Chalfie M (2002) MEC-2 regulates *C. elegans* DEG/ENaC channels needed for mechanosensation. Nature 415: 1039–1042.

Goodyear R, Richardson G (1999) The ankle-link antigen: an epitope sensitive to calcium chelation associated with the hair-cell surface and the calycal processes of photoreceptors. J Neurosci 19:3761–3772.

Grunder S, Muller A, Ruppersberg JP (2001) Developmental and cellular expression pattern of epithelial sodiuim channel alpha, beta and gamma subunits in the inner ear of the rat. Eur J Neurosci 13:641–648.

Hackney CM, Furness DN (1995) Hair cell ultrastructure and mechanotransduction: morphological effects of low extracellular calcium levels on stereociliary bundles in the turtle cochlea. In: Flock A, Ottoson D, Ulfendahl M (eds), Active Hearing. Oxford: Pergamon, pp. 103–111.

Hackney CM, Fettiplace R, Furness DN (1993) The functional morphology of stereociliary bundles on turtle cochlear hair cells. Hear Res 69:163–175.

Hackney CM, Furness DN, Benos DJ, Woodley JF, Barratt J (1992) Putative immunolocalization of the mechanoelectrical transduction channels in mammalian cochlear hair cells. Proc R Soc Lond B 248:215–221.

Hackney CM, Mahendrasingam S, Jones EMC, Fettiplace R (2003) The distribution of calcium buffering proteins in the turtle cochlea. J Neurosci 23:4577–4589.

Hacohen N, Assad JA, Smith WJ, Corey DP (1989) Regulation of tension on hair-cell transduction channels: displacement and calcium dependence. J Neurosci 9:3988–3997.

Hasson T, Gillespie PG, Garcia JA, MacDonald RB, Zhao Y, Yee AG, Mooseker MS, Corey DP (1997) Unconventional myosins in inner-ear sensory epithelia. J Cell Biol 137:1287–1307.

He DZ, Jia S, Dallos P (2004) Mechanoelectrical transduction of adult outer hair cells studied in a gerbil hemicochlea. Nature 429:766–770.

Heller S, Bell AM, Denis CS, Choe Y, Hudspeth AJ (2002) Parvalbumin 3 is an abundant Ca^{2+} buffer in hair cells. J Assoc Res Otolaryngol 3:488–498.

Hiel H, Navaratnam D, Oberholtzer JO, Fuchs PA (2001) Topological and developmental gradients of calbindin expression in the chick's inner ear. J Assoc Res Otolaryngol 3:1–15.

Holt JR, Corey DP, Eatock RA (1997) Mechanoelectrical transduction and adaptation in hair cells of the mouse utricle, a low frequency vestibular organ. J Neurosci 17:8739–8748.

Holt JR, Gillespie SK, Provance DW, Shah K, Shokat KM, Corey DP, Mercer JA, Gillespie PG (2002)A chemical-genetic strategy implicates myosin-1c in adaptation by hair cells. Cell 108:371–381.

Holton T, Hudspeth AJ (1986) The transduction channel of hair cells from the bull-frog characterized by noise analysis. J Physiol 375:195–227.

Howard J (2001) Mechanics of Motor Proteins and the Cytoskeleton. Sunderland, MA: Sinauer.

Howard J, Ashmore JF (1986) Stiffness of sensory hair bundles in the sacculus of the frog. Hear Res 23:93–104.

Howard J, Hudspeth AJ (1987) Mechanical relaxation of the hair bundle mediates adaptation in mechanoelectrical transduction by the bullfrog's saccular hair cell. Proc Natl Acad Sci USA 84:3064–3068.

Howard J, Hudspeth AJ (1988) Compliance of the hair bundle associated with gating of mechanoelectrical transduction channels in the bullfrog's saccular hair cell. Neuron 1:189–199.

Huang M, Chalfie M (1994) Gene interactions affecting mechanosensory transduction in *Caenorhabditis elegans.* Nature 367:467–470.

Huber A (2001) Scaffolding proteins organize multimolecular protein complexes for sensory signal transduction. Eur J Neurosci 14:769–776.

Hudspeth AJ (1982) Extracellular current flow and the site of transduction by vertebrate hair cells. J Neurosci 2:1–10.

Hudspeth AJ (1985) The cellular basis of hearing: the biophysics of hair cells. Science 230:745–752.

Hudspeth A (1997) Mechanical amplification of stimuli by hair cells. Curr Opin Neurobiol 7:480–486.

Hudspeth AJ, Corey DP (1977) Sensitivity, polarity, and conductance change in the response of vertebrate hair cells to controlled mechanical stimuli. Proc Natl Acad Sci USA 74:2407–2411.

Hudspeth AJ, Choe Y, Mehta AD, Martin P (2000) Putting ion channels to work: mechanoelectrical transduction, adaptation, and amplification by hair cells. Proc Natl Acad Sci USA 97:11765–11772.

Ikeda K, Kusakari JU, Takasaka T, Saito Y (1987) The Ca^{2+} activity of cochlear endolymph of the guine pig and the effect of inhibitors. Hear Res 26:117–125.

Ismailov II, Awayda MS, Berdiev BK, Bubien JK, Lucas JE, Fuller CM, Benos DJ (1996) Triple-barrel organization of ENaC, a cloned epithelial Na^+ channel. J Biol Chem 271:807–816.

Jaramillo F, Hudspeth AJ (1991) Localization of the hair cell's transduction channels at the hair bundle's top by iontophoretic application of a channel blocker. Neuron 7: 409–420.

Jaramillo F, Hudspeth AJ (1993) Displacement-clamp measurement of the forces exerted by gating springs in the hair bundle. Proc Natl Acad Sci USA 90:1330–1334.

Jaramillo F, Howard J, Hudspeth AJ (1990) Calcium ions promote rapid mechanically evoked movements of hair bundles. In: Dallos P, Geisler, CD, Matthews JW, Ruggero MA, Steele CR (eds), The Mechanics and Biophysics of Hearing. Berlin: Springer-Verlag, pp. 26–33.

Jones EM, Gray-Keller M, Fettiplace R (1999) The role of Ca^{2+}-activated K^+ channel spliced variants in the tonotopic organization of the turtle cochlea. J Physiol 518:653–665.

Jørgensen F, Kroese ABA (1995) Ca selectivity of the transduction channels in hair cell of the frog sacculus. Acta Physiol Scand 155:363–376.

Jørgensen F, Ohmori H (1988) Amiloride blocks the mechanoelectrical transduction channel of hair cells of chick. J Physiol 403:577–588.

Kachar B, Parakkal M, Kurc M, Zhao Y, Gillespie PG (2000) High-resolution structure of hair-cell tip links. Proc Natl Acad Sci USA 97:13336–13341.

Kemp DT (1978) Stimulated acoustic emissions from within the human auditory system. J Acoust Soc Am 64:1386–1391.

Kennedy HJ, Evans MG, Crawford AC, Fettiplace R (2003) Fast adaptation of mechanoelectrical transducer channels in mammalian cochlear hair cells. Nat Neurosci 6: 832–836.

Kennedy HJ, Crawford AC, Fettiplace R (2005) Force generation by mammalian hair bundles supports a role in cochlear amplification. Nature 433:880–883.

Kharkovets T, Hardelin JP, Safieddine S, Schweizer M, El-Amraoui A, Petit C, Jentsch TJ (2000) KCNQ4, a K^+ channel mutated in a form of dominant deafness, is expressed in the inner ear and the central auditory pathway. Proc Natl Acad Sci USA 97:4333–4338.

Kikuchi T, Kimura RS, Paul DL, Takasaka T, Adams JC (2000) Gap junction systems in the mammalian cochlea. Brain Res Brain Res Rev 32:163–166.

Kim J, Chung YD, Park D-Y, Choi S, Shin, DW, Soh H, Lee HW., Son W, Yim J, Park C-S, Kernan MJ, Kim CA (2003) TRPV family ion channel required for hearing in *Drosophila*. Nature 424:81–84.

Kimitsuki T, Ohmori H (1992) The effect of caged calcium release on the adaptation of the transduction current in chick hair cells. J Physiol 458:27–40.

Kimitsuki T, Ohmori H (1993) Dihydrostreptomycin modifies adaptation and blocks the mechano-electric transducer in chick cochlear hair cells. Brain Res 624:143–150.

Kimitsuki T, Nakagawa T, Hisashi K, Komune S, Komiyama S (1996) Gadolinium blocks mechano-electric transducer current in chick cochlear hair cells. Hear Res 101:75–80.

Kolston PJ (1999) Comparing in vitro, in situ, and in vivo experimental data in a three-dimensional model of mammalian cochlear mechanics. Proc Natl Acad Sci USA 96: 3676–3681.

Köppl C (1995) Otoacoustic emissions as an indicator for active cochlear mechanics: a primitive property of vertebrate hearing organs. In: Manley GA (ed), Advances in Hearing Research. Singapore: World Science Publishers, pp. 207–216.

Köppl C, Manley GA, Konishi M (2000) Auditory processing in birds. Curr Opin Neurobiol 10:474–481.

Kössl M (1994) Otoacoustic emissions from the cochlea of the 'constant frequency' bats, *Pteronotus parnellii* and *Rhinolophus rouxi*. Hear Res 72:59–72.

Kozel PJ, Friedman RA, Erway LC, Yamoah EN, Liu LH, Riddle T, Duffy JJ, Doetschman T, Miller ML, Cardell EL, Shull GE (1998) Balance and hearing deficits in mice with a null mutation in the gene encoding plasma membrane Ca^{2+}-ATPase isoform 2. J Biol Chem 273:18693–18696.

Kozel PJ, Davis RR, Krieg EF, Shull GE, Erway LC (2002) Deficiency in the plasma membrane calcium ATPase isoform 2 increases susceptibility to noise-induced hearing loss. Hear Res 164:231–239.

Kroese AB, Das A, Hudspeth AJ (1989) Blockage of the transduction channels of hair cells in the bullfrog's sacculus by aminoglycoside antibiotics. Hear Res 37:203–217.

Kros CJ, Rüsch A, Richardson GP (1992) Mechano-electrical transducer currents in hair cells of the cultured neonatal mouse cochlea. Proc R Soc Lond B 249:185–193.

Kros CJ, Marcotti W, van Netten SM, Self TJ, Libby RT, Brown SD, Richardson GP, Steel KP (2002) Reduced climbing and increased slipping adaptation in cochlear hair cells of mice with *Myo7a* mutations. Nat Neurosci 5:41–47.

Kubisch C, Schroeder BC, Friedrich T, Lutjohann B, El-Amraoui A, Marlin S, Petit C, Jentsch TJ (1999) KCNQ4, a novel potassium channel expressed in sensory outer hair cells, is mutated in dominant deafness. Cell 96:437–446.

Langer MG, Fink S, Koitschev A, Rexhausen U, Horber JK, Ruppersberg JP (2001) Lateral mechanical coupling of stereocilia in cochlear hair bundles. Biophys J 80: 2608–2621.

Lim DJ (1986) Functional structure of the organ of Corti: a review. Hear Res 22:117–146.

Little KF, Neugebauer D-C (1985) Interconnections between the stereovilli of the fish inner ear. II. Systematic investigation of saccular hair bundles of *Rutilus rutilus* (Teleostei). Cell Tissue Res 284:473–479.

Lumpkin EA, Hudspeth AJ (1995) Detection of Ca^{2+} entry through mechanosensitive channels localizes the site of mechanoelectrical transduction in hair cells. Proc Natl Acad Sci USA 92:10297–10301.

Lumpkin EA, Marquis RE, Hudspeth AJ (1997) The selectivity of the hair cell's mechanoelectrical-transduction channel promotes Ca^{2+} flux at low Ca^{2+} concentrations. Proc Natl Acad Sci USA 94:10997–11002.

Mammano F, Ashmore JF (1996) Differential expression of outer hair cell potassium currents in the isolated cochlea of the guinea-pig. J Physiol 496:639–646.

Manley GA (2001) Evidence for an active process and a cochlear amplifier in nonmammals. J Neurophysiol. 86:541–549.

Manley GA, Kirk DL, Koppl C, Yates GK (2001) In vivo evidence for a cochlear amplifier in the hair-cell bundle of lizards. Proc Natl Acad Sci USA 98:2826–2831.

Marcotti W, Kros CJ (1999) Developmental expression of the potassium current $I_{K,n}$ contributes to maturation of mouse outer hair cells. J Physiol 520:653–660.

Markin VS, Hudspeth AJ (1995) Gating-spring models of mechanoelectrical transduction by hair cells of the internal ear. Annu Rev Biophys Biomol Struct 24:59–83.

Martin P, Hudspeth AJ (1999) Active hair-bundle movements can amplify a hair cell's response to oscillatory mechanical stimuli. Proc Natl Acad Sci USA 96:14306–14311.

Martin P, Mehta AD, Hudspeth AJ (2000) Negative hair-bundle stiffness betrays a mechanism for mechanical amplification by the hair cell. Proc Natl Acad Sci USA 97: 12026–12031.

Martin P, Bozovic D, Choe Y, Hudspeth AJ (2003) Spontaneous oscillations by hair bundles of bullfrog's sacculus. J Neurosci 23:4533–4548.

Matulef K, Zagotta WN (2003) Cyclic nucleotide-gated ion channels. Annu Rev Cell Dev Biol 19:23–44.

Meyer J, Furness DN, Zenner HP, Hackney CM, Gummer AW (1998) Evidence for opening of hair-cell transducer channels after tip-link loss. J Neurosci 18:6748–6756.

Meyers JR, MacDonald RB, Duggan A, Lenzi D, Standaert DG, Corwin JT, Cory DP (2003) Lighting up the senses: FM1-43 loading of sensory cells through nonslelective ion channels. J Neurosci 23:4054–4065.

Mhatre AN, Li J, Kim Y, Coling DE, Lalwani AK (2004) Cloning and developmental expression of nonmuscle myosin IIA (Myh9) in the mammalian inner ear. J Neurosci Res 76:296–305.

Minke B, Cook B (2002) TRP channel proteins and signal transduction. Physiol Rev 82:429–472.

Mountain DC, Hubbard AE (1989) Rapid force production in the cochlea. Hear Res 42: 195–202.

Mulroy MJ (1974) Cochlear anatomy of the alligator lizard. Brain Behav Evol 10:69–87.

Neely ST, Kim DO (1983) An active cochlear model showing sharp tuning and high sensitivity. Hear Res 9:123–130.

Nilius B, Vennekens R, Prenen J, Hoenderop JG, Bindels RJ, Droogmans G (2000) Whole-cell and single channel monovalent cation currents through the novel rabbit epithelial Ca^{2+} channel ECaC. J Physiol 527:239–248.

Nobili R, Mammano F (1996) Biophysics of the cochlea II: stationary nonlinear phenomenology. J Acoust Soc Am 99:2244–2255.

Ohmori H (1985) Mechano-electrical transduction currents in isolated vestibular hair cells of the chick. J Physiol 359:189–217.

Ostap EM, Pollard TD (1996) Biochemical kinetic characterization of the Acanthamoeba myosin-I ATPase. J Cell Biol 132:1053–1060.

Perrault-Micale C, Shushan AD, Coluccio LM (2000) Truncation of mammalian myosin I results in loss of Ca^{2+}-sensitive motility. J Biol Chem 275:21618–21623.

Petit C, Levilliers J, Hardelin J-P (2001) Molecular genetics of hearing loss. Annu Rev Genet 35:589–646.

Pickles JO (1993) A model for the mechanics of the stereociliar bundle on acousticolateral hair cells. Hear Res 68:159–172.

Pickles JO, Comis SD, Osborne MP (1984) Cross-links between stereocilia in the guinea pig organ of Corti, and their possible relation to sensory transduction. Hear Res 15: 103–112.

Pollak G, Henson OW, Novick A (1972) Cochlear microphonic audiograms in the pure tone bat, *Chilonycteris parnelli parnelli*. Science 176:66–68.

Preyer P, Renz S, Hemmert, W, Zenner, H-P, and Gummer AW (1996) Receptor potential of outer hair cells isolated from base to apex of the adult guinea pig cochlea: implications for cochlear tuning mechanisms. Aud Neurosci 2:145–157.

Probst R, Lonsbury-Martin BL, Martin GK (1991) A review of otoacoustic emissions. J Acoust Soc Am 89:2027–2067.
Ramanathan K, Michael TH, Jiang GJ, Hiel H, Fuchs PA. (1999) A molecular mechanism for electrical tuning of cochlear hair cells. Science 283:215–217.
Ren T, Nuttall AL (1996) Extracochlear electrically evoked otoacoustic emissions: a model for in vivo assessment of outer hair cell electromotility. Hear Res 92:178–183.
Ricci AJ (2002) Differences in mechano-transducer channel kinetics underlie tonotopic distribution of fast adaptation in auditory hair cells. J Neurophysiol 87:1738–1748.
Ricci AJ, Fettiplace R (1997) The effects of calcium buffering and cyclic AMP on mechano-electrical transduction in turtle auditory hair cells. J Physiol 501:111–124.
Ricci AJ, Fettiplace R (1998) Calcium permeation of the turtle hair cell mechanotransducer channel and its relation to the composition of endolymph. J Physiol 506:159–173.
Ricci AJ, Wu YC, Fettiplace R (1998) The endogenous calcium buffer and the time course of transducer adaptation in auditory hair cells. J Neurosci 18:8261–8277.
Ricci AJ, Crawford AC, Fettiplace R (2000) Active hair bundle motion linked to fast transducer adaptation in auditory hair cells. J Neurosci 20:7131–7142.
Ricci AJ, Crawford AC, Fettiplace R (2002) Mechanisms of active hair bundle motion in auditory hair cells. J Neurosci 22:44–52.
Ricci AJ, Crawford AC, Fettiplace R (2003) Tonotopic variation in the conductance of the hair cell mechanotransducer channel. Neuron 40:983–990.
Ricci AJ, Kennedy HJ, Crawford AC, Fettiplace R (2005) The transduction channel filter in auditory hair cells. J Neurosci 25:7831–7839.
Richardson GP, Forge A, Kros CJ, Fleming J, Brown SDS, Steel KP (1997) Myosin VIIA is required for aminoglycoside accumulation in cochlear hair cells. J Neurosci 17:9506–9519.
Robles L, Ruggero MA (2001) Mechanics of the mammalian cochlea. Physiol Rev 81:1305–1352.
Rüsch A, Kros CJ, Richardson GP (1994) Block by amiloride and its derivatives of mechano-electrical transduction in outer hair cells of mouse cochlear cultures. J Physiol 474:75–86.
Russell IJ, Richardson GP (1987) The morphology and physiology of hair cells in organotypic cultures of the mouse cochlea. Hear Res 31:9–24.
Russell IJ, Kössl M, Richardson GP (1992) Nonlinear mechanical responses of mouse cochlear hair bundles. Proc Roy Soc Lond Ser B 250:217–227.
Rzadzinska AK, Schneider ME, Davies C, Riordan GP, Kachar B (2004) An actin molecular treadmill and myosins maintain stereocilia functional architecture and self-renewal. J Cell Biol 164:887–897.
Salt AN, Inamura N, Thalmann R, Vora A (1989) Calcium gradients in inner ear endolymph. Am J Otolaryngol 10:371–375.
Santos-Sacchi J (1992) On the frequency limit and phase of outer hair cell motility: effects of the membrane filter. J Neurosci 12:1906–1916.
Sariban-Sohraby S, Benos DJ (1986) The amiloride-sensitive sodium channel. Am J Physiol 250:C175–190.
Saunders JC, Cohen YE, Szymko YM (1991) The structural and functional consequences of acoustic injury in the cochlea and peripheral auditory system: a five year update. J Acoust Soc Am 90:136–146.
Shera CA (2003) Mammalian spontaneous otoacoustic emissions are amplitude-stabilized cochlear standing waves. J Acoust Soc Am 114:244–262.

Shotwell SL, Jacobs R, Hudspeth AJ (1981) Directional sensitivity of individual vertebrate hair cells to controlled deflection of their hair bundles. Ann NY Acad Sci 374: 1–10.

Sidi S, Friedrich RW, Nicolson T (2003) NompC TRP channel required for vertebrate sensory hair cell mechanotransduction. Science 301:96–99.

Siemens J, Lillo C Dumont RA, Reynolds A, Williams DS, Gillespie PG, Muller U (2004) Cadherin 23 is a component of the tip link in hair cell stereocilia. Nature 428: 901–903.

Söllner C, Rauch GJ, Siemens J, Geisler R, Schuster SC, Muller U, Nicolson T (2004) Mutations in cadherin 23 affect tip links in zebrafish sensory hair cells. Nature 428: 955–959.

Steyger PS, Gillespie PG, Baird RA (1998) Myosin Ibeta is located at tip link anchors in vestibular hair bundles. J Neurosci 18:4603–4615.

Strassmaier M, Gillespie PG (2002) The hair cell's transduction channel. Curr Opin Neurobiol 12:380–386.

Street VA, McKee-Johnson JW, Fonseca RC, Tempel BL, Noben-Trauth K (1998) Mutations in a plasma membrane Ca^{2+}-ATPase gene cause deafness in *deafwaddler* mice. Nat Genet 19:390–394.

Strelioff D, Flock A (1984) Stiffness of sensory-cell hair bundles in the isolated guinea pig cochlea. Hear Res 15:19–28.

Sukharev SI, Martinac B, Arshavsky VY, Kung C (1993) Two types of mechanosensitive channels in the Escherichia coli cell envelope: solubilization and functional reconstitution. Biophys J 65:177–183.

Szymko YM, Dimitri PS, Saunders JC (1992) Stiffness of hair bundle in the chick cochlea. Hear Res 59:241–249.

Tilney LG, Saunders JC (1983) Actin filaments, stereocilia, and hair cells of the bird cochlea. I. Length, number, width, and distribution of stereocilia of each hair cell are related to the position of the hair cell on the cochlea. J Cell Biol 96:807–821.

Tucker T, Fettiplace R (1995) Confocal imaging of calcium mocrodomains and calcium extrusion in turtle hair cells. Neuron 15:1323–1335.

van Dijk P, Narins PM, Wang J (1996) Spontaneous otoacoustic emissions in seven frog species. Hear Res 101:102–112.

van Netten SM, Kros CJ (2000) Gating energies and forces of the mammalian hair cell transducer channel and related hair bundle mechanics. Proc R Soc Lond B 267:1915–1923.

van Netten SM, Dinklo T, Marcotti W, Kros CJ (2003) Channel gating forces govern accuracy of mechano-electrical transduction in hair cells. Proc Natl Acad Sci USA 100:15510–15515.

Vennekens R, Hoenderop JG, Prenen J, Stuiver M, Willems PH, Droogmans G, Nilius B, Bindels RJ (2000) Permeation and gating properties of the novel epithelial Ca^{2+} channel. J Biol Chem 275:3963–3969.

Vollrath MA, Eatock RA (2003)Time course and extent of mechanotransducer adaptation in mouse utricular hair cells: comparison with frog saccular hair cells. J Neurophysiol 90:2676–2689.

Walker RG, Hudspeth AJ (1996) Calmodulin controls adaptation of mechanoelectrical transduction by hair cells of the bullfrog's sacculus. Proc Natl Acad Sci USA 93: 2203–2207.

Walker RG, Willingham AT, Zuker CS (2000) A *Drosophila* mechanosensory transduction channel. Science 287:2229–2234.

Wu YC, Ricci AJ, Fettiplace R (1999) Two components of transducer adaptation in auditory hair cells. J Neurophysiol 82:2171–2181.

Xu H, Ramsey IS, Kotecha SA, Moran MM, Chong JA, Lawson D, Ge P, Lilly J, Silos-Santiago I, Xie Y, DiStefano PS, Curtis R, Clapham DE (2002) TRPV3 is a calcium-permeable temperature-sensitive cation channel. Nature 418:181–186.

Yamoah EN, Gillespie PG (1996) Phosphate analogs block adaptation in hair cells by inhibiting adaptation-motor force production. Neuron 17:523–533.

Yamoah EN, Lumpkin EA, Dumont RA, Smith PJ, Hudspeth AJ, Gillespie PG (1998) Plasma membrane Ca^{2+}-ATPase extrudes Ca^{2+} from hair cell stereocilia. J Neurosci 18:610–624.

Yates GK, Kirk DL (1998) Cochlear electrically evoked emissions modulated by mechanical trnasduction channels. J Neurosci 18:1996–2003.

Zhao Y, Yamoah EN, Gillespie PG (1996) Regeneration of broken tip links and restoration of mechanical transduction in hair cells. Proc Natl Acad Sci USA 93:15469–15474.

Zheng J, Shen W, He DZ, Long KB, Madison LD, Dallos P (2000) Prestin is the motor protein of cochlear outer hair cells. Nature 405:149–155

Zhu T, Beckingham K, Ikebe M (1998) High affinity Ca^{2+} binding sites of calmodulin are critical for regulation of myosin Iβ motor function. J Biol Chem 273:20481–20486.

5
Contribution of Ionic Currents to Tuning in Auditory Hair Cells

JONATHAN J. ART AND ROBERT FETTIPLACE

1. Introduction

The vertebrate hearing organ is a spectrum analyzer that in transducing sound stimuli also separates the frequency constituents of the sound along its length. At each place on the basilar membrane the hair cell receptor potential is tuned at low sound levels to one particular frequency, the characteristic frequency (CF). The auditory periphery largely dichotomizes between those in which frequency selectivity is a property local to each hair cell, and those, like mammals, in which the mechanical input to the hair cell is tuned prior to the generation of the transduction current. Local hair cell tuning can be further subdivided into micromechanical tuning attributable to the properties of the apical stereociliary bundle and electrical tuning by basolateral voltage-dependent conductances (Fettiplace and Fuchs 1999; Manley 2000). No taxonomy is without its complications and there are examples, epitomized by birds, in which cochlear frequency selectivity results from mechanical mechanisms preceding and electrical mechanisms following transduction (von Békésy 1960; Fuchs et al. 1988). In other cases, such as the alligator lizard, *Gerrhonotus multicarinatus*, different mechanisms may be used to encode different parts of the acoustic spectrum: electrical tuning for low frequencies and mechanical resonances of free-standing hair bundles for high frequencies (Holton and Weiss 1983a,b; Eatock et al. 1993).

In non-mammalian vertebrates the role of voltage-dependent Ca^{2+} channels and K^+ channels in cochlear frequency selectivity has been analyzed functionally, the mechanisms modeled, and the molecular and structural basis of channel variations are now being addressed Wu et al. 1995; Fettiplace and Fuchs 1999). The story is far less clear in the mammalian cochlea, in which both inner and outer hair cells possess the same types of ion channel as those found in non-mammals but where there is no evidence that these channels are used for electrical tuning. Nevertheless, these channels must make an important contribution to auditory transduction because mutations in either Ca^{2+} channels or K^+ channels result in hair cell degeneration that can produce hearing loss or deafness (Kubisch et al. 1999; Platzer et al. 2000; Rüttiger et al. 2004). While such results are intriguing, they have yet to be assembled into a coherent picture in which the critical roles of these channels are fully understood.

A review of the role of ionic conductances in hair cell tuning is necessarily uneven with respect to members of the vertebrate phylum. The material will be largely retrospective concerning tuning in non-mammalian vertebrates, where much of the work dissecting the components and reconstructing them into a coherent scheme has been accomplished. The material with respect to mammals is necessarily more prospective, because even though additional data have been collected since the previous review in this series (Kros 1996), it has not yet been possible to fully fathom the cellular nuances and the precise role of ionic conductances in the mammalian cochlea.

2. Characterization of the Receptor Potential

2.1 Response to Pure Tones

Successful intracellular recordings were reported from sensory epithelia in amphibians, reptiles and mammals in the late 1970s using ultrafine glass microelectrodes (Hudspeth and Corey 1977; Russell and Sellick 1978; Crawford and Fettiplace 1980). Many of the features common to the receptor potential are illustrated in Figure 5.1 for recordings from hair cells in the basilar papilla of the red-eared turtle, *Trachemys scripta elegans*. In response to a tone burst at the cell's CF, the receptor potential is an analog representation of the pressure wave at the eardrum (Fig. 5.1A). For low intensity stimuli the response increases transiently over the first few cycles, and then the amplitude remains constant for the duration of the tone. At the end of the stimulus the amplitude of the periodic response decays exponentially toward the resting level. As the intensity of the stimulus increases, the amplitude of the receptor potential increases proportionally. At higher intensities two nonlinearities become apparent. The first is that the proportionality between the amplitudes of the input and output is lost, and with an order of magnitude increase in the input, the amplitude of the receptor potential increases more modestly and eventually saturates at close to 50 mV peak to peak. The other nonlinearity is a progressive distortion in the receptor potential waveform, with the depolarizing phase noticeably larger and sharper than the corresponding hyperpolarizing phase of each cycle. These nonlinearities reflect both the saturating and asymmetric transducer function as well as the voltage-dependent properties of the hair cell membrane. For stimuli significantly higher than the CF, the response in the turtle resembles that in mammals where filtering by the membrane capacitance results in a sustained depolarization known as the summating potential accompanied by attenuation of the periodic response (Russell and Sellick 1978; Crawford and Fettiplace 1980; Palmer and Russell 1986).

For a given hair cell the response amplitude also varies with frequency (Fig. 5.1B), displaying a V-shaped tuning curve with maximal response at the CF.

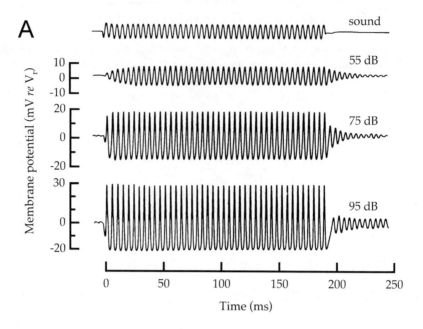

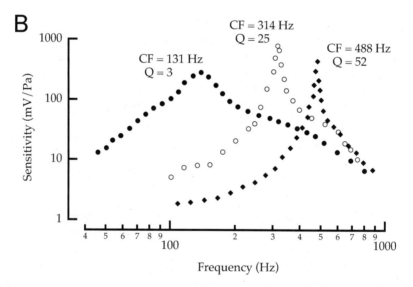

FIGURE 5.1. Microelectrode recordings of hair cell responses evoked by pure tones in the intact turtle cochlea. (**A**) Average responses to tone bursts at 55, 75, and 95 dB (re 20 μPa) at 220 Hz, the characteristic frequency (CF) of the hair cell. Ordinate plotted with respect to a resting potential (V_r) of −45 mV. (**B**) Frequency responses of three hair cells derived from sound frequency sweeps between 40 Hz and 1 kHz. The ordinate, the linear sensitivity of the cell near threshold, was calculated by dividing the amplitude of the fundamental component of the receptor potential by the sound pressure at the eardrum. The CF and quality factor Q (CF divided by 3 dB bandwidth) of each cell are indicated near the peak of the tuning curve. The details of stimulation and the analysis can be found in Art et al. (1985).

The tuning curve can be determined from responses to low-level tones at different frequencies that produce receptor potentials in the linear range. At each frequency point linear acoustic sensitivity was calculated by scaling the amplitude of the receptor potential by the sound pressure. Several features of the tuning curves common to the three cells in Figure 5.1B should be noted. First, each has a remarkably high acoustic sensitivity, in some cases nearly 1 V/Pa, at its CF. Second, at frequencies above or below CF, the sensitivity declines dramatically, and thus viewed through the lens of the receptor potential, the epithelium acts as a series of acoustic filters, with a broad stimulus spectrum detected by an array of cells, each of which is narrowly tuned to a small frequency range. Third, the narrowness of the tuning curve, which can be quantified by the quality factor or Q (defined as the CF divided by the half-power bandwidth) increases with CF. Because the receptor potential is highly tuned, the response of the cell to transient stimuli will be prolonged, and its temporal resolution degraded. This accounts for the slow buildup and decay of the response to CF tones (Fig. 5.1A).

2.2 Response to Acoustic Transients and Current Injection

In the turtle the notion that frequency selectivity might be a locally addressable function of each hair cell resulted from the observation of a clear correspondence between the features of the receptor potential in response to sound and that of the potential following injection of exogenous current through the recording electrode (Fig. 5.2A). The intimate relationship between the hair cell and its tuning mechanism is illustrated in the similarity between the receptor potential in response to a brief acoustic "click" and the membrane oscillations following injection of current through the recording electrode. The frequency and duration of the oscillations with the two types of stimuli are comparable. Moreover, the frequency of the voltage oscillations is identical to the CF of the hair cell determined with tonal stimuli (Crawford and Fettiplace 1981a). This implies that there is a functional link between the filtering of an acoustic signal and the electrical properties of the cell in the turtle. As might be expected from the tonotopic organization of the cochlea (Crawford and Fettiplace 1980), hair cells at different positions along the turtle basilar membrane resonate at different frequencies (Fig. 5.2B).

Electrical tuning has been found in the auditory end organs of a variety of non-mammals besides the turtle. These include the saccules of the goldfish, *Carassius auratus* (Sugihara and Furukawa 1989) and the bullfrog, *Rana catesbeiana* (Lewis and Hudspeth 1983; Hudspeth and Lewis 1988a), the amphibian papilla of the frogs *Rana temporaria* and *Rana pipiens* (Pitchford and Ashmore 1987; Smotherman and Narins 1999), the alligator lizard cochlea (Eatock et al. 1993), and the tall hair cells of the chick basilar papilla (Fuchs et al. 1988; Pantelias et al. 2001). However it is remarkable that little other experimental evidence has been acquired that directly compares the acoustic and electrical responses of a given hair cell within an intact auditory organ. All that follows for species other than the turtle is therefore based on the untested assumption

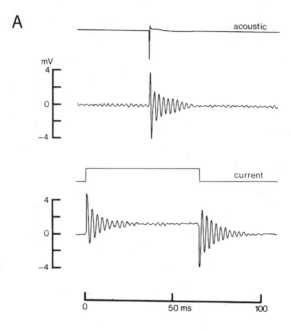
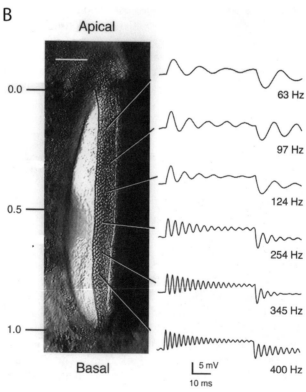

that the properties elicited electrically in reduced preparations, usually isolated hair cells, accurately reflects their contributions to tuning in the intact ear. The fact that this idea remains untested in many preparations may be the major reason that it has been difficult to characterize the roles of ionic currents in tuning the auditory periphery.

2.3 Theoretical Considerations

Sound-evoked vibration of the stereociliary bundle gates the mechanotransducer channels and generates a transducer current that at low levels is a faithful representation of the acoustic waveform. This current then flows out across the basolateral membrane to produce a receptor potential. In turtles and frogs the mechanotransducer channels activate with sub-millisecond kinetics (Corey and Hudspeth 1983; Crawford et al. 1989) and are probably not rate limiting over the auditory range of the animal concerned (less than 1 kHz). Therefore for a given transducer current, the amplitude of the receptor potential depends on the filtering characteristics of the basolateral membrane. If the membrane is passive, being composed of a capacitance, C, and an instantaneous leak conductance G_L, it will behave as a low-pass filter with a time constant, τ_m, equal to C/G_L; this causes the receptor potential to decline in amplitude above a corner frequency F_C ($F_C = 1/2\pi\tau_m$). In practice, τ_m is typically on the order of 1 ms, which results in attenuation above an F_C of 160 Hz. An increase in the bandwidth of the hair

FIGURE 5.2. Electrical tuning in turtle auditory hair cells (**A**) Comparison of the changes in membrane potential for an acoustic click (*top*) and an extrinsic current pulse (*bottom*). At top, a brief acoustic impulse at the eardrum evokes a damped oscillation at the cell's CF (370 Hz). At bottom, a current pulse injected through the microelectrode produces damped oscillations at beginning and end of the step reflecting the cell's inherent resonant properties. Note that the frequency of oscillation and their rate of decay in response to the current pulse are virtually identical to the cell's response to the acoustic stimulus, implying that the electrical resonance accounts for most of the cochlear frequency selectivity. The oscillation frequency is at the CF of the cell and the rate of decay of the oscillations reflects the quality factor, Q: the more prolonged the oscillation, the higher the Q. Changes in membrane potential are relative to the resting potential (-50 mV). (**B**) Tonotopic organization of the turtle cochlea. *Left*: Surface view of the turtle basilar papilla with the strip of hair cells on the right-hand side of the basilar membrane. Numbers indicate the fractional distance along the epithelium from the low-frequency end. Scale bar, 100 μm. *Right*: Examples of the electrical resonance recorded in hair cells at different positions along the epithelium. Each trace is the averaged response to 25 presentations of a small depolarizing current step. Standing depolarizing currents were added where necessary to bring the cell to a resting potential at which it was optimally tuned. Resonant frequencies are given next to each trace. Membrane potentials prior to the current step were: -51 mV (63 Hz), -47 mV (97 Hz), -47 mV (124 Hz), -39 mV (254 Hz), -44 mV (345 Hz), -44 mV (400 Hz). (B) is reproduced from Ricci et al. (2000b).

cell could in principle be achieved by reducing the capacitance, but there is a limit to how small the cell capacitance can be made given that the stereociliary bundle contributes nearly 40% of the hair-cell membrane area. A second approach to expanding the bandwidth would be to alter the passive basolateral conductance G_L. An increase in G_L would increase the corner frequency by an equivalent amount, allowing the cell to follow high-frequency stimuli with less attenuation. However, the amplitude of the receptor potential in the steady state would necessarily be reduced (Fig. 5.3). Clearly use of a fixed conductance, of whatever size, leads to unacceptable tradeoffs between the bandwidth and the amplitude of the receptor potential (Fig. 5.3B). Furthermore irrespective of the magnitude of G_L, the membrane impedance, and hence receptor potential, declines at high frequencies due to the capacitance.

An alternative approach is to introduce a time- and voltage-dependent K^+ conductance into the basolateral membrane that can behave in a negative feedback manner. By appropriate choice of the activation time constant for this conductance it is possible to offset the effects of the capacitance and maximize the membrane impedance over a limited frequency range. Because the current flowing through the capacitor is proportional to the rate of change of the membrane potential, dV/dt, it leads the voltage by a quarter of a cycle, whereas the K^+ current, because of its delayed activation, progressively lags the voltage. Thus there will exist a limited range of frequencies where currents in the two elements balance out, and where a minimum transducer current is needed to charge the membrane capacitance. Although it is difficult to satisfy these criteria across a wide frequency range in any one cell, a high impedance membrane at a given frequency can be created by choosing the time constant for activation of the conductance, τ_K, to be proportional to the square of the conventional membrane time constant at the desired frequency, that is, $\tau_K = \kappa \, (C/G)^2$, where κ is a constant, and C and G are the membrane capacitance and conductance respectively. If the activation time constant of the ionic current is chosen in this way, the membrane behaves not as a simple low-pass filter, but as an electrical resonator at the desired frequency. From a theoretical standpoint, we have created a receptor that is maximally sensitive to a narrow range of frequencies (Fig. 5.3B). The resonant frequency F_0 is then given by (Art and Fettiplace 1987):

$$F_0^2 = [\partial G/\partial V \cdot (V - E_K)]/[4\pi^2 C \tau_K] \tag{5.1}$$

where $\partial G/\partial V$ is the slope conductance at the membrane potential V and E_K is the K^+ equilibrium potential. To extend the bandwidth of the system as a whole requires producing a collection of cells with different kinetics to span the frequency range. Assuming a constant cell capacitance, to create an array of receptors that cover a decade in frequency, as occurs in the turtle, requires an order of magnitude change in the size of the conductance and two orders of magnitude change in the kinetics of its activation.

In the analysis described above, the high impedance at the resonant frequency depends upon the capacitive current being balanced out by the current flowing through voltage-dependent K^+ channels having simple kinetics with a single activation time constant. Introduction of a second K^+ conductance, as an in-

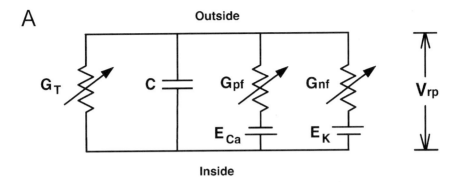

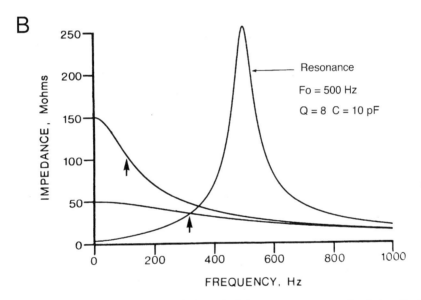

FIGURE 5.3. Equivalent circuit and theoretical frequency response of a hair cell. (**A**) Equivalent circuit is the parallel combination of a mechanotransducer conductance modulated by hair bundle deflections (G_T), cell capacitance (C) and two voltage-dependent conductances giving positive feedback (G_{pf}) and negative feedback (G_{nf}). G_{pf} primarily represents Ca^{2+} channels (see text for other channels that can produce positive feedback) and G_{nf} represents K^+ channels. The battery E_{Ca} is the Ca^{2+} equilibrium potential which is positive to the resting potential (-50 mV) whereas the battery E_K is the K^+ equilibrium potential negative to the resting potential. (**B**) Variation of the cell impedance with frequency for three conditions: a low-pass membrane filter with resistance of 150 MΩ and capacitance 10 pF; a low-pass membrane with resistance of 50 MΩ and capacitance 10 pF; and an electric resonance due to a voltage–dependent conductance and a capacitance. Arrows indicate half-power frequencies of 106 Hz and 318 Hz. Note that for the two low-pass filters, reducing the resistance increases bandwidth but reduces steady-state sensitivity; for both, the impedance declines along a common curve at high frequency due to membrane capacitance.

stantaneous leak in parallel with the voltage-dependent conductance, reduces the impedance at resonance and broadens the tuning curve. This is known as shunt damping and largely explains the loss of frequency selectivity that occurs on stimulation of efferent nerve fibers to the turtle cochlea (Art et al. 1984). The efferent axons make synapses on the hair cells where release of acetylcholine elicits slow hyperpolarizing synaptic potentials due to an increase in the K^+ conductance. Another type of shunt damping may arise by addition of a second set of voltage-dependent K^+ channels. If these channels have significantly different activation kinetics from the primary voltage-dependent K^+ channels, they will also offset the high impedance at resonance because their quarter cycle phase lag occurs at a different frequency. Thus if a hair cell possesses two types of voltage-dependent K^+ channel, they will need to have comparable activation time constants to generate sharply tuned resonance. This is an important consideration in mixing K^+ channel subtypes to generate a smooth gradation of resonant frequencies.

3. The Hair Cell Current–Voltage Relationship

In the turtle, alterations in the receptor potential in the presence of pharmacological agents such as the K^+-channel blocker tetraethylammonium (TEA) suggested that the hair cell response was dominated by voltage- and time-dependent conductances similar to those that shape responses in other excitable cells (Crawford and Fettiplace 1981b). By analogy with techniques used to understand mechanisms of excitability in nerve and muscle, a more complete analysis of hair cell function required experiments to characterize the voltage-dependent gating of the underlying currents. The highly nonlinear steady-state current-voltage (I–V) relationship (Fig. 5.4A) comprises a central region of high impedance near the resting potential (about -50 mV) flanked by two relatively low impedance regions of outward and inward rectification. Extensive studies using pharmacological blocking agents and ionic substitution have demonstrated that the net I–V curve is the result of a superposition of at least three types of current, for which the representative steady-state curves are illustrated in Figs. 5.4B and C. The high impedance and the outward rectification near the resting level are usually due to the combined effects of inward Ca^{2+} currents and outwardly rectifying K^+ currents. The inward rectification at membrane potentials hyperpolarized to the resting potential is due to additional voltage-activated conductances permeable to K^+. Functionally, all these currents fall into two classes: those which upon depolarization give positive feedback and work in concert with the transduction current to depolarize the cell further, and those that give negative feedback, work against the transduction current and repolarize the cell to its resting level (Fig. 5.3A). Because the Ca^{2+} equilibrium potential is positive to the resting potential, an increase in the Ca^{2+} conductance, G_{pf}, evoked by depolarization produces an inward current that causes further depolarization. In contrast, an increase in the K^+ conductance, G_{nf}, generates an outward current

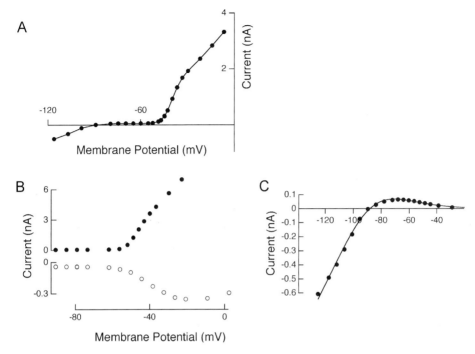

FIGURE 5.4. Whole-cell current-voltage (*I–V*) relationships in turtle hair cells. (**A**) Steady-state net *I–V* curve in an isolated hair cell with resonant frequency of 13 Hz. Cell bathed in normal saline solution containing (in m*M*): NaCl, 125; KCl, 4; MgCl$_2$, 2.8; CaCl$_2$, 2.2; HEPES, 10. Patch electrode was filled with a K$^+$-based solution (in m*M*): KCl, 125; MgCl$_2$, 2.8; CaCl$_2$, 0.45; K$_2$.EGTA, 5; Na$_2$.ATP, 2.5; KHEPES, 5. Each point is the average current flowing during the last 5 ms of a 140-ms voltage step. (**B**) Comparison of the steady-state net *I–V* curve (filled circles) and the inward Ca^{2+} current (*open circles*) measured in a solitary 234-Hz hair cell with a −54 mV resting potential. The net *I–V* curve determined as before, with a patch pipette containing a K$^+$-based intracellular solution. The inward Ca^{2+} current was subsequently measured while superfusing with a modified saline solution containing 65mM TEA substituted on an equimolar basis for NaCl, and a Cs$^+$-based intracellular solution containing (in m*M*): CsCl, 125; MgCl$_2$, 2.8; CaCl$_2$, 0.45; Cs$_2$.EGTA, 5; Na$_2$.ATP, 2.5; CsHEPES, 5. (**C**) The inward rectifier (K_{IR}) current in a 6-Hz hair cell. The current was isolated by superfusion with a solution containing 65 mM TEA, 5 mM 4–AP and 50 μM nisoldipine to block other hair cell currents carried by K$^+$ and Ca^{2+}. This cocktail also blocks K_{IR} by 35% in a voltage-independent manner. Conventional K$^+$-based intracellular solution in the recording pipette. Each point is an average current at the end of a 120ms command pulse from a holding potential, −90 mV.

that tries to repolarize the cell because the K$^+$ equilibrium potential is negative to the resting potential.

In general, currents giving positive feedback are rapidly activating inward currents that closely follow the membrane potential and add to the depolarizing current generated by the transducer. Currents giving negative feedback may be partially activated at the resting potential, are more sluggish, and their delay in activation ultimately leads to the resonant behavior in the hair cell response to injected current. Negative feedback systems are well known to be capable of generating resonance at a frequency that is dictated by the speed and strength of the feedback (Murray 1989). Indeed, in electrically tuned cells, it is the size and kinetics of the dominant outward current, flowing through either large conductance Ca^{2+}-activated K^+ (BK_{Ca}) channels or voltage-dependent K^+ (K_V) channels which largely determines the hair cell CF. By comparison the inward current, for example the voltage-dependent Ca^{2+} current, if it activates on a fast time scale, increases the Q or sharpness of tuning. Clearly this only works if the inward current operates in subthreshold regime (Hodgkin and Huxley 1952). Once the net current is inward, the response becomes regenerative and leads to action potentials. Whether the cell produces action potentials or resonance depends on the relative magnitude of the inward and outward currents. Models employing experimentally derived descriptions of the Ca^{2+} current and Ca^{2+}-activated K^+ (BK_{Ca}) current have been shown to correctly predict the electrical tuning in responses to applied current pulses, including the nonlinear dependences of resonant frequency and Q on membrane potential (Hudspeth and Lewis 1988b; Wu et al. 1995).

4. Positive Feedback: Identification of Currents that Enhance Depolarization

Currents that provide positive feedback and augment the depolarization produced by the inward transduction current include inward Ca^{2+} and Na^+ currents that are activated by depolarization, and a K^+-selective inward rectifier whose outward current is blocked by depolarization at voltages near the resting potential. Of the inward currents, the most important is the voltage-dependent Ca^{2+} current. This current not only directly contributes to membrane excitability as a charge carrier, but also elevates intracellular calcium concentration leading to activation of BK_{Ca} channels in the basolateral membrane (as well as exocytosis during synaptic transmission). Less prominent in generating positive feedback are the inward Na^+ currents, which are transient compared to the calcium currents. Na^+ currents in conjunction with the Ca^{2+} current develop strong and rapid depolarization leading to either spiking behavior or action potentials. In mature auditory hair cells, Na^+ currents are relatively rare, being observed in fish (Sugihara and Furukawa 1989) and crocodilians (Evans and Fuchs 1987), but they are transiently expressed in the mammalian cochlea where they may contribute to spontaneous firing during development (Oliver et al. 1997; Marcotti et al. 2003). The third current contributing positive feedback is the inwardly rectifying K^+ current near the resting potential. The significance of the inward rectifiers in hair cell function stems from the positive feedback afforded by the negative-slope region of the I–V curve positive to E_K (Goodman and Art 1996a).

4.1 Calcium Currents

Voltage-dependent Ca^{2+} currents in hair cells have been characterized in auditory and vestibular epithelia in a variety of species (Lewis and Hudspeth 1983; Art and Fettiplace 1987; Fuchs and Evans 1988; Hudspeth and Lewis 1988a; Fuchs et al. 1990; Roberts et al. 1990; Prigioni et al. 1992; Steinacker and Romero 1992; Art et al. 1993; Zidanic and Fuchs 1995; Smotherman and Narins 1999; Rispoli et al. 2000; Engel et al. 2002). The hair cell Ca^{2+} current has a half activation voltage of -30 to -40 mV, and activates and deactivates rapidly with a time constant less than 0.5 ms at room temperature. It shows little inactivation on the millisecond time scale, can be blocked by extracellular Cd^{2+} or Ni^{2+} and is enhanced when Ba^{2+} is substituted for Ca^{2+} as the charge carrier (Art and Fettiplace 1987; Fuchs et al. 1990; Zidanic and Fuchs 1995; Smotherman and Narins 1999; Schnee and Ricci 2003). This specification, combined with sensitivity to dihydropyridines (Fuchs et al. 1990; Schnee and Ricci, 2003), suggests that the major species of Ca^{2+} current in hair cells is of the L-type. However, the hair cell Ca^{2+} current has faster kinetics, a more hyperpolarized activation voltage, and reduced sensitivity to block by dihydropyridines such as nimodipine compared to the L-type Ca^{2+} current originally described in cardiac myocytes (Tsien et al. 1987; Lipscombe et al. 2004). Based on the pharmacological profile it has been argued that only L-type Ca^{2+} channels are present in turtle cochlear hair cells (Schnee and Ricci 2003) but a subpopulation of N-type channels may be present in frog saccular hair cells (Rodriguez-Contreras and Yamoah 2001).

Molecular analysis of Ca^{2+} channels has demonstrated that they are multimeric proteins, composed of α, β, α2δ, and sometimes γ subunits (Hille 2001). L-type Ca^{2+} channels are dihydropyridine-sensitive channels that are divided into four subtypes, based on differences in the pore-forming α subunits, as α1S (CaV1.1), α1C (CaV1.2), α1D (CaV1.3), or α1F (CaV1.4). The α1S and α1C subtypes are found mainly in skeletal and cardiac muscle respectively, α1D is prevalent in neurons and endocrine cells and α1F is a retinal isoform. When examined in a heterologous expression system, the α1D most closely resembles the native currents in hair cells, with fast kinetics, hyperpolarized activation curves, and lower sensitivity to dihydropyridines than the other L-type channels (Xu and Lipscombe 2001; Lipscombe et al. 2004). The α1D subtype has been identified as the major molecular species in cochlear hair cells of chicken based on the greater abundance of mRNA for α1D compared to α1C, (Kollmar et al. 1997). A similar conclusion was reached for mouse inner hair cells where knockout of the α1D subtype (α1D$^{-/-}$) produced more than 90% reduction in the mean Ca^{2+} current (Platzer et al. 2000), thus arguing for conservation of the α1D type across a wide range of species.

Ca^{2+} influx through voltage-dependent Ca^{2+} channels drives synaptic transmission and the Ca^{2+} channels are clustered, probably at the synaptic release sites on the basolateral membrane of the hair cell (Roberts et al. 1990; Issa and

Hudspeth 1994; Tucker and Fettiplace 1995). In non-mammals, this Ca^{2+} influx also gates the BK_{Ca} channels that underlie electrical tuning. In turtle cochlear hair cells, the numbers of Ca^{2+} channels increase systematically with CF in parallel with the numbers of release sites and BK_{Ca} channels (Art and Fettiplace 1987; Ricci et al. 2000). The significance of this tonotopic variation for synaptic transmission is not fully understood. However, it is likely to be important for electrical tuning because it will ensure uniform activation of the increased number of BK_{Ca} channels and it will also provide more positive feedback to maintain the sharpness of tuning in the face of a larger outward K^+ current.

4.2 Sodium Currents

Voltage-dependent Na^+ currents sensitive to tetrodotoxin (TTX) were first described in the tall, low frequency cochlear hair cells in the alligator, *A. mississippiensis* (Evans and Fuchs 1987; Fuchs and Evans 1988). Na^+ currents have since been reported in auditory and vestibular hair cells in fish, reptiles and mammals (Sugihara and Furukawa 1989; Witt et al. 1994; Kros 1996; Oliver et al. 1997; Rüsch and Eatock 1997; Lennan et al. 1999; Chabbert et al. 2003; Masetto et al. 2003; Marcotti et al. 2003). Classically the rapid TTX-sensitive current is known to give positive feedback and to produce action potentials in nerve (Hodgkin and Huxley 1952; Hille 2001). Evidence in support of such a role in hair cells is clearest in fish (Sugihara and Furukawa 1989) and crocodilian (Evans and Fuchs 1987; Fuchs and Evans 1988) preparations, where superfusion with TTX or removal of Na^+ either changes or eliminates spiking potentials in a distinct population of hair cells within the epithelia. Similar data are available in the immature inner hair cells in the mammalian cochlea (Marcotti et al. 2003). A surprising feature, however, is the wide range of voltages for the steady-state inactivation of the Na^+ current. The extremely hyperpolarized potentials reported for half-inactivation bring into question the physiological role of the current at the resting potential. Indeed, there is no evidence that Na^+ currents contribute either to setting the resonant frequency or improving the Q in hair cells that possess electrical tuning.

4.3 Potassium Inward Rectifiers

The existence of an inwardly rectifying K^+ current has been demonstrated in a variety of auditory and vestibular hair cells (Corey and Hudspeth 1979; Ohmori 1984; Art and Fettiplace 1987; Fuchs and Evans 1990; Holt and Eatock 1995; Marcotti et al. 1999). The most frequently cited role for the K^+-selective inward rectifier (K_{IR}) relates to its ability to maintain a cell's resting membrane potential near E_K (see Hille 2001). Inward rectifiers are unlikely to play this type of role in auditory hair cells however, since the average zero-current potential of -50 mV is 30 mV positive to E_K (Crawford and Fettiplace 1980). Nevertheless, the finding that the magnitude of K_{IR} conductance varies systematically with CF suggests that it makes some contribution to the tuning process. The K_{IR} conductance differs from other voltage-dependent or transduction conductances in the turtle in that it shows a reversed tonotopic gradient, being largest in cells tuned to the lowest CF and declining systematically with increasing CF (Goodman and Art 1996a). A decrease in the expression of K_{IR} channels with CF also occurs in the chick cochlea as indicated both by variation in K_{IR} current with hair cell resonant frequency (Fuchs and Evans 1990) and the distribution of cIRK1 transcripts (Navaratnam et al. 1995).

The inward rectifier is so named because it carries large inward currents negative to E_K but small outward currents positive to E_K. The inward rectification arises because the channel is subject to rapid, sub-millisecond, voltage-dependent block by positively charged intracellular particles, divalent cations or polyamines, as the channel is depolarized (Lopatin et al. 1994). As a consequence of this voltage-dependent block, the steady-state I–V relationship can have a negative slope near the resting potential. The increasing block of outward current is functionally equivalent to activating an inward current and implies the channel would be capable of generating positive feedback. It may therefore serve a function similar to activation of the voltage-gated Ca^{2+} current for which depolarization leads to a rapid activation of an inward current. Interestingly both the magnitude and kinetics of the two inward-going currents are similar in turtle cochlear hair cells with low CF (Fig. 5.5). The K_{IR} current and the Ca^{2+} current sum to give a combined inward current that spans a range of membrane potentials from -60 to -20 mV. As noted earlier, positive feedback does not directly contribute to the frequency of the electrical resonance, but instead boosts the sharpness of tuning or Q. That K_{IR} can perform this role has been verified by showing that application of 5 mM extracellular Cs^+ both blocks the inward rectifier and also reduces or abolishes electrical tuning in hair cells (Goodman and Art 1996a). This effect is most pronounced in cells tuned to low frequencies where the negative feedback is generated by a purely voltage dependent K^+ current. The contribution of K_{IR} in augmenting the Q in the low-frequency range has been confirmed by modeling of electrical tuning with and without the inward rectifier (Wu et al. 1995; Goodman and Art 1996a).

Experimental study and modeling of the contribution of K_{IR} has also been done for frog saccular hair cells (Holt and Eatock 1995; Catacuzzeno et al. 2004). In those cells interaction between K_{IR} and a hyperpolarization-activated current carrying both Na^+ and K^+ (I_h) causes spontaneous low-frequency (approximately 5 Hz) oscillations in the membrane potential (Catacuzzeno et al. 2004). The contributions of the two currents were established by using barium to block K_{IR} and the selective inhibitor ZD-7288 to block I_h. If K_{IR} provides the positive feedback, the negative feedback is generated by I_h (depolarization closes channels with a reversal potential positive to rest resulting in hyperpolarization), which has a slow activation time constant of approximately 100 ms and sets the oscillation frequency. The significance of these oscillations in not fully clear, but they may act in concert with similarly timed spontaneous hair bundle oscillations (Martin et al. 2003) to amplify the sound stimuli in the low frequency range.

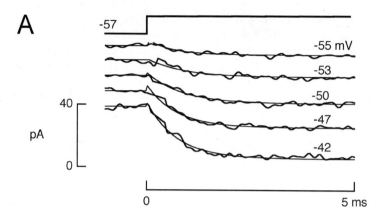

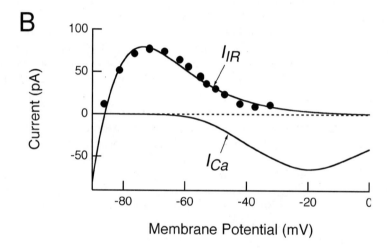

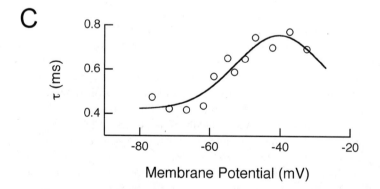

5. Negative Feedback: Repolarizing the Membrane with K$^+$ Conductances

5.1 Comparison of Hair Cell Response in Current- and Voltage-Clamp

To understand how the kinetics of the basolateral current might vary between cells tuned to different frequencies, it is useful to compare the resonant behavior in current clamp (Fig. 5.6A) with the membrane current seen in voltage clamp (Fig. 5.6B) for an isolated turtle hair cell. Figure 5.6A illustrates the voltage oscillations and damping in response to a small current step reminiscent of that seen for current injection in hair cells in the intact cochlea (Fig. 5.2). In the corresponding voltage-clamp experiment, small voltages steps around the resting potential evoke a net K$^+$ outward current that activates with a delay at the onset of depolarization, but at the termination of the step, the current deactivates exponentially back to the resting level. Cells tuned to higher frequencies have both a larger current and faster time constants for activation and deactivation (Art and Fettiplace 1987). The conductance is directly proportional to resonant frequency, and there is an inverse relation between the dominant time constant for current deactivation and resonant frequency (Fig. 5.6C), the fitted line being predicted by Eq. (5.1). Similar results have been obtained in the frog amphibian papilla (Smotherman and Narins 1999), showing that the size and speed of the K$^+$ currents increase with CFs up to nearly 400 Hz. The amplitude of the K$^+$

◀───────────────────────

FIGURE 5.5. Behavior of I_{KIR} and I_{Ca} in the physiological voltage range. (**A**) Rapid block produced by small depolarizations from -57 mV. Current carried by K$_{IR}$ was isolated by superfusion with the modified saline solution specified in the Figure 5.4C. Each trace is the average of four to six presentations and was digitally filtered with a three-point Gaussian filter. Single-exponential functions (thin lines) were fit to the onset of the block. (**B**) Comparison of the negative slope region of K$_{IR}$ with the average calcium *I–V* curve. *Filled symbols* are the steady-state current measured in the blocking cocktail, scaled to account for 35% inhibition of K$_{IR}$. The smooth curve was calculated using

$$I = g_{max} \cdot (V - E_K) \frac{1}{1 + B\exp\left(\dfrac{V - V_B}{V_e}\right)}$$

with fitting parameters: $g_{max} = 35$ nS, $E_K = -86$ mV, $B = 2.0$, $V_B = -88$ mV, $V_e = 11$ mV. Calcium current was calculated following Wu et al. (1995), according to $I_{Ca} = Ni_{Ca}p_{Ca}$, where N is the number of channels, i_{Ca} is the single channel current given by the constant field equation, and p_{Ca} is the steady-state open probability, calculated from an m^2 Hodgkin–Huxley scheme. N was adjusted to give the peak I_{Ca} estimated from the relationship between Ca^{2+} current size and F_0 determined previously (Art et al. 1993). (**C**) Time constant, τ, of K$_{IR}$ block exhibits a modest voltage-sensitivity about the resting level comparable to that seen for the Ca^{2+} current.

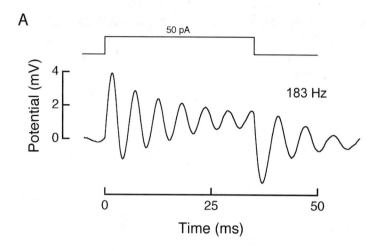
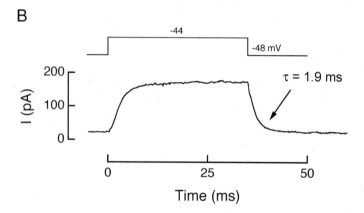
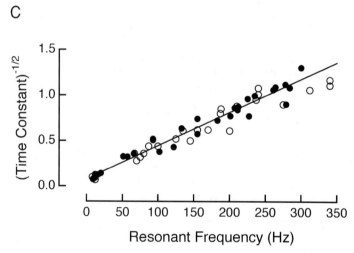

currents was also found to increase in progressing toward the high-frequency end of the chick cochlea (Pantelias et al. 2001).

Given these results, a fundamental question is the extent to which the kinetic variation is explained by a single class of K^+ channels, or by different members of the channel superfamily. The evidence for a contribution by large Ca^{2+}-activated K^+ (BK_{Ca}) channels to resonance was established early on (Lewis and Hudspeth 1983; Art and Fettiplace 1987) and similar channels are found in hair cells in most species. Other types of K^+ channel, akin to delayed rectifiers and inactivating, "A" type channels, have been reported in a variety of cochlear and vestibular hair cells (Hudspeth and Lewis 1988a; Housley et al. 1989; Lang and Correia, 1989; Sugihara and Furukawa 1989; Fuchs and Evans 1990; Kros and Crawford 1990; Masetto et al. 1994; Murrow 1994; Rennie and Correia 1994; Rüsch and Eatock 1996; Smotherman and Narins 1999), but their functional role was initially unclear. One possibility was that rather than a single class of channels with variable kinetics being used across the acoustic spectrum, the range might be fractionated, perhaps for functional reasons relating to the range of kinetics that can be achieved within a given channel type. For example, the notion that low frequency cells are tuned by a purely voltage-sensitive K^+ (K_V) current, rather than a BK_{Ca} current, was supported by the observation that outward currents persist in these cells when exposed to a saline containing only 1 μM Ca^{2+} (Art et al. 1993), a concentration that fails to activate the BK_{Ca} current.

5.2 Identification and Dissection of the Outward Currents

To determine the ensemble of currents underlying resonance at different frequencies in the turtle, wide-spectrum K^+ channel antagonists, 4-aminopyridine (4-AP) and TEA were used to analyze whether hair cells tuned to different

◄───

FIGURE 5.6. The behavior of turtle hair cells in current-clamp and voltage-clamp. (**A**) Damped oscillations at 183 Hz elicited by a depolarizing 50-pA current step. Voltage is expressed relative to the mean resting potential, -48 mV. (**B**) Net current elicited in the same cell by a 4-mV depolarization from -48 mV holding potential. Following activation, the current is maintained for the duration of the depolarization. On returning to the resting level, the current deactivates exponentially with time constant of 1.9ms. (**C**) Time constant of current decay is correlated with resonant frequency of hair cell. Measurements were made in solitary hair cells, isolated either in low calcium (*filled circles*) or using enzymatic dissociation with papain (*open circles*). There was no evidence that the method of isolation and the use of papain affected the results as previously claimed (Armstrong and Roberts 1998). The pooled data were fit with a straight line, ($R = 0.99$ for $n = 62$) indicating that the over the entire spectrum the resonant frequency is proportional to the (time constant of K^+ channel activation)$^{-1/2}$ (see eq.[5.1] of the text). The group of points at less than 50 Hz represents hair cells with K_V channels whereas the rest of the points are for BK_{Ca} channels. Modifed from Goodman and Art (1996b).

frequencies express more than a single type of K$^+$ channel (Goodman and Art 1996b). The design relies on the fact that most, if not all K$^+$ channels are sensitive to these agents, but at widely varying concentrations. Peptide toxins are now available for some of the K$^+$ channel subtypes (iberiotoxin for the BK$_{Ca}$ channel or dendrotoxins for certain K$_V$ channel isoforms) but in general these have not been used to characterize hair cell K$^+$ currents. The virtue of using 4-AP and TEA is that together these cover the spectrum of K$^+$ channel subtypes. Pharmacological dissection of the contributions of multiple K$^+$ currents relies on estimating the blocker affinities for each channel type, which in turn requires that a dose–response relationship can be determined for each current and blocker in isolation. In the turtle this requirement is met at the extremes of the acoustic spectrum, where the outward currents are carried by a single species of K$^+$ channel with very different pharmacological profiles.

5.2.1 Inhibition of Outward Current by 4-AP

An example of the dose-dependent inhibition of the outward current in a low-frequency hair cell is illustrated in Figure 5.7A. The inhibition of K$^+$ channels by 4-AP is complicated since it is known to exert its block on the internal face of the channel in most preparations, as originally demonstrated for the K$_V$ current in squid axon (Kirsch and Narahashi 1983). Construction of dose–response curves based on external superfusion of 4-AP thus relies on the fact that an uncharged hydrophobic molecule at physiological pH (7.4 to 7.6) has access to both sides of the plasma membrane. Moreover, since 4-AP ionization within the cell is required to block the channel, the intracellular concentration of the relevant species at the binding site is largely unknown. The apparent affinity constant for block of the outward current in Figure 5.7A was 70 µM, but in other cells, it ranged from 26 to 102 µM, and was correlated with the time constant of current deactivation upon repolarization to -50 mV: cells with slower deactivation time constant required larger 4-AP concentrations for half-

FIGURE 5.7. Pharmacological and voltage sensitivity of the different K$^+$ channels underlying electrical tuning in turtle hair cells of low and high resonant frequency (F_0). (**A**) Inhibition of the outward current in a low-frequency hair cell ($F_0 = 6$ Hz) by progressively larger concentrations of 4-aminopyridine (4–AP) which blocks K$_V$ channels. Estimated inward currents have been subtracted from the average traces. (**B**) Inhibition of the outward current in a medium-frequency hair cell ($F_0 = 93$ Hz) by progressively larger concentrations of tetraethylammonium (TEA). Note that with highest TEA concentration the K$_{Ca}$ channels are completely blocked leaving the residual inward Ca^{2+} current. In both (A) and (B) the outward current was evoked by depolarizing voltage steps to -40 mV from a holding potential of -60 mV. Note the differences in time scale between (A) and (B). (**C**) Steady-state voltage activation curves for K$_{Ca}$ channels (open circles) and K$_V$ channels (filled symbols). Results derived from tail current measurements scaled to their maximum value (I/I_{max}). Modifed from Goodman and Art (1996b).

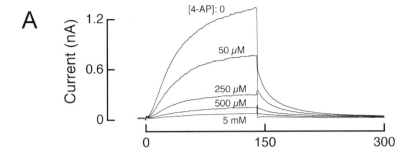

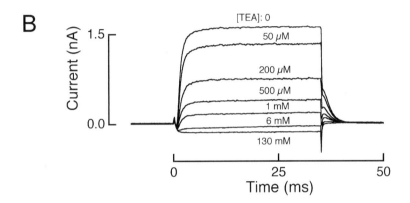

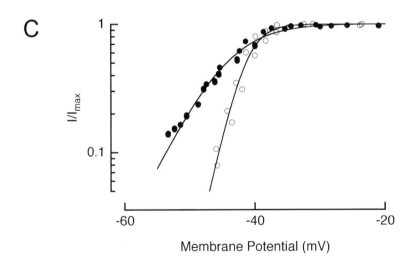

block. The mechanism linking 4-AP affinity to kinetics is unknown, though it is worth noting that a similar correlation has been described in cloned K_V channels (Kirsch et al. 1993); furthermore small changes in the amino acid residues flanking the interior of the pore region are sufficient to modulate deactivation rate and 4-AP affinity (Judge et al. 2002) in tandem (Shieh and Kirsch 1994).

By contrast, the K^+ currents in hair cells tuned to high frequencies were relatively insensitive to 4-AP, and dose–response relationships indicated a single class of low-affinity binding site with a Hill coefficient of 1 and an average half-blocking concentration of 13.8 mM. In most cells, 4-AP block did not alter the time course of K^+ current activation. In a minority of high-CF cells the 4-AP block gave slightly lower affinity, modified K^+ current kinetics, and was best fit with a Hill coefficient of approximately 1.4. These results suggest that in some instances inhibition of BK_{Ca} by 4-AP is a complex process, and may include indirect actions on the Ca^{2+} current in addition to a direct effect on the K^+ channel. Nevertheless, the K^+ current in high-CF cells is 100-fold less sensitive to block by 4-AP than is the corresponding current in low-CF cells. Cells with intermediate CF, between 60 and 80 Hz, had dose–response relationships consistent with at least two classes of binding sites, suggesting that they may express a mixture of K^+ channels. However, only a single deactivation time constant was evident in the tail currents.

5.2.2 Inhibition of Outward Current by TEA

The effects of varying the concentration of TEA on outward currents are illustrated for a moderately high frequency hair cell in Figure 5.7B. In such cells the TEA was remarkably potent, with a block of half the current at 215 μM. At higher concentrations, TEA completely blocked the K^+ current, revealing the rapid inward calcium current activated simultaneously with depolarization. In contrast, in low-frequency cells a TEA concentration of 36 mM was needed to block half the current. All TEA dose–response curves for cells of low- and high-CF can be fit with single-site equilibrium binding functions. The apparent TEA affinity for the outward current in cells of low CF is nearly 200-fold lower than in cells of high CF, a result that supports the case for different K^+ channel types being expressed at low and high frequencies.

5.2.3 Voltage Sensitivity of the K^+ Currents

The prior pharmacological dissection indicated that 25 mM TEA is sufficient to block 99% of the K^+ current in cells of high CF, and 40% of the current in cells of low CF (Goodman and Art 1996b). Similarly 0.8 mM 4-AP blocks 90% of the current in low-CF cells, and virtually none of the current in those of high CF. Use of these blocker concentrations was therefore adequate to determine the voltage dependence of the activation of the dominant K^+ currents, with different voltage dependences supporting the expression of different channel subtypes. Activation curves were estimated from tail current amplitude at -50 mV, and the results were fit with Boltzmann functions in both low- and high-frequency cells (Fig. 5.7C). In three low-CF cells the current was resistant to external TEA, increased e-fold in 4.0 ± 0.3 mV, and reached a half-maximum at −44.3 ± 0.8 mV. Similarly, for two high-CF cells, the normalized tail current was half-maximal at −41.3 mV and increased e-fold in just 1.9 mV. This remarkable voltage dependence in cells of high CF far exceeds that of any known purely voltage-sensitive conductance and is likely due to the combined effects of membrane potential and the voltage-dependent elevation of intracellular Ca^{2+} to activate BK_{Ca} channels. Thus the voltage dependence of Ca^{2+} entry (e-fold in approximately 6 mV) is steepened by the cooperative binding of multiple Ca^{2+} ions to the BK_{Ca} channel. With a Hill coefficient of 3.5 for Ca^{2+} binding (Art et al. 1995), this produces a voltage dependence for BK_{Ca} channel activation of e-fold in 6/3.5 = 1.7 mV, similar to that observed experimentally.

5.2.4 Variations in the Relative Contribution of Different Types of K^+ Current

Because of the large difference between the blocking concentrations of 4-AP and TEA on the different K^+ channels, these blockers could be used to analyze the relative contributions of BK_{Ca} and K_V channels to the total K^+ current in hair cells along the tonotopic axis. The K^+ current I_K was calculated from the saturating tail current amplitude and assumed to represent only the sum of TEA- and 4-AP-sensitive currents, $I(_K) = I(BK_{Ca}) + I(K_V)$, in each cell. In addition to the response in normal saline, currents were examined in combinations of 4-AP and TEA. By using the reduction in the total current in two test solutions, it was possible to estimate the contributions of the two unknown currents, $I(BK_{Ca})$ and $I(K_V)$, to the total. This notion can be generalized, and for any concentration of the two blocking agents, the equation from Goodman and Art (1996b), is given by

$$I_K(\alpha,\beta) = \sum_\alpha \left(\prod_\beta \left(\frac{K_i(\alpha,\beta)}{K_i(\alpha,\beta) + [\beta]} \right) \cdot I_\alpha \right), \quad (5.2)$$

where the summation is over the set $\alpha = (BK_{Ca}, K_V)$, and the product is over the set $\beta = $ (4-AP, TEA). Thus the total current, $I_K(\alpha, \beta)$, is a function of the 4-AP and TEA concentrations, and the apparent affinities, $K_i(\alpha, \beta)$, of each reagent for each channel type. The affinity constants derived from single-cell dose–response curves were used. This technique of deconstructing the relative contributions of the two major currents assumes that external TEA and 4-AP inhibit K^+ channels independently, since they act at nonoverlapping sites, with TEA lodged near the external mouth of the ion pore (Heginbotham and MacKinnon 1992), and 4-AP binding to the internal face (Kirsch and Narahashi 1983).

Block of both K^+ currents by larger doses of 4-AP and TEA, combined with block of the Ca^{2+} current using nisoldipine, could be used also to determine the size of the K_{IR} in each cell. The conclusion from these studies is that there is a tonotopic expression of all three currents, and, to a first approximation, there are three categories of cell as illustrated in Figure 5.8. In the lowest frequency cells, with CF up to about 60 Hz, the primary outward current providing negative feedback and determining the kinetics of the response is the 4-AP-sensitive K_V. The presence of a rapidly deactivating K_{IR} in this range augments the positive feedback provided by the small Ca^{2+} currents in these cells. In the next higher range of CF, up to about 250 Hz, the BK_{Ca} current becomes the dominant outward current determining frequency, and the contribution of K_{IR} to positive feedback is progressively decreased. In the highest frequency cells, up to 600 Hz, only the tightly coupled BK_{Ca} and Ca^{2+} currents remain to contribute negative and positive feedback to tune the hair cell.

The distributions of outward and inward currents have been examined along the tonotopic axis in other species, and in some, such as the chick, across the width of the sensory epithelium as well (Murrow 1994). In the chick, variation in channel expression along the major axis in some reports resembles that outlined for the turtle, with voltage-sensitive K^+ channels providing the negative feedback for tuning at low CF and the BK_{Ca} channels underlying tuning at high frequencies (Fuchs and Evans 1990). However, in another study in which measurements were made in the intact chick basilar papilla rather than in dissociated hair cells, increased expression of both K_V and BK_{Ca} channels occurred with

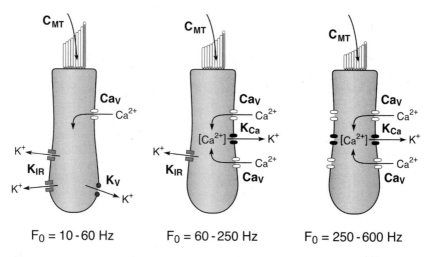

FIGURE 5.8. Distribution of ionic currents in turtle auditory hair cells with different resonant frequencies (F_0) All cells contain cationic mechanotransducer channels in the hair bundle (C_{MT}) and voltage-dependent Ca^{2+} channels (Ca_V) in the basolateral membrane. Voltage-dependent K^+ channels (K_V), which have slow activation kinetics, occur only in cells tuned to low frequency whereas Ca^{2+}-activated K^+ channels (K_{Ca}) with faster kinetics occur in middle and high frequency cells. With increasing F_0, there are increases in the number of Ca_V channels and K_{Ca} channels per cell at a stoichiometry of 2:1. Inwardly rectifying K^+ channels (K_{IR}) occur in cells tuned to low and middle frequencies.

higher CF (Pantelias et al. 2001). Similarly, molecular (Navaratnam et al. 1995; Mason et al. 1998) and physiological (Fuchs and Evans 1990; Pantelias et al. 2001) evidence demonstrates that the K_{IR} is preferentially expressed in the low-frequency half of the chick epithelium, suggesting that in chick K_{IR} may enhance positive feedback, as had been proposed in turtle (Goodman and Art 1996a). This pattern of preferential expression of K_{IR} at low CF is not universal, however, since it is the high-frequency region of the frog amphibian papilla in which K_{IR} is expressed (Smotherman and Narins 1999). The significance of other variations in channel expression across the width of the epithelium, such as the presence of an inactivating K_V, or "A" current, in short hair cells of the chick cochlea (Murrow and Fuchs 1990) suggests that variation in the ensemble of expressed currents may have different, as yet unexplored roles in hair cell function, in addition to determining the characteristic frequency.

6. Possible Mechanisms of Kinetic Variation

6.1 Analysis of Single-Channel Biophysics

Measurements of the macroscopic current show a clear correlation of the time constant of the dominant outward K^+ current with the resonant frequency of the cell. For a BK_{Ca} channel, variations in the whole-cell current kinetics might result from two distinct, but equally plausible mechanisms. Since the activation occurs by increasing $[Ca^{2+}]$, the variation might reflect differences in the time course of the cytoplasmic Ca^{2+} transients, attributable to differences either in Ca^{2+} channel kinetics or in diffusion and buffering of cytoplasmic calcium. Alternatively BK_{Ca} channels in cells tuned to different frequencies might possess different kinetic properties even at constant levels of intracellular Ca^{2+}. By isolating BK_{Ca} channels from cells of known resonant frequency and analyzing voltage and Ca^{2+} sensitivity of the channel kinetics, it was possible to show a relationship between the time course of activation and deactivation of the macroscopic current, and the properties of the BK_{Ca} channel. Membrane patches excised from both low- and high-frequency hair cells contained BK_{Ca} channels with a mean conductance of 320 pS in symmetrical 150 mM K^+ (Art et al. 1995). Under the same steady voltage and Ca^{2+} conditions, the mean open times of BK_{Ca} channels in different cells ranged from 0.25 to 12ms. Similarly, following depolarization, the time constant of relaxation of average single-channel currents at -50 mV in 4 μM Ca^{2+} varied between 0.4 and 13 ms, and was inversely correlated with the resonant frequency estimated for the hair cell (Fig. 5.9). This range of time constants is roughly equivalent to the relaxation time constants for the whole-cell BK_{Ca} current in cells tuned between 60 and 600 Hz, covering most of the range of turtle hearing. In addition the voltage dependence of the single-channel time constant (an e-fold decrease for approximately 25 mV of hyperpolarization) was found to be similar to that of the macroscopic BK_{Ca} current. Taken together these results support the notion that

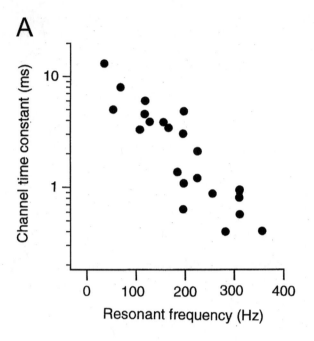
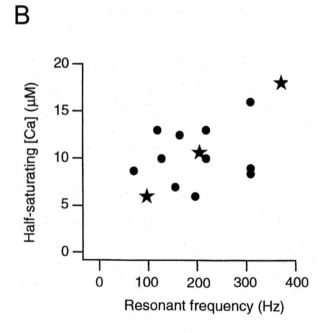

the behavior of the macroscopic BK_{Ca} current in turtle auditory hair cells is limited by the BK_{Ca} channel kinetics rather than by the rate of accumulation and decay of intracellular Ca^{2+} (Art et al. 1995).

A sixfold variation in the deactivation time constants with hair-cell location along the tonotopic axis has also been reported for BK_{Ca} channels in chick cochlear hair cells (Duncan and Fuchs 2003). In both turtle and chick hair cells, the differences in BK_{Ca} channel kinetics with CF are accompanied by smaller changes in the Ca^{2+} sensitivity of the channel. Recordings from macropatches excised from precise locations along the turtle cochlea indicated that the Ca^{2+} sensitivity of the BK_{Ca} channel is threefold greater in cells of low CF than those of high CF (Fig. 5.8; Ricci et al. 2000). A tonotopic gradient in the Ca^{2+} sensitivity of BK_{Ca} channels was also observed in chick cochlear hair cells (Duncan and Fuchs 2003). Although there was considerable scatter at a single cochlear location, the average Ca^{2+} sensitivity of BK_{Ca} channels in chick hair cells declined with CF.

Loose-patch recording (Roberts et al. 1990) and Ca^{2+} imaging (Issa and Hudspeth 1994) in frog saccular hair cells had suggested that the voltage-sensitive Ca^{2+} channels and BK_{Ca} channels are colocalized at the synaptic release sites, where they occur in clusters of about 90 Ca^{2+} channels to 40 BK_{Ca} channels. Each cell has multiple channel clusters and multiple release sites, identified by the electron-dense synaptic bodies. A similar conclusion was reached in turtle hair cells where the Ca^{2+} channels and BK_{Ca} channels were found to occur in the same membrane patch (Art et al. 1995) with a an approximately 2:1 stoichiometry of Ca^{2+} channels to BK_{Ca} channels independent of the CF of the hair cell (Art et al. 1993, 1995). Thus even though the sizes of both BK_{Ca} and Ca^{2+} currents increase with CF, the relative proportions of channels remain constant, with every active BK_{Ca} channel influenced by the Ca^{2+} influx through a pair of adjacent Ca^{2+} channels. The fixed stoichiometry might be achieved if the Ca^{2+} and BK_{Ca} channels are assembled and inserted into the membrane as trichannel complexes. Evidence in support of such an arrangement includes the relative insensitivity of BK_{Ca} channel activation to the concentration of intracellular calcium buffer (Ricci et al. 2000). At least 10 mM BAPTA was needed to alter

FIGURE 5.9. Properties of Ca^{2+}-activated K^+ (BK_{Ca}) channels in turtle hair cells. (**A**) Correlation between the BK_{Ca} channel time constant and the resonant frequency of the hair cell. The channel were recorded in inside-out patches and the time constants were derived from fitting the relaxation of the ensemble-average current following repolarization to -50 mV with 4 μM Ca^{2+} perfused over the intracellular face of the patch. (**B**) Ca^{2+} sensitivities of BK_{Ca} channels inferred from the half-saturation Ca^{2+} concentration at -50 mV in cells of different resonant frequency. Estimates from whole-cell patch recording in solitary hair cells (*filled circles*; Art et al. 1995), with measurements at three specific locations in the intact hair cell epithelium (*stars*). Resonant frequency was estimated from the height of the stereociliary bundle in isolated cells or from the location of the hair cell along the cochlea.

significantly the activation range of the BK_{Ca} channel, implying that its distance from the Ca^{2+} source is less than 20 nm (Ricci et al. 2000). Immunoprecipitation experiments have suggested coassembly of L-type Ca^{2+} channels and BK_{Ca} channels in the brain (Grunnet and Kaufmann 2004). Furthermore a possible linker protein for the two channel types was recently identified as β-catenin from yeast two-hybrid screens of chick cochleas with the BK_{Ca} channel C-terminus as bait (Lesage et al. 2004).

The trichannel complexes are not uniformly distributed over the hair cell basolateral membrane but are aggregated into larger assemblies. Recordings from hair cells in conjunction with intracellular calcium imaging have demonstrated a few hotspots of Ca^{2+} entry indicative of clustering of Ca^{2+} channels (Tucker and Fettiplace 1995; Ricci et al. 2000). From measurements at specific locations in the intact turtle cochlea, it was found that the numbers of hotspots per cell increase with CF, suggesting that the complement of Ca^{2+} channels is changed by addition of clusters at a constant channel density (Ricci et al. 2000). The codistribution of Ca^{2+} and BK_{Ca} channels has also been studied in chick hair cells using specific antibodies to the two channel types (Samaranayake et al. 2004). Both channels occurred as clusters mainly around the basal pole of hair cell. Furthermore, the number of clusters per cell increased with hair cell location towards the high-frequency end of the papilla, implying an increase in channel numbers as found from electrical recordings (Martinez-Dunst et al. 1997). Although the majority of Ca^{2+} channels were colocalized with BK_{Ca} channels, a proportion of BK_{Ca} channel clusters did not contain Ca^{2+} channels (Samaranayake et al. 2004). This lack of colocalization might reflect an error in the mechanism for targeting the two channels, as it is not clear what role such orphan BK_{Ca} channels could serve or how they would be activated.

Recordings of single BK_{Ca} channels in turtle and chick hair cells have shown that there are systematic variations of the channel properties with CF. These results raise two related questions: What is the source of the diversity and how many different variants are needed to produce the apparently smooth gradation in tuning along the cochlea? Modeling of electrical tuning using the measured properties of the native channels demonstrated that is possible to reproduce most of the frequency range in the turtle (40 to 600 Hz) with just five subtypes of BK_{Ca} channel with different kinetics (Wu and Fettiplace 1996). To achieve the appropriate degree of tuning the modeling required variation with CF in the numbers of Ca^{2+} channels, BK_{Ca} channels, and K_{IR} channels, as found experimentally. It was also necessary to assume tonotopic gradients in the maximum mechanotransducer current and in the cell's cytoplasmic calcium buffer, assumptions that were later justified experimentally (Ricci et al. 1998). The mechanotransducer channels, although not directly involved in electrical tuning, provide an important leak current to set the hair cell resting potential in the range where the voltage-dependent channels can be activated. Because the number of BK_{Ca} channels increases with CF, an increased leak through the mechanotransducer channels was needed to maintain an approximately constant resting potential (Wu and Fettiplace 1996). A smooth variation in electrical tuning was achieved by individual hair cells possessing multiple kinetic variants of the BK_{Ca} channel. Interestingly, mixing of these variants did not show up as distinct components in the voltage clamped currents for the cell. The model could be extended (Wu and Fettiplace 1996) to the chick cochlea by using nine subtypes of channel with different kinetics. After applying temperature corrections for the kinetic processes, the model generated a frequency range from 130 Hz to 4000 Hz, similar to that observed experimentally. The modeling revealed that only a few channel variants are needed to produce a smooth frequency map, but the molecular origin of the variation required isolating the hair cell BK_{Ca} channels.

6.2 Properties of Cloned Ca^{2+}-Activated K^+ (BK_{Ca}) Channels

The BK_{Ca} channel is a voltage- and Ca^{2+}-activated channel that consists minimally of four pore-forming α subunits (Quirk and Reinhart 2001) encoded by a single gene (KCNMA1). Each α subunit has seven N-terminal transmembrane domains (S0 to S6), including a positively charged S4 region conferring voltage sensitivity, a pore region between S5 and S6, and a long intracellular C-terminus (Fig. 5.10; Meera et al. 1997). In many respects it resembles structurally the voltage-dependent K^+ channels to which it is related. The major difference lies in the extended C-terminus, which contains four additional hydrophobic segments as well as an aspartate-rich "Ca^{2+} bowl" thought to be the high-affinity Ca^{2+} binding site (Schreiber and Salkoff 1997). The BK_{Ca} channel can be activated by depolarization in the absence of intracellular Ca^{2+}, but the divalent ion is a modulator that decreases the energy to open the channel (Stefani et al. 1997). It has been proposed that the four C-termini of the α subunits contribute eight RCK (regulator of K^+ channel; Jiang et al. 2002) domains that form an intracellular gating ring, and that expansion of the ring by conformational changes due to Ca^{2+} binding reduces the energy for channel opening by exerting force on the S6 gate (Niu et al. 2004).

Variation of BK_{Ca} channel properties in vivo can be achieved by several mechanisms, including phosphorylation, usually by cAMP-protein kinase (PKA) (Hall and Armstrong 2000; Schubert and Nelson 2001; Tian et al. 2001); binding of accessory β subunits (McManus et al. 1995; Meera et al. 1997; Orio et al. 2002); and alternative mRNA splicing that changes the length of the C-terminus (Lagrutta et al. 1994; Tseng-Crank et al. 1994; Saito et al. 1997; Xie and McCobb 1998; Jones et al. 1999). Although PKA phosphorylation had a modest effect on BK_{Ca} channel activity in turtle hair cells, there was no evidence for significant changes in channel kinetics (Art et al. 1995). By contrast, a number of studies in both turtle (Jones et al. 1998, 1999) and chick (Navaratnam et al. 1997; Rosenblatt et al. 1997; Ramanathan et al. 2000) demonstrate that not only are different splice variants of the α-subunit expressed in the basilar papilla, but that (at least in chick) cells in the low-frequency region of the epithelium may express accessory β subunits as well (Ramanathan et al. 1999).

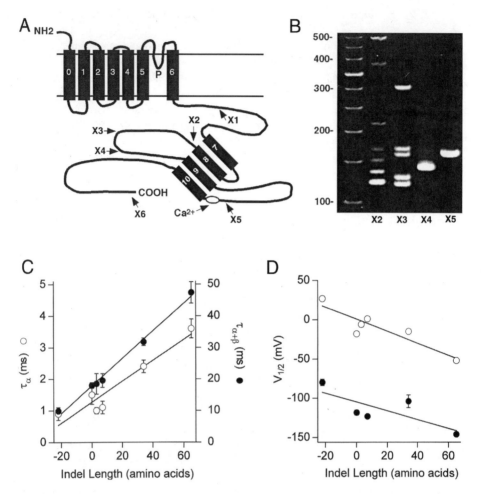

FIGURE 5.10. Properties of cloned BK_{Ca} channels. (**A**) Schematic of the BK_{Ca} channel α subunit, showing seven transmembrane domains, a pore region (P) and an extended intracellular C-terminus. Position of high-affinity Ca^{2+} binding site (Ca^{2+} bowl) and six sites in the intracellular C-terminus where residues are inserted because of alternative splicing of mRNA are indicated. Other splice sites may exist in the transmembrane region. (**B**) Polyacrylamide gel of PCR products from the turtle hair cell epithelium amplified with primers constructed around splice sites X2, X3, X4, and X5. No inserts were detected in sites X4, X5 (or in X1 and X6; not shown). However multiple products were isolated for sites X2 and X3. *Left-hand lane* contains size markers. (**C**) Variation of kinetics of cloned BK_{Ca} channel isoforms as a function of the number of amino acids inserted by alternative splicing at X2 and X3 (the indel length). Kinetics determined from relaxation time constant of ensemble current at −50 mV in 5 μM Ca^{2+} (*open circles*, α subunit only, left-hand ordinate) or in 2.5 μM Ca^{2+} (*filled circles*, α+β subunit, right-hand ordinate). (**D**) Voltage for half-activation of BK_{Ca} current with 5 μM Ca^{2+} bathing intracellular face of detached inside-out patch for different BK channel isoforms, α subunit only (*open circles*), α+β subunit (*filled circles*). More negative half-activation voltage is equivalent to higher Ca^{2+} sensitivity. In (C) and (D) cloned channels were expressed in *Xenopus* oocytes and coexpressed β subunit was the mammalian $β_1$.

6.2.1 Alternative Splicing of BK_{Ca} Channels

The C-terminus of the a subunit contains at least six sites for alternative splicing (X1 to X6; Fig. 5.10), two of which are used in the turtle. Cloned exons that may be inserted at the second site (X2) have sizes of 4, 9, 13, and 31 amino acids. The third site (X3) can have inserts of 3, 14, 17, or 61 amino acids (Jones et al. 1998) or a -26 deletion. In theory, a large number of combinations are possible but only eight were detected by reverse transcriptase polymerase chain reaction (RT-PCR) of mRNA from turtle hair cells using primers that bracket the two splice sites. Expression in *Xenopus* oocytes of those isoforms detected demonstrated up to a fourfold variation in BK_{Ca} channel kinetics and Ca^{2+} sensitivity depending on the size of the insert: larger inserts were slower and more Ca^{2+} sensitive (Fig. 5.10; Jones et al. 1999). Increasing the size of the C-terminus may affect the ease with which Ca^{2+} can expand the intracellular gating ring, thus enhancing the Ca^{2+} sensitivity. The distribution of channel time constants was between 1 and 4 ms, values in the middle of those obtained from recordings of BK_{Ca} channels in turtle hair cells (0.4 to 13 ms) under similar experimental conditions. In chick hair cells, some of the same inserts at X2 and X3 were detected as well as insert at sites X4 (0 or 8 amino acids), X5 (0 or 28 amino acids), and X6 (7, 8 or 60 amino acids) with sites numbered as in Fig. 5.10 (Navaratnam et al. 1997; Rosenblatt et al. 1997; Ramanathan et al. 2000). Single-cell RT-PCR indicated that multiple isoforms may be present in a given hair cell (Navaratnam et al. 1997).

In both turtle and chick, there were suggestions that different splice variants were differentially distributed along the tonotopic axis. Thus both the 4 (SRKR) insert and the 31 insert at X2 were found preferentially at the apical low-frequency end of the turtle cochlea (Jones et al. 1999); a similar apical location for the SRKR insert was reported in chick (Rosenblatt et al. 1997). The 28 insert at X5 is distributed in the basal high-frequency half of the chick cochlea whereas the zero insert at this site occurs in the apex (Navaratnam et al. 1997). However, these results must be taken with some caution because of the difficulties in quantifying PCR. Furthermore, it is not clear that the distributions fit with expected differences in channel properties. For example, the 31 insert (X2) is slower than the insertless, which fits with its apical location in turtle (Jones et al. 1999); however, the 28 insert (X5) is also slower than the insertless (Ramamanthan et al. 2000) but this does not fit with its more basal location in chick (Navaratnam et al. 1997).

Apart from differences in Ca^{2+} sensitivity and kinetics, an extra property of the splice variants is their differential susceptibility to phosphorylation. Thus, for example, PKA activates BK_{Ca} channels with no inserts (α_0) but inhibits channels with a 58-amino acid insert (and presumably a 61; α_{61}) at X3 (Tian et al. 2001). Whether this feature is exploited in hair cells is unknown, but it could provide a mechanism for equalizing the voltage and Ca^{2+} sensitivity of α_0 and α_{61} while preserving the kinetic differences between these two isoforms. Alternative splicing in the distal part of the C-terminus (site X6) may also influence the localization of the channel. The sequence in this region has been shown to be involved in the membrane targeting of the BK_{Ca} channel in renal epithelial cells (Kwon and Guggino 2004).

6.2.2 BK_{Ca} Channel β subunits

The variation in BK_{Ca} channel properties can also be achieved with accessory β subunits. The N-terminal S0 of the α subunit may serve as a point of interaction with an accessory β subunit, and in its native state the channel may exist maximally as an octomer of four α subunits with four associated β subunits (Wang et al. 2002). Four β subunits have been cloned in mammals (encoded by genes *KCNMB1–4*; Orio et al. 2002; Lippiat et al. 2003). $β_1$ was first isolated from smooth muscle (Knaus et al. 1994) and shown to produce a large leftward-shift in the voltage of half-activation of the channel and a slowing of channel kinetics (McManus et al. 1995). $β_2$ and $β_3$ confer inactivation on the channel whereas the neuronal $β_4$ reduces Ca^{2+} sensitivity but slows the channel activation kinetics (Wallner et al. 1999; Brenner et al. 2000; Wang et al. 2002). The non-mammalian equivalent of the $β_1$ subunit (Oberst et al. 1997) has been found in chick hair cells and in situ hybridization showed it was preferentially expressed at the apical low-frequency end of the chick cochlea (Ramanathan et al. 1999). It is thus appropriately localized to extend the low frequency range of BK_{Ca} channels in chick, but it is not known which of the β subunits is expressed in the turtle cochlea. Coexpression of the mammalian $β_1$ subunit with the various isoforms of the a subunit found in the turtle produced an approximately 12-fold slowing of the channel time constant and approximately 100 mV leftward shift in the voltage activation at constant Ca^{2+} (Fig. 5.10; Jones et al. 1999). If instead the avian $β_1$ was coexpressed, the effects were qualitatively similar though the leftward shift in the voltage activation at constant Ca^{2+} was only 50 mV, half that with the mammalian $β_1$.

While there exist mechanisms, attributable to alternative splicing and β subunit association, that can generate substantial diversity in BK_{Ca} channel kinetics, more detailed information is needed on which are employed in different cochlear regions. For example, the distributions of splice variants along the cochlea derived from PCR (Navaratnam et al. 1997; Rosenblatt et al. 1997; Jones et al. 1999) are likely to be unreliable and need to be confirmed by more quantitative methods such as immunolabeling for specific exons. Similar techniques should also be applied to quantify the distribution of the β subunit. Surprisingly, it has not been possible to clone any β subunit from the turtle cochlea. Moreover, native BK_{Ca} channels in turtle hair cells appear to be insensitive to dehydrosoyasaponin-I (Gray-Keller, Ricci, and Fettiplace, unpublished), an alkaloid that preferentially activates BK_{Ca} channels containing $β_1$ subunits (McManus et al. 1995; Tanaka et al. 1997). It is possible an accessory subunit other than $β_1$ is employed in turtle hair cells. This notion is strengthened by the recent discovery of $β_2$ and $β_4$ subunits in chick cochlear hair cells (Samaranayake et al. 2005).

6.2.3 Modeling the Role of BK_{Ca} Channel α and β Subunits in Electrical Tuning

Whether the kinetics and Ca^{2+} sensitivity of the cloned variants can explain the range of properties of the native channels is unclear. To address this question, Ramanathan and Fuchs (2002) modeled electrical tuning in chick hair cells using four types of BK_{Ca} channel: alternatively spliced isoforms α_0 (zero insert) and α_{61} (a 61-amino acid insert at X2) and the same α subunits with β subunits: $\alpha_0\beta$ and $\alpha_{61}\beta$. Channel time constants were in the order: $\alpha_{61}\beta > \alpha_0\beta > \alpha_{61} > \alpha_0$. Using properties for the four subunits obtained from expression studies in HEK293 cells, and assuming an increase in channel numbers/cell with frequency, the modeling yielded resonant frequencies at 22°C from 88 Hz with the $\alpha_{61}\beta$ channel to 282 Hz with α_0 which had the most rapid kinetics. This variation was consistent with the distribution of the β subunit, which, in the chick, was confined to the apical low-frequency end of the papilla. When temperature corrections were applied to the various kinetic processes, the frequency range extrapolated to the avian body temperature (40°C) was 307 to 1133 Hz. Mixing of two subunit types in a given cell produced resonant frequencies intermediate between those of the two components alone.

A defect of the modeling based on cloned channels is the absence of sufficiently fast gating kinetics to produce the highest frequency hair cells (Jones et al. 1999; Ramanathan and Fuchs 2002). For example, in modeling of electrical tuning in chick hair cells, the highest resonant frequency predicted (1133 Hz) is less than the highest CF in the papilla (4000 Hz). There are a number of possible explanations for this discrepancy: (1) the existence of undiscovered BK_{Ca} channel isoforms with faster kinetics, especially those with deletions such as the 26-amino-acid deletion at X3 in the turtle (Jones et al. 1999); (2) differences in membrane composition and fluidity, due to cholesterol or anionic lipids, between the expression system and the native hair cell; (3) the modifying effects of coupler proteins linking the BK_{Ca} channel to the Ca^{2+} channel or the cytoskeleton. However, it must also be the case that the maximum CF achievable by electrical tuning is restricted by the kinetic limitations of the BK_{Ca} channel.

Models of electrical tuning have shown it is possible to generate resonant frequencies up to a few kilohertz (Wu et al. 1995), but to achieve higher frequencies with this mechanism requires unrealistic assumptions about the density and kinetic properties of the BK_{Ca} channels. In other reptiles (e.g., crocodiles), and especially in birds and mammals, pressure to extend the upper frequency limit of the auditory range has driven the evolution of new tuning mechanisms. In mammals, where some species hear above 100 kHz, electrical tuning has probably been abandoned in favor of mechanical tuning of the basilar membrane, partly passive but enhanced by electromechanical feedback from a new class of outer hair cell (Robles and Ruggero 2001). Birds and crocodiles, with a more limited frequency range less than 5 kHz in most species, have retained electrical tuning, but have still found the need for an alternative tuning mechanism. In a remarkable case of parallel evolution to mammals, both have developed mechanical tuning of the basilar membrane (von Békésy 1960) and acquired a second class of hair cell (short or outer hair cells) that may be capable of providing electromechanical feedback and amplification (Fettiplace and Fuchs 1999; Manley 2000).

7. Tuning Conferred by the Mechanotransducer Channels

The electrical resonance is driven in vivo by changes in membrane potential elicited by opening and closing of the mechanotransducer channels in response to deflections of the hair bundle. Although the mechanotransducer channels are not directly involved in electrical tuning, they are part of the electrical profile of the hair cell and participate in tuning both by setting the resting potential and by contributing an initial filtering stage. These effects depend upon CF because (at least in the turtle) the properties of the mechanotransducer channels vary tonotopically in tandem with the BK_{Ca} channels. Both the maximum transducer conductance and the channel kinetics increase with CF, changes which stem at least partly from a variation in single-channel conductance (Ricci et al. 2003). The mechanotransducer channels have a significant open probability in the absence of stimulation (approximately 0.3 under in vivo conditions in the turtle; Ricci et al. 1998). They will therefore provide an important inward "leak" current to depolarize the hair cell from E_K and set the resting potential in the range where the voltage-dependent channels can be activated and where the Q of the resonance is maximal. Because of changes in the maximum transducer conductance, this leak will increase with CF to counter the increase in BK_{Ca} conductance in order to maintain an approximately constant resting potential of -50 mV. Given the narrow voltage range over which the BK_{Ca} channels are activated (e-fold in approximately 2 mV), and the fact that the resonant frequency and sharpness of the electrical tuning are voltage dependent (Crawford and Fettiplace 1981a), the contribution of the mechanotransducer channels to the resting potential is a crucial part of hair cell performance.

In response to a step deflection of the hair bundle, the mechanotransducer channels open and quickly close again in a process of fast adaptation mediated by elevation in stereociliary Ca^{2+} (Fettiplace and Ricci 2003). In the turtle, the time constants of activation and adaptation of the mechanotransducer channels change with CF, suggesting these kinetic steps may play a part in hair cell frequency selectivity (Fettiplace and Ricci 2003; Ricci et al. 2005). If channel activation and adaptation are both approximated as first-order processes, the two together produce a band-pass transduction filter. Under some conditions, fast adaptation can even display resonance in the turtle's auditory range (Ricci et al. 1998), which will make the adaptation filter higher order. If the kinetics of the mechanotransducer channels are adjusted for the ionic conditions in vivo, the transduction filter appears to be centered on the CF of the hair cell (Ricci et al. 2005). However, the degree of tuning, evaluated by the quality factor, Q, lies between 1 and 3, significantly less than that of the electrical resonance, which taken at face value implies only a minor contribution to hair cell frequency selectivity. The transduction filter could play some other role: for example, improving the signal to noise in the transduction mechanism itself by confining the frequency components in the input to those encoded by the cell's electrical filter. But it may be misleading to consider the two filters in isolation especially when both are effectively voltage-controlled feed back processes relying on Ca^{2+} influx. The exact contribution of the transduction filter to the overall frequency selectivity of the cochlea may be determined only by a quantitative model of the hair cell that takes into account both the attributes of the mechanotransducer channels in the apical pole as well as the voltage-dependent channels in the basal pole of the cell.

A complication in assessing the involvement of the mechanotransducer channels in tuning is that there are active hair bundle movements linked to opening and closing of the channels and hence to adaptation (Benser et al. 1996; Ricci et al. 2000a). In the turtle, the time course of these movements, synchronous with fast adaptation, varies tontopically. The bundle movements may act to boost the mechanical input to the hair cell, but it is at present unclear whether they significantly sharpen the tuning curve of turtle hair cells. They may be more important in the avian and mammalian cochleas, where fast electromechanical feedback from the outer (or short in the bird) hair cells has been proposed to augment the stimulus to the inner (or tall) hair cells (for review see Fettiplace and Ricci, Chapter 4).

8. K^+ Channels in the Mammalian Cochlea

The same types of K^+ channel found in non-mammals, both K_V and BK_{Ca}, are also expressed in mammalian hair cells but there is no evidence in mammals for sharp electrical tuning. Nevertheless they are important for the proper functioning of the mammalian cochlea. Deletion of the BK_{Ca} channel leads to progressive hearing loss (Rüttiger et al. 2004) and a point mutation in KCNQ4, a voltage-dependent K_V channel, causes deafness (Kubisch et al. 1999). KCNQ4 channels are a major but relatively slowly activating type of K^+ channel most conspicuous in outer hair cells. They also occur in inner hair cells where they probably aid with setting the cell's resting potential (Oliver et al. 2003). BK_{Ca} channels are found in both inner and outer hair cells, but their role is most clear in inner hair cells.

Outward currents in mature inner hair cells consist of two voltage-dependent K^+ currents: a slowly activating current, I_{Ks}, and a rapidly activating one, I_{Kf}, which together shape the receptor potentials (Kros and Crawford 1990). The slow current is a voltage-dependent K_V which has an activation time constant of approximately 5 ms and is blocked by intracellular 4-AP. However, the fast current, with about 10-fold faster kinetics, probably flows through BK_{Ca} channels (Kros et al. 1998; Skinner et al. 2003; Pyott et al. 2004) because it is blocked by iberiotoxin, a specific inhibitor of large-conductance Ca^{2+}-activated K^+ channels, as well as by TEA with a half-blocking concentration of 0.3 mM, similar to BK_{Ca} channels in turtle hair cells. Furthermore, in situ hybridization against transcripts of the BK_{Ca} channels gives particularly strong labeling in inner hair cells (Langer et al. 2003; Skinner et al. 2003). The channels associated with I_{Kf} appear at the onset of hearing in mammals (Kros et al. 1998; Pyott et al. 2004; Hafidi et al. 2005) as do BK_{Ca} channels in chick hair cells. Interestingly, these channels are of the inactivating variety which implies they contain a β_2 or β_3 subunit. The large size and the fast kinetics of I_{Kf} will reduce the membrane time constant to approximately 0.5 ms and thereby extend the frequency range of the cells, which may increase the precision of phase-locking in high-frequency responses. However, this also reduces the low-frequency impedance and hence the size of the receptor potential (Fig. 5.3B).

Why do inner hair cells, with a similar complement of voltage-dependent Ca^{2+} and BK_{Ca} channels to turtle hair cells, not produce electrical tuning? First, the amplitude of the Ca^{2+} current is less than half that in the turtle and may be too small to generate enough positive feedback to augment the electrical quality factor, Q. Second, the K_V conductance with kinetics much slower than the BK_{Ca} conductance may short out the high impedance at the resonant frequency (see Section 2.3). Third, although the BK_{Ca} conductance is large, it has a voltage dependence sixfold less steep than in the turtle, only e-fold in about 12 mV (Marcotti et al. 2004), which undermines electrical tuning (Ashmore and Attwell 1985). The shallow voltage sensitivity also raises questions about how the BK_{Ca} channels are activated, and implies that this does not arise by Ca^{2+} influx through nearby voltage-dependent Ca^{2+} channels (see Section 5.2.3). Indeed, activation of the inner hair cell BK_{Ca} current is largely unaffected by removal of external Ca^{2+} or by addition of L-type Ca^{2+} channel blockers (Kros and Crawford 1990; Marcotti et al. 2004). Furthermore, immunolabeling for BK_{Ca} channels has shown that the majority of them are clustered around the neck of the inner hair cells away from the Ca^{2+} channels at the synaptic release sites near the base (Pyott et al. 2004; Hafidi et al. 2005). Nevertheless, the BK_{Ca} channels are sensitive to manipulations of intracellular Ca^{2+}. They are activated by intracellular release of caged Ca^{2+} (Dulon et al. 1995) and their voltage-activation range is shifted in the depolarized direction by increasing intracellular calcium buffering (Marcotti et al. 2004) just as it is in the turtle (Ricci et al. 2000b). These observations imply that the BK_{Ca} channels in inner hair cells operate mainly as voltage-dependent channels under conditions of constant intracellular Ca^{2+}. This explains the shallow voltage dependence, only slightly more than that in detached membrane patches at constant Ca^{2+} (e-fold activation in 15 to 20 mV). Modulation of BK_{Ca} channels may also occur by Ca^{2+} release from intracellular stores (Marcotti et al. 2004).

The most puzzling observation about BK_{Ca} channels in the mammalian cochlea is that deletion of the BK_{Ca} α subunit in knockout mice causes a more than 30 dB hearing loss, not by affecting inner hair cells but by promoting degeneration of outer hair cells (Rüttiger et al. 2004). Hair cell degeneration develops after about 8 weeks and proceeds from the cochlear base to apex. It is unclear what part BK_{Ca} channels play in outer hair cell operation but this seems to be somehow linked to the KCNQ4 channels which in BK_{Ca} α-null mice disappear prior to any sign of outer hair cell apoptosis (Rüttiger et al. 2004). The KCNQ4 channels underlie a slow voltage-dependent K^+ current referred to as $I_{K,n}$ (Housley and Ashmore 1992; Nenov et al. 1997) which is activated at negative membrane potentials (half-activation −80 to −90 mV) and is therefore substantially turned on at a resting potential of −70 mV. Possible roles for this K^+ conductance include setting the resting potential, reducing the membrane time constant, and acting as a shunt to allow K^+ ions that enter the cell via the mechanotransducer channels to exit into the perilymph. $I_{K,n}$ in outer hair cells is specifically blocked by linopirdine (Marcotti and Kros 1999; Wong et al. 2004). Chronic administration of linopirdine elicits a phenotype similar to BK_{Ca} channel deletion, with progressive degeneration of outer hair cells proceeding from base to apex (Nouvian et al. 2003). Both KCNQ4 and BK_{Ca} channels in outer hair cells display a tonotopic gradient with channel density increasing toward the base (Mammano and Ashmore 1996; Rüttiger et al. 2004). This will decrease the membrane time constant in basal hair cells (Preyer et al. 1996) and help with high-frequency signaling. However, how BK_{Ca} channel deletion leads to KCNQ4 channel loss is currently obscure.

9. Summary and Future Directions

Extensive work in the auditory end organs of non-mammalian vertebrates, including fish, frogs, reptiles, and birds, has demonstrated electrical tuning of the hair cell receptor potential as an important means of frequency selectivity. The mechanism of electrical tuning relies on negative feedback produced by voltage- and/or Ca^{2+}-dependent K^+ channels, with variations in the density and activation kinetics of the K^+ channels generating a range of resonant frequencies. Over much of the frequency range, BK_{Ca} channels are employed but in both turtle and chick, purely voltage-activated (delayed rectifier) K^+ channels, with slow kinetics and small unitary conductance, are recruited for tuning at the lowest frequencies. The narrowness of tuning is enhanced by positive feedback from voltage-dependent Ca^{2+} channels assisted at low frequencies by inwardly rectifying K^+ channels, which provide positive feedback when they are deactivated (blocked) by depolarization. The origin of the kinetic variation among BK_{Ca} channels is still not firmly established. Studies on cloned channels show possible mechanisms that include use of different splice variants of the α subunit together with an expression gradient in a β subunit. However, more refined mappings of the α subunits are needed to complete this picture and define which exons are included at each position along the tonotopic axis. It will also be important to provide a more detailed map for β subunits and to determine whether accessory subunits other than $β_1$ contribute to hair cell BK_{Ca} channels.

A second area is to identify proteins involved in clustering of BK_{Ca} channels in association with Ca^{2+} channels and anchoring to the cytoskeleton at synaptic release sites. These scaffolding proteins may also modify the properties of the BK_{Ca} channel, either by directly interacting with the channel or by coassociating with other components such as protein kinases. It is possible that such interactions alter the kinetics of the different BK_{Ca} channel isoforms and may account for the missing high frequency resonance. However, it is clear that even the most rapid activation time constants achievable with BK_{Ca} channels are too slow to produce electrical tuning at the frequencies above 10 kHz required for mammalian hearing (Wu et al. 1995). This is probably why an electrical tuning mechanism was abandoned in mammals in favor of mechanical filtering. Despite the absence of electrical tuning in mammals, BK_{Ca} channels are expressed in both inner and outer cochlear hair cells, though neither their role nor mode of activation is fully understood.

Another important but largely unexplored area pertains to how the precise mapping of density and subtypes of K^+ channels along the tonotopic axis is established and maintained. It is of interest that during development of the chick cochlea, the BK_{Ca} channels are late to appear around E19 (Fuchs and Sokolowski 1990). In contrast, voltage-dependent Ca^{2+} channels and purely voltage-dependent K^+ channels are already present at E13. It would be instructive to know whether the appropriate BK_{Ca} channel splice variants are expressed immediately or whether these variants change with maturation. Evidence on the SRKR-insert at X2 indicates a different distribution in embryonic and adult chick cochleas (Rosenblatt et al. 1997). Changes in intracellular Ca^{2+}, initiated by the onset of hair cell transduction in the embryo, may themselves regulate the splicing. For example, in GH3 pituitary cells, expression of the 58 insert at X3 can be repressed by an increase in intracellular Ca^{2+} via activation of Ca^{2+}/calmodulin-dependent kinase IV (Xie and Black 2001). Do changes in intracellular Ca^{2+} resulting from hair cell activity control other aspects of the channel expression during development? Use of embryonic cochlear cultures, incubated in different media to manipulate intracellular Ca^{2+}, may give insight to these control mechanisms. Cataloging the properties of regenerating hair cells following destruction of sections of the auditory epithelium in adult chicks (Bermingham-McDonogh and Rubel 2003) may also provide clues to the signal gradients that determine tonotopic organization. Such signals may be common to all auditory end organs where they may control cochlear gradients in the numbers and isoforms of K^+ channels in mammals as well as non-mammals.

Acknowledgments. Work in the authors' laboratories was supported by Grants RO1 DC 03443 (J.J.A) and RO1 DC 01362 (R.F.) from the National Institutes on Deafness and other Communication Disorders.

References

Armstrong CE, Roberts WM (1998) Electrical properties of frog saccular hair cells: distortion by enzymatic dissociation. J Neurosci 18:2962–2973.

Art JJ, Fettiplace R (1987) Variation of membrane properties in hair cells isolated from the turtle cochlea. J Physiol 385:207–242.

Art JJ, Fettiplace R, Fuchs PA (1984) Synaptic hyperpolarization and inhibition of turtle cochlear hair cells. J Physiol 356:525–1550.

Art JJ, Crawford AC, Fettiplace R, Fuchs PA (1985) Efferent modulation of hair cell tuning in the cochlea of the turtle. J Physiol 360:397–421.

Art JJ, Fettiplace R, Wu YC (1993) The effects of low calcium on the voltage-dependent conductances involved in tuning of turtle hair cells. J Physiol 470:109–126.

Art JJ, Wu YC, Fettiplace R (1995) The calcium-activated potassium channels of turtle hair cells. J Gen Physiol 105:49–72.

Ashmore JF, Attwell D (1985) Models for electrical tuning in hair cells. Proc R Soc Lond Ser B 226:325–344.

Benser ME, Marquis RE, Hudspeth AJ (1996) Rapid, active hair bundle movements in hair cells from the bullfrog's sacculus. J Neurosci 16:5629–5643.

Bermingham-McDonogh O, Rubel EW (2003) Hair cell regeneration: winging our way towards a sound future. Curr Opin Neurobiol 13:119–126.

Brenner R, Jegla TJ, Wickenden A, Liu Y, Aldrich RW (2000) Cloning and functional characterization of novel large conductance calcium-activated potassium channel beta subunits, hKCNMB3 and hKCNMB4. J Biol Chem 275:6453–6461.

Catacuzzeno L, Fioretti B, Perin P, Franciolini F (2004) Spontaneous low-frequency voltage oscillations in frog saccular hair cells. J Physiol 561:685–701.

Chabbert C, Mechaly I, Sieso V, Giraud P, Brugeaud A, Lehouelleur J, Couraud F, Valmier J, Sans A (2003) Voltage-gated Na^+ channel activation induces both action potentials in utricular hair cells and brain-derived neurotrophic factor release in the rat utricle during a restricted period of development. J Physiol 553:113–123.

Corey DP, Hudspeth AJ (1979) Ionic basis of the receptor potential in a vertebrate hair cell. Nature 281:675–677.

Corey DP, Hudspeth AJ (1983) Kinetics of the receptor current in bullfrog saccular hair cells. J Neurosci 3:962–976.

Crawford AC, Fettiplace R (1980) The frequency selectivity of auditory nerve fibers and hair cells in the cochlea of the turtle. J Physiol 306:79–125.

Crawford AC, Fettiplace R (1981a) An electrical tuning mechanism in turtle cochlear hair cells. J Physiol 312:377–412.

Crawford AC, Fettiplace R (1981b) Non-linearities in the responses of turtle hair cells. J Physiol 315:317–338.

Crawford AC, Evans MG, Fettiplace R (1989) Activation and adaptation of transducer currents in turtle hair cells. J Physiol 419:405–434.

Dulon D, Sugasawa M, Blanchet C, Erostegui C (1995) Direct measurements of Ca^{2+}-activated K^+ currents in inner hair cells of the guinea-pig cochlea using photolabile Ca^{2+} chelators. Pflugers Arch 430:365–373.

Duncan RK, Fuchs PA (2003) Variation in the large-conductance calcium-activated potassium channels from hair cells along the chicken basilar papilla. J Physiol 547:357–371.

Eatock RA, Saeki M, Hutzler MJ (1993) Electrical resonance of isolated hair cells does not account for acoustic tuning in the free-standing region of the alligator lizard's cochlea. J Neurosci 13:1767–1783.

Engel J, Michna M, Platzer J, Striessnig J. (2002) Calcium channels in mouse hair cells: function, properties and pharmacology. Adv Otorhinolaryngol 59:35–41.

Evans MG, Fuchs PA (1987) Tetrodotoxin-sensitive, voltage-dependent sodium currents in hair cells from the alligator cochlea. Biophys J 52:649–652.

Fettiplace R, Fuchs PA (1999) Mechanisms of hair cell tuning. Annu Rev Physiol 61: 809–34.

Fettiplace R, Ricci AJ (2003) Adaptation in auditory hair cells. Curr Opin Neurobiol 13:446–451.

Fuchs PA, Evans MG (1988) Voltage oscillations and ionic conductances in hair cells isolated from the alligator cochlea. J Comp Physiol [A] 164:151–163.

Fuchs PA, Evans MG (1990) Potassium currents in hair cells isolated from the cochlea of the chick. J Physiol 429:529–551.

Fuchs PA, Nagai T, Evans MG (1988) Electrical tuning in hair cells isolated from the chick cochlea. J Neurosci. 8:2460–2467.

Fuchs PA, Sokolowski BH (1990) The acquisition during development of Ca-activated potassium currents by cochlear hair cells of the chick. Proc R Soc Lond B Biol Sci 241:122–1260.

Fuchs PA, Evans MG, Murrow BW (1990) Calcium currents in hair cells isolated from the cochlea of the chick. J Physiol 429:553–568.

Goodman MB, Art JJ (1996a) Positive feedback by a potassium-selective inward rectifier enhances tuning in vertebrate hair cells. Biophys J 71:430–442.

Goodman MB, Art JJ (1996b) Variations in the ensemble of potassium currents underlying resonance in turtle hair cells. J Physiol 497:395–412.

Grunnet M, Kaufmann WA (2004) Coassembly of big conductance Ca^{2+}-activated K^+ channels and L-type voltage-gated Ca^{2+} channels in rat brain. J Biol Chem 279: 36445–36453.

Hafidi A, Beurg M, Dulon D (2005) Localization and developmental expression of BK channels in mammalian cochlear hair cells. Neuroscience 130:475–484.

Hall SK, Armstrong DL (2000) Conditional and unconditional inhibition of calcium-activated potassium channels by reversible protein phosphorylation. J Biol Chem 275: 3749–3754.

Heginbotham L, MacKinnon R (1992) The aromatic binding site for tetraethylammonium ion on potassium channels. Neuron 8:483–491.

Hille B (2001) Ion Channels of Excitable Membranes, 3rd ed. Sunderland, MA: Sinauer Press.

Hodgkin AL, Huxley AF (1952) A quantitative description of membrane current and its application to conduction and excitation in nerve. J Physiol 117:500–544.

Holt JR, Eatock RA (1995) The inwardly rectifying currents of saccular hair cells from the leopard frog. J Neurophysiol 73:1484–1502.

Holton T, Weiss TF (1983a) Frequency selectivity of hair cells and nerve fibres in the alligator lizard cochlea. J Physiol 345:241–260.

Holton T, Weiss TF (1983b) Receptor potentials of lizard cochlear hair cells with free-standing stereocilia in response to tones. J Physiol 345:205–240.

Housley GD, Ashmore JF (1992) Ionic currents of outer hair cells isolated from the guinea-pig cochlea. J Physiol 448:73–98.

Housley GD, Norris CH, Guth PS (1989) Electrophysiological properties and morphology of hair cells isolated from the semicircular canal of the frog. Hear Res 38:259–276.

Hudspeth AJ, Corey DP (1977) Sensitivity, polarity, and conductance change in the response of vertebrate hair cells to controlled mechanical stimuli. Proc Natl Acad Sci USA 74:2407–2411.

Hudspeth AJ, Lewis RS (1988a) Kinetic analysis of voltage- and ion-dependent conductances in saccular hair cells of the bull-frog, *Rana catesbeiana*. J Physiol 400:237–274.

Hudspeth AJ, Lewis RS (1988b) A model for electrical resonance and frequency tuning in saccular hair cells of the bull-frog, *Rana catesbeiana*. J Physiol 400:275–297.

Issa NP, Hudspeth AJ (1994) Clustering of Ca^{2+} channels and Ca^{2+}-activated K^+ channels at fluorescently labeled presynaptic active zones of hair cells. Proc Natl Acad Sci USA 91:7578–7582.

Jiang GJ, Zidanic M, Michaels RL, Michael TH, Griguer C, Fuchs PA (1997) CSlo encodes calcium-activated potassium channels in the chick's cochlea. Proc R Soc Lond B Biol Sci 264:731–737.

Jiang Y, Lee A, Chen J, Cadene M, Chait BT, MacKinnon R (2002) Crystal structure and mechanism of a calcium-gated potassium channel. Nature 417:515–522.

Jones EMC, Laus C, Fettiplace R (1998) Identification of Ca^{2+}-activated K^+ channel splice variants and their distribution in the turtle cochlea. Proc R Soc Lond B Biol Sci 265:685–692.

Jones EMC, Gray-Keller M, Fettiplace R (1999) The role of Ca^{2+}-activated K^+ channel spliced variants in the tonotopic organization of the turtle cochlea. J Physiol 518:653–665.

Judge SI, Yeh JZ, Goolsby JE, Monteiro MJ, Bever CT, Jr. (2002) Determinants of 4-aminopyridine sensitivity in a human brain kv1.4 $K^{(+)}$ channel: phenylalanine substitutions in leucine heptad repeat region stabilize channel closed state. Mol Pharmacol 61:913–920.

Kirsch GE, Narahashi T (1983) Site of action and active form of aminopyridines in squid axon membranes. J Pharmacol Exp Ther 226:174–1179.

Kirsch GE, Shieh CC, Drewe JA, Vener DF, Brown AM (1993) Segmental exchanges define 4-aminopyridine binding and the inner mouth of K^+ pores. Neuron 11:503–512.

Knaus HG, Folander K, Garcia-Calvo M, Garcia ML, Kaczorowski GJ, Smith M, Swanson R (1994) Primary sequence and immunological characterization of beta-subunit of high conductance Ca^{2+}-activated K^+ channel from smooth muscle. J Biol Chem 269:17274–8.

Kollmar R, Montgomery LG, Fak J, Henry LJ, Hudspeth AJ (1997) Predominance of the alpha1D subunit in L-type voltage-gated Ca^{2+} channels of hair cells in the chicken's cochlea. Proc Natl Acad Sci USA 94:14883–14888.

Kros CJ (1996) Physiology of Mammalian Cochlear Hair Cells. In: Dallos PP, Popper AN, Fay RR (eds), The Cochlea. New York: Springer-Verlag, pp. 318–385.

Kros CJ, Crawford AC (1990) Potassium currents in inner hair cells isolated from the guinea-pig cochlea. J Physiol 421:263–291.

Kros CJ, Ruppersberg JP, Rüsch A (1998) Expression of a potassium current in inner hair cells during development of hearing in mice. Nature 394:281–284.

Kubisch C, Schroeder BC, Friedrich T, Lutjohann B, El-Amraoui A, Marlin S, Petit C, Jentsch TJ (1999) KCNQ4, a novel potassium channel expressed in sensory outer hair cells, is mutated in dominant deafness. Cell 96:437–446.

Kwon SH, Guggino WB (2004) Multiple sequences in the C terminus of MaxiK channels are involved in expression, movement to the cell surface, and apical localization. Proc Natl Acad Sci USA 101:15237–42.

Lagrutta A, Shen K-Z, North RA, Adelman JP (1994) Functional differences among alternatively spliced variants of Slowpoke, a Drosophila calcium-activated potassium channel. Biol Chem 269:20347–20351.

Lang DG, Correia MJ (1989) Studies of solitary semicircular canal hair cells in the adult pigeon. II. Voltage-dependent ionic conductances. J Neurophysiol 62:935–945.

Langer P, Grunder S, Rüsch A (2003) Expression of Ca^{2+}-activated BK channel mRNA and its splice variants in the rat cochlea. J Comp Neurol 455:198–209.

Lennan GW, Steinacker A, Lehouelleur J, Sans A (1999) Ionic currents and current-clamp depolarisations of type I and type II hair cells from the developing rat utricle. Pflugers Arch 438:40–46.

Lesage F, Hibino H, Hudspeth AJ (2004) Association of beta-catenin with the alpha-subunit of neuronal large-conductance Ca^{2+}-activated K^+ channels. Proc Natl Acad Sci USA 101:671–675.

Lewis RS, Hudspeth AJ (1983) Voltage- and ion-dependent conductances in solitary vertebrate hair cells. Nature 304:538–541.

Lippiat JD, Standen NB, Harrow ID, Phillips SC, Davies NW (2003) Properties of BK(Ca) channels formed by bicistronic expression of hSloα and β1–4 subunits in HEK293 cells. J Membr Biol 192:141–148.

Lipscombe D, Helton TD, Xu W. (2004) L-type calcium channels: the low down. J Neurophysiol. 92:2633–2641.

Lopatin AN, Makhina EN, Nichols CG (1994) Potassium channel block by cytoplasmic polyamines as the mechanism of intrinsic rectification. Nature 372:366–369.

Mammano F, Ashmore JF (1996) Differential expression of outer hair cell potassium currents in the isolated cochlea of the guinea-pig. J Physiol 496:639–646.

Manley GA (2000) Cochlear mechanisms from a phylogenetic viewpoint. Proc Natl Acad Sci USA 97:11736–11743.

Marcotti W, Kros CJ (1999) Developmental expression of the potassium current $I_{K,n}$ contributes to maturation of mouse outer hair cells. J Physiol 520:653–660.

Marcotti W, Geleoc GS, Lennan GW, Kros CJ (1999) Transient expression of an inwardly rectifying potassium conductance in developing inner and outer hair cells along the mouse cochlea. Pflugers Arch 439:113–122.

Marcotti W, Johnson SL, Rüsch A, Kros CJ (2003) Sodium and calcium currents shape action potentials in immature mouse inner hair cells. J Physiol 552:743–761.

Marcotti W, Johnson SL, Kros CJ (2004) Effects of intracellular stores and extracellular Ca^{2+} on Ca^{2+}-activated K^+ currents in mature mouse inner hair cells. J Physiol 557: 613–633.

Martin P, Bozovic D, Choe Y, Hudspeth AJ (2003) Spontaneous oscillation by hair bundles of the bullfrog's sacculus. J Neurosci 23:4533–4548.

Martinez-Dunst C, Michaels RL, Fuchs PA (1997) Release sites and calcium channels in hair cells of the chick's cochlea. J Neurosci 17:9133–9144.

Masetto S, Russo G, Prigioni I (1994) Differential expression of potassium currents by hair cells in thin slices of frog crista ampullaris. J Neurophysiol 72:443–455.

Masetto S, Bosica M, Correia MJ, Ottersen OP, Zucca G, Perin P, Valli P (2003) Na^+ currents in vestibular type I and type II hair cells of the embryo and adult chicken. J Neurophysiol 90:1266–1278.

Mason K, Peale FV Jr, Stone JS, Rubel EW, Bothwell M (1998) Expression of novel potassium channels in the chick basilar papilla. Hear Res 125:120–130.

McManus OB, Helms LM, Pallanck L, Ganetzky B, Swanson R, Leonard RJ (1995) Functional role of the beta subunit of high conductance calcium-activated potassium channels. Neuron 14:645–650.

Meera P, Wallner M, Song M, Toro L (1997) Large conductance voltage- and calcium-dependent K$^+$ channel, a distinct member of voltage-dependent ion channels with seven N-terminal transmembrane segments (S0-S6), an extracellular N terminus, and an intracellular (S9-S10) C terminus. Proc Natl Acad Sci USA 94:14066–14071.

Murray JD (1989) Mathematical Biology. Berlin: Springer-Verlag.

Murrow BW (1994) Position-dependent expression of potassium currents by chick cochlear hair cells. J Physiol 480:247–259. (Erratum in: J Physiol 482:725).

Murrow BW, Fuchs PA (1990) Preferential expression of transient potassium current (IA) by 'short' hair cells of the chick's cochlea. Proc R Soc Lond B Biol Sci 242:189–195.

Navaratnam DS, Escobar L, Covarrubias M, Oberholtzer JC (1995) Permeation properties and differential expression across the auditory receptor epithelium of an inward rectifier K$^+$ channel cloned from the chick inner ear. J Biol Chem 270:19238–19245.

Navaratnam DS, Bell TJ, Tu TD, Cohen EL, Oberholtzer JC (1997) Differential distribution of Ca^{2+}-activated K$^+$ channel splice variants among hair cells along the tonotopic axis of the chick cochlea. Neuron 19:1077–1085.

Nenov AP, Norris C, Bobbin RP (1997) Outwardly rectifying currents in guinea pig outer hair cells. Hear Res 105:146–158.

Niu X, Qian X, Magleby KL (2004) Linker-gating ring complex as passive spring and Ca^{2+}-dependent machine for a voltage- and Ca^{2+}-activated potassium channel. Neuron 42:745–756.

Nouvian R, Ruel J, Wang J, Guitton MJ, Pujol R, Puel JL (2003) Degeneration of sensory outer hair cells following pharmacological blockade of cochlear KCNQ channels in the adult guinea pig. Eur J Neurosci 17:2553–2562.

Oberst C, Weiskirchen R, Hartl M, Bister K (1997) Suppression in transformed avian fibroblasts of a gene (CO6) encoding a membrane protein related to mammalian potassium channel regulatory subunits. Oncogene 14:1109–1116.

Ohmori H (1994) Studies of ionic currents in the isolated vestibular hair cell of the chick. J Physiol (London) 350:561–581.

Oliver D, Plinkert P, Zenner HP, Ruppersberg JP (1997) Sodium current expression during postnatal development of rat outer hair cells. Pflugers Arch 434:772–778.

Oliver D, Knipper M, Derst C, Fakler B (2003) Resting potential and submembrane calcium concentration of inner hair cells in the isolated mouse cochlea are set by KCNQ-type potassium channels. J Neurosci 23:2141–2149.

Orio P, Rojas P, Ferreira G, Latorre R (2002) New disguises for an old channel: MaxiK channel beta-subunits. News Physiol Sci 17:156–161.

Palmer AR, Russell IJ (1986) Phase-locking in the cochlear nerve of the guinea-pig and its relation to the receptor potential of inner hair-cells. Hear Res 24:1–15.

Pantelias AA, Monsivais P, Rubel EW (2001) Tonotopic map of potassium currents in chick auditory hair cells using an intact basilar papilla. Hear Res 156:81–94.

Pitchford S, Ashmore JF (1987) An electrical resonance in hair cells of the amphibian papilla of the frog *Rana temporaria*. Hear Res 27:75–83.

Platzer J, Engel J, Schrott-Fischer A, Stephan K, Bova S, Chen H, Zheng H, Striessnig J (2000) Congenital deafness and sinoatrial node dysfunction in mice lacking class D L-type Ca^{2+} channels. Cell 102:89–97.

Preyer P, Renz S, Hemmert, W, Zenner, H-P, and Gummer AW (1996) Receptor potential of outer hair cells isolated from base to apex of the adult guinea pig cochlea: implications for cochlear tuning mechanisms. Aud Neurosci 2:145–157.

Prigioni I, Masetto S, Russo G, Taglietti V (1992) Calcium currents in solitary hair cells isolated from frog crista ampullaris. J Vestib Res 2:31–39.

Pyott SJ, Glowatzki E, Trimmer JS, Aldrich RW (2004) Extrasynaptic localization of inactivating calcium-activated potassium channels in mouse inner hair cells. J Neurosci 24:9469–9474.

Quirk JC, Reinhart PH (2001) Identification of a novel tetramerization domain in large conductance K(ca)channels. Neuron 32:13–23.

Ramanathan K, Fuchs PA (2002) Modeling hair cell tuning by expression gradients of potassium channel beta subunits. Biophys J 82:64–75.

Ramanathan K, Michael TH, Jiang GJ, Hiel H, Fuchs PA (1999) A molecular mechanism for electrical tuning of cochlear hair cells. Science 283:215–217.

Ramanathan K, Michael TH, Fuchs PA (2000) Beta subunits modulate alternatively spliced, large conductance, calcium-activated potassium channels of avian hair cells. J Neurosci 20:1675–1684.

Rennie KJ, Correia MJ (1994) Potassium currents in mammalian and avian isolated type I semicircular canal hair cells. J Neurophysiol 71:317–329.

Ricci AJ, Wu YC, Fettiplace R (1998) The endogenous calcium buffer and the time course of transducer adaptation in auditory hair cells. J Neurosci 18:8261–8277.

Ricci AJ, Crawford AC, Fettiplace R (2000a) Active hair bundle motion linked to fast transducer adaptation in auditory hair cells. J Neurosci 20:7131–7142.

Ricci AJ, Gray-Keller M, Fettiplace R (2000b) Tonotopic variations of calcium signalling in turtle auditory hair cells. J Physiol 524:423–436.

Ricci AJ, Crawford AC, Fettiplace R (2003) Tonotopic variation in the conductance of the hair cell mechanotransducer channel. Neuron 40:983–990.

Ricci AJ, Kennedy HJ, Crawford AC, Fettiplace R (2005) The transduction channel filter in auditory hair cells. J Neurosci 25:7831–7839.

Rispoli G, Martini M, Rossi ML, Rubbini G, Fesce R (2000) Ca^{2+}-dependent kinetics of hair cell Ca^{2+} currents resolved with the use of cesium BAPTA. NeuroReport 11: 2769–2774.

Roberts WM, Jacobs RA, Hudspeth AJ (1990) Colocalization of ion channels involved in frequency selectivity and synaptic transmission at presynaptic active zones of hair cells. J Neurosci 10:3664–3684.

Robles L, Ruggero MA (2001) Mechanics of the mammalian cochlea. Physiol Rev 81: 1305–1352.

Rodriguez-Contreras A, Yamoah EN (2001) Direct measurement of single-channel Ca^{2+} currents in bullfrog hair cells reveals two distinct channel subtypes. J Physiol 534: 669–89.

Rosenblatt KP, Sun ZP, Heller S, Hudspeth AJ (1997) Distribution of Ca^{2+}-activated K^+ channel isoforms along the tonotopic gradient of the chicken's cochlea. Neuron 19: 1061–1075.

Rüsch A, Eatock RA (1996) A delayed rectifier in type I hair cells of the mouse utricle. J Neurophysiol 76:995–1004.

Rüsch A, Eatock RA (1997) Sodium currents in hair cells of the mouse utricle. In: Lewis ER, Long GR, Lyon RF, Steele CR, Narins PM, Hecht-Poinar E (eds), Diversity in Auditory Mechanics Singapore: World Scientific, pp. 549–555.

Russell IJ, Sellick PM (1978) Intracellular studies of hair cells in the mammalian cochlea. J Physiol 284:261–290.

Rüttiger L, Sausbier M, Zimmermann U, Winter H, Braig C, Engel J, Knirsch M, Arntz

C, Langer P, Hirt B, Müller M, Köpschall I, Pfister M, Münkner S, Rohbock K, Pfaff I, Rüsch A, Ruth P, Knipper M (2004) Deletion of the Ca^{2+}-activated potassium (BK) α-subunit but not the BK β1-subunit leads to progressive hearing loss. Proc Natl Acad Sci USA 101:12922–12927.

Saito M, Nelson C, Salkoff L, Lingle CJ (1997) A cysteine-rich domain defined by a novel exon in a slo variant in rat adrenal chromaffin cells and PC12 cells. J Biol Chem 272:11710–11717.

Samaranayake H, Saunders JC, Greene MI, Navaratnam DS (2004) Ca^{2+} and K^+ (BK) channels in chick hair cells are clustered and colocalized with apical-basal and tonotopic gradients. J Physiol 560:13–20.

Samaranayake H, Santos-Sacchi J, Navaratnam D (2005) The effects of beta subunits on the kineitc properties of BK channels in chick hair cells. Assoc Res Otolaryngol Abstr 28:367.

Schnee ME, Ricci AJ (2003) Biophysical and pharmacological characterization of voltage-gated calcium currents in turtle auditory hair cells. J Physiol 549:697–717.

Schreiber M, Salkoff L (1997) A novel calcium-sensing domain in the BK channel. Biophys J 73:1355–1363.

Schubert R, Nelson MT (2001) Protein kinases: tuners of the BKCa channel in smooth muscle. Trends Pharmacol Sci 22:505–512.

Shieh C-C, Kirsch GE (1994) Mutational analysis of ion conduction and drug binding sites in the inner mouth of voltage-gated K channels. Biophys J 67:2316–2325.

Skinner LJ, Enee V, Beurg M, Jung HH, Ryan AF, Hafidi A, Aran JM, Dulon D (2003) Contribution of BK Ca^{2+}-activated K^+ channels to auditory neurotransmission in the guinea pig cochlea. J Neurophysiol 90:320–332.

Smotherman MS, Narins PM (1999) The electrical properties of auditory hair cells in the frog amphibian papilla. J Neurosci 19:5275–5292.

Stefani E, Ottolia M, Noceti F, Olcese R, Wallner M, Latorre R, Toro L (1997) Voltage-controlled gating in a large conductance Ca^{2+}-sensitive K^+ channel (hslo). Proc Natl Acad Sci USA 94:5427–5431.

Steinacker A, Romero A (1992) Voltage-gated potassium current and resonance in the toadfish saccular hair cell. Brain Res 574:229–236.

Sugihara I, Furukawa T (1989) Morphological and functional aspects of two different types of hair cells in the goldfish sacculus. J Neurophysiol 62:1330–1343.

Tanaka Y, Meera P, Song M, Knaus HG, Toro L (1997) Molecular constituents of maxi KCa channels in human coronary smooth muscle: predominant alpha + beta subunit complexes. J Physiol 502:545–557.

Tian L, Duncan RR, Hammond MS, Coghill LS, Wen H, Rusinova R, Clark AG, Levitan IB, Shipston MJ (2001) Alternative splicing switches potassium channel sensitivity to protein phosphorylation. J Biol Chem 276:7717–7720.

Tseng-Crank J, Foster CD, Krause JD, Mertz R, Godinot N, DiChiara TJ, Reinhart PH (1994) Cloning, expression, and distribution of functionally distinct Ca^{2+}-activated K^+ channel isoforms from human brain. Neuron 13:1315–1330.

Tsien RW, Hess P, McCleskey EW, Rosenberg RL (1987) Calcium channels: mechanisms of selectivity, permeation, and block. Annu Rev Biophys Biophys Chem 16:265–290.

Tucker T, Fettiplace R (1995) Confocal imaging of calcium microdomains and calcium extrusion in turtle hair cells. Neuron 15:1323–1335.

von Békésy G (1960) Experiments in Hearing. New York: McGraw-Hill.

Wallner M, Meera P, Toro L (1999) Molecular basis of fast inactivation in voltage and

Ca^{2+}-activated K^+ channels: a transmembrane beta-subunit homolog. Proc Natl Acad Sci USA 96:4137–4142.

Wang YW, Ding JP, Xia XM, Lingle CJ (2002) Consequences of the stoichiometry of Slo1 alpha and auxiliary beta subunits on functional properties of large-conductance Ca^{2+}-activated K^+ channels. J Neurosci 22:1550–1561.

Wong WH, Hurley KM, Eatock RA (2004) Differences between the negatively activating potassium conductances of mammalian cochlear and vestibular hair cells. J Assoc Res Otolaryngol 5:270–284.

Witt CM, Hu HY, Brownell WE, Bertrand D (1994) Physiologically silent sodium channels in mammalian outer hair cells. J Neurophysiol 72:1037–1040.

Wu YC, Fettiplace R (1996) A developmental model for generating frequency maps in the reptilian and avian cochleas. Biophys J 70:2557–2570.

Wu YC, Art JJ, Goodman MB, Fettiplace R (1995) A kinetic description of the calcium-activated potassium channel and its application to electrical tuning of hair cells. Prog Biophys Mol Biol 63:131–158.

Xie J, Black DL (2001) A CaMK IV responsive RNA element mediates depolarization-induced alternative splicing of ion channels. Nature 410:936–939.

Xie J, McCobb DP (1998) Control of alternative splicing of potassium channels by stress hormones. Science 280:443–446.

Xu W, Lipscombe D. (2001) Neuronal Ca(V)1.3alpha(1) L-type channels activate at relatively hyperpolarized membrane potentials and are incompletely inhibited by dihydropyridines. J Neurosci 21:5944–5951.

Zidanic M, Fuchs PA (1995) Kinetic analysis of barium currents in chick cochlear hair cells. Biophys J 68:1323–1336.

6
The Synaptic Physiology of Hair Cells

Paul A. Fuchs and Thomas D. Parsons

1. Introduction

Mechanosensory hair cells of the inner ear convert sound waves or head motion into bioelectrical signals for propagation throughout the nervous system. Mechanotransduction begins at the hair bundle where gated ionic flux changes the hair cell's membrane potential. This initial bioelectrical event is communicated centrally through a chemical synapse between the hair cell and an associated afferent fiber of the eighth nerve. Afferent, or centripetal flow is modulated by centrifugal, or efferent feedback provided by cholinergic efferent neurons from the auditory brainstem that synapse with hair cells or primary afferent fibers in the inner ear. Most vertebrate hair cells have efferent as well as afferent synaptic contacts (Fig. 6.1A). A notable exception is the mammalian cochlea, where afferent and efferent synapses are largely segregated to inner and outer hair cells, respectively. Even here, however, efferent and afferent synapses can be found in close proximity on the basolateral surface of hair cells. In particular, inner hair cells in the apical turn of the mouse cochlea have numerous efferent, as well as afferent, synaptic contacts before the onset of hearing at postnatal day 12 (Fig. 6.1B) and similar arrangements persist into adulthood to a variable degree in different species.

Specialized organelles are found at afferent and efferent synapses of hair cells. Afferent transmission is mediated by a synaptic ribbon, an electron-dense ball or disc to which are tethered small clear vesicles thought to contain the neurotransmitter substance. Unlike most neuronal synapses, transmitter release from the hair cell is continuously modulated with sub-millisecond temporal precision, over a significant dynamic range, and all without benefit of action potentials! Vesicular release is mediated by voltage-gated calcium channels, as in neurons; however, the synaptic ribbon undoubtedly helps to provide some of these hair-cell specific features. At efferent synapses, acetylcholine (ACh) is released to cause hyperpolarization and inhibition of the hair cells. The surprise here is that the hair cell's ACh receptor (AChR) is nicotinic, like those that cause excitation of skeletal muscle. Inhibition occurs because of associated calcium-dependent potassium flux that hyperpolarizes the membrane. A near membrane

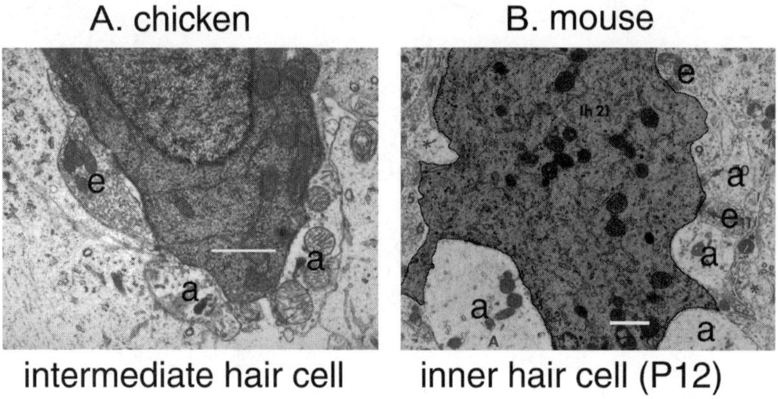

A. chicken

B. mouse

intermediate hair cell   inner hair cell (P12)

FIGURE 6.1. Afferent and efferent synapses on hair cells. Electron micrographs of an (**A**) intermediate hair cell from the chicken basilar papilla (longitudinal section) and (**B**) a mouse inner hair cell (horizontal section, apical turn of the cochlea, postnatal day 12). Afferent (a) and efferent (e) synapses can occur in close proximity on hair cells (scale bars-1 µm). (A) reproduced from Robin Michaels with permission. (B) modified with permission from Figure 1 in Sobkowicz HM, Slapnick SM (1994) The efferents interconnecting auditory inner hair cells. Hear Res 75:81–92, with permission from Elsevier.

endoplasmic reticulum (the synaptic cistern) may act as a calcium store that modulates or regulates the efferent calcium signal.

A complete understanding of inner ear function requires a thorough characterization of these synaptic mechanisms. At the most fundamental level, signal processing by the synapse determines the signal processing capacity of the system. These synaptic mechanisms also confer a significant biosynthetic and metabolic load onto the hair cell. Continuous, life-long transmitter release from the hair cell is driven by continuous calcium entry into the cell through voltage-gated channels, requiring energy-dependent sequestration and removal. Synaptic vesicles are synthesized, transported, fused with plasma membrane, and recycled, again representing a continuous drain on cellular resources of energy and material. Similar arguments apply to the efferent synapse that also triggers intracellular calcium signals to exert its effect. It is reasonable to suppose that metabolic trauma to the hair cell must also impact its synaptic function. Indeed, synaptic dysfunction could be instrumental in pathogenesis. For example, overstimulation by excess glutamate (glutamate excitotoxicity) is a well-established cause of neuronal death in the brain, and may arise in the ear from hair cell release of glutamate. Also, the calcium influx that is central to both afferent and efferent synaptic function could become cytotoxic if buffering and extrusion mechanisms failed.

This chapter is divided into a consideration of afferent synapses and efferent synapses of hair cells. The emphasis will be on cellular physiology, with an aim of discussing molecular mechanisms where possible.

2. The Afferent Synapse

The afferent neurons that innervate hair cells have cell bodies in peripheral ganglia, from which a neurite extends to form a postsynaptic contact on the hair cell, while a central axon enters the brainstem to make presynaptic contacts with neurons of target nuclei. Both the peripheral and central processes are myelinated and conduct tetrodotoxin-sensitive action potentials. The number and extent of postsynaptic contacts varies markedly among afferent neurons in different end organs. For example, many vestibular afferents arborize extensively, contacting numerous hair cells (Eatock and Lysakowski, Chapter 8), while spiral ganglion neurons contact only one inner hair cell in the mammalian cochlea, and typically receive their input from a single synaptic ribbon (Slepecky 1996). It is intriguing (and perhaps alarming) to realize that the perception of sound depends on such a limited synaptic contact, albeit repeated for each of the 10 to 30 neurons that contact each hair cell.

A great deal of knowledge has been gained in the last 25 years about mechanoelectric transduction, the process of converting the acoustic wave into a change in membrane potential (Fettiplace and Ricci, Chapter 4; Art and Fettiplace, Chapter 5). However, comparatively little is known about how the graded membrane voltages or receptor potentials in the hair cell result in the release of a chemical messenger capable of generating a spike train in the afferent fiber of a ganglion cell.

Several recent technical advances have provided the opportunity to gain new insight into the mechanism of neurotransmission between the cochlear hair cell and the afferent dendrite of the eighth nerve ganglion cells. These advances now allow us to begin to compare basic synaptic mechanisms of the cochlea to those of other well-studied synapses. This chapter focuses on three general questions: What are the requirements placed on synaptic mechanisms by audition? What is known about mechanisms of transmitter release in other fast synaptic systems? What do we know about how these or other mechanisms might be used by hair cells to meet the demands of audition?

2.1 Signaling Requirements of the Afferent Synapse

The cochlea encodes information about the amplitude, duration, and frequency of the acoustic wave. Both amplitude and duration of an acoustic signal are encoded in the spike trains of cochlear nerve recordings—amplitude as the frequency of firing and duration as the duration of firing. The persistent nature of acoustic signals, as well as ongoing spontaneous activity, requires that the afferent synapse maintain the release of neurotransmitter for prolonged periods of time. Continued pure tone stimulation results in an eighth nerve response that features several types of adaptation that relax to a sustained firing rate that is above the basal rate of firing (Fig. 6.2A). This type of high-frequency firing can persist unabated for several minutes (Kiang 1965). The ability of the hair cell synapse to maintain sustained, or tonic, neurotransmitter release distin-

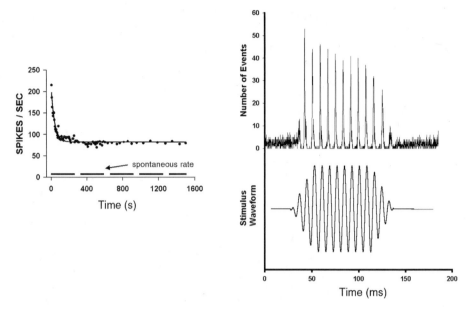

FIGURE 6.2. Response properties of auditory afferent neurons of the chicken basilar papilla. (*Left*) Mean firing rate of a single afferent fiber during presentation of a sustained suprathreshold CF tone. Firing rate dropped approximately 50% in the first 30 s, but then was maintained at 80 Hz for over 25 min! (*Right*) Phase-locking in the response of a single afferent fiber to a tone burst at 400 Hz. Unpublished data from James Saunders, with permission.

guishes it from phasic glutamatergic synapses in the central nervous system (CNS), where spike-driven transmission usually fails completely after several hundred milliseconds of high-frequency stimulation (Grover and Teyler 1990; Kapur et al. 1998). The auditory system also places unusual demands on synaptic timing. Much is known about how the cochlea acts to discriminate the frequency of acoustic signals via mechanical (von Békésy 1960), electrical, and electromotile/active mechanisms (Art and Fettiplace, Chapter 5; Brownell, Chapter 7). In many instances, the frequency components of acoustical waves also are encoded in the temporal pattern of the spike train—and thus by inference also encoded by the hair cell synapse. Two examples of the demands placed on phasic signaling by the hair cell synapse occur in binaural hearing and phase locking. In the case of the former, minute differences (tens to hundreds of microseconds) in the arrival time of acoustic waves at each ear are used by higher centers in the brain for azimuthal sound localization (Moiseff and Konishi 1981). Precise temporal fidelity is required to prevent the degradation of these small timing differences by synaptic jitter. At conventional chemical synapses, variation in the arrival of postsynaptic potentials relative to the presynaptic action potential can be 1 to 2 ms (Van der Kloot 1988; Isaacson and Walmsley

1995; Borst and Sakmann 1996); somehow the hair cell's synapse performs far more exactly.

Phase-locking is another phenomenon of the auditory system that places temporal demands on the hair cell synapse. Cochlear nerve fibers can follow high-frequency signals on a cycle-by-cycle basis with great precision. Depending on the species, the ability of an afferent fiber to phase-lock to an acoustic signal fails between 0.9 kHz (frog: Hillery and Narins 1984) and 9 kHz (barn owl: Sullivan and Konishi 1984; Koppl 1997). This is truly remarkable when synaptic latencies at most synapses are considered to be at least 1 ms. Of course, the eighth nerve does not follow the acoustic signal at high frequencies on a one-to-one basis (one cycle–one spike), but rather drops out spikes as frequency increases. However, those spikes that are recorded are in phase (Rose et al. 1967) showing that the precise timing of transmitter release is somehow maintained (Fig. 6.2B). Such precision requires rapid deactivation, as well as activation, of transmitter release. Another remarkable feature of phase locking is its intensity invariance—as the amplitude of a best frequency stimulus increases, the phase relationship between the stimulus and the response does not change (Anderson et al. 1971). This is in contradiction to widely accepted models of calcium-dependent neurotransmitter release at phasic synapses where a phase advance is predicted with increased stimulus intensity (Bollmann et al. 2000; Schneggenburger and Neher 2000).

Thus, the hair cell synapse is faced with unusual demands on both tonic and phasic neurotransmitter release. What specialized synaptic mechanisms have evolved to enable the hair cell to meet the extraordinary demands of audition? Before exploring this question further, it will be useful to review briefly the mechanisms of transmitter release established for the generic chemical synapse.

2.2 The Canonical View of Fast Chemical Neurotransmission

The basic tenets of chemical neurotransmission are the calcium and the vesicle hypotheses, both proposed by Bernard Katz and his colleagues (Katz 1969). This canonical model includes the arrival of presynaptic action potentials that result in the activation of voltage-dependent calcium channels. The resulting influx of calcium triggers the fusion of docked synaptic vesicles to the plasmalemma and the release of neurotransmitter into the synaptic cleft. Transmitter diffuses across the cleft and binds to receptor–ionophore complexes that alter postsynaptic membrane potential. Continued intensive study of synaptic physiology and molecular biology has provided several embellishments to the original model of calcium-dependent neurotransmitter release as it applies to fast glutamatergic synaptic transmission (Fig. 6.3).

2.2.1 Calcium Microdomains

The realization of distinct types of calcium channels (Hagiwara 1975; Llinas et al. 1992; Dunlap et al. 1995) immediately suggested the possibility of functional

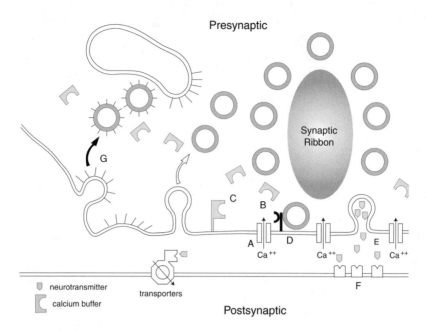

FIGURE 6.3. Hypothesized functional schematic of the hair cell ribbon synapse. This schematic is a compilation of functional attributes based in part on direct evidence from hair cells, but also on inferences from other neuronal synapses. Numerous small, clear-core synaptic vesicles are tethered around an electron-dense ribbon, or synaptic body, demarcating the active zone for transmitter release. Vesicles located under the synaptic body and docked to the plasmalemma are thought to comprise an ultrafast release pool while those tethered to the synaptic ribbon may constitute the "readily releasable pool." Synaptic vesicles needed to sustain tonic neurotransmitter release may also include other cytoplasmic vesicles not tethered to the synaptic ribbon that require translocation prior to exocytosis. The synaptic vesicles in the active zone are triggered to fuse to the plasma membrane following the entry of calcium via voltage-gated calcium channels (**A**). A calcium microdomain (**B**) results from both mobile and fixed calcium buffers (**C**) that restrict elevations in calcium concentration to the vicinity of open calcium channels. Putative interactions between the calcium channels and core complex proteins such as syntaxin hold the low-affinity calcium receptor of exocytosis (**D**) within the high-calcium plume of the microdomain. Cooperative binding of calcium to this receptor (synaptotagmin?) results in vesicle exocytosis and the release of neurotransmitter into the synaptic cleft (**E**). There diffusion, degradation, and active removal all shape the arrival of neurotransmitter at the postsynaptic receptors (**F**). Receptor affinity, kinetics, and desensitization in turn determine the postsynaptic potential. After fusion and release of neurotransmitter, vesicular membrane is recaptured in the process of endocytosis (**G**) that can occur through clathrin-mediated and non-clathrin–mediated pathways.

specialization—with different classes of calcium channels subserving different cellular functions. Subsequent kinetic, pharmacological, and now molecular biological studies have recognized a plethora of different calcium channel types that appear to be differentially expressed both between soma and synapse, and between synapses of different neurons. Numerous studies have described preferential roles for different types of calcium channels at different synapses, for example, Meir et al. (1999a).

Elegant studies on intracellular calcium concentration dynamics in the restricted area close to the mouth of the calcium channel suggest that at some synapses the effectiveness of calcium to trigger neurotransmitter may be very limited in both time and space (Adler et al. 1991; Llinas et al. 1992). Such experiments have inspired the concept of a calcium microdomain that requires the calcium sensing portion of the release machinery to be within molecular distances of the calcium channel (Zucker 1996; Neher 1998). Interacting protein domains of calcium channels and syntaxin, a member of a highly conserved protein complex that has a putative role in membrane fusion, provide a possible physical basis for such a spatial relationship (Sheng et al. 1994; Rettig et al. 1997). The microdomain model is an attractive concept for fast synaptic transmission as it provides a mechanism for the rapid onset of neurotransmitter release, and if coupled to a low-affinity calcium sensor in the release machinery, also provides a mechanism for the rapid cessation of neurotransmitter release. However, there is mounting evidence at some fast synapses in the CNS that neurotransmitter release cannot be accounted for solely by single channel calcium domains (Borst and Sakmann 1996; Augustine 2001).

Critical to the spatial–temporal profiles of intracellular calcium that regulate neurotransmitter release are a variety of calcium sinks that reduce free calcium and may ultimately remove it from the cytoplasm. Both fixed and mobile calcium buffers act on a fast time scale to bind the influx of calcium entering through the calcium channels, allowing only a small fraction (1% to 2%) of the total calcium entering the cell to act to trigger exocytosis (Neher and Augustine 1992). Mobile buffers tend to shorten the time course of intracellular calcium transients at the synapse whereas the fixed calcium buffers will tend to lengthen the duration of calcium transients (Zhou and Neher 1993; Roberts 1994). Mobile buffers bind calcium and may diffuse away to other parts of the cell before releasing their calcium ions and effectively remove the calcium ion from the site of action. Conversely the fixed buffers will also rapidly bind calcium, but remain in a dynamic equilibration with the microdomain. Fixed buffers tend to release calcium as the local concentration drops, acting then to prolong the duration of elevated calcium.

On a slower time scale, calcium also is sequestered into intracellular compartments or removed from the cell by a variety of pumps and exchangers. The relative contributions of these different calcium clearance mechanisms appear to vary between different types of synapse and according to activity level. They include the plasma membrane Ca-ATPase (Tucker and Fettiplace 1995; Krizaj and Copenhagen 1998), the Na–Ca exchanger (Mulkey and Zucker 1992; Regehr

1997), and uptake into mitochondria (Tang and Zucker 1997; David et al. 1998); but see (Zenisek and Matthews 2000) or endoplasmic reticulum (Llano et al. 2000; Emptage et al. 2001) but see (Carter et al. 2002). Subsequent release of calcium from intracellular stores also may mediate neurotransmitter release (Llano et al. 2000; Emptage et al. 2001).

How does calcium influx trigger vesicle fusion and transmitter release? Physiological evidence indicates that these processes are mediated by a low-affinity calcium receptor that has a K_D ranging from 10 μM (Bollmann et al. 2000; Schneggenburger and Neher 2000) to 100 μM (Heidelberger et al. 1994). Furthermore, the binding of the calcium ion to its sensor has a high order of cooperativity (Dodge and Rahamimoff 1967). The molecular identity or identities of the calcium sensor remains a source of robust scientific inquiry, although synaptotagmin I has been proposed as the calcium sensor in fast exocytosis (Bommert et al. 1993; Fernandez-Chacon et al. 2001).

2.2.2 The Vesicle Cycle

The cellular and molecular details that follow calcium binding and mediate vesicle fusion are less well understood. Numerous other synapse-associated proteins have been identified, but their functions mostly remain elusive (Chen and Scheller 2001). The severe temporal requirements of calcium-exocytosis coupling in fast transmission (1 to 2 ms) leave little time for multiple protein–protein interactions and conformational changes, suggesting that most of the action of these molecules must occur prior to calcium's arrival.

Synaptic vesicles reside in functional pools defined by differing stages of fusion competency as revealed by multiple kinetic components of exocytosis (von Gersdorff and Matthews 1994; Stevens and Tsujimoto 1995) and presumably reflect differing states of molecular readiness resulting from the actions and interactions of the numerous synapse-associated proteins (Parsons et al. 1996), or differing physical locations relative to the release site (von Gersdorff et al. 1996). The anatomical substrate of these different functional pools appears to vary from preparation to preparation. Synaptic vesicles docked to the plasmalemma are thought to comprise the readily releasable pool of hippocampal synapses (Schikorski and Stevens 2001), whereas in neuroendocrine cells, only a subset of the docked vesicles correspond to the readily releasable pool (Parsons et al. 1995; Plattner et al. 1997). In retinal bipolar cells, the docked vesicles also comprise the readily releasable pool, but vesicles tethered to a specialized presynaptic structure, the synaptic ribbon, appear to comprise an additional rapidly releasable pool of vesicles (Mennerick and Matthews 1996; Neves and Lagnado 1999; Palmer et al. 2003).

Once the vesicle is fused, its contents equilibrate with the synaptic cleft located between the pre- and postsynaptic cell. The spatial–temporal pattern of neurotransmitter in the cleft is shaped by diffusion (Eccles and Jaeger 1958), and degradative enzymes (Eccles et al. 1942) or transport mechanisms that return the transmitter to either the presynaptic terminal or surrounding glia cells (reviewed in Bergles et al. 1999). However, neurotransmitter concentration likely reaches millimolar levels in the cleft before binding to postsynaptic receptors (Clements et al. 1992; Diamond and Jahr 1997).

To sustain the calcium-dependent exocytosis of synaptic vesicles and the release of neurotransmitter, the synaptic vesicle cycle must be maintained (Kosaka and Ikeda 1983; Sudhof 1995). Membrane added to the plasmalemma during exocytosis of synaptic vesicles must be recovered by the process of endocytosis (Ceccarelli et al. 1973; Heuser and Reese 1973). The appropriate membrane proteins must be sorted, before synaptic vesicles are reformed and then reloaded with neurotransmitter (Mundigl and De Camilli 1994; Cremona and De Camilli 1997; Jarousse and Kelly 2001). There remains some controversy over how many pathways for membrane retrieval are used by neurons (Pyle et al. 2000; Stevens and Williams 2000; Sankaranarayanan and Ryan 2001). The fastest pathway may result from synaptic vesicles reforming immediately after neurotransmitter release by a mechanism called "kiss and run" (Fesce et al. 1994; Valtorta et al. 2001). Although evidence for this energetically attractive form of vesicle cycling continues to accumulate (Aravanis et al. 2003; Gandhi and Stevens 2003; Staal et al. 2004), its physiological importance remains under debate (Fernandez-Alfonso and Ryan 2004).

2.2.3 The Postsynaptic Response

Ultimately, transmitter binding to the postsynaptic receptors and the subsequent depolarization of the postsynaptic cell lead to spike initiation and continued transmission of the signal. Both the amplitude and time course of the postsynaptic potential play an important role in spike initiation as multiple postsynaptic potentials can summate in time to dictate postsynaptic cell activation. Several properties of the glutamate receptor–ionophore complex are known to shape the time course of the postsynaptic potential change including kinetics (Lester et al. 1990), affinity (Lester and Jahr 1992), and desensitization (Jones and Westbrook 1996). Compared to the neuromuscular junction, the number of postsynaptic receptors clustered at the release site of a fast central synapse appears to be far fewer and may become saturated by the release of a few synaptic vesicles (Auger and Marty 1997). Such a high degree of receptor occupancy has important implications for the function of the synapse. If receptor occupancy following the exocytosis of a single vesicle is *low*, then the synapse functions in a graded mode where the postsynaptic response is proportional to the number of quanta released. However if receptor occupancy is *high*, then the synapse functions in essentially a binary mode—any release, even just one vesicle, leads to a postsynaptic spike. The exocytosis of additional vesicles contributes sublinearly to the postsynaptic response (Auger and Marty 2000).

2.2.4 Summary of the Neo-Katzian Synapse

The amount of information on synaptic mechanisms is expanding exponentially as biochemical, molecular biological, genetic, and physiological analyses converge on the details of synaptic transmission. Thus, the classical Katzian view of fast synaptic transmission has been enriched with: the identification of specific calcium channel types that mediate release, the concept of calcium microdomains regulating intraterminal calcium dynamics, recognition of a low-affinity calcium receptor mediating release, multiple functional pools of vesicles, neurotransmitter cleft dynamics, postsynaptic receptor affinity, kinetics, desensitization, and saturation, and the existence of multiple endocytic pathways. Acquisition of similar knowledge about the hair cell synapse has lagged behind due to technical challenges of the inner ear as an experimental system. However, the last 5 years have provided significant advances in hair cell synaptic physiology. These findings are reviewed now to determine whether the hair cell utilizes novel synaptic mechanisms to meet the demands of audition.

2.3 Hair Cell Synaptic Mechanisms

2.3.1 The Synaptic Ribbon

The hair cell-afferent fiber synapse falls into a general class of synapses based on the presence of a characteristic electron-dense cytoplasmic structure localized at the release site. A number of sensory receptor cells that utilize graded membrane potentials (so-called "nonspiking") are found in the auditory, vestibular, and visual systems and share this common anatomical feature. The structure was termed a "ribbon" (Sjostrand 1953, 1958), and synapses made by these cells, "ribbon synapses." All ribbons are surrounded by a halo of synaptic vesicles that appear to be tethered to it by fine filaments (Fig. 6.4). Recent reviews

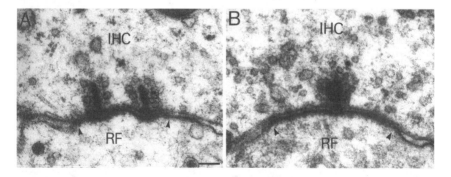

FIGURE 6.4. Diversity of synaptic body shape and size. The size and shape of the ribbon seen in transmission electron micrographs varies between and within hair cells, and may reflect functional attributes of that synaptic contact. The example shows ribbons from the modiolar (**A**) and pillar (**B**) side of inner hair cells in the cat cochlea. Afferent fibers with low rates of spontaneous activity tend to make modiolar synapses, high spontaneous rate fibers tend to make pillar synapses. Reproduced with permission from Figure 2 in Liberman et al. (1990).

are recommended for a general description of the ribbon synapse (Lenzi and von Gersdorff 2001; von Gersdorff 2001).

The molecular composition of the ribbon synapse appears to be generally similar to that of a conventional synapse, although unique isoforms of common synaptic proteins often are found at ribbon synapses (Morgan et al. 2000; Lenzi and von Gersdorff 2001). A notable exception is the synapsin family. These molecules that link vesicles to the actin cytoskeleton appear to be absent from ribbon synapses (Favre et al. 1986; Mandell et al. 1990). More and more information is available about the molecular composition of the ribbon itself. Several proteins have been identified via immunolocalization as possible constituents of the ribbon. These proteins include the antigen recognized by the ribbon-specific antibody B16, thought to be a clathrin-adaptor-like protein (Balkema and Rizkalla 1996; Nguyen and Balkema 1999), RIM, a putative Rab3-interacting protein (Wang et al. 1997), KIF3A, a kinesin molecular motor subunit (Muresan et al. 1999); bassoon and piccolo, putative scaffolding proteins of presynaptic terminals (Brandstatter et al. 1999; Dick et al. 2001), and RIBEYE, a novel protein that shares homology with both a transcriptional repressor factor and a NAD dehydrogenase (Schmitz et al. 2000). Antibodies to RIBEYE have recently been shown to label frog saccular hair cells (Zenisek et al. 2003) but the other putative ribbon proteins have been immunolocalized only to the synapses of retinal cells and await confirmation in hair cells.

Electron-dense ribbons were identified in various acousto-lateralis sensory systems (Smith and Sjostrand 1961; Lysakowski and Goldberg 1997), and the general terms "dense body" or "synaptic body" have often been used to describe the ribbon equivalent structure in hair cells (Fig. 6.4). The dense body may appear spherical, ellipsoidal, or bar shaped (Liberman 1980; Martinez-Dunst et al. 1997), ranges in size from approximately 100 to 400 nm depending on species and cell type (Martinez-Dunst et al. 1997; Lenzi et al. 1999), and may vary depending on developmental stage (Sobkowicz et al. 1982) or position within the cell (Liberman 1980). The number of dense bodies per hair cell also can vary depending on location in the end organ (Lysakowski and Goldberg 1997; Martinez-Dunst et al. 1997), or even vary dynamically in response to deafferentation (Sobkowicz et al. 1998) or weightlessness (Ross 2000).

The proximity of the synaptic body to the putative release site and its decoration with synaptic vesicles renders it likely to be involved in synaptic transmission. Ultrastructural studies in the developing cochlea suggest that the dense body descends from the cytoplasm to the basolateral membrane to initiate the formation of a synaptic specialization (Sobkowicz et al. 1986; Goodyear, Kros and Richardson, Chapter 2). Afferent nerve spontaneous rate correlates with hair cell synaptic body electron density in the frog sacculus (Guth et al. 1993) and dense body area in the cat cochlea (Merchan-Perez and Liberman 1996). The synaptic ribbon is likely important for the maintenance of high rates of synaptic vesicle release characteristic of the graded, tonic synapse where it is found. However, exact function of the synaptic ribbon remains controversial. It may serve as a "conveyer belt" to deliver vesicles to release sites (von Gersdorff 2001) or as a "safety belt" to tether vesicles in place for multivesicular release (Parsons and Sterling 2003).

2.3.2 Calcium Channels

The peculiar demands of the hair cell for continuous, but rapidly modulated transmitter release near the resting membrane potential appear to be met by unusually configured voltage gated calcium channels. Calcium current kinetics and voltage-sensitivity have been characterized in hair cells of amphibians (Lewis and Hudspeth 1983; Rodriguez-Contreras et al. 2002), reptiles (Art and Fettiplace, 1987; Schnee and Ricci 2003), birds (Zidanic and Fuchs, 1995; Spassova et al. 2001), and mammals (Beutner and Moser 2001; Engel et al. 2002; Art, and Fettiplace Chapter 5), showing in all cases a negative voltage activation range (half-activation voltage near -25 mV), very rapid gating (activation time constants less than 0.5 ms), and little or no inactivation. Similar results are obtained from studies in retinal cells. This combination of properties is not found in voltage-gated calcium channels common to most neuronal synapses, but is well suited for transmitter release from the ribbon.

At the neuromuscular junction and many CNS synapses, action potentials trigger transmitter release by calcium entry through a variety of voltage-gated calcium channels, including so-called N, P, Q, and R subtypes (Meir et al. 1999b). Some evidence has been reported for heterogeneous calcium channel types in vestibular hair cells of guinea pigs (Rennie and Ashmore 1991), frogs (Su et al. 1995) (Rodriguez-Contreras and Yamoah 2001; Eatock and Lysakowski, Chapter 8), and cochlear hair cells in chickens (Kimitsuki et al. 1994). However, most evidence suggests that a homogeneous class of "L-type" calcium channels is critical for hair cell neurotransmitter release. These calcium channels are defined in part by their pharmacological sensitivity to dihydropyridines (DHPs). Cochlear perfusion with DHP antagonists and agonists modulated both sound-evoked compound action potentials and single neuron activity in the mammalian auditory nerve (Sueta et al. 2004). Spassova et al. (2001) reported that in the chicken tall ("inner") hair cell, calcium-dependent vesicular fusion (measured as an increase in membrane capacitance) is mediated completely by low-voltage activated, noninactivating calcium channels modulated by both DHP agonists and antagonists (Fig. 6.5). The principal calcium channel subunit gene expressed in chicken cochlear hair cells is the DHP-sensitive L-type channel α_{1D} (Kollmar et al. 1997) A similar but less complete dependence on the DHP antagonist, nifedipine, was observed for both calcium current and capacitance changes in mouse inner hair cells (Moser and Beutner 2000). Finally, knockout of the gene that encodes the L-type calcium channel pore-forming subunit α_{1D} in mice results in deafness. The deafness is presumably mediated by the reduction in voltage-gated calcium currents and elimination of evoked exocytosis observed in the knockout cochlear hair cells (Platzer et al. 2000; Brandt et al. 2003). Interestingly, vestibular function is relatively unaffected in the α_{1D} knockout. In these animals, utricular hair cell calcium current is only partially

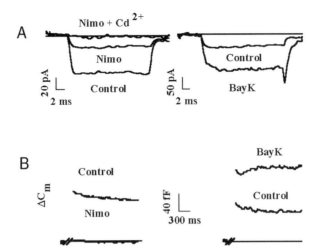

FIGURE 6.5. L-type voltage-gated calcium channels mediate hair cell neurotransmitter release. (**A**) Voltage-gated calcium current in hair cells of the chicken basilar papilla was reduced by dihydropyridine antagonists (nimodipine) and enhanced by dihydropyridines agonists that facilitate gating (Bay K 8644), providing evidence for a large L-type component. (**B**) Changes in membrane capacitance give an electrical measure of cell surface area and can be used as a proxy for transmitter release following the fusion of synaptic vesicles. Dihydropyridines also modulate voltage-dependent increases in hair cell membrane capacitance. (A) and (B) reproduced with permission from Figures 3D and E in Spassova et al. (2001).

inhibited (Dou et al. 2004), suggesting that vestibular afferent synaptic transmission is mediated by additional gene products.

2.3.3 Calcium Buffering and Extrusion

Rapid synaptic signaling requires rapidly activating calcium channels, as described above. Equally important to high-frequency synaptic transmission however, is the ability to *terminate rapidly* the calcium signals that trigger release. Consistent with this requirement, the hair cell boasts several specialized buffering mechanisms that provide a tight temporal and spatial regulation of intracellular calcium.

The bullfrog saccular hair cell has proven to be an informative model in which to study calcium buffering. High concentrations of mobile calcium buffers (Roberts 1993) such as calretinin (Edmonds et al. 2000) or parvalbumin (Heller et al. 2002) are thought to maintain highly localized concentration gradients or microdomains of calcium near the site of influx (Roberts 1994). This elegant work involved the use of calcium-activated potassium channels that colocalize with calcium channels in clusters on the basolateral surface of the hair cell (Roberts et al. 1990; Issa and Hudspeth 1994) as detectors of calcium concentration in spatial and temporal domains too small to be measured with then available optical methods. By comparing the effects of known amounts of buffer in the patch pipette to the physiological effect of endogenous buffers on calcium-activated channels, Roberts was able to predict that the endogenous buffer in the bullfrog saccular hair cells was equivalent in buffering capacity to approximately 1 mM of the fast buffer BAPTA, consistent with the notion that tight regulation of calcium does occur in hair cells. A similar value was obtained for endogenous buffers in the stereocilia of these same cells (Hall et al. 1997).

Estimates of endogenous mobile buffers also have been reported in turtle and mouse cochlear hair cells using various physiological probes of calcium dynamics. The endogenous buffer value in turtle hair cells was 0.1 to 0.4 mM BAPTA in the hair bundle (Ricci et al. 1998), based on comparisons to a calcium-dependent rapid adaptation of mechanotransduction. Intriguingly, hair cells at higher frequency locations in the tonotopically organized papilla had higher levels of buffer and an expression gradient for calbindin-D28k has been shown by quantitative molecular histology in this preparation (Hackney et al. 2003). Analysis based on the gating of large-conductance, calcium-sensitive potassium channels located in the basolateral membrane yielded similar values ranging from 0.2 to 0.5 mM BAPTA in the hair cell soma (Ricci et al. 2000). A value of 0.9 mM BAPTA was reported in turtle cochlear hair cells when the small-conductance calcium-activated potassium channel was used as the calcium sensor (Tucker and Fettiplace 1996).

Thus, estimates of mobile buffering in frog and turtle hair cells appear to range over an order of magnitude, but share the property of being mimicked by a fast exogenous buffer, BAPTA. Interestingly, only 200 µM of the slow buffer EGTA (150-fold slower on rate than BAPTA for binding calcium) was needed to mimic the effect of endogenous buffer on exocytosis from mouse cochlear cells, measured as a change in membrane capacitance (Moser and Beutner 2000). Since the cytoplasmic spread of incoming calcium is determined by the speed and concentration of the buffer, vesicular fusion measured as a capacitance increase in mouse hair cells may occur at a greater distance from the site of calcium influx than predicted from measurements in turtle and frog hair cells. The origin of this discrepancy remains to be determined. Overall, the microdomain model predicts localized increases in intracellular calcium "hotspots" along the basolateral surface of the hair cell. Such calcium hotspots have been observed in hair cells (Tucker and Fettiplace 1995; Hudspeth and Issa 1996; Rispoli et al. 2001). Indeed, fluorescent calcium indicators show a localized and rapid change in calcium immediately around synaptic ribbons in frog hair cells (Zenisek et al. 2003) providing further support for specialized calcium dynamics at sites of transmitter release.

Both mobile and fixed calcium buffers are thought to bind calcium near the mouth of the channel within microseconds of its entry (Roberts 1994). Estimates of fixed buffer equivalents in the hair bundle are similar to those for mobile buffers, approximately 0.5 mM of a fast BAPTA-like buffer (Lumpkin and Hudspeth 1998). However, similar somatic estimates for fixed buffers have

6. The Synaptic Physiology of Hair Cells   263

not been attempted and thus their role in afferent synaptic transmission remains unknown. Saturation of the fixed buffer can be prevented by a mobile buffer that is in rapid equilibrium with the rest of the cytoplasm. The calcium-loaded mobile buffer can diffuse away from the microdomain created by the calcium channel and be replaced by unbound buffer. Once away, the calcium ion will reequilibrate with the cytoplasm, releasing from the buffer. Therefore, the mobile buffer delimits calcium transients by shuttling calcium away from release sites. These ions, however, are not finally rendered biologically inactive until they are removed from the cytoplasm.

In addition to mobile buffers, both plasma membrane efflux and uptake into intracellular stores (Tucker and Fettiplace 1995; Yamoah et al. 1998; Boyer et al. 1999) act to remove calcium ions from the cell cytoplasm. Calcium-ATPases are the dominant mechanism of calcium efflux by the plasmalemma, and different isoforms of the plasma membrane calcium ATPase (PMCA) appear to be expressed in different regions of these hair cells (Dumont et al. 2001). The isoenzyme PMCA2 is expressed specifically in the hair bundles and is required for auditory and vestibular function (Kozel et al. 1998). A critical role for PMCA1, the isoform localized to the basolateral membrane, remains to be determined. However, it is interesting to speculate that PMCA1 may have a specialized role in afferent transmission as it is not found in outer hair cells (Crouch and Schulte 1995).

2.3.4 The Calcium Receptor

The original work in the squid giant synapse suggested that fast synaptic transmission should be mediated by a low-affinity calcium receptor (Adler et al. 1991; Augustine et al. 1991). The use of photolabile calcium chelators to trigger exocytosis has allowed for direct measurements of the intracellular calcium dependence of neurotransmitter release in a number of cell types. Transmitter release from the glutamatergic calyx of Held synapse in the auditory brainstem has a calcium affinity (K_D) of 25 µM (Bollmann et al. 2000; Schneggenburger and Neher 2000). Capacitance measurements of vesicular fusion in mouse hair cells (Fig. 6.6) give a similar K_D, 45 µM (Beutner et al. 2001), while the rod bipolar cell appears to be considerably less sensitive, with a K_D of 194 µM (Heidelberger et al. 1994). Recent studies of the rod photoreceptor ribbon synapse identified a highly calcium-sensitive pool of vesicles with a K_D of approximately 1 µM (Thoreson et al. 2004). The significance of these dramatic differences in calcium requirements for exocytosis between ribbon synapses of different cell types remains to be realized. Interestingly, synaptotagmins I and II, putative calcium-sensitive mediators of synchronous and asynchronous release in the brain (Geppert et al. 1994; Fernandez-Chacon et al. 2001), have not been found in the cochlea (Safieddine and Wenthold 1999).

The affinity and cooperativity of calcium binding to its receptor for exocytosis determines the minimal level of calcium needed to mediate neurotransmitter release. Matthews (2000) has suggested that the low calcium affinity observed

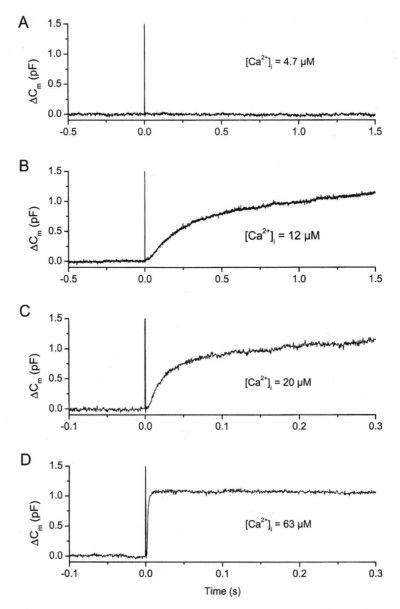

FIGURE 6.6. The calcium dependence of exocytosis from inner hair cells. Flash photolysis of caged calcium was used to increase membrane capacitance through vesicular fusion. Calcium concentration was recorded with an indicator dye. (**A**) No increase in capacitance occurred for a calcium increase to approximately 5 µM. At higher calcium levels, capacitance increased more and more rapidly (**B, C**), achieving a time constant of 1.4 ms in 63 µM calcium (**D**). A low-affinity calcium sensor triggers rapid transmitter release from hair cells. Reproduced with permission from Figure 2 in Beutner et al. (2001).

at the ribbon bipolar cell could be a hallmark for sustained neurotransmitter release at ribbon synapses, preventing a too rapid exhaustion of releasable vesicles. The hair cell ribbon synapse calcium sensor has almost an order of magnitude higher affinity for calcium than the bipolar cell, but is also not activated in submicromolar calcium levels (Beutner et al. 2001), likely owing to its high-order dependence on calcium. Interestingly, the rod photoreceptor ribbon synapse features both a higher affinity for calcium and a low-order cooperativity, thus allowing it to be activated by low levels of intracellular calcium (Rieke and Schwartz 1996; Thoreson et al. 2004). The rod photoreceptor is tonically active under dark conditions and decreases its transmitter output in response to light stimuli. While the rod bipolar and hair cell ribbon synapse are also considered tonic releasers, both possess the ability to respond with phasic exocytosis (Protti et al. 2000; Glowatzki and Fuchs 2002). Thus perhaps the low affinity and high cooperativity of calcium binding characteristic of some ribbon synapses is actually a signature of phasic neurotransmitter release. The ability to minimize tonic release of transmitter under resting conditions provides the opportunity for a greater dynamic range of graded phasic signaling.

2.3.5 Multiple Functional Pools of Vesicles

Direct postsynaptic measurements of hair cell transmitter release are extremely difficult to obtain. Thus, most studies of hair cell transmitter release have employed the capacitance method to detect the addition of membrane as synaptic vesicles fuse during release. The principles underlying this technique were proven originally in studies of large dense-core vesicular release from chromaffin cells, and have been used extensively to study release from ribbon synapses in retinal cells (Matthews 2000). Capacitance measurements have provided significant insights into hair cell function as well. Single unit recordings of eighth-nerve activity suggest that hair cells must be able to accommodate both phasic and sustained release of neurotransmitter. Experiments performed on three different hair cell preparations reveal multiple kinetic components of exocytosis and provide possible mechanisms to mediate the varied demands of audition. Multiple, distinct kinetic components of exocytosis are often interpreted as the exhaustion of different functional pools of vesicles (Neher and Zucker 1993; Thomas et al. 1993). In chicken hair cells, repeated stimulation results in exocytotic depression as each subsequent stimulus triggers a smaller and smaller jump in capacitance (Eisen et al. 2004). Several possible mechanisms that might mediate this form of synaptic depression have been ruled out, including calcium channel inactivation (Wu and Saggau 1997), presynaptic metabotropic glutamate receptor (mGluR) activation (Scanziani et al. 1997; von Gersdorff et al. 1997) and rundown of neurotransmitter release (Hay and Martin 1992), It was thus concluded that the exocytotic depression must be attributable to the exhaustion of a functional pool of vesicles (Eisen et al. 2004). Given this interpretation, subsequent analysis then provides the opportunity to make quantitative estimates of the size and kinetics of these functional pools. Both fast and slow components have been observed in different hair cell preparations, although quantitative differences exist and are detailed in Table 6.1.

The goldfish retinal bipolar cell remains the gold standard for interpretation of kinetic components of ribbon synapse exocytosis. There three components have been documented and are ascribed to the following functional vesicle pools: a readily releasable pool, a releasable pool, and a reserve pool. The readily releasable pool ($\tau = 1.5$ ms) numbering approximately 1100 vesicles is suggested to correspond to those vesicles docked at release sites between the ribbon and the plasma membrane (Mennerick and Matthews 1996; Neves and Lagnado 1999); the releasable pool ($\tau = 125$ ms) numbering approximately 6000 vesicles, corresponds to the "halo" of vesicles attached to the ribbon but not docked at the plasma membrane (von Gersdorff and Matthews 1994; von Gersdorff et al. 1996); and finally the reserve pool that consists of vesicles recruited from nearby cytoplasm (von Gersdorff and Matthews 1997; Gomis et al. 1999). Different kinetic components of exocytosis also can be revealed with the use of photolabile chelators of calcium to raise calcium uniformly throughout the cytoplasm of the cell (Neher and Zucker 1993; Thomas et al. 1993). Here exocytosis is not limited by calcium influx via voltage-gated calcium channels and so a pool of approximately 6000 vesicles can be released with a time constant of < 1 ms (Heidelberger 1998; Heidelberger et al. 1994). This functional pool corresponds to the number of vesicles tethered to the ribbon and previously defined as the releasable pool.

TABLE 6.1. Parameters of synaptic ribbons.

Ribbon synapse cell type	Kinetic component			Putative anatomical substrate	Reference
	Tau (ms)	Size (vesicles)	Max rate (vesicles/s)		
Goldfish retinal bipolar cell	< 2	1100	7×10^5	Vesicles docked at base of ribbon	1
	125	6000	5×10^4	Vesicles tethered to ribbons	2, 3
		12,000	2×10^3	Cytoplasmic vesicles	4, 5
Chick cochlear tall hair cell	< 50	1000	$> 2 \times 10^4$	Vesicles tethered to ribbons	6
	600	8000	8×10^3	Cytoplasmic vesicles	7
Frog saccular hair cell	3	6000	5×10^5	Vesicles tethered to ribbons	8, 9, 10
			1×10^4	Unknown	11
			3×10^4	Unknown	8
Mouse cochlear IHC	10	300	3×10^4	Unknown	12
			6×10^3	Unknown	12

1, Mennerick and Matthews 1996; 2, von Gersdorff and Matthews 1994; 3, von Gersdorff et al. 1996; 4, von Gersdorff and Matthews 1997; 5, Heidelberger et al. 2002; 6, Spassova et al. 2004; 7, Eisen 2000; 8, Edmonds and Schweizer 2000; 9, Lenzi et al. 1999; 10, Lenzi et al. 2002; 11, Parsons et al. 1994; 12, Moser and Beutner 2000.
SOURCE: Edmonds B, Schweizer FE. (2000) Two phases of exocytosis from vestibular hair cells. Soc Neurosci Abstr 30, 34.1.

As for the bipolar cell, quantitative electron microscopic studies on frog saccular and chicken cochlear hair cells have yielded estimates of both the number of dense bodies and their relative sizes (Roberts et al. 1990; Martinez-Dunst et al. 1997; Lenzi et al. 1999) and provide a basis for comparison with physiological studies. The fast kinetic component of release observed in the chicken cochlear hair cell (Spassova et al. 2004) exhausts with a time constant faster than 50 ms and yields a maximum rate of >20,000 vesicles/s. The size of this functional pool corresponds well to the number of vesicles tethered to dense bodies in this preparation. Furthermore, this fast component is not prevented by moderate concentrations of EGTA, whereas a slower component can be reduced. Consistent with the interpretation that the fast pool equates to vesicles on the dense body, the calculated mean free pathway of calcium diffusion in the cell with EGTA is approximately equivalent to the diameter of the dense bodies. The slower component of exocytosis that is blocked by EGTA has a rate of 8000 vesicles/s and is interpreted to result from the recruitment of cytoplasmic vesicles. In the frog, the fast component exhausts with a maximum time constant of 3 ms and yields a maximum rate of 300,000 vesicles/s (Edmonds, Gregory, and Schweizer, personal communication). The size of this functional pool (5000 vesicles) also corresponds to the total number of vesicles tethered to the dense bodies (Lenzi et al. 1999). A slow phase of exocytosis also has been observed that ranges from 10,000 to 30,000 vesicles/s (Parsons et al. 1994).

Moser and his colleagues have extensively studied the kinetics of exocytosis from mouse cochlear inner hair cells. They found a small pool of approximately 300 vesicles that exhausts with a time constant of 10 ms following a voltage step, and yields a maximum rate of release of 30,000 vesicles/s; and a sustained response to depolarizing stimulation that is an order of magnitude slower than the fast response (Moser and Beutner 2000). Interestingly, when exocytosis is triggered by the release of caged calcium, a functional pool is identified that exhausts with a time constant less than 10 ms, but numbers approximately 28,000 vesicles (Beutner et al. 2001). Successive vesicles would have to fuse within microseconds at each ribbon to sustain this rate, leading the authors to propose that significant release can take place away from ribbons. Whether such "ectopic" release requires specialized fusion sites is unknown.

What emerges is that in each of the hair cells studied both fast and slow components of exocytosis have been observed that differ by about one order of magnitude. Although the absolute rates of exocytosis can vary between different hair cells, the relationship between the fast and the slow components varies proportionately. Based on correlation with anatomical studies, the fast component of exocytosis appears to represent the release of vesicles (or a subset of vesicles) tethered to the ribbon and the slower components reflect recruitment of vesicles from cytoplasm. Given the hypothesis that the transmitter released from a single vesicle would be adequate to trigger an action potential (Geisler 1981; Siegel and Dallas 1986), the fast component of exocytosis would be more than adequate to meet the requirements of phasic release in the auditory system.

The question remains: How does the hair cell meet its need to sustain release for prolonged periods of time? Evidence suggests that the hair cell is capable of rapidly reloading its readily releasable pools, perhaps to maintain transmitter release. In mouse, the readily releasable pool is refilled with a time constant of 140 ms (Moser and Beunter 2000). In chick, where the readily releasable pool is thought to be comprised of vesicles tethered to the synaptic ribbon, the fast exocytotic component is 95% recovered in less than 200 ms. Interestingly, the time course of the readily releasable pool recovery corresponds to the recovery time of adaptation by single afferent fibers in the chick cochlear nerve (Spassova et al. 2004). These observations support the proposal that adaptation results from the exhaustion of a functional vesicle pool (Furukawa et al. 1978; Moser and Beutner 2000), and suggests that adaptation and dis-adaptation of the eight nerve is specifically mediated by the depletion and reloading of the synaptic ribbon with vesicles (Spassova et al. 2004). Such ability to rapidly refill functional pools is in distinct contrast to the retinal bipolar cell. There even the fastest exocytic component recovers relatively slowly ($\tau = 600$ ms [Gomis et al. 1999]) and putative reloading of the ribbon is even slower (approximately 20 s [von Gersdorff and Matthews 1997]). This functional difference may reflect different physiological requirements for the processes of hearing and vision, and portends the possibility of underlying molecular differences between auditory and visual synaptic ribbons.

2.3.6 Neurotransmitter Release and Postsynaptic Potentials

As evident from the foregoing discussion, much of our understanding of hair cell synaptic function is inferred from recordings of membrane capacitance that reflect the exocytosis of synaptic vesicles. These measurements act as a proxy for neurotransmitter release measured by intracellular recordings from the afferent dendrites. Conventional, sharp electrode "current-clamp" recordings of postsynaptic potentials are extremely difficult to obtain, and less revealing than voltage-clamp recordings. Nonetheless, recordings from vestibular afferent fibers in cold-blooded vertebrates have suggested that the release of neurotransmitter from hair cells occurs in quantal packets, perhaps representing single vesicles (Furukawa et al. 1978; Rossi et al. 1994). In addition, a number of hard-won recordings from mammalian cochlear afferent fibers in vivo (Palmer and Russell 1986; Siegel and Dallos 1986; Siegel 1992) showed that the majority of spontaneous postsynaptic potentials resulted in an action potential, consistent with the notion that cochlear afferent fibers have a high input impedance that could be driven to threshold with minimal excitatory currents, perhaps even those resulting from the release of a single vesicle's worth of transmitter.

A recent step forward has been made using an excised organ of Corti preparation from neonatal rat or mouse. Taking advantage of the superior optics that can be used with isolated tissue, it is possible to visualize afferent boutons at the base of inner hair cells, and so to apply "patch" pipettes for whole-cell tight-seal voltage-clamp recording (Glowatzki and Fuchs 2002). In this way, postsynaptic currents could be recorded with high resolution directly at the site of release from the ribbon synapse. As expected from the temporal demands of auditory signaling, the majority of these excitatory postsynaptic currents (EPSCs) had impressively brief waveforms, rising and falling in a few milliseconds at room temperature (Fig. 6.7A). A surprise was that EPSC amplitudes were quite variable about an average of approximately 150 pA. The amplitude distribution had a prominent mode at 36 pA, but was highly skewed, with the largest EPSCs reaching 800 pA (Fig. 6.7B). This pattern is not expected if the ribbon released single vesicles of transmitter. Rather, it was as if a variable number of vesicles were being released at once to produce different-sized EPSCs. If the modal peak of the amplitude distribution results from the release

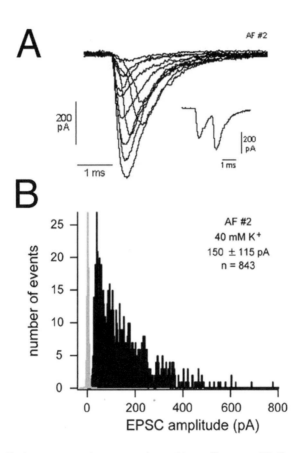

FIGURE 6.7. Excitatory synaptic currents in cochlear afferents. (**A**) Synaptic currents produced in afferent terminals by spontaneous release of transmitter at the hair cell ribbon synapse. The majority of synaptic currents had monophasic waveforms with submillisecond kinetics. (**B**) The amplitude distribution of spontaneous currents had a single defined peak at approximately 35 pA, but was markedly skewed, encompassing 20-fold larger amplitudes. This pattern suggests that multiple synaptic vesicles are coordinately released at the ribbon synapse. The light gray peak indicates the noise level. Reproduced with permission from Figures 2C and 4C in Glowatzki and Fuchs (2002).

of a single vesicle (as suggested by the observation that the equivalent conductance is the same as that for glutamatergic "minis" in the CNS) then the average EPSC represented three to six vesicles, and the largest EPSCs more than 20.

It is not known why the quantum content of the cochlear ribbon synapse appears to differ from that found in the vestibular studies. However, Siegel also deduced that synaptic potentials in cochlear afferents arose from multivesicular release (Siegel 1992). Multivesicular release may reflect the requirement for the ribbon to release vesicles rapidly and continuously. At low release frequencies the ability of the ribbon rapidly to shepherd vesicles to fusion sites ensures that the probability of multiple site occupancy is high. The mechanism for multivesicular release remains to be determined, but it might involve the coordinated release of multiple docked vesicles at the active zone (Auger and Marty 2000; Glowatzki and Fuchs 2002) or the compound exocytosis (fusion) of a synaptic vesicle "stack" tethered along the synaptic ribbon (Parsons and Sterling 2003).

Encouragingly, the extent and time course of neurotransmitter release reported here were consistent with capacitance recordings made on mice of the same developmental age (Beutner and Moser 2001). Before the onset of hearing, neonatal IHCs can generate calcium action potentials and the associated capacitance change corresponds to the release of approximately 40 vesicles onto each of the 25 or so afferent contacts. Bursts of EPSCs thought to result from spontaneous calcium action potentials were recorded from single afferent boutons and were calculated to include 47 vesicles on average (Glowatzki and Fuchs 2002), very near to the aforementioned capacitance measurement.

Afferent bouton recordings also have provided direct tests of neurotransmitter pharmacology at this synapse. The EPSCs could be blocked entirely by an α-amino-3-hydroxy-5-methyl-4-isoxazolepropionic acid (AMPA)/kainate receptor antagonist (10 μM 5-cyano-7-nitroquinoxaline-2,3-dione [CNQX]), and were enhanced by cyclothiazide that reduces desensitization of AMPA, but not kainate, receptors. These results were in good agreement with earlier in vivo studies showing that cyclothiazide increased spontaneous activity in guinea pig afferent fibers (Ruel et al. 1999) and a selective AMPA receptor antagonist blocked action potential generation and glutamate excitotoxicity in the cochlea (Ruel et al. 2000). Another strategy was to use glutamate "sniffer" patches to detect the release of glutamate from isolated chicken hair cells (Kataoka and Ohmori 1994; Kataoka and Ohmori 1996). In this technique an outside-out membrane patch is excised from a cultured cerebellar granule cell that expresses glutamate receptors. Individual hair cells are placed into the same culture dishes so that the outside-out patch can be brought near to a hair cell that is being depolarized by a second patch pipette. N-methyl-D-aspartate (NMDA) receptors in the granule cell patch are activated by transmitter released during hair cell depolarization, suggesting that the hair cell releases an NMDA receptor agonist (i.e., glutamate).

2.3.7 Endocytosis

Much emphasis has been placed on the necessity of cellular mechanisms to mediate both the phasic and sustained aspects of hair cell transmitter release.

However, the exocytosed synaptic vesicle membrane ultimately needs also to be recovered and recycled. At a classic fast synapse, the recovery of this exocytosed membrane is critical to synapse function (Koenig et al. 1989), and resupply of synaptic vesicles to the synapse is local (Heuser and Reese 1973). An unusual aspect of the somatic release of neurotransmitter by the hair cell is the close proximity of other intracellular membrane trafficking networks. Thus, these other subcellular membrane compartments could provide, in theory at least, another source of vesicles to sustain exocytosis at the hair cell synapse.

We have gained much insight in the last several years with regard to the process of membrane retrieval or endocytosis following exocytosis in the hair cell. The original membrane capacitance recordings of hair cells (Parsons et al. 1994) demonstrated the ability of the hair cell to maintain its cell surface area with great fidelity. Following stimulation, endocytosis returned the frog saccular hair cell to within 10% of its original cell surface area with a time constant of 14 s. Furthermore, this mechanism of membrane retrieval was shown to depend on the presence of a low molecular weight compound in the cytosol (endocytosis failed during ruptured patch recordings, presumably because necessary components were dialyzed from the cytoplasm). Endocytosis with nearly identical kinetics has been reported for mouse cochlear hair cells ($\tau = 16$ seconds (Beutner et al. 2001). However, these authors also describe a fast component of endocytosis ($\tau = 300$ ms) that is promoted by high intracellular calcium concentrations (> 15 μM). Apparently, hair cell synaptic vesicle membrane can be retrieved by one of two mechanisms. Both presumably represent examples of "rapid endocytosis," reviewed by Henkel and Almers (1996) that have been reported for neurons and endocrine cells and are faster than conventional clathrin-mediated endocytosis. The faster form recruited by high calcium may reflect a specialization to sustain transmitter release during times of intense or prolonged stimulation.

The strong exo–endocytosis coupling in hair cells can rapidly recover synaptic vesicle membrane from the plasmalemma, but it is not clear how fast the endocytosed membrane is then recycled into synaptic vesicles and if the recycling is fast enough to sustain exocytosis. Hair cell synapses are characterized by large aggregates of cytoplasmic vesicles that in the electron microscope appear similar to synaptic vesicles (Martinez-Dunst et al. 1997; Lenzi et al. 1999; Spicer et al. 1999) and may contribute to sustaining neurotransmitter release. Recent studies of ribbon synapses in the retina suggest that this large reservoir of cytoplasmic synaptic vesicles functions as a "buffer" between exocytosis and endocytosis and allow sustained bouts of rapid transmitter release (Lagnado et al. 1996; Holt et al. 2004; Rea et al. 2004).

Photoreceptors tonically release neurotransmitter in the dark and appear to utilize a continuous, "streamlined" cycle of exocytosis and endocytosis (Rieke and Schwartz 1996; Rea et al. 2004). Following neurotransmitter release, the exocytosed plasma membrane is retrieved via the endocytosis of small vesicles that are immediately mobile in the cytoplasm. Essentially all the vesicles remain mobile to insure that a sufficient number of vesicles encounter a synaptic ribbon and that the high tonic rates of release are maintained in the dark. The situation in the bipolar cells is similar but different. There membrane retrieval is mediated by bulk endocytosis (Holt et al. 2003), but this compartment rapidly degrades into synaptic vesicles creating a similar large reservoir of highly mobile synaptic vesicles (Holt et al. 2004). However, the photoreceptor synaptic vesicles are an order of magnitude more mobile than those in bipolar cells, likely subserving the increased requirements for tonic release necessitated by the sensory cell compared to the phasic-tonic requirements of the second-order bipolar neuron.

Both phasic and tonic demands for neurotransmitter release are placed on the hair cell, but the mobility of synaptic vesicles is unknown. There are several other membranous subcellular compartments identified in hair cells that could serve as immobilizing sinks for endocytosed membrane. This includes a dense network of smooth endoplasmic reticulum and Golgi complexes in the perinuclear region (Spicer et al. 1999) and a large endosomal network in the apical portion of the cell (Kachar et al. 1997). Tracer studies have demonstrated communication between these different subcellular compartments (Siegel and Brownell 1986; Leake and Snyder 1987; Griesinger et al. 2002). In particular, hair cells appear to have a robust, constitutive membrane turnover at the apical surface (Seiler and Nicolson 1999; Self et al. 1999; Meyer et al. 2001). Apical endocytosis by guinea pig cochlear inner hair cells is purported to provide membrane for trafficking to the basal synaptic zone and is putatively regulated by synaptic activity (Griesinger et al. 2002). These provocative findings suggest the intriguing notion that hair cell synapses may rely on sources of vesicles in addition to those recycled from the basolateral membrane to maintain a large reservoir of synaptic vesicles necessary for exocytosis, albeit possibly at the expense of vesicle mobility.

Trafficking between apical and basal membrane compartments of hair cells is not surprising given their epithelial origins and may represent transcytotic mechanisms. Griesinger and colleagues (Griesinger et al. 2002) showed that the upregulation of apical endocytosis was blocked by agents such as calcium channel blockers that inhibit hair cell exocytosis. However, the stimulation protocols used consisted of high extracellular potassium for several minutes and are likely to drive intracellular calcium throughout the cell to levels not normally experienced during mechanotransduction. This has the potential to trigger other cellular events in addition to synaptic exocytosis that might regulate apical endocytosis, and thus the physiological relevance of coupling between basolateral synaptic activity and apical endocytosis in hair cells remains open to further experimentation.

2.3.8 Ontogeny of Afferent Transmission

The onset of hearing varies from species to species, occurring during the last week in ovo for chickens (Saunders et al. 1973; Rebillard and Rubel 1981) and during the second postnatal week for rats and mice (Ehret and Merzenich 1985; Goodyear, Kros, and Richardson, Chapter 23). Dramatic changes in the electrical properties of the hair cell (Fuchs and Sokolowski 1990; Griguer and Fuchs

1996; Kros et al. 1998) precede functional maturation of hearing and are thought possibly to mediate patterned spontaneous activity that might be important for the refinement and maintenance of synaptic connections (Jones and Jones 2000; Jones et al. 2001). This is also a time of dynamic changes in synaptic morphology (Sobkowicz et al. 1982; Ard et al. 1985); however, far less is known about the onset of neurotransmitter release from hair cells.

Mice provide a convenient model to explore the ontogeny of presynaptic mechanisms in cochlear hair cells. Both calcium-dependent exocytosis and endocytosis are present shortly after birth. However, prior to the onset of hearing, the inner hair cells pass through a stage of strong exocytotic activity attributable to increased expression of calcium channels (Beutner and Moser 2001) at a time when the highest number of synaptic ribbons are present (Sobkowicz et al. 1986). Around the onset of hearing, the number of ribbons at each synaptic contact falls, the shape of the synaptic bodies also changes from spherical to planar, calcium channel density falls, but the efficiency of stimulus–secretion coupling improves—a greater capacitance change occurs for a given level of calcium influx. Release kinetics also change during postnatal development as the size of the fast component, but not the slow component, is markedly reduced in the hearing animals. It is interesting to speculate whether this change in size of the readily releasable pool is in part mediated by the conversion from a spherical to a plate-like dense body. Much larger readily releasable pools are observed in the chicken and the frog where the dense bodies are more spherical than planar (Martinez-Dunst et al. 1997; Lenzi et al. 1999).

2.4 Summary of the Afferent Synapse

The hair cell afferent synapse is remarkable. Throughout the lifetime of the cell, continuous release of neurotransmitter is modulated with sub-millisecond precision to encode the frequency composition and intensity of sound or head movements. And, all this is accomplished by the hair cell without the benefit of action potentials to trigger release. To carry out these tasks the hair cell employs the synaptic ribbon, an electron-dense body 0.5 µm or less in diameter to which are tethered a number of clear-core vesicles presumed to contain the neurotransmitter glutamate, or a functionally equivalent molecule. The molecular composition of the synaptic ribbon is essentially unknown but is distinguished from the presynaptic density of neuronal synapses by its much greater size, and the apparent absence of at least some of the identified proteins of the synaptic complex, notably the synapsins and synaptotagmins I and II. The functional consequences of these differences remain to be determined, although synapsin's absence is logical given its role in restricting vesicle movement by anchoring them to the actin cytoskeleton.

The relatively high rates of "spontaneous" transmitter release from hair cells demonstrates another marked difference from most neuronal synapses, and in particular from the canonical neuromuscular junction. The voltage-gated calcium channels (VGCCs) of hair cells are encoded by different genes than those that provide for neuronal synapses. Rather than the relatively high-threshold, dihydropyridine-insensitive VGCCs serving neuronal transmitter release, hair cells employ low threshold, noninactivating, dihydropyridine-sensitive VGCCs, related to the VGCCs of cardiac muscle. Thus, some of the calcium channels will be open at the hair cell's resting membrane potential, and drive "spontaneous" eighth nerve activity. These channels also activate very rapidly, inactivate minimally, and therefore allow hair cell calcium signaling to track the periodicity of acoustic stimuli.

The mechanisms of vesicular release and recycling in hair cells differ quantitatively from those of other synapses. There seems to be a nearly inexhaustible supply of vesicles feeding the ribbon, and initial studies suggest that multiple vesicles can be released simultaneously. Parallel activation of multiple fusion sites at individual ribbons would help to explain the high rates of release observed in capacitance recordings. What remains unanswered is how such multivesicular release can occur, particularly in the absence of action potentials to coordinate the opening of multiple VGCCs. This suggests the possibility that multivesicular release can be driven by the opening of a single VGCC. Presumably the synaptic ribbon also plays a role, perhaps to load multiple vesicles into parallel fusion sites on the plasma membrane or perhaps to facilitate compound fusion within stacks of vesicles running up the ribbon; the alternatives at present are bounded only by the limits of the imagination.

3. The Efferent Synapse

The transmission of afferent signals from the ear is modulated by neural feedback from the brain. Such efferent, or centrifugal, feedback is a general aspect of neural organization of sensory pathways, with functional effects including blocking self-stimulation and/or adaptation during motor output, preservation of dynamic range, discriminating signals from noise, and protection from damaging stimuli. In the auditory periphery, activation of the efferent pathway restores dynamic range in the presence of background noise (Winslow and Sachs 1987), leading to enhanced discrimination of threshold differences (May and McQuone 1995). Also, efferent inhibition of the mammalian ear can confer a protective effect during exposure to loud sound (Rajan and Johnstone 1988; Liberman 1991). In the vestibular periphery, the overall role of efferent control may be to convert from postural to dynamic coding (Highstein 1991; Brichta and Goldberg 2000). Another attractive idea has been that efferents modulate afferent gain during volitional head movements. However, recent studies cast doubt on this function at least for a certain range of head movements in primates (Cullen and Minor 2002). An important distinction is that efferent effects in the auditory periphery are inhibitory (so far as is known) while vestibular afferents can be inhibited, excited, or both (Goldberg and Fernandez 1980; Eatock and Lysakowski, Chapter 8). Thus, the cellular mechanisms of efferent feedback must vary significantly between different end organs of the inner ear. How does synaptic feedback alter the response of peripheral mechanoreceptors? Although efferent synapses also are made on afferent nerve endings in the cochlea and vestibule, this chapter focuses particularly on the cholinergic inhibition of cochlear hair cells. Before discussing the cellular mechanisms of inhibition, it will be useful briefly to review the anatomy and pharmacology of the efferents to the inner ear.

The efferent innervation pattern in the mammalian ear has been examined by many, with a hallmark analysis of cochlear efferents by Warr and colleagues (Warr 1975), reviewed in Warr (1992) and Guinan (1996). Vestibular efferent innervation has been summarized by Guth and colleagues (Guth et al. 1998). Briefly stated, efferent axons to the cochlea arise from neurons located within the superior olivary complex of the brainstem. Degeneration studies by Rasmussen first identified this olivocochlear pathway (Rasmussen 1946). Olivocochlear axons exit the cranium in the anterior branch of the eighth nerve, then decussate at the peripheral anastomosis of Oort to travel with afferent fibers to the cochlea. Within the mammalian cochlea one finds larger efferent synapses on the bases of outer hair cells, and small endings on the afferent fibers beneath the inner hair cells. Large-diameter, myelinated axons of medial olivocochlear neurons give rise to the larger synapses on outer hair cells, while lateral olivocochlear neurons with smaller unmyelinated axons contact inner hair cell afferents. Efferent synapses are found on inner hair cells in the neonatal cochlea and persist to a variable degree in the cochlear apex of some species. In the vestibular periphery efferent synapses are made directly onto type II hair cells, and onto afferent contacts beneath both type II and type I hair cells (Eatock and Lysakowski, Chapter 8).

Efferent innervation of the ear in birds, reptiles, and amphibians differs somewhat from that of mammals. The chief difference is that most hair cells in the non-mammalian hearing organs receive both efferent and afferent contacts. An exception is the avian basilar papilla, in which short hair cells, like outer hair cells of mammals, receive predominantly or exclusively efferent contacts (Takasaka and Smith 1971).

Histological, biochemical, and pharmacological studies have been used to demonstrate that many, if not most, of the efferent fibers to the inner ear are cholinergic (reviewed in (Guth et al. 1998). Other immunohistological studies have identified a smaller fraction of efferent fibers that may use γ-aminobutyric acid (GABA) as their neurotransmitter (Vetter et al. 1991; Nitecka and Sobkowicz 1996), while some evidence points to neuropeptides such as calcitonin gene-related peptide (CGRP), opioids, the enkephalins, and dynorphins in efferent neurons. This chapter emphasizes cholinergic effects on hair cells since these have been studied most comprehensively to date.

Synaptic ultrastructure has been reviewed by Slepecky (1996). Of particular interest to a following discussion is the unusual, but consistent, appearance of the synaptic cistern at efferent synapses on hair cells (Fig. 6.8A). This near membrane endoplasmic organelle (sometimes referred to as the "subsynaptic cistern") is closely aligned (20 to 30-nm spacing) with the postsynaptic plasma

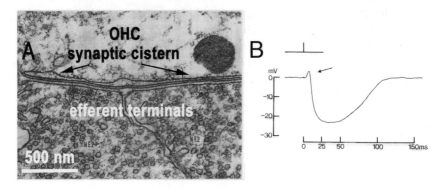

FIGURE 6.8. Efferent synapses in the mammalian cochlea. (**A**) Two vesiculated nerve endings contact an outer hair cell. A "subsynaptic" (or synaptic) cistern is found closely apposed to the postsynaptic membrane of the hair cell, coextensive with the efferent contacts. Reproduced from Figure 7 in Saito (1980), with permission from Elsevier Science. (**B**) Inhibitory synaptic potentials in turtle hair cells. The averaged response to a single shock to the efferent axons consisted of a large, long-lasting hyperpolarization from the resting membrane potential (-45 mV in this cell). The predominant hyperpolarization was preceded by a smaller, briefer depolarization, indicated by the arrow. From Art, Fettiplace, and Fuchs, unpublished, with permission.

membrane of the hair cell and is essentially coextensive with the efferent contact (Saito 1980). Based on its appearance, several investigators have suggested that this cistern may be a synaptically activated calcium store (see Section 3.9).

The function of the efferent pathway has been studied extensively, with techniques ranging from behavioral to molecular biological. Excellent and comprehensive reviews of this body of literature can be found in Wiederhold (1986) and Guinan (1996). Here we focus on the cellular physiology of efferent inhibition, based largely on intracellular recordings from hair cells.

3.1 Cellular Studies of Synaptic Inhibition

The first intracellular recordings of efferent inhibition were obtained from hair cells in the lateral line of the burbot (*Lota lota*, a cod-like fish) by Flock and Russell (1976). The authors described inhibitory postsynaptic potentials (IPSPs) that hyperpolarized the hair cell by as much as 10 mV and outlasted the efferent shock train by 150 to 200 ms. The IPSPs were blocked by cholinergic antagonists. During inhibition, synaptic excitation of the associated afferent fiber was diminished, indicating that hair cell transmitter release was reduced. Hyperpolarizing synaptic potentials also were reported to inhibit the receptor potentials of frog saccular hair cells (Ashmore and Russell 1982).

3.1.1 IPSPs in Turtle Hair Cells

An extensive study of efferent inhibition was carried out by Crawford, Fettiplace, and colleagues (Art and Fettiplace 1984; Art et al. 1984, 1985) who made intracellular recordings from hair cells in the auditory papilla of the turtle. Brief trains of efferent shocks produced maximal hyperpolarizing IPSPs that ranged from 12 to 30 mV in different hair cells, with a half-amplitude duration of 150 to 200 ms. The amplitude and waveform of the IPSP varied when the hair cell's membrane potential was altered by current injection through the microelectrode. The prominent hyperpolarization of the IPSP reversed in sign at -80 mV. The reversal potential varied as a function of the external potassium concentration, leading to the conclusion that the hyperpolarizing phase of the IPSP was due to an increase in the potassium conductance of the hair cell membrane. A depolarizing component that preceded the larger hyperpolarization was revealed when the hyperpolarizing phase was minimized at its reversal potential. The early depolarizing component of single IPSPs also could be seen at the resting potential in some hair cells (Fig. 6.8B). The ionic basis of the early depolarization was not determined, but this component was believed to result from the efferent neurotransmitter as well. Application of the nicotinic antagonist curare or the muscarinic antagonist atropine blocked both components of the efferent IPSP reversibly.

Intracellular microelectrode recording was used later to demonstrate similar hyperpolarizing IPSPs in hair cells of the frog's saccule (Sugai et al. 1992), which also were thought to arise by the release of ACh from the efferent terminals (see Section 3.8). Thus, in hair cells of a fish lateral line, an amphibian vestibular organ, and the hearing organ of a reptile, the electrophysiological features of efferent inhibition appeared similar. Electrical stimulation of efferent axons produced large hyperpolarizing IPSPs whose waveform decayed over the course of 100 ms or more. In each case the pharmacology suggested a cholinergic mechanism, although the character of the ACh receptor remained undefined since both nicotinic and muscarinic antagonists were effective blockers.

3.2 Inhibition of the Receptor Potential

In the turtle inner ear, as in many vertebrates, afferent and efferent synapses converge on the same hair cell. The efferent effect is to open ion channels to hyperpolarize and shunt the basolateral membrane, so that the sound-evoked receptor potential fails to release neurotransmitter. The conductance increase itself can account for several aspects of efferent inhibition, including the enlarged extracellular microphonic potential seen in the mammalian cochlea that results from the increased flow of current through the hair cell. Other considerations, however, suggest that hair cell inhibition might rely specifically on the associated hyperpolarization as well. For example, synaptic inhibition provided by GABA- or glycine-gated chloride channels (common in the vertebrate CNS) produces only small changes in membrane potential (since E_{Cl} lies near the resting membrane potential in most cells), and depends largely on the conductive shunt. Thus, one might ask why hair cell inhibition involves such a pronounced hyperpolarization. Does this method of inhibition reflect additional requirements unique to hair cell function? Some insights can be provided by studying the effect of inhibition on the acoustic response of turtle hair cells.

To explain the inhibitory effect it is necessary first to describe the response of turtle hair cells to sound and the role played by electrical tuning (reviewed in Fettiplace and Fuchs, 1999). The voltage response to a tone is a sinusoid whose amplitude, phase, and linearity depend on sound frequency. During a constant intensity frequency sweep the peak-to-peak amplitude of the receptor potential is largest at the characteristic frequency (Fig. 6.9A, B), revealing the acoustic tuning of the hair cell response (Crawford and Fettiplace 1980). This acoustic tuning arises from a mechanism of electrical tuning that is intrinsic to each hair cell, and is derived from the cell-specific expression of voltage-gated ion channels (Crawford and Fettiplace 1981; Lewis and Hudspeth 1983; Art and Fettiplace 1987; Hudspeth and Lewis 1988a,b; Art and Fettiplace, Chapter 5).

The receptor potential to a tone at the characteristic frequency is markedly reduced by efferent inhibition (Fig. 6.9A, B). Not only is the peak-to-peak amplitude of the sinusoid many times smaller, but the hair cell is also strongly hyperpolarized. This combined effect makes inhibition of turtle hair cells very powerful, requiring 50 to 75 dB additional sound pressure to overcome (Art et al. 1984). Thus, one consequence of the efferent hyperpolarization is to fortify the inhibitory effect beyond that arising from a conductive shunt alone. A still more interesting consequence is found when considering tones that are *not* at the characteristic frequency.

3.2.1 "De-tuning" the Hair Cell

Remarkably, inhibitory efficacy varies as a function of the frequency of acoustic stimulation, and is maximal at the characteristic, or best frequency of the cell. Thus, efferent activity not only inhibits the cochlear response, but it does so in a frequency-specific way, in essence de-tuning the cochlear filter (Art and Fettiplace 1984). This can be seen in Figure 6.8 where the peak-to-peak voltage sinusoid is reduced at characteristic frequency (CF), but is unchanged at high frequencies, while the sinusoidal response to low frequency tones is actually enhanced! The net result is to convert the hair cell from a band-pass to a low-pass detector (Art et al. 1985).

Efferent de-tuning in the turtle can be explained by considering the effect of the synaptic input on the electrical tuning of the hair cell. The hair cell behaves analogously to a resonant (RLC—resistor, inductor, capacitor) circuit (Fig. 6.9C) whose impedance varies with frequency, being highest at the characteristic frequency (Fig. 6.9D). Synaptic inhibition alters the hair cell in two ways, as a parallel shunt of the membrane impedance, and by hyperpolarizing the membrane. In the shunt mechanism, the synaptic input is a parallel resistance that reduces the net impedance of the circuit. This effect is largest where the circuit's impedance is highest—at the CF (Fig. 6.9D, curve c).

Clearly, however, shunt damping does not produce the boost in low-frequency sensitivity seen during inhibition (Fig. 6.9B). Rather, this low-frequency boost would occur in the model circuit if the membrane resistance in series with the inductance had been increased—an effect referred to as "series damping" (Fig.

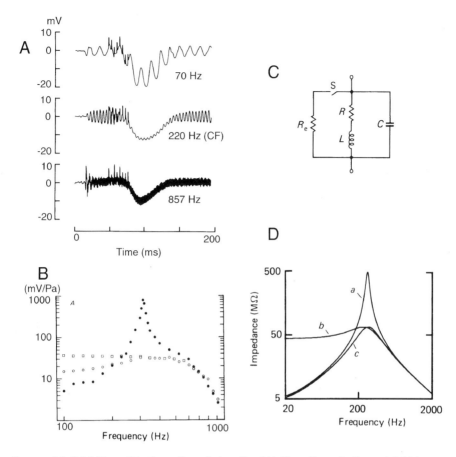

FIGURE 6.9. Inhibition of turtle auditory hair cells. (**A**) The effect of efferent inhibition on the receptor potential depends on the frequency of acoustic stimulation. Inhibition severely attenuates the peak to peak amplitude of the receptor potential at the characteristic or best frequency (CF, *middle record*). The receptor sinusoid is unchanged at higher frequencies (*bottom record*) and actually enhanced at frequencies below CF (*top record*). (**B**) A frequency "sweep" (acoustic intensity remains constant while frequency changes gradually) caused changes in hair cell membrane potential that were measured with a phase-lock amplifier (*filled circles*). The sweep was repeated while efferent axons were stimulated electrically at 50/s, causing hair cell hyperpolarization greater than 8 mV (*open circles*), or between 5 and 8 mV (*open squares*). The effect of inhibition is to convert the hair cell from a tuned resonance to a simple low-pass filter. (A) and (B) reproduced with permission from Figures 1 and 2A in Art et al. (1985). (**C**) The equivalent circuit used to describe electrical resonance in the hair cell. Voltage-gated potassium channels can act as an equivalent inductance (generating current flow opposite in effect to imposed voltage changes). The switch, S, represents the synaptic conductance change. (**D**) Voltage drop generated by frequency sweep of constant current, expressed as equivalent impedance. Sharply tuned resonance curve, *a*, is with switch, S, open. Curve *c* is with switch closed, "shunt-damping" of the resonance. Curve *b* arises when resistance in series with the inductance is increased, illustrating "series damping." Both changes reduce the overall tuning, but series damping additionally increases the low frequency impedance. (C) and (D) reproduced with permission from Figure 3 in Art et al. (1985).

6.9D, curve b). Inhibition could produce series damping if the associated hyperpolarization deactivated already active voltage-dependent conductances. Indeed, the collection of voltage-gated calcium and potassium channels that confer electrical tuning is partially activated and contributes significantly to the resting membrane conductance of the hair cell (Art and Fettiplace, Chapter 5). Hyperpolarizing IPSPs deactivate these channels, causing input resistance to rise (Art et al. 1984). Consequently, the slope resistance of the membrane is actually higher at the peak of small IPSPs than at the resting potential.

Both features of inhibition, conductance increase and hyperpolarization, are needed to explain entirely the loss of tuning. The conductance increase alone cannot provide the low-frequency enhancement. On the other hand, a modest hyperpolarization of the hair cell (produced by current injection through the microelectrode) lowers the resonant frequency, while an equivalent synaptic hyperpolarization does not, presumably because the accompanying synaptic conductance increase neutralizes the series damping effect. It is important to note that the enhancement of low-frequency sensitivity seen during inhibition of the hair cell is not reflected in the afferent fiber tuning curve, which is either suppressed or unchanged at frequencies below CF (Art and Fettiplace 1984). This is because even the largest low-frequency receptor potentials do not overcome the synaptic hyperpolarization that reduces transmitter release from the hair cell.

Thus, the efferent synaptic hyperpolarization deactivates voltage-dependent processes that confer tuning and sensitivity to the turtle hair cells. We will consider an analogous voltage effect in the later discussion on inhibition of the mammalian cochlea.

3.2.2 The Functional Significance of Inhibition

The functional consequences of hair cell inhibition in the turtle, as in other species, can be assessed by recording the response of single afferent nerve fibers. Action potentials can be evoked by pure tone stimuli, with the threshold sound pressure varying as a function of frequency. The complete frequency–threshold plot for an individual afferent fiber describes a V-shaped tuning curve, with best sensitivity at CF (Crawford and Fettiplace, 1980). When the tuning curve of the afferent fiber is examined during inhibition, there is preferential reduction of sensitivity at the characteristic frequency (Art and Fettiplace 1984). The strength of inhibition can be as much as 80 dB equivalent sound pressure. That is, it takes a sound that is nearly 10,000 times louder to overcome maximal inhibition at the CF. Inhibition produces a lesser effect on sensitivity at frequencies above and below the CF, with the net effect being that frequency selectivity, or tuning, is reduced. This result is predicted by the combined effects of inhibitory shunting and hyperpolarization on the hair cell (Art et al. 1985) as described in the previous section.

A reduction in sensitivity during inhibition could serve to extend dynamic range, or protect from loud sound damage, as mentioned earlier. But, could the reduction of tuning, per se, offer some additional functional benefit? One possibility arises from the recognition that sharp frequency selectivity and temporal acuity are negatively correlated. This is because the transient response of a tuned system is prolonged (it "rings") to a greater extent the more sharply tuned it is. The "click response" of a sharply tuned hair cell or afferent fiber consists of a decaying oscillation in membrane potential (in the case of the hair cell) or action potential firing probability (in the afferent fiber) that far outlasts the duration of the original stimulus, and obscures the response to a second click. Thus, if the ear is required to analyze a rapidly occurring series of sounds, as produced by an approaching predator for instance, sharp tuning may actually be a hindrance. Low levels of inhibition can markedly reduce tuning, thus shortening response time, with only modest losses of sensitivity (Art and Fettiplace 1984).

3.3 Effects of ACh on Isolated Hair Cells

Various lines of evidence support the identity of ACh as the predominant neurotransmitter of olivocochlear efferents (e.g., Norris and Guth 1974, reviewed in Sewell 1996). Thus, it should be possible to reproduce the effects of efferent inhibition by application of ACh to hair cells isolated from the cochlea. Such an experiment was first performed on hair cells isolated from the chicken's inner ear (Shigemoto and Ohmori 1990). These investigators used calcium indicator dyes and whole-cell, tight-seal recording to show that application of ACh raised intracellular calcium and hyperpolarized hair cells. Both effects lasted for many minutes, and the hyperpolarization was attributed to the activation of calcium-dependent potassium channels. These authors proposed that a muscarinic ACh receptor caused release of calcium from intracellular stores.

The involvement of a calcium-dependent potassium current in cholinergic inhibition was further elucidated in voltage-clamp recordings of hair cells isolated from the chicken (Shigemoto and Ohmori 1991) and the guinea pig (Housley and Ashmore 1991). In contrast to the metabotropic, G-protein-coupled mechanism proposed by Shigemoto and Ohmori, Housley and Ashmore concluded that the ACh receptor in guinea pig hair cells was more likely to be a ligand-gated cation channel that provided calcium entry to activate calcium-dependent potassium channels. However, neither study reported a biphasic change in membrane current (equivalent to the biphasic IPSPs seen during efferent inhibition of turtle and frog hair cells), as would be expected if ACh did first gate cation channels. An interesting feature of both these studies was the relatively long time course of ACh-evoked currents, in some cases outlasting ACh application by seconds to minutes.

3.3.1 A Rapid Biphasic Response to ACh

Later studies on hair cells of the chicken basilar papilla (Fuchs and Murrow 1992a) were able to demonstrate that ACh could cause a biphasic change in membrane conductance essentially identical to that found during efferent inhibition of turtle hair cells. Further, these experiments identified the cation current that flowed through hair cell ACh receptors themselves. Whole-cell, tight-seal recordings were used to reveal a biphasic change in membrane potential, or membrane current, during brief (50 to 100 ms) application of 100 μM ACh (Fig. 6.10). At a membrane potential of −40 mV, ACh evoked a small (less than 50 pA) inward current followed within a few milliseconds by a much larger outward potassium current that rose to a peak within 50 to 70 ms. This ACh-evoked potassium current ($I_{K(ACh)}$) was thought to flow through small-conductance, calcium-activated potassium (SK) channels, as confirmed in later experiments on mammalian and avian hair cells using channel-specific blockers (Nenov et al. 1996; Yuhas and Fuchs 1999).

The calcium dependence of $I_{K(ACh)}$ is based on two features of the current. First, it was possible to prevent activation of $I_{K(ACh)}$ by buffering cytoplasmic calcium with 10 mM of the rapid calcium buffer BAPTA, leaving only a smaller, earlier inward current. However, $I_{K(ACh)}$ was not prevented by 11 mM of the slower calcium buffer EGTA. This differential effect of fast and slow calcium buffers means not only that $I_{K(ACh)}$ activation depends on a rise of intracellular calcium, but also that calcium cannot diffuse far from its site of origin to activate potassium channels, perhaps as little as 100 to 200 nm (see, e.g., Tucker and Fettiplace 1995) for discussion of spatially limited calcium signals in hair cells). A second indication of the calcium-dependence of $I_{K(ACh)}$ derives from its steady-state voltage dependence, as seen in the "bell-shaped" current-voltage relation (Fig. 6.11). Maximal $I_{K(ACh)}$ occurs between −40 and 0 mV (at −40 in the illustrated experiment), but essentially disappears at positive membrane potentials. Calcium-dependent potassium currents that rely on calcium influx have

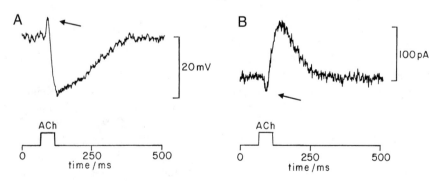

FIGURE 6.10. Whole-cell, tight-seal recording of the effect of acetylcholine (ACh) on short (outer) hair cells isolated from the chicken basilar papilla. (**A**) A brief (100 ms) puff of 100 μM ACh caused the membrane potential of a short hair cell to be briefly depolarized (*arrow*) then hyperpolarized by 20 mV for approximately 100 ms (resting membrane potential −40 mV). (**B**) In voltage clamp at −40 mV, the same treatment generates a combination of inward (*arrow*) and outward current. (A) and (B) reproduced with permission from Figure 1 in Fuchs and Murrow (1992b).

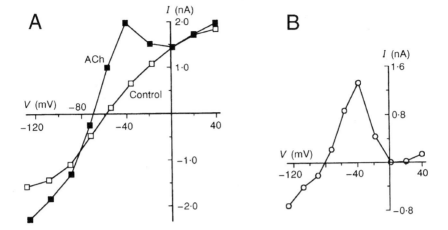

FIGURE 6.11. The voltage dependence of membrane current evoked by ACh. (**A**) Current-voltage curves in the presence (*solid squares*) and absence (*open squares*) of ACh. (**B**) The difference between these curves is the ACh-dependent current that reverses in sign near the potassium equilibrium potential, −80 mV, reaches a peak near −40, and reduces to zero at positive membrane potentials. (A) and (B) reproduced with permission from Figure 2B+C in Evans (1996).

just such a bell-shaped current-voltage relation because calcium driving force, and hence its influx, are reduced at positive membrane potentials. This effect of membrane potential on calcium entry can account quantitatively for the steady-state current-voltage relation of $I_{K(ACh)}$ (Martin and Fuchs 1992). As predicted by this hypothesis, $I_{K(ACh)}$ is eliminated by removal of extracellular calcium. However, this last experiment is not, in fact, conclusive, since ligand-gating of the hair cell AChR itself depends on extracellular calcium, as will be discussed next.

The early inward component of the biphasic current response to ACh could be examined in hair cells buffered with BAPTA to prevent activation of $I_{K(ACh)}$. Under these conditions the remaining current reversed in sign near 0 mV, and was carried equally by sodium and potassium ions. Does this current also include calcium ions, and thereby trigger the subsequent calcium-dependent potassium current? The available data suggest that the channels carrying the early inward current are approximately equipermeant to sodium and calcium ions (McNiven et al. 1996). The exact permeability ratio has been difficult to obtain because extracellular calcium has several effects in addition to its role as a charger carrier. Thus, cation currents evoked by ACh in hair cells of chickens (McNiven et al. 1996) and guinea pigs (Blanchet et al. 1996; Evans 1996) disappear entirely when extracellular calcium is eliminated. In addition, calcium at concentrations above 10 mM produced a voltage-dependent block of the channels, as evident in the calcium-dependent rectification of the current–voltage relationship (McNiven et al. 1996). These effects are not predicted from consideration of independent ionic flux through a nonselective cation channel. However, calcium is known to be a necessary cofactor for gating of some neuronal nicotinic receptors (Vernino et al. 1992) and can retard monovalent ionic flux through some cation channels (see, e.g., Mulle et al. 1992). A more thorough description of hair cell AChR gating and permeability has been obtained by functional expression of cloned receptors (see Section 3.7).

3.4 Synaptic Inhibition of Hair Cells in the Mammalian Cochlea

How accurately do the findings on isolated hair cells account for the actual synaptic effect observed in the auditory periphery? IPSPs in hair cells of turtles and frogs are biphasic, and to the extent tested, demonstrate the same pharmacological profile. More recently it has become possible to use an excised organ of Corti preparation from postnatal rats and mice to reexamine this issue in mammalian hair cells. In this preparation it is possible to make intracellular voltage-clamp recordings from hair cells and to observe inhibitory postsynaptic currents (IPSCs) arising by spontaneous release of neurotransmitter from sur-

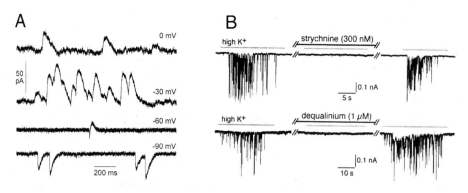

FIGURE 6.12. Cholinergic synaptic currents in hair cells of the mammalian cochlea. (**A**) Synaptic currents resulting from spontaneous release of ACh from efferent terminals were recorded by voltage-clamp of inner hair cells in the excised organ of Corti of mouse (membrane potential listed beside each record). In this preparation, efferent terminals are cut off from their cell bodies in the brainstem, but remain viable and able to release neurotransmitter for hours. The predominant outward current reversed in sign near the potassium equilibrium potential (-80 mV). Biphasic currents like those produced by ACh in hair cells of lower vertebrates could be observed at -60 mV (see expanded version in Fig. 13). Unpublished data from Elisabeth Glowatzki, with permission. (**B**) Spontaneous synaptic currents in outer hair cells of the rat cochlea were increased in frequency by high potassium to depolarize the efferent nerve terminals. The synaptic response was blocked by strychnine, a potent antagonist of some neuronal nicotinic receptors and by dequalinium ions that block small conductance calcium-activated potassium channels. (B) reproduced with permission from Figure 1C in Oliver et al. (2000).

viving efferent terminals (Glowatzki and Fuchs 2000; Oliver et al. 2000). IPSC frequency can be increased using high potassium saline to depolarize the efferent terminals. In the neonatal cochlea the efferent neurons synapse with inner as well as outer hair cells. This efferent innervation of IHCs diminishes or disappears after the onset of hearing to a variable extent depending on the region of the cochlea and species examined (Simmons 2002). IPSCs in IHCs reversed in sign near the potassium equilibrium potential (Fig. 6.12A), showing that the majority of current flowed through potassium channels (Glowatzki and Fuchs 2000). At −60 to −70 mV it was possible to observe biphasic currents with an earlier inward, and later outward component (Fig. 6.13A). The IPSCs, as

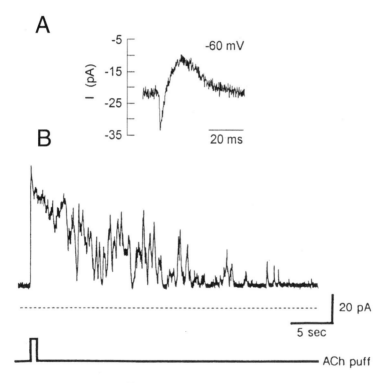

FIGURE 6.13. Fast and slow effects of ACh on hair cells. (**A**) Biphasic synaptic current recorded from a mouse inner hair cell voltage-clamped to −60 mV. Synaptic release of ACh produces rapid membrane currents compared to those arising from bath application of ACh. The outward component of this current decays with a time constant of approximately 11 ms. The inward current rises to a peak within approximately 1 ms, at room temperature. The mechanisms that couple the inward and outward component of the ACh response must act within one ms of the binding of ACh to its membrane receptor on the hair cell (expanded version of record from Fig. 12A). (**B**) A 1-s application of ACh causes a slowly decaying outward current in frog saccular hair cells. Prominent fluctuations occur during the decay phase. Reproduced with permission from Figure 1 in Yoshida et al. (1994).

well as the IHC's response to applied ACh, were blocked by strychnine and α-bungarotoxin, antagonists of AChRs in hair cells of non-mammalian vertebrates (Fuchs 1996). The outward component of the synaptic current, and the potassium currents caused by ACh both were inhibited by apamin, the SK channel blocker. Thus, the synaptic contacts made on neonatal inner hair cells appear to release ACh to activate a postsynaptic mechanism like that seen in isolated hair cells: a biphasic conductance change involving ligand-gated cation channels, and calcium-activated potassium channels.

These synaptic events are not an anomaly unique to the IHCs. Spontaneous synaptic currents also were seen in OHCs of the postnatal rat (Fig. 6.12B) and shown to have similar pharmacological sensitivities (Oliver et al. 2000). Careful comparisons of the synaptic currents in OHCs to those produced by cloned and expressed *SK2* genes suggested that the time course of synaptic current was determined by the gating kinetics of these small conductance calcium-activated potassium channels. This result supports the hypothesis that the potassium channels carrying $I_{K(ACh)}$ are *SK2* gene products, (Dulon et al. 1998) and implies tight coupling between gating of the AChR and activation of the SK channels.

3.5 Inhibition of the Response to Sound in the Mammalian Cochlea

What is the effect of inhibition on the hair cell's response to sound in the mammalian cochlea? This topic has received extensive treatment in earlier reviews (Guinan 1996) and is summarized only briefly here. Before answering this question it is helpful first to recapitulate the synaptic organization of the cochlea. The medial olivocochlear efferents (whose influence has been studied most thoroughly) innervate OHCs predominantly in the mid-frequency region of the cochlea (Liberman et al. 1990). Activation of this pathway suppresses the response of the cochlea to sound (Galambos 1956). What remained unknown for many years was how a synaptic effect on OHCs could alter the activity of afferent fibers that receive synaptic excitation from IHCs. The most promising idea, based on the appearance of distortion products and other indications of mechanical feedback in the inner ear, was that OHCs are mechanically active, somehow amplifying the vibration of the cochlear partition to enhance the sensitivity of the IHC (Mountain et al. 1980; Siegel and Kim 1982). A cellular basis for such mechanical amplification was first reported by Brownell and colleagues who found that OHCs isolated from the guinea pig cochlea "contracted" during depolarization (Brownell et al. 1985). Substantial effort from a number of laboratories has gone into characterizing "electromotility"—the term coined for the voltage-driven movement of OHCs (Brownell, Chapter 7). The voltage-sensitive motor protein has been cloned and shown to be a member of a sulfate ion transporter family (Zheng et al. 2000).

The electromotility mechanism provides a promising target for synaptic inhibition by olivocochlear efferents. It might be expected that the synaptic conductance increase would shunt sound-evoked receptor potentials that normally drive electromotility. Perhaps more importantly, the relatively long-lasting hyperpolarization could clamp the OHC in an extended state, preventing its normal mechanical feedback. Unfortunately, it has not appeared to be so straightforward in studies made to date.

3.5.1 IHC Recording During Efferent Inhibition

Intracellular studies of hair cell inhibition in mammals began with Brown and colleagues, who used sharp microelectrodes to record from IHCs in the adult guinea pig (Brown and Nuttall 1984). A tone burst at the characteristic frequency (16 to 22 kHz) gave rise to a depolarization whose duration matched that of the tone burst. Electrical stimulation of the efferent axons in the floor of the fourth ventricle reduced the amplitude of the DC receptor potential, but produced no hyperpolarization beyond the resting membrane potential. Such an effect is consistent with the expectation that inhibitory synapses made on the OHCs reduce their contribution to the overall cochlear response encoded by the IHC, without altering the IHC directly. Unfortunately, intracellular recordings from OHCs during combined acoustic and efferent stimulation have not yet been achieved, largely because recording from these cells in the intact cochlea is an extremely difficult undertaking, made still more so when combined with electrical stimulation of the efferent axons.

3.5.2 ACh and Electromotility

An alternative approach has been to examine the effects of the efferent neurotransmitter, ACh, on the electromotility of isolated OHCs produced by applied electrical fields. Unexpectedly, prolonged application of ACh caused an *increase* in somatic motility by reducing somatic stiffness (Dallos et al. 1997). Increased motility (to a standard electrical stimulus) began 5 to 10 s after exposure to ACh, and is reminiscent of the ACh-induced increases in length reported in the original description of electromotility (Brownell et al. 1985). The slow increase in motility was accompanied by an equally slow rise in the global intracellular calcium. The proposal was made that calcium-dependent alteration of cytoskeletal elements reduced somatic stiffness, facilitating electromotility.

However, it remains to be determined how this enhanced motion ultimately results in inhibition of afferent fiber sensitivity. It also is unclear why ACh had no effect on motility for many seconds, despite the fact that other studies showed that potassium currents evoked in OHCs by ACh rise to their peak within 300 ms or less. Perhaps more importantly, these slow cholinergic effects on the motility of isolated OHCs also differed from the time course of inhibition as measured in the IHC in vivo, which rises and falls within 100 ms of the onset and offset of electrical stimulation of the efferent axons (Brown and Nuttall 1984). One possible explanation for these discrepancies may be that cholinergic inhibition acts on motility through both fast and slow cellular mechanisms, but that only the slower mechanism survives in isolated OHCs.

3.5.3 Fast and Slow Inhibition

Fast and slow inhibitory effects of efferent activity have been described in the intact guinea pig cochlea by Brown, Sewell and colleagues (Sridhar et al. 1997). Both fast and slow inhibitory components were blocked by antagonists of the hair cell AChR and were dependent on intracellular calcium signaling. Addition of ryanodine (a facilitator of calcium-induced calcium release from intracellular stores) to the perilymphatic fluid enhanced both fast and slow efferent suppression of the compound action potential. Compounds that block the smooth endoplasmic reticulum Ca-ATPase (SERCA) selectively enhanced the slow efferent effect. The authors proposed that intracellular calcium stores mediate some or all of the efferent effects. Further, the preferential effect of the SERCA blockers on slow efferent effects may imply that two classes of calcium stores are at work, perhaps the synaptic cisterns versus the lateral cisterns that lie further from the synapse. Another possibility that remains to be tested is whether ryanodine acts presynaptically on the efferent terminal to enhance the magnitude of transmitter release, which of course also depends on intracellular calcium signaling. For example, ryanodine affects short-term plasticity and transmitter release in the hippocampus (Emptage et al. 2001).

In summary, the known effect of efferent activation is to cause inhibition in the mammalian cochlea and in the analogous auditory end organ of non-mammalian vertebrates. However, the mechanism of inhibition in mammals, particularly how OHC electromotility is affected, remains uncertain. Still less is known about those "lateral" efferent contacts that are directed against afferent fibers at the base of IHCs, or of those few efferent synapses on IHCs that persist into the adult. This picture becomes murkier still with recognition of the host of additional transmitter substances that are thought to modulate, or in some cases perhaps to mediate, lateral efferent effects (Sewell 1996). For example, a recent study from Ruel and colleagues demonstrated that dopamine acts to suppress spontaneous and driven afferent activity in the guinea pig cochlea (Ruel et al. 2001). Also, peptide neurotransmitters have been detected in lateral efferents and could serve a neuormodulatory role (Lioudyno et al. 2002).

3.6 Efferent Inhibition and Cochlear Development

It has been suggested that efferent innervation of the cochlea may play a role in development of the auditory pathway. Functional maturation of the auditory periphery is thought to depend on intact efferent innervation (Walsh et al. 1998). The transient innervation of inner hair cells during development could play a role in shaping synapse formation by modulating spontaneous activity prior to the onset of hearing. This would be analogous to visual pathway development where it is known that the maturation of synaptic organization depends on continued spontaneous activity prior to the onset of vision (Katz and Shatz 1996). Immature auditory hair cells can generate spontaneous calcium-dependent action potentials (Fuchs and Sokolowski 1990; Marcotti et al. 2003), and immature neurons in the auditory nerve and central nuclei fire low-frequency bursts spontaneously (Gummer and Mark 1994; Lippe 1994). The bursting pattern of afferent activity is eliminated when the efferent input is cut in neonatal cats (Walsh and McGee 1997) implying that these inhibitory inputs impose rhythmicity onto the immature afferent auditory pathway. Afferent synapses of the IHCs undergo significant functional changes while efferent synapses are present (Beutner and Moser 2001). The patterned spontaneous activity of the eighth nerve may be important for modulating or synchronizing activity-dependent changes in brainstem nuclei, or in the IHCs themselves. Efferent feedback could shape emerging activity in the auditory system by cholinergic inhibition of IHCs.

3.7 Molecular Components of the Cholinergic Mechanism

The pharmacology of cochlear inhibition is remarkable for the number and variety of antagonists that have been found to have some effect, as reviewed in Guth and Norris (1996), and Sewell (1996), with the most common conclusion being that a cholinergic receptor with mixed nicotinic/muscarinic properties mediates inhibition. Studies of isolated hair cells provided some refinement of this picture, showing that nicotinic antagonists such as α-bungarotoxin and curare are highly effective, while muscarinic antagonists such as atropine are less so (Housley and Ashmore 1991; Fuchs and Murrow 1992b; Erostegui et al. 1994). Nonetheless, the hair cell ACh receptor (AChR) clearly is not a typical nicotinic receptor since it is highly sensitive to block by strychnine—normally considered a glycinergic antagonist—and the blocking effect of α-bungarotoxin is readily reversed (in contrast to the irreversible block of nicotinic AChRs at the neuromuscular junction). Finally and rather awkwardly, this putative nicotinic receptor also is not activated by nicotine! Instead, nicotine, and muscarine, serve as weak antagonists and the only effective agonists are suberyldicholine, carbachol, and dimethylphenylpiperazinium (DMPP) in order of decreasing potency (Erostegui et al. 1994; McNiven et al. 1996).

3.7.1 The α9 ACh Receptor Gene

In 1994, Elgoyhen and colleagues cloned a new member of the ligand-gated cation channel superfamily that had 33% to 39% amino acid identity with neuronal nicotinic alpha subunits (Elgoyhen et al. 1994). When expressed in *Xenopus* oocytes the sequentially named α9 subunit generated ligand-gated cation currents as well as calcium-dependent chloride currents in response to application of ACh. These latter chloride currents are native to the oocyte, but serve as a convenient amplifier of the response, and indicate that some fraction of the α9 cation current must be carried by calcium. The pharmacology of the α9-dependent currents was essentially identical to that of the hair cell AChRs. It was antagonized by nicotine and muscarine, and blocked by strychnine, α-bungarotoxin, curare, and atropine. No other nicotinic receptor possesses this combined pharmacology. The mRNA for α9 has been shown by in situ hybridization and reverse transcriptase-polymerase chain reaction (RT-PCR) to be expressed in both cochlear (Elgoyhen et al. 1994) and vestibular hair cells (Zuo et al. 1999) of rodents. The orthologous gene was cloned from a chicken cochlear cDNA library and similarly shown to be expressed by auditory (Hiel et al. 2000) and vestibular hair cells (Lustig et al. 1999).

The role of α9 as a mediator of cochlear inhibition was convincingly demonstrated by generation of a transgenic mouse in which the functional gene was disabled—an α9 knockout (Vetter et al. 1999). Such animals had apparently normal hearing, and efferent innervation of the cochlea was essentially intact. However, electrical stimulation of the efferent fibers failed to suppress the compound action potential arising from IHC excitation, or distortion product otoacoustic emissions generated by OHCs. Normal suppression of these signals did occur in heterozygous littermates. Cochlear hair cells in the transgenic mice did not respond to exogenous ACh and had no efferent synaptic currents (E. Glowatzki, personal communication). An implication of these findings is that no other gene product appears able to substitute for α9 in forming functional AChRs in hair cells. This conclusion has particular significance since the discovery of another hair cell AChR gene, α10.

3.7.2 The α10 ACh Receptor Gene

Although α9 largely replicated the pharmacological profile of the hair cell AChR, the channels it generated in oocytes differed biophysically, most notably by remaining conductive when extracellular calcium was eliminated (Katz et al. 2000); conditions in which hair cell AChRs fail to activate. While this distinction may arise from differences in the synthetic machinery of the oocyte versus that of the hair cell, another possibility is that the native hair cell receptor contains subunits in addition to α9 that contribute to its functional properties. This latter possibility was strengthened by the discovery of α10 (Elgoyhen et al. 2001). This gene product shares 57% overall amino acid identity with α9 and like it is detected in cochlear and vestibular hair cells of the rat with in situ hybridization. Interestingly, α10 on its own produces no detectable current when injected into *Xenopus* oocytes, reinforcing the conclusion from knockout mice that α9 is required for hair cell AChR function. However, when α9 and α10 are coinjected, the resulting currents were hundreds of times larger than those due to α9 alone, as though α10 facilitated the formation of functional channels by α9. Furthermore, α9α10-injected oocytes produced channels whose activation required extracellular calcium, as do native AChRs in hair cells (Weisstaub et al. 2002). This suggests that α10 does combine with α9 and thereby forms a channel with biophysical properties more like those of the native hair cell AChR. Human α10 and α9 (*CHRNA10*, *CHRNA9*) also have been cloned and are found on chromosomes 11p15.5 and 4p14.5, respectively (Lustig et al. 2001). Both genes are expressed in tissues of the inner ear, and at lower levels in blood lymphocytes, where their functional significance is unknown.

The enhanced expression of α9/α10 in oocytes has provided the opportunity for closer comparison of their biophysical properties with those of the native AChR in hair cells. The expressed α9/α10 channels conducted calcium approximately 10 times better than sodium. Thus, the α9/α10 heteromer falls into the category of ligand-gated cation channels with high calcium permeability, and indeed might be considered formally as a calcium channel. Relatively small net ionic flux through such a calcium-selective channel should be sufficient to trigger SK channels, or other calcium-dependent processes.

3.7.3 The ACh-Sensitive Potassium Channel in Hair Cells

A candidate gene also has been identified for the potassium channels that carry $I_{K(ACh)}$. The *SK2* mRNA has been demonstrated in cochlear hair cells of the rat by in situ hybridization (Dulon et al. 1998), and by specific antibody labeling (Oliver et al. 2000). *SK2* is one of a family of genes that encode small-conductance, calcium-activated potassium channels (Kohler et al. 1996). All these channels are activated by micromolar concentrations of calcium, but are voltage insensitive. They are blocked by apamin, a peptide toxin of bee venom. $I_{K(ACh)}$ in hair cells, or synaptically gated potassium currents, likewise are blocked by apamin (Doi and Ohmori 1993; Nenov et al. 1996; Yuhas and Fuchs 1999; Glowatzki and Fuchs 2000) and have an equivalent sensitivity and binding kinetic for calcium (Oliver et al. 2000). A puzzling discrepancy is that fast efferent effects in the guinea pig cochlea appeared to be resistant to apamin (Yoshida et al. 2001), in contrast to the evidence that SK channels mediate fast inhibition of isolated hair cells as well as synaptic inhibition of IHCs and OHCs in excised cochlear coils (Glowatzki and Fuchs 2000; Oliver et al. 2000).

3.8 Vestibular Efferent Inhibition

This section will touch briefly on efferent effects in vestibular hair cells for the purpose of comparison; a fuller description can be found in Chapter 8 (Eatock and Lysakowski). The structure and function of vestibular organs has been reviewed extensively by Guth and colleagues (Guth et al. 1998) and by Lysakowski and Goldberg (Lysakowski 2003) and so will be summarized only briefly. Efferent neurons reside in a loose assembly near the vestibular nuclei of the brainstem. Their peripheral axons ramify extensively, in some cases to multiple vestibular end organs, with a minority having a bilateral projection to both labyrinths (Dechesne et al. 1984). Efferent synapses are made directly onto type II hair cells, and onto the calyx endings of type I hair cells that are found in mammals, birds, and reptiles. Efferent function in the vestibular periphery resists simple generalizations; in part because of the widely spread innervation pattern, and because efferent effects can be inhibitory, excitatory, or both, depending on afferent fiber type, end organ, and species. For example, efferent activation inhibits posterior nerve afferents in the frog (Rossi et al. 1980) and saccular afferents of the toad (Sugai et al. 1991) and reduces the rate of transmitter release from hair cells in the vestibular lagena of the toadfish (Locke et al. 1999), but facilitates afferent firing in semicircular canal organs of mammals and the toadfish (Goldberg and Fernandez 1980; Boyle and Highstein 1990). Pharmacological studies in the amphibian periphery have led to the proposition that facilitation may be mediated by an "atropine-preferring," presumably muscarinic cholinergic receptor, and inhibition by a "strychnine-preferring" nicotinic cholinergic receptor, presumably the α9/α10 containing receptor (Guth et al. 1998).

3.8.1 Synaptic Potentials

While efferent effects on the vestibular periphery are complex, to date the limited number of intracellular recordings from vestibular hair cells have revealed only hyperpolarization by efferent stimulation. Ashmore and Russell described hyperpolarizing IPSPs in frog saccular hair cells produced by electrical stimulation of the eighth nerve root (Ashmore and Russell 1982). Sugai and colleagues also used sharp microelectrodes to record the effect of efferent activity on membrane potential in hair cells of the frog's saccule (Sugai et al. 1992). Brief trains of electrical shocks to the eighth nerve stump innervating the excised saccular epithelium caused hyperpolarizing IPSPs of 10 to 30 mV amplitude and duration of up to 500 ms. A faster, smaller hyperpolarizing component preceded the prominent slower hyperpolarization. This early component reversed in sign at −65 mV, differing in that respect from the early depolarization found in turtle hair cells. The IPSPs were blocked by 0.5 μM curare and 2.0 μM atropine. In keeping with the pharmacology of α9/α10 containing receptors described above, the IPSPs also were reduced or eliminated in the presence of nicotinic agonists such as nicotine, while ACh and carbachol both hyperpolarized the hair cells and desensitized the synaptic receptors. In a subsequent study it was shown that ACh activates a calcium-dependent potassium current in frog saccular hair cells (Yoshida et al. 1994).

3.8.2 ACh Receptors in Vestibular Hair Cells

Application of ACh evokes a potassium current, $I_{K(ACh)}$, in hair cells isolated from the sacculus, but not from the crista ampullaris of the frog (Holt et al. 2001). $I_{K(ACh)}$ in saccular hair cells can be blocked by apamin, supporting the identification of SK-type channels as in cochlear hair cells. The nicotinic receptor α9 is expressed by saccular hair cells, and the cholinergic response can be blocked by many of the same antagonists that block AChRs in cochlear hair cells. An interesting additional feature is that the saccular hair cell AChR appears to be inhibited directly by morphine. This opioid antagonism also can be demonstrated on currents produced by rat α9 expressed in *Xenopus* oocytes (Lioudyno et al. 2000). Still more recent work suggests that opioid binding sites are found on α9/α10-containing receptors in cochlear hair cells as well (Lioudyno et al. 2002).

Messenger RNA for the candidate hair cell receptor genes α9 and α10 are expressed by many vestibular hair cells, including both type II and type I hair cells (Elgoyhen et al. 1994, 2001). This last observation presents a small mystery since the type I hair cell is surrounded entirely by the afferent calyx and does not receive direct efferent synaptic contact (Lysakowski and Goldberg 1997). It remains to be seen whether the expression of α9 and α10 mRNA has any functional significance for the type I hair cell.

3.9 Calcium Stores in Cholinergic Inhibition of Hair Cells

Most of the experimental evidence given above can be incorporated into an "ionotropic" mechanism of hair cell inhibition. According to this hypothesis, ACh binds to and opens a ligand-gated cation channel (the α9/α10 containing receptor). Calcium entry through that channel triggers the activation of small-conductance, calcium-activated potassium channels whose current flow hyperpolarizes and inhibits the hair cell. Biphasic changes in membrane potential or current consistent with this "two-channel" hypothesis have been observed in hair cells of frogs (Sugai et al. 1992), turtles (Art et al. 1984), chickens (Fuchs and Murrow 1992a), mice and rats (Dulon and Lenoir 1996; Glowatzki and Fuchs 2000; Oliver et al. 2000), guinea pigs (Blanchet et al. 1996; Evans 1996), and gerbils (He and Dallos 1999), although the early cation current often is not seen (presumably because it is obscured by the larger outward potassium current). Such an ionotropic mechanism acts by altering membrane potential and conductance. In keeping with this proposed mechanism, the combined conductance change activates rapidly. Timed application of ACh to isolated hair cells produces inward currents that are followed by outward currents with a delay of only a few milliseconds. Indeed, the fastest possible agonist application, that owing to release from the efferent synapse on cochlear hair cells, causes IPSCs in which the outward current begins only 1 ms after the inward current, and rises to a peak within 10 to 20 ms at room temperature in mammalian IHCs (Fig. 6.13A).

A relatively direct, ionotropic mechanism of inhibition accounts for many, but not all features of inhibition that have been described. For instance, although $I_{K(ACh)}$ can rise within milliseconds, in some cases it persists for seconds after a brief application of ACh. $I_{K(ACh)}$ described in frog saccular hair cells was particularly long-lasting, with a highly irregular, almost oscillatory, waveform that decayed for more than 30 s after a 1-s puff of 100 μM ACh (Fig. 6.13B). Irregularly decaying $I_{K(ACh)}$ also was noted during prolonged application of ACh to hair cells of the chicken (Fuchs and Murrow 1992a) and guinea pig (Housley and Ashmore 1991) and may contribute to the low-frequency voltage noise seen during inhibition of turtle (Art et al. 1984) and chicken (Shigemoto and Ohmori 1991) hair cells.

This prolonged and irregular time course suggests slower, perhaps calcium-store–dependent mechanisms such as those associated with metabotropic, G-protein-coupled receptors. Consistent with this possibility, perfusion with GTPγS, or inositol triphosphate (IP_3), could activate outward currents in frog saccular hair cells, while the IP_3 receptor blocker heparin caused a relatively rapid loss (compared to normal "rundown") of ACh-evoked outward currents (Yoshida et al. 1994). Similarly, GTPγS or IP_3 included in the patch pipette activated a potassium current in chick hair cells (Shigemoto and Ohmori 1991). Also, G-protein inhibitors pertussis toxin, GDPβS, or heparin all reduced $I_{K(ACh)}$ in guinea pig OHCs (Kakehata et al. 1993). In other experiments, however, photoreleased IP_3 had no detectable effect on $I_{K(ACh)}$ in OHCs (Blanchet et al. 1996).

3.9.1 Calcium-Induced Calcium Release

How can these disparate hypotheses be reconciled? One possibility is that ACh could act through more than one pathway to alter hair cell function. Thus the faster, immediate effects of inhibition may be on membrane conductances and membrane potential, mediated by calcium influx directly activating SK channels. Slower, longer-lasting changes could result from G-protein–coupled cascades that include activation of intracellular calcium stores, perhaps to alter membrane potassium conductance, but then also to initiate other biochemical changes such as those proposed to modulate OHC electromotility. The efferent synapse on hair cells is endowed with a synaptic cistern that could be just such a calcium store.

Many of the observations listed above would be consistent with the hypothesis that sustained activation of $I_{K(ACh)}$ depends on calcium-induced calcium release (CICR) from the synaptic cistern. In this scheme, prolonged calcium entry through the α9/α10 containing AChR accumulates to trigger a calcium release channel in the synaptic cistern, thereby providing larger, longer-lasting cytoplasmic calcium transients (Fig. 6.14). In many cell types, CICR is regenerative, leading to calcium sparks, waves, and oscillations, reminiscent of the prolonged and variable $I_{K(ACh)}$ observed in some hair cells.

CICR is mediated by calcium-sensitive ryanodine receptors in most cell types. The dependence of $I_{K(ACh)}$ in hair cells on membrane potential is best explained by the voltage dependence of calcium influx through AChRs (Martin and Fuchs 1992), supporting an hypothesis of store release triggered by that influx. It is even possible that the fastest SK activation in hair cells could result from CICR. For example, the ryanodine receptors (RyRs) that mediate excitation-contraction coupling of heart muscle can activate within 200 μs at high calcium concentrations (Zahradnikova et al. 1999), easily within the time requirements of the efferent synapse. The alternative possibility of an IP_3-dependent calcium store seems considerably less likely at the hair cell's efferent synapse, both on the basis of activation speed and the requirement for calcium influx. However, it remains possible that IP_3-sensitive calcium stores become involved during longer-term calcium signaling in hair cells. Molecular and histological studies will be helpful in resolving the identity and subcellular localization of calcium release channels at the hair cell's efferent synapse.

While the molecular components of a synaptic calcium store remain to be determined, sufficient evidence has accumulated to conclude that the hair cell's

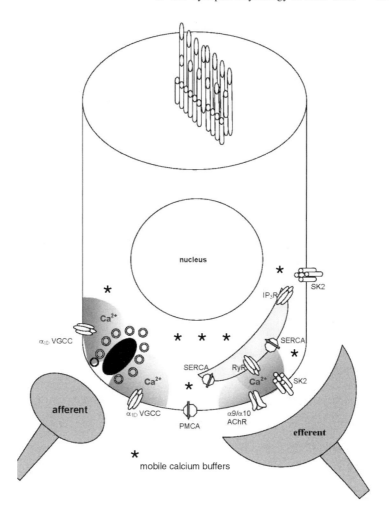

FIGURE 6.14. Schematic diagram of synaptic calcium domains. Voltage-gated calcium channels (VGCCs) trigger vesicular fusion and release at the ribbon synapse. Fast-acting mobile calcium buffers restrict the temporal and spatial profile of cytoplasmic calcium. At the efferent synapse, a synaptic cistern establishes a restricted diffusion space for calcium that enters through the α9/α10-containing AChRs. Calcium entry may act directly on nearby SK channels, or via calcium-induced calcium release from the synaptic cistern. The synaptic cistern takes up calcium via an ATP-requiring calcium pump (SERCA). Ryanodine receptors and IP3 receptors may effect calcium release on different time scales.

efferent synapse is at least modulated by its associated cistern; not a surprising thought given the obligatory and close (20 to 30 nm) association of the cistern with the postsynaptic plasma membrane. This small distance, and the enclosed diffusion space, make ionic, chemical, and even mechanical coupling feasible for postsynaptic signaling. Finally it should be pointed out that the synaptic cistern may be there to absorb calcium, rather than to release it. Since efferent and afferent synapses can occur in close proximity to one another on the basolateral surface of many hair cells (Fig. 6.1), it is likely that intracellular buffering and sequestration is necessary to maintain discrete calcium signals for these contradictory synaptic functions (Fig. 6.14).

3.10 Summary of the Efferent Synapse

Considerable progress has been made in determining the cellular and molecular details of cholinergic inhibition of hair cells. The identity of the hair cell's AChR, and the basic ionic mechanism it supports are now known. These discoveries have added to the neuronal repertoire of synaptic mechanisms and established one of the few certified postsynaptic functions for non-muscle or "neuronal" nicotinic receptors. Electrophysiological studies, especially in the non-mammalian vertebrates, have provided working hypotheses for how inhibition alters the sensitivity and tuning of hair cells to extend dynamic range and improve temporal resolution. In many respects, these data from hair cells complement the far more extensive view of efferent inhibition obtained from eighth nerve single units and other higher-level recordings. In other cases the predicted effects of efferent feedback were not found (Cullen and Minor 2002). The ultimate hope, however, is to use our growing knowledge of molecular mechanisms for a fuller understanding of hair cell inhibition in the context of behavior. A step in this direction has been made with the genesis of "α9 knockout" mice (Vetter et al. 1999) described earlier that can be examined with behavioral testing. For example, one effect of efferent inhibition is to extend the dynamic range of cochlear afferent fibers so as to improve intensity discrimination (May and McQuone 1995). However, when tested with a behavioral task for intensity discrimination in noise, the knockout mice showed no deficit compared to controls (May et al. 2002). The authors proposed that central collaterals of the efferents (acting through non-α9–containing receptors) mediate compensatory central feedback mechanisms for these highly trained listening tasks. Thus, the utilization of genetic models is not necessarily straightforward. Future progress will depend on more subtle design at both the molecular and behavioral levels to elucidate the functional role of individual gene products.

4. Chapter Summary

The mechanosensory hair cells of vertebrates present a fascinating array of functional problems for physiological study. These small cells serve as detectors, filters, and transmitters of essential sensory information for audition and body and eye position. This chapter has detailed some of the particular solutions hair cells have found for forward afferent transmission, and efferent feedback. A central fact, and perhaps the crux to understanding these processes, is the role of cytoplasmic calcium. Calcium influx across the plasma membrane triggers afferent transmitter release, and may serve analogously to trigger efferent inhibition (Fig. 6.14). The sustained release of chemical neurotransmitter requires that the hair cell continuously spend its metabolic resources to sequester and remove the underlying calcium load. This observation, as well as the temporal precision of release, demands that calcium signals be spatially, as well as temporally restricted. While the temporal requirements on efferent synaptic calcium signals are not as severe, the necessity for spatial limitation may be even more rigorous. As shown in Figure 6.1, efferent and afferent synaptic contacts can occur cheek-by-jowl on hair cells. Calcium signals triggered by ACh released from efferent terminals would be contradictory if they caused transmitter release from the hair cell while at the same time trying to prevent it through hair cell hyperpolarization. Likewise, the timing and efficacy of voltage changes driving afferent synaptic function would be severely degraded if calcium influx at the ribbon synapse spread to activate SK channels at efferent synapses.

Mobile calcium buffers such as calbindin and calretinin occur at high concentrations in hair cells and are essential to the segregation of synaptic functions as just described. Membranous calcium stores like the efferent synaptic cistern can provide additional levels of calcium amplification and/or sequestration. The roles of such intracellular calcium stores in efferent and afferent synaptic function remains incompletely resolved. It seems certain too that the life-long stability of hair cell synaptic function must depend on the calcium homeostasis provided by intracellular stores, as well as by plasma membrane transport. It will not be a surprise, therefore, if calcium store function is found at the center of some inner ear pathologies. Therapeutic agents aimed at calcium stores in cardiac pathology may offer important leads for treatments of vertigo, tinnitus, hyperacusis, and progressive hearing loss. Finally, the intriguing interplay of afferent and efferent synaptic contacts on IHCs during development of the mammalian cochlea implies some causal role in the maturation of hair cell function. It is possible that some forms of inherited deafness could result from problems at this critical developmental juncture. If we understand better the synaptic choreography at and just before the onset of hearing, we may achieve the ability to repair those processes that go awry in a given disease state.

Acknowledgments. Research in the authors' laboratories is supported by Grants NIDCD DC00276, NIDCD DC01508, and NIDCD DC05211 to P.A.F. and NIDCD DC03763 and a Sloan Research Fellowship Award to T.D.P. We thank R. Eatock for helpful criticism of the manuscript and Phyllis Taylor for Figure 6.3.

References

Adler EM, Augustine GJ, Duffy SN, Charlton MP (1991) Alien intracellular calcium chelators attenuate neurotransmitter release at the squid giant synapse. J Neurosci 11: 1496–1507.

Anderson DJ, Rose JE, Hind JE, Brugge JF (1971) Temporal position of discharges in single auditory nerve fibers within the cycle of a sine-wave stimulus: frequency and intensity effects. J Acoust Soc Am 49:Suppl 2:1131–1139.

Aravanis AM, Pyle JL, Tsien RW (2003) Single synaptic vesicles fusing transiently and successively without loss of identity. Nature 423:643–647.

Ard MD, Morest DK, Hauger SH (1985) Trophic interactions between the cochleovestibular ganglion of the chick embryo and its synaptic targets in culture. Neuroscience 16:151–170.

Art JJ, Crawford AC, Fettiplace R, Fuchs PA (1985) Efferent modulation of hair cell tuning in the cochlea of the turtle. J Physiol 360:397–421.

Art JJ, Fettiplace R (1984) Efferent desensitization of auditory nerve fibre responses in the cochlea of the turtle Pseudemys scripta elegans. J Physiol 356:507–523.

Art JJ, Fettiplace R (1987) Variation of membrane properties in hair cells isolated from the turtle cochlea. J Physiol 385:207–242.

Art JJ, Fettiplace R, Fuchs PA (1984) Synaptic hyperpolarization and inhibition of turtle cochlear hair cells. J Physiol 356:525–550.

Ashmore J, Russell J (1982) Effect of efferent stimulation on hair cells of the frog sacculus. J Physiol 329:25p.

Auger C, Marty A (1997) Heterogeneity of functional synaptic parameters among single release sites. Neuron 19:139–150.

Auger C, Marty A (2000) Quantal currents at single-site central synapses. J Physiol 526 Pt 1:3–11.

Augustine GJ (2001) How does calcium trigger neurotransmitter release? Curr Opin Neurobiol 11:320–326.

Augustine GJ, Adler EM, Charlton MP (1991) The calcium signal for transmitter secretion from presynaptic nerve terminals. Ann NY Acad Sci 635:365–381.

Balkema GW, Rizkalla R (1996) Ultrastructural localization of a synaptic ribbon protein recognized by antibody B16. J Neurocytol 25:565–571.

Bergles DE, Diamond JS, Jahr CE (1999) Clearance of glutamate inside the synapse and beyond. Curr Opin Neurobiol 9:293–298.

Beutner D, Moser T (2001) The presynaptic function of mouse cochlear inner hair cells during development of hearing. J Neurosci 21:4593–4599.

Beutner D, Voets T, Neher E, Moser T (2001) Calcium dependence of exocytosis and endocytosis at the cochlear inner hair cell afferent synapse. Neuron 29:681–690.

Blanchet C, Erostegui C, Sugasawa M, Dulon D (1996) Acetylcholine-induced potassium current of guinea pig outer hair cells: its dependence on a calcium influx through nicotinic-like receptors. J Neurosci 16:2574–2584.

Bollmann JH, Sakmann B, Borst JG (2000) Calcium sensitivity of glutamate release in a calyx-type terminal. Science 289:953–957.

Bommert K, Charlton MP, DeBello WM, Chin GJ, Betz H, Augustine GJ (1993) Inhibition of neurotransmitter release by C2-domain peptides implicates synaptotagmin in exocytosis. Nature 363:163–165.

Borst JG, Sakmann B (1996) Calcium influx and transmitter release in a fast CNS synapse. Nature 383:431–434.

Boyer C, Sans A, Vautrin J, Chabbert C, Lehouelleur J (1999) K^+-dependence of Na^+-Ca^{2+} exchange in type I vestibular sensory cells of guinea-pig. Eur J Neurosci 11: 1955–1959.

Boyle R, Highstein SM (1990) Efferent vestibular system in the toadfish: action upon horizontal semicircular canal afferents. J Neurosci 10:1570–1582.

Brandstatter JH, Fletcher EL, Garner CC, Gundelfinger ED, Wassle H (1999) Differential expression of the presynaptic cytomatrix protein bassoon among ribbon synapses in the mammalian retina. Eur J Neurosci 11:3683–3693.

Brandt A, Striessnig J, Moser T (2003) Cav1.3 channels are essential for development and presynaptic activity of cochlear inner hair cells. J Neurosci 23:10832–10840.

Brichta AM, Goldberg JM (2000) Responses to efferent activation and excitatory response-intensity relations of turtle posterior-crista afferents. J Neurophysiol 83: 1224–1242.

Brown MC, Nuttall AL (1984) Efferent control of cochlear inner hair cell responses in the guinea-pig. J Physiol 354:625–646.

Brownell WE, Bader CR, Bertrand D, de Ribaupierre Y (1985) Evoked mechanical responses of isolated cochlear outer hair cells. Science 227:194–196.

Carter AG, Vogt KE, Foster KA, Regehr WG (2002) Assessing the role of calcium-induced calcium release in short-term presynaptic plasticity at excitatory central synapses. J Neurosci 22:21–28.

Ceccarelli B, Hurlbut WP, Mauro A (1973) Turnover of transmitter and synaptic vesicles at the frog neuromuscular junction. J Cell Biol 57:499–524.

Chen YA, Scheller RH (2001) SNARE-mediated membrane fusion. Nat Rev Mol Cell Biol 2:98–106.

Clements JD, Lester RA, Tong G, Jahr CE, Westbrook GL (1992) The time course of glutamate in the synaptic cleft. Science 258:1498–1501.

Crawford AC, Fettiplace R (1980) The frequency selectivity of auditory nerve fibres and hair cells in the cochlea of the turtle. J Physiol 306:79–125.

Crawford AC, Fettiplace R (1981) An electrical tuning mechanism in turtle cochlear hair cells. J Physiol 312:377–412.

Cremona O, De Camilli P (1997) Synaptic vesicle endocytosis. Curr Opin Neurobiol 7: 323–330.

Crouch JJ, Schulte BA (1995) Expression of plasma membrane Ca-ATPase in the adult and developing gerbil cochlea. Hear Res 92:112–119.

Cullen KE, Minor LB (2002) Semicircular canal afferents similarly encode active and passive head-on-body rotations: implications for the role of vestibular efference. J Neurosci 22:RC226.

Dallos P, He DZ, Lin X, Sziklai I, Mehta S, Evans BN (1997) Acetylcholine, outer hair cell electromotility, and the cochlear amplifier. J Neurosci 17:2212–2226.

David G, Barrett JN, Barrett EF (1998) Evidence that mitochondria buffer physiological Ca^{2+} loads in lizard motor nerve terminals. J Physiol 509:59–65.

Dechesne C, Raymond J, Sans A (1984) The efferent vestibular system in the cat: a horseradish peroxidase and fluorescent retrograde tracers study. Neuroscience 11: 893–901.

Diamond JS, Jahr CE (1997) Transporters buffer synaptically released glutamate on a submillisecond time scale. J Neurosci 17:4672–4687.

Dick O, Hack I, Altrock WD, Garner CC, Gundelfinger ED, Brandstatter JH (2001) Localization of the presynaptic cytomatrix protein Piccolo at ribbon and conventional synapses in the rat retina: comparison with Bassoon. J Comp Neurol 439:224–234.

Dodge FA, Jr., Rahamimoff R (1967) Co-operative action a calcium ions in transmitter release at the neuromuscular junction. J Physiol 193:419–432.

Doi T, Ohmori H (1993) Acetylcholine increases intracellular Ca^{2+} concentration and hyperpolarizes the guinea-pig outer hair cell. Hear Res 67:179–188.

Dou H, Vazquez AE, Namkung Y, Chu H, Cardell EL, Nie L, Parson S, Shin HS, Yamoah EN (2004) Null mutation of alpha(1D) $Ca(^{2+})$ channel gene results in deafness but no vestibular defect in mice. J Assoc Res Otolaryngol (epub ahead of print).

Dulon D, Lenoir M (1996) Cholinergic responses in developing outer hair cells of the rat cochlea. Eur J Neurosci 8:1945–1952.

Dulon D, Luo L, Zhang C, Ryan AF (1998) Expression of small-conductance calcium-activated potassium channels (SK) in outer hair cells of the rat cochlea. Eur J Neurosci 10:907–915.

Dumont RA, Lins U, Filoteo AG, Penniston JT, Kachar B, Gillespie PG (2001) Plasma membrane Ca^{2+}-ATPase isoform 2a is the PMCA of hair bundles. J Neurosci 21:5066–5078.

Dunlap K, Luebke JI, Turner TJ (1995) Exocytotic Ca^{2+} channels in mammalian central neurons. Trends Neurosci 18:89–98.

Eccles JC, Jaeger JC (1958) The relationship between the mode of operation and the dimensions of the junctional regions at synapses and motor end-organs. Proc R Soc Lond B Biol Sci 148:38–56.

Eccles JC, Katz B, Kuffler SW (1942) Effects of eserine on neuromuscular transmission. J Neurophysiol 5:211–230.

Edmonds B, Reyes R, Schwaller B, Roberts WM (2000) Calretinin modifies presynaptic calcium signaling in frog saccular hair cells. Nat Neurosci 3:786–790.

Ehret G, Merzenich MM (1985) Auditory midbrain responses parallel spectral integration phenomena. Science 227:1245–1247.

Eisen M (2000) Neurotransmitter release at the chick cochlear hair cell afferent synapse. Doctoral thesis, Department of Neuroscience, University of Pennsylvania, Philadelphia.

Eisen M, Spassova M, Parsons TD (2004) Large releasable pool of synaptic vesicles in chick cochlear hair cells. J Neurophysiol 91:2422–2428.

Elgoyhen AB, Johnson DS, Boulter J, Vetter DE, Heinemann S (1994) Alpha 9: an acetylcholine receptor with novel pharmacological properties expressed in rat cochlear hair cells. Cell 79:705–715.

Elgoyhen AB, Vetter DE, Katz E, Rothlin CV, Heinemann SF, Boulter J (2001) alpha 10: A determinant of nicotinic cholinergic receptor function in mammalian vestibular and cochlear mechanosensory hair cells. Proc Natl Acad Sci USA 98:3501–3506.

Emptage NJ, Reid CA, Fine A (2001) Calcium stores in hippocampal synaptic boutons mediate short-term plasticity, store-operated Ca^{2+} entry, and spontaneous transmitter release. Neuron 29:197–208.

Engel J, Michna M, Platzer J, Striessnig J (2002) Calcium channels in mouse hair cells: function, properties and pharmacology. Adv Otorhinolaryngol 59:35–41.

Erostegui C, Norris CH, Bobbin RP (1994) In vitro pharmacologic characterization of a cholinergic receptor on outer hair cells. Hear Res 74:135–147.

Evans MG (1996) Acetylcholine activates two currents in guinea-pig outer hair cells. J Physiol 491:563–578.

Favre D, Scarfone E, Di Gioia G, De Camilli P, Dememes D (1986) Presence of synapsin I in afferent and efferent nerve endings of vestibular sensory epithelia. Brain Res 384:379–382.

Fernandez-Alfonso T, Ryan TA (2004) The kinetics of synaptic vesicle pool depletion at CNS synaptic terminals. Neuron 41:943–953

Fernandez-Chacon R, Konigstorfer A, Gerber SH, Garcia J, Matos MF, Stevens CF, Brose N, Rizo J, Rosenmund C, Sudhof TC (2001) Synaptotagmin I functions as a calcium regulator of release probability. Nature 410:41–49.

Fesce R, Grohovaz F, Valtorta F, Meldolesi J (1994) Neurotransmitter release: fusion or 'kiss-and-run'? Trends Cell Biol 4:1–4.

Fettiplace R, Fuchs PA (1999) Mechanisms of hair cell tuning. Annu Rev Physiol 61: 809–834.

Flock A, Russell I (1976) Inhibition by efferent nerve fibres: action on hair cells and afferent synaptic transmission in the lateral line canal organ of the burbot *Lota lota*. J Physiol 257:45–62.

Fuchs PA (1996) Synaptic transmission at vertebrate hair cells. Curr Opin Neurobiol 6: 514–519.

Fuchs PA, Murrow BW (1992a) Cholinergic inhibition of short (outer) hair cells of the chick's cochlea. J Neurosci 12:800–809.

Fuchs PA, Murrow BW (1992b) A novel cholinergic receptor mediates inhibition of chick cochlear hair cells. Proc R Soc Lond B Biol Sci 248:35–40.

Fuchs PA, Sokolowski BH (1990) The acquisition during development of Ca-activated potassium currents by cochlear hair cells of the chick. Proc R Soc Lond B Biol Sci 241:122–126.

Furukawa T, Hayashida Y, Matsuura S (1978) Quantal analysis of the size of excitatory post-synaptic potentials at synapses between hair cells and afferent nerve fibres in goldfish. J Physiol 276:211–226.

Galambos R (1956) Suppression of auditory nerve activity by stimulation of efferent fibers to the cochlea. J Neurophysiol 19:424–437.

Gandhi SP, Stevens CF (2003) Three modes of synaptic vesicular recycling revealed by single vesicle imaging. Nature 423:607–613.

Geisler CD (1981) A model for discharge patterns of primary auditory-nerve fibers. Brain Res 212:198–201.

Geppert M, Goda Y, Hammer RE, Li C, Rosahl TW, Stevens CF, Sudhof TC (1994) Synaptotagmin I: a major Ca^{2+} sensor for transmitter release at a central synapse. Cell 79:717–727.

Glowatzki E, Fuchs PA (2000) Cholinergic synaptic inhibition of inner hair cells in the neonatal mammalian cochlea. Science 288:2366–2368.

Glowatzki E, Fuchs PA (2002) Transmitter release at the hair cell ribbon synapse. Nat Neurosci 5:147–154.

Goldberg JM, Fernandez C (1980) Efferent vestibular system in the squirrel monkey: anatomical location and influence on afferent activity. J Neurophysiol 43:986–1025.

Gomis A, Burrone J, Lagnado L (1999) Two actions of calcium regulate the supply of releasable vesicles at the ribbon synapse of retinal bipolar cells. J Neurosci 19:6309–6317.

Griesinger CB, Richards CD, Ashmore JF (2002) Fm1-43 reveals membrane recycling in adult inner hair cells of the mammalian cochlea. J Neurosci 22:3939–3952.

Griguer C, Fuchs PA (1996) Voltage-dependent potassium currents in cochlear hair cells of the embryonic chick. J Neurophysiol 75:508–513.

Grover LM, Teyler TJ (1990) Two components of long-term potentiation induced by different patterns of afferent activation. Nature 347:477–479.

Guinan JJ (1996) Physiology of olivocochlear efferents. In: Dallos P, Popper AN, Fay RR (eds), The Cochlea. New York: Springer-Verlag, pp. 435–502.

Gummer AW, Mark RF (1994) Patterned neural activity in brain stem auditory areas of a prehearing mammal, the tammar wallaby (*Macropus eugenii*). NeuroReport 5: 685–688.

Guth PS, Norris CH (1996) The hair cell acetylcholine receptors: a synthesis. Hear Res 98:1–8.

Guth P, Norris C, Fermin CD, Pantoja M (1993) The correlated blanching of synaptic bodies and reduction in afferent firing rates caused by transmitter-depleting agents in the frog semicircular canal. Hear Res 66:143–149.

Guth PS, Perin P, Norris CH, Valli P (1998) The vestibular hair cells: post-transductional signal processing. Prog Neurobiol 54:193–247.

Hackney CM, Mahendrasingam S, Jones EM, Fettiplace R (2003) The distribution of calcium buffering proteins in the turtle cochlea. J Neurosci 23:4577–4589.

Hagiwara S (1975) Ca-dependent action potential. Membranes 3:359–381.

Hall JD, Betarbet S, Jaramillo F (1997) Endogenous buffers limit the spread of free calcium in hair cells. Biophys J 73:1243–1252.

Hay JC, Martin TF (1992) Resolution of regulated secretion into sequential MgATP-dependent and calcium-dependent stages mediated by distinct cytosolic proteins. J Cell Biol 119:139–151.

He DZ, Dallos P (1999) Development of acetylcholine-induced responses in neonatal gerbil outer hair cells. J Neurophysiol 81:1162–1170.

Heidelberger R (1998) Adenosine triphosphate and the late steps in calcium-dependent exocytosis at a ribbon synapse. J Gen Physiol 111:225–241.

Heidelberger R, Heinemann C, Neher E, Matthews G (1994) Calcium dependence of the rate of exocytosis in a synaptic terminal. Nature 371:513–515.

Heller S, Bell AM, Denis CS, Choe Y, Hudspeth AJ (2002) Parvalbumin 3 is an abundant Ca^{2+} buffer in hair cells. J Assoc Res Otolaryngol 3:488–498.

Henkel AW, Almers W (1996) Fast steps in exocytosis and endocytosis studied by capacitance measurements in endocrine cells. Curr Opin Neurobiol 6:350–357.

Heuser JE, Reese TS (1973) Evidence for recycling of synaptic vesicle membrane during transmitter release at the frog neuromuscular junction. J Cell Biol 57:315–344.

Hiel H, Luebke AE, Fuchs PA (2000) Cloning and expression of the alpha9 nicotinic acetylcholine receptor subunit in cochlear hair cells of the chick. Brain Res 858:215–225.

Highstein SM (1991) The central nervous system efferent control of the organs of balance and equilibrium. Neurosci Res 12:13–30.

Hillery CM, Narins PM (1984) Neurophysiological evidence for a traveling wave in the amphibian inner ear. Science 225:1037–1039.

Holt JC, Lioudyno M, Athas G, Garcia MM, Perin P, Guth PS (2001) The effect of proteolytic enzymes on the alpha9-nicotinic receptor-mediated response in isolated frog vestibular hair cells. Hear Res 152:25–42.

Holt M, Cooke A, Wu MM, Lagnado L (2003) Bulk membrane retrieval in the synaptic terminal of retinal bipolar cells. J Neurosci 23:1329–1339.

Holt M, Cooke A, Neef A, Lagnado L (2004) High mobility of vesicles supports continuous exocytosis at a ribbon synapse. Curr Biol 14:173–183.

Housley GD, Ashmore JF (1991) Direct measurement of the action of acetylcholine on isolated outer hair cells of the guinea pig cochlea. Proc R Soc Lond B Biol Sci 244:161–167.

Hudspeth AJ, Issa NP (1996) Confocal-microscopic visualization of membrane addition during synaptic exocytosis at presynaptic active zones of hair cells. Cold Spring Harb Symp Quant Biol 61:303–307.

Hudspeth AJ, Lewis RS (1988a) Kinetic analysis of voltage- and ion-dependent conductances in saccular hair cells of the bull-frog, *Rana catesbeiana*. J Physiol 400: 237–274.

Hudspeth AJ, Lewis RS (1988b) A model for electrical resonance and frequency tuning in saccular hair cells of the bull-frog, *Rana catesbeiana*. J Physiol 400:275–297.

Isaacson JS, Walmsley B (1995) Counting quanta: direct measurements of transmitter release at a central synapse. Neuron 15:875–884.

Issa NP, Hudspeth AJ (1994) Clustering of Ca^{2+} channels and $Ca^{(2+)}$-activated K^+ channels at fluorescently labeled presynaptic active zones of hair cells. Proc Natl Acad Sci USA 91:7578–7582.

Jarousse N, Kelly RB (2001) Endocytotic mechanisms in synapses. Curr Opin Cell Biol 13:461–469.

Jones MV, Westbrook GL (1996) The impact of receptor desensitization on fast synaptic transmission. Trends Neurosci 19:96–101.

Jones TA, Jones SM (2000) Spontaneous activity in the statoacoustic ganglion of the chicken embryo. J Neurophysiol 83:1452–1468.

Jones TA, Jones SM, Paggett KC (2001) Primordial rhythmic bursting in embryonic cochlear ganglion cells. J Neurosci 21:8129–8135.

Kachar B, Battaglia A, Fex J (1997) Compartmentalized vesicular traffic around the hair cell cuticular plate. Hear Res 107:102–112.

Kakehata S, Nakagawa T, Takasaka T, Akaike N (1993) Cellular mechanism of acetylcholine-induced response in dissociated outer hair cells of guinea-pig cochlea. J Physiol 463:227–244.

Kapur A, Yeckel MF, Gray R, Johnston D (1998) L-Type calcium channels are required for one form of hippocampal mossy fiber LTP. J Neurophysiol 79:2181–2190.

Kataoka Y, Ohmori H (1994) Activation of glutamate receptors in response to membrane depolarization of hair cells isolated from chick cochlea. J Physiol 477 (Pt 3):403–414.

Kataoka Y, Ohmori H (1996) Of known neurotransmitters, glutamate is the most likely to be released from chick cochlear hair cells. J Neurophysiol 76:1870–1879.

Katz B (1969) The Release of Neural Transmitter Substances. Liverpool: Liverpool University Press.

Katz LC, Shatz CJ (1996) Synaptic activity and the construction of cortical circuits. Science 274:1133–1138.

Katz E, Verbitski M, Rothlin CV, Vetter DE, Heinemann SF, Elgoyhen AB (2000) High calcium permeability and calcium block of the alpha9 nicotinic acetylcholine receptor. Hear Res 141:117–128.

Kiang NY-S (1965) Discharge patterns of single fibers in the cat's auditory nerve. Cambridge, MA: MIT Press.

Kimitsuki T, Nakagawa T, Hisashi K, Komune S, Komiyama S (1994) Single channel recordings of calcium currents in chick cochlear hair cells. Acta Otolaryngol 114: 144–148.

Koenig JH, Kosaka T, Ikeda K (1989) The relationship between the number of synaptic vesicles and the amount of transmitter released. J Neurosci 9:1937–1942.

Kohler M, Hirschberg B, Bond CT, Kinzie JM, Marrion NV, Maylie J, Adelman JP

(1996) Small-conductance, calcium-activated potassium channels from mammalian brain. Science 273:1709–1714.

Kollmar R, Montgomery LG, Fak J, Henry LJ, Hudspeth AJ (1997) Predominance of the alpha1D subunit in L-type voltage-gated Ca^{2+} channels of hair cells in the chicken's cochlea. Proc Natl Acad Sci USA 94:14883–14888.

Koppl C (1997) Phase locking to high frequencies in the auditory nerve and cochlear nucleus magnocellularis of the barn owl, *Tyto alba*. J Neurosci 17:3312–3321.

Kosaka T, Ikeda K (1983) Possible temperature-dependent blockage of synaptic vesicle recycling induced by a single gene mutation in *Drosophila*. J Neurobiol 14:207–225.

Kozel PJ, Friedman RA, Erway LC, Yamoah EN, Liu LH, Riddle T, Duffy JJ, Doetschman T, Miller ML, Cardell EL, Shull GE (1998) Balance and hearing deficits in mice with a null mutation in the gene encoding plasma membrane Ca^{2+}-ATPase isoform 2. J Biol Chem 273:18693–18696.

Krizaj D, Copenhagen DR (1998) Compartmentalization of calcium extrusion mechanisms in the outer and inner segments of photoreceptors. Neuron 21:249–256.

Kros CJ, Ruppersberg JP, Rusch A (1998) Expression of a potassium current in inner hair cells during development of hearing in mice. Nature 394:281–284.

Lagnado L, Gomis A, Job C (1996) Continuous vesicle cycling in the synaptic terminal of retinal bipolar cells. Neuron 17:957–967.

Leake PA, Snyder RL (1987) Uptake of horseradish peroxidase from perilymph by cochlear hair cells. Hear Res 25:153–171.

Lenzi D, von Gersdorff H (2001) Structure suggests function: the case for synaptic ribbons as exocytotic nanomachines. Bioessays 23:831–840.

Lenzi D, Runyeon JW, Crum J, Ellisman MH, Roberts WM (1999) Synaptic vesicle populations in saccular hair cells reconstructed by electron tomography. J Neurosci 19:119–132.

Lenzi D, Crum J, Ellisman MH, Roberts WM (2002) Depolarization redistributes synaptic membrane and creates a gradient of vesicles on the synaptic body at a ribbon synapse. Neuron 36:649–659.

Lester RA, Jahr CE (1992) NMDA channel behavior depends on agonist affinity. J Neurosci 12:635–643.

Lester RA, Clements JD, Westbrook GL, Jahr CE (1990) Channel kinetics determine the time course of NMDA receptor-mediated synaptic currents. Nature 346:565–567.

Lewis RS, Hudspeth AJ (1983) Voltage- and ion-dependent conductances in solitary vertebrate hair cells. Nature 304:538–541.

Liberman MC (1980) Morphological differences among radial afferent fibers in the cat cochlea: an electron-microscopic study of serial sections. Hear Res 3:45–63.

Liberman MC (1991) The olivocochlear efferent bundle and susceptibility of the inner ear to acoustic injury. J Neurophysiol 65:123–132.

Liberman MC, Dodds LW, Pierce S (1990) Afferent and efferent innervation of the cat cochlea: quantitative analysis with light and electron microscopy. J Comp Neurol 301:443–460.

Lioudyno MI, Verbitsky M, Holt JC, Elgoyhen AB, Guth PS (2000) Morphine inhibits an alpha9-acetylcholine nicotinic receptor-mediated response by a mechanism which does not involve opioid receptors. Hear Res 149:167–177.

Lioudyno MI, Verbitsky M, Glowatzki E, Holt JC, Boulter J, Zadina JE, Elgoyhen AB, Guth PS (2002) The alpha9/alpha10-containing nicotinic ACh receptor is directly modulated by opioid peptides, endomorphin-1, and dynorphin B, proposed efferent cotransmitters in the inner ear. Mol Cell Neurosci 20:695–711.

Lippe WR (1994) Rhythmic spontaneous activity in the developing avian auditory system. J Neurosci 14:1486–1495.

Llano I, Gonzalez J, Caputo C, Lai FA, Blayney LM, Tan YP, Marty A (2000) Presynaptic calcium stores underlie large-amplitude miniature IPSCs and spontaneous calcium transients. Nat Neurosci 3:1256–1265.

Llinas R, Sugimori M, Silver RB (1992) Microdomains of high calcium concentration in a presynaptic terminal. Science 256:677–679.

Locke R, Vautrin J, Highstein S (1999) Miniature EPSPs and sensory encoding in the primary afferents of the vestibular lagena of the toadfish, *Opsanus tau*. Ann NY Acad Sci 871:35–50.

Lumpkin EA, Hudspeth AJ (1998) Regulation of free Ca^{2+} concentration in hair-cell stereocilia. J Neurosci 18:6300–6318.

Lustig LR, Hiel H, Fuchs PA (1999) Vestibular hair cells of the chick express the nicotinic acetylcholine receptor subunit alpha 9. J Vestib Res 9:359–367.

Lustig LR, Peng H, Hiel H, Yamamoto T, Fuchs PA (2001) Molecular cloning and mapping of the human nicotinic acetylcholine receptor alpha10 (CHRNA10). Genomics 73:272–283.

Lysakowski A (2003) Morphophysiology of the vestibular sensory periphery. In: Highstein S, Popper AN, Fay RR (eds), Anatomy and Physiology of the Central and Peripheral Vestibular System. New York: Springer-Verlag, pp. 503–534.

Lysakowski A, Goldberg JM (1997) A regional ultrastructural analysis of the cellular and synaptic architecture in the chinchilla cristae ampullares. J Comp Neurol 389: 419–443.

Mandell JW, Townes-Anderson E, Czernik AJ, Cameron R, Greengard P, De Camilli P (1990) Synapsins in the vertebrate retina: absence from ribbon synapses and heterogeneous distribution among conventional synapses. Neuron 5:19–33.

Marcotti W, Johnson SL, Holley MC, Kros CJ (2003) Developmental changes in the expression of potassium currents of embryonic, neonatal and mature mouse inner hair cells. J Physiol 548:383–400.

Martin AR, Fuchs PA (1992) The dependence of calcium-activated potassium currents on membrane potential. Proc R Soc Lond B Biol Sci 250:71–76.

Martinez-Dunst C, Michaels RL, Fuchs PA (1997) Release sites and calcium channels in hair cells of the chick's cochlea. J Neurosci 17:9133–9144.

Matthews G (2000) Vesicle fiesta at the synapse. Nature 406:835–836.

May B, McQuone SJ (1995) Effects of bilateral olivocochlear lesions on pure-tone intensity discrimination in cats. Aud Neurosci 1:385–400.

May BJ, Prosen CA, Weiss D, Vetter D (2002) Behavioral investigation of some possible effects of the central olivocochlear pathways in transgenic mice. Hear Res 171: 142–157.

McNiven AI, Yuhas WA, Fuchs PA (1996) Ionic dependence and agonist preference of an acetylcholine receptor in hair cells. Aud Neurosci 2:63–77.

Meir A, Ginsburg S, Butkevich A, Kachalsky SG, Kaiserman I, Ahdut R, Demirgoren S, Rahamimoff R (1999a) Ion channels in presynaptic nerve terminals and control of transmitter release. Physiol Rev 79:1019–1088.

Meir A, Ginsburg S, Butkevich A, Kachalsky SG, Kaiserman I, Ahdut R, Demirgoren S, Rahamimoff R (1999b) Ion channels in presynaptic nerve terminals and control of transmitter release. Physiol Rev 79:1019–1088.

Mennerick S, Matthews G (1996) Ultrafast exocytosis elicited by calcium current in synaptic terminals of retinal bipolar neurons. Neuron 17:1241–1249.

Merchan-Perez A, Liberman MC (1996) Ultrastructural differences among afferent synapses on cochlear hair cells: correlations with spontaneous discharge rate. J Comp Neurol 371:208–221.
Meyer J, Mack AF, Gummer AW (2001) Pronounced infracuticular endocytosis in mammalian outer hair cells. Hear Res 161:10–22.
Moiseff A, Konishi M (1981) Neuronal and behavioral sensitivity to binaural time differences in the owl. J Neurosci 1:40–48.
Morgan JR, Prasad K, Hao W, Augustine GJ, Lafer EM (2000) A conserved clathrin assembly motif essential for synaptic vesicle endocytosis. J Neurosci 20:8667–8676.
Moser T, Beutner D (2000) Kinetics of exocytosis and endocytosis at the cochlear inner hair cell afferent synapse of the mouse. Proc Natl Acad Sci USA 97:883–888.
Mountain DC, Geisler CD, Hubbard AE (1980) Stimulation of efferents alters the cochlear microphonic and the sound-induced resistance changes measured in scale media of the guinea pig. Hear Res 3:231–240.
Mulkey RM, Zucker RS (1992) Monensin can transport calcium across cell membranes in a sodium independent fashion in the crayfish *Procambarus clarkii*. Neurosci Lett 143:115–118.
Mulle C, Choquet D, Korn H, Changeux JP (1992) Calcium influx through nicotinic receptor in rat central neurons: its relevance to cellular regulation. Neuron 8:135–143.
Mundigl O, De Camilli P (1994) Formation of synaptic vesicles. Curr Opin Cell Biol 6:561–567.
Muresan V, Lyass A, Schnapp BJ (1999) The kinesin motor KIF3A is a component of the presynaptic ribbon in vertebrate photoreceptors. J Neurosci 19:1027–1037.
Neher E (1998) Vesicle pools and Ca^{2+} microdomains: new tools for understanding their roles in neurotransmitter release. Neuron 20:389–399.
Neher E, Augustine GJ (1992) Calcium gradients and buffers in bovine chromaffin cells. J Physiol 450:273–301.
Neher E, Zucker RS (1993) Multiple calcium-dependent processes related to secretion in bovine chromaffin cells. Neuron 10:21–30.
Nenov AP, Norris C, Bobbin RP (1996) Acetylcholine response in guinea pig outer hair cells. II. Activation of a small conductance $Ca(^{2+})$-activated K^+ channel. Hear Res 101:149–172.
Neves G, Lagnado L (1999) The kinetics of exocytosis and endocytosis in the synaptic terminal of goldfish retinal bipolar cells. J Physiol (Lond) 515:181–202.
Nguyen TH, Balkema GW (1999) Antigenic epitopes of the photoreceptor synaptic ribbon. J Comp Neurol 413:209–218.
Nitecka LM, Sobkowicz HM (1996) The GABA/GAD innervation within the inner spiral bundle in the mouse cochlea. Hear Res 99:91–105.
Norris CH, Guth PS (1974) The release of acetylcholine (ACH) by the crossed olivocochlear bundle (COCB). Acta Otolaryngol 77:318–326.
Oliver D, Klocker N, Schuck J, Baukrowitz T, Ruppersberg JP, Fakler B (2000) Gating of Ca^{2+}-activated K^+ channels controls fast inhibitory synaptic transmission at auditory outer hair cells. Neuron 26:595–601.
Palmer AR, Russell IJ (1986) Phase-locking in the cochlear nerve of the guinea-pig and its relation to the receptor potential of inner hair-cells. Hear Res 24:1–15.
Palmer MJ, Hull C, Vigh J, von Gersdorff H (2003) Synaptic cleft acidification and modulation of short-term depression by exocytosed protons in retinal bipolar cells. J Neurosci 23:11332–11341.

Parsons TD, Sterling P (2003) Synaptic ribbon. Conveyor belt or safety belt? Neuron 37:379–382.
Parsons TD, Lenzi D, Almers W, Roberts WM (1994) Calcium-triggered exocytosis and endocytosis in an isolated presynaptic cell: capacitance measurements in saccular hair cells. Neuron 13:875–883.
Parsons TD, Coorssen JR, Horstmann H, Almers W (1995) Docked granules, the exocytic burst, and the need for ATP hydrolysis in endocrine cells. Neuron 15:1085–1096.
Parsons TD, Ellis-Davies GC, Almers W (1996) Millisecond studies of calcium-dependent exocytosis in pituitary melanotrophs: comparison of the photolabile calcium chelators nitrophenyl-EGTA and DM-nitrophen. Cell Calcium 19:185–192.
Plattner H, Artalejo AR, Neher E (1997) Ultrastructural organization of bovine chromaffin cell cortex-analysis by cryofixation and morphometry of aspects pertinent to exocytosis. J Cell Biol 139:1709–1717.
Platzer J, Engel J, Schrott-Fischer A, Stephan K, Bova S, Chen H, Zheng H, Striessnig J (2000) Congenital deafness and sinoatrial node dysfunction in mice lacking class D L-type Ca^{2+} channels. Cell 102:89–97.
Protti DA, Flores-Herr N, von Gersdorff H (2000) Light evokes Ca^{2+} spikes in the axon terminal of a retinal bipolar cell. Neuron 25:215–227
Pyle JL, Kavalali ET, Piedras-Renteria ES, Tsien RW (2000) Rapid reuse of readily releasable pool vesicles at hippocampal synapses. Neuron 28:221–231.
Rajan R, Johnstone BM (1988) Electrical stimulation of cochlear efferents at the round window reduces auditory desensitization in guinea pigs. I. Dependence on electrical stimulation parameters. Hear Res 36:53–73.
Rasmussen G (1946) The olivary peduncle and other fiber projections of the superior olivary complex. J Comp Neurol 84:141–219.
Rea R, Li J, Dharia A, Levitan ES, Sterling P, Kramer RH (2004) Streamlined synaptic vesicle cycle in cone photoreceptor terminals. Neuron 41:755–766.
Rebillard G, Rubel EW (1981) Electrophysiological study of the maturation of auditory responses from the inner ear of the chick. Brain Res 229:15–23.
Regehr WG (1997) Interplay between sodium and calcium dynamics in granule cell presynaptic terminals. Biophys J 73:2476–2488.
Rennie KJ, Ashmore JF (1991) Ionic currents in isolated vestibular hair cells from the guinea-pig crista ampullaris. Hear Res 51:279–291.
Rettig J, Heinemann C, Ashery U, Sheng ZH, Yokoyama CT, Catterall WA, Neher E (1997) Alteration of Ca^{2+} dependence of neurotransmitter release by disruption of Ca^{2+} channel/syntaxin interaction. J Neurosci 17:6647–6656.
Ricci AJ, Wu YC, Fettiplace R (1998) The endogenous calcium buffer and the time course of transducer adaptation in auditory hair cells. J Neurosci 18:8261–8277.
Ricci AJ, Gray-Keller M, Fettiplace R (2000) Tonotopic variations of calcium signalling in turtle auditory hair cells. J Physiol 524 Pt 2:423–436.
Rieke F, Schwartz EA (1996) Asynchronous transmitter release: control of exocytosis and endocytosis at the salamander rod synapse. J Physiol 493:1–8.
Rispoli G, Martini M, Rossi ML, Mammano F (2001) Dynamics of intracellular calcium in hair cells isolated from the semicircular canal of the frog. Cell Calcium 30: 131–140.
Roberts WM (1993) Spatial calcium buffering in saccular hair cells. Nature 363:74–76.
Roberts WM (1994) Localization of calcium signals by a mobile calcium buffer in frog saccular hair cells. J Neurosci 14:3246–3262.

Roberts WM, Jacobs RA, Hudspeth AJ (1990) Colocalization of ion channels involved in frequency selectivity and synaptic transmission at presynaptic active zones of hair cells. J Neurosci 10:3664–3684.

Rodriguez-Contreras A, Yamoah EN (2001) Direct measurement of single-channel $Ca^{(2+)}$ currents in bullfrog hair cells reveals two distinct channel subtypes. J Physiol 534: 669–689.

Rodriguez-Contreras A, Nonner W, Yamoah EN (2002) Ca^{2+} transport properties and determinants of anomalous mole fraction effects of single voltage-gated Ca^{2+} channels in hair cells from bullfrog saccule. J Physiol 538:729–745.

Rose JE, Brugge JF, Anderson DJ, Hind JE (1967) Phase-locked response to low-frequency tones in single auditory nerve fibers of the squirrel monkey. J Neurophysiol 30:769–793.

Ross MD (2000) Changes in ribbon synapses and rough endoplasmic reticulum of rat utricular macular hair cells in weightlessness. Acta Otolaryngol 120:490–499.

Rossi ML, Prigioni I, Valli P, Casella C (1980) Activation of the efferent system in the isolated frog labyrinth: effects on the afferent EPSPs and spike discharge recorded from single fibres of the posterior nerve. Brain Res 185:125–137.

Rossi ML, Martini M, Pelucchi B, Fesce R (1994) Quantal nature of synaptic transmission at the cytoneural junction in the frog labyrinth. J Physiol 478 (Pt 1):17–35.

Ruel J, Chen C, Pujol R, Bobbin RP, Puel JL (1999) AMPA-preferring glutamate receptors in cochlear physiology of adult guinea-pig. J Physiol 518 (Pt 3):667–680.

Ruel J, Bobbin RP, Vidal D, Pujol R, Puel JL (2000) The selective AMPA receptor antagonist GYKI 53784 blocks action potential generation and excitotoxicity in the guinea pig cochlea. Neuropharmacology 39:1959–1973.

Ruel J, Nouvian R, Gervais d'Aldin C, Pujol R, Eybalin M, Puel JL (2001) Dopamine inhibition of auditory nerve activity in the adult mammalian cochlea. Eur J Neurosci 14:977–986.

Safieddine S, Wenthold RJ (1999) SNARE complex at the ribbon synapses of cochlear hair cells: analysis of synaptic vesicle- and synaptic membrane-associated proteins. Eur J Neurosci 11:803–812.

Saito K (1980) Fine structure of the sensory epithelium of the guinea pig organ of Corti: afferent and efferent synapses of hair cells. J Ultrastruct Res 71:222–232.

Sankaranarayanan S, Ryan TA (2001) Calcium accelerates endocytosis of vSNAREs at hippocampal synapses. Nat Neurosci 4:129–136.

Saunders JC, Coles RB, Gates GR (1973) The development of auditory evoked responses in the cochlea and cochlear nuclei of the chick. Brain Res 63:59–74.

Scanziani M, Salin PA, Vogt KE, Malenka RC, Nicoll RA (1997) Use-dependent increases in glutamate concentration activate presynaptic metabotropic glutamate receptors. Nature 385:630–634.

Schikorski T, Stevens CF (2001) Morphological correlates of functionally defined synaptic vesicle populations. Nat Neurosci 4:391–395.

Schmitz F, Konigstorfer A, Sudhof TC (2000) RIBEYE, a component of synaptic ribbons: a protein's journey through evolution provides insight into synaptic ribbon function. Neuron 28:857–872.

Schnee M, Ricci A (2003) Biophysical and pharmacological characterization of voltage-gated calcium currents in turtle auditory hair cells. J Physiol 549:697–717.

Schneggenburger R, Neher E (2000) Intracellular calcium dependence of transmitter release rates at a fast central synapse. Nature 406:889–893.

Schneider SW (2001) Kiss and run mechanism in exocytosis. J Membr Biol 181:67–76.
Seiler C, Nicolson T (1999) Defective calmodulin-dependent rapid apical endocytosis in zebrafish sensory hair cell mutants. J Neurobiol 41:424–434.
Self T, Sobe T, Copeland NG, Jenkins NA, Avraham KB, Steel KP (1999) Role of myosin VI in the differentiation of cochlear hair cells. Dev Biol 214:331–341.
Sewell WF (1996) Neurotransmitters and synaptic transmission. In: Dallos P, Popper A, N, Fay RR (eds), The Cochlea. New York: Springer-Verlag, pp. 503–534.
Sheng ZH, Rettig J, Takahashi M, Catterall WA (1994) Identification of a syntaxin-binding site on N-type calcium channels. Neuron 13:1303–1313.
Shigemoto T, Ohmori H (1990) Muscarinic agonists and ATP increase the intracellular Ca^{2+} concentration in chick cochlear hair cells. J Physiol 420:127–148.
Shigemoto T, Ohmori H (1991) Muscarinic receptor hyperpolarizes cochlear hair cells of chick by activating $Ca^{(2+)}$-activated K^+ channels. J Physiol 442:669–690.
Siegel JH (1992) Spontaneous synaptic potentials from afferent terminals in the guinea pig cochlea. Hear Res 59:85–92.
Siegel JH, Brownell WE (1986) Synaptic and Golgi membrane recycling in cochlear hair cells. J Neurocytol 15:311–328.
Siegel JH, Dallos P (1986) Spike activity recorded from the organ of Corti. Hear Res 22:245–248.
Siegel JH, Kim DO (1982) Efferent neural control of cochlear mechanics? Olivocochlear bundle stimulation affects cochlear biomechanical nonlinearity. Hear Res 6:171–182.
Simmons DD (2002) Development of the inner ear efferent system across vertebrate species. J Neurobiol 53:228–250.
Sjostrand F (1953) The ultrastructure of retinal rod synapse of the guinea pig eye. J Appl Phys 24:1422.
Sjostrand F (1958) Ultrastructure of retinal rod synapses of the guinea pig as revealed by three dimensional reconstruction from serial sections. J Ultrastruct Res 2:122.
Slepecky NB (1996) Structure of the mammalian cochlea. In: Dallos P, Popper AN, Fay RR (eds), The Cochlea. New York: Springer-Verlag, pp. 44–129.
Smith C, Sjostrand F (1961) A synaptic structure in the hair cells of the guinea pig cochlea. J Ultrastruct Res 5:184–192.
Sobkowicz HM, Rose JE, Scott GE, Slapnick SM (1982) Ribbon synapses in the developing intact and cultured organ of Corti in the mouse. J Neurosci 2:942–957.
Sobkowicz HM, Rose JE, Scott GL, Levenick CV (1986) Distribution of synaptic ribbons in the developing organ of Corti. J Neurocytol 15:693–714.
Sobkowicz HM, August BK, Slapnick SM, Luthy DF (1998) Terminal dendritic sprouting and reactive synaptogenesis in the postnatal organ of Corti in culture. J Comp Neurol 397:213–230.
Spassova M, Eisen MD, Saunders JC, Parsons TD (2001) Chick cochlear hair cell exocytosis mediated by dihydropyridine-sensitive calcium channels. J Physiol 535: 689–696.
Spassova M, Avissar M, Furman AC, Crumling MA, Saunders JC, Parsons TD (2004) Evidence that rapid vesicle replenishment of the synaptic ribbon mediates recovery from short-term adaptation at the hair cell afferent synapse. J Assoc Res Otolaryngol 5:376–390.
Spicer SS, Thomopoulos GN, Schulte BA (1999) Novel membranous structures in apical and basal compartments of inner hair cells. J Comp Neurol 409:424–437.
Sridhar TS, Brown MC, Sewell WF (1997) Unique postsynaptic signaling at the hair cell

efferent synapse permits calcium to evoke changes on two time scales. J Neurosci 17: 428–437.

Staal RG, Mosharov EV, Sulzer D (2004) Dopamine neurons release transmitter via a flickering fusion pore. Nat Neurosci 7:341–346.

Stevens CF, Tsujimoto T (1995) Estimates for the pool size of releasable quanta at a single central synapse and for the time required to refill the pool. Proc Natl Acad Sci USA 92:846–849.

Stevens CF, Williams JH (2000) "Kiss and run" exocytosis at hippocampal synapses. Proc Natl Acad Sci USA 97:12828–12833.

Su ZL, Jiang SC, Gu R, Yang WP (1995) Two types of calcium channels in bullfrog saccular hair cells. Hear Res 87:62–68.

Sudhof TC (1995) The synaptic vesicle cycle: a cascade of protein–protein interactions. Nature 375:645–653.

Sueta T, Zhang SY, Sellick PM, Patuzzi R, Robertson D (2004) Effects of a calcium channel blocker on spontaneous neural noise and gross action potential waveforms in the guinea pig cochlea. Hear Res 188: 117–125.

Sugai T, Sugitani M, Ooyama H (1991) Effects of activation of the divergent efferent fibers on the spontaneous activity of vestibular afferent fibers in the toad. Jpn J Physiol 41:217–232.

Sugai T, Yano J, Sugitani M, Ooyama H (1992) Actions of cholinergic agonists and antagonists on the efferent synapse in the frog sacculus. Hear Res 61:56–64.

Sullivan WE, Konishi M (1984) Segregation of stimulus phase and intensity coding in the cochlear nucleus of the barn owl. J Neurosci 4:1787–1799.

Takasaka T, Smith CA (1971) The structure and innervation of the pigeon's basilar papilla. J Ultrastruct Res 35:20–65.

Tang Y, Zucker RS (1997) Mitochondrial involvement in post-tetanic potentiation of synaptic transmission. Neuron 18:483–491.

Thomas P, Wong JG, Lee AK, Almers W (1993) A low affinity Ca^{2+} receptor controls the final steps in peptide secretion from pituitary melanotrophs. Neuron 11:93–104.

Thoreson WB, Rabl K, Townes-Anderson E, Heidelberger R (2004) A highly Ca^{2+}-sensitive pool of vesicles contributes to linearity at the rod photoreceptor ribbon synapse. Neuron 42:595–605.

Tucker T, Fettiplace R (1995) Confocal imaging of calcium microdomains and calcium extrusion in turtle hair cells. Neuron 15:1323–1335.

Tucker TR, Fettiplace R (1996) Monitoring calcium in turtle hair cells with a calcium-activated potassium channel. J Physiol 494 (Pt 3):613–626.

Valtorta F, Meldolesi J, Fesce R (2001) Synaptic vesicles: is kissing a matter of competence? Trends Cell Biol 11:324–328.

Van der Kloot W (1988) The kinetics of quantal releases during end-plate currents at the frog neuromuscular junction. J Physiol 402:605–626.

Vernino S, Amador M, Luetje CW, Patrick J, Dani JA (1992) Calcium modulation and high calcium permeability of neuronal nicotinic acetylcholine receptors. Neuron 8: 127–134.

Vetter DE, Adams JC, Mugnaini E (1991) Chemically distinct rat olivocochlear neurons. Synapse 7:21–43.

Vetter DE, Liberman MC, Mann J, Barhanin J, Boulter J, Brown MC, Saffiote-Kolman J, Heinemann SF, Elgoyhen AB (1999) Role of alpha9 nicotinic ACh receptor subunits in the development and function of cochlear efferent innervation. Neuron 23:93–103.

von Békésy G (1960) Experiments in Hearing. New York: McGraw-Hill.
von Gersdorff H (2001) Synaptic ribbons: versatile signal transducers. Neuron 29:7–10.
von Gersdorff H, Matthews G (1994) Dynamics of synaptic vesicle fusion and membrane retrieval in synaptic terminals. Nature 367:735–739.
von Gersdorff H, Matthews G (1997) Depletion and replenishment of vesicle pools at a ribbon-type synaptic terminal. J Neurosci 17:1919–1927.
von Gersdorff H, Vardi E, Matthews G, Sterling P (1996) Evidence that vesicles on the synaptic ribbon of retinal bipolar neurons can be rapidly released. Neuron 16:1221–1227.
von Gersdorff H, Schneggenburger R, Weis S, Neher E (1997) Presynaptic depression at a calyx synapse: the small contribution of metabotropic glutamate receptors. J Neurosci 17:8137–8146.
Walsh EJ, McGee J (1997) Does activity in the olivocochlear bundle affect development of the auditory periphery? In: Lewis ER (ed), Diversity in Auditory Mechanics. Singapore: World Scientific, pp. 376–385.
Walsh EJ, McGee J, McFadden SL, Liberman MC (1998) Long-term effects of sectioning the olivocochlear bundle in neonatal cats. J Neurosci 18:3859–3869.
Wang Y, Okamoto M, Schmitz F, Hofmann K, Sudhof TC (1997) Rim is a putative Rab3 effector in regulating synaptic-vesicle fusion. Nature 388:593–598.
Warr WB (1975) Olivocochlear and vestibular efferent neurons of the feline brain stem: their location, morphology and number determined by retrograde axonal transport and acetylcholinesterase histochemistry. J Comp Neurol 161:159–181.
Warr WB (1992) Organization of olivocochlear efferent systems in mammals. In: Douglas W, Popper AN, Fay RR (eds), The Mammalian Auditory Pathway: Neuroanatomy. New York: Springer-Verlag, pp. 410–448.
Weisstaub N, Vetter DE, Elgoyhen AB, Katz E (2002) The alpha9alpha10 nicotinic acetylcholine receptor is permeable to and is modulated by divalent cations. Hear Res 167:122–135.
Wiederhold M (1986) Physiology of the olivocochlear system. In: Altschuler RA, Bobbin RP (eds), The Cochlea. New York: Raven Press, pp. 349–370.
Winslow RL, Sachs MB (1987) Effect of electrical stimulation of the crossed olivocochlear bundle on auditory nerve response to tones in noise. J Neurophysiol 57:1002–1021.
Wu LG, Saggau P (1997) Presynaptic inhibition of elicited neurotransmitter release. Trends Neurosci 20:204–212.
Yamoah EN, Lumpkin EA, Dumont RA, Smith PJ, Hudspeth AJ, Gillespie PG (1998) Plasma membrane Ca^{2+}-ATPase extrudes Ca^{2+} from hair cell stereocilia. J Neurosci 18:610–624.
Yoshida N, Shigemoto T, Sugai T, Ohmori H (1994) The role of inositol trisphosphate on ACh-induced outward currents in bullfrog saccular hair cells. Brain Res 644:90–100.
Yoshida N, Liberman MC, Brown MC, Sewell WF (2001) Fast, but not slow, effects of olivocochlear activation are resistant to apamin. J Neurophysiol 85:84–88.
Yuhas WA, Fuchs PA (1999) Apamin-sensitive, small-conductance, calcium-activated potassium channels mediate cholinergic inhibition of chick auditory hair cells. J Comp Physiol [A] 185:455–462.
Zahradnikova A, Zahradnik I, Gyorke I, Gyorke S (1999) Rapid activation of the cardiac ryanodine receptor by submillisecond calcium stimuli. J Gen Physiol 114:787–798.

Zenisek D, Matthews G (2000) The role of mitochondria in presynaptic calcium handling at a ribbon synapse. Neuron 25:229–237.

Zenisek D, Davila V, Wan L, Almers W (2003) Imaging calcium entry sites and ribbon structures in two presynaptic cells. J Neurosci 23:2538–2548.

Zheng J, Shen W, He DZ, Long KB, Madison LD, Dallos P (2000) Prestin is the motor protein of cochlear outer hair cells. Nature 405:149–155.

Zhou Z, Neher E (1993) Mobile and immobile calcium buffers in bovine adrenal chromaffin cells. J Physiol 469:245–273.

Zidanic M, Fuchs PA (1995) Kinetic analysis of barium currents in chick cochlear hair cells. Biophys J 68:1323–1336.

Zucker RS (1996) Exocytosis: a molecular and physiological perspective. Neuron 17:1049–1055.

Zuo J, Treadaway J, Buckner TW, Fritzsch B (1999) Visualization of alpha9 acetylcholine receptor expression in hair cells of transgenic mice containing a modified bacterial artificial chromosome. Proc Natl Acad Sci USA 96:14100–14105.

7
The Piezoelectric Outer Hair Cell

WILLIAM E. BROWNELL*

1. Overview: High-Frequency Force Production in the Ear

Auditory nerve fibers originating at a given location in vertebrate hearing organs respond most vigorously to sounds at a specific (or best) frequency. This frequency is determined by the mechanical properties of vibrating structures at the location in the auditory epithelium innervated by the nerve fiber. A systematic progression of mechanical properties along the epithelium results in a topographic mapping of best frequencies and the resulting "tonotopic" map is retained in the central nervous system (CNS). Factors that impose limits on the frequency at which inner ear structures can vibrate will define the frequency limits for a given species. Vertebrate hearing organs are liquid filled (most likely reflecting the aquatic origins of the vertebrates), and the fluid environment imposes a damping force on vibrations. Because fluid or viscous damping is directly proportional to the velocity of the vibrating structures, the force resisting movement increases proportionally with frequency, making it more difficult for animals to hear as the frequency increases.

The ability to localize predator or prey is improved by analyzing sounds over a wide range of frequencies, resulting in an evolutionary selection pressure for detecting ever higher frequencies. Diverse strategies to either get around or mechanically counteract fluid damping and increase the upper limit of hearing are found in animal ears. The electrically tuned membranes described by Art and Fettiplace in Chapter 5 may represent a nonmechanical strategy to improve hair cell performance. The mechanical strategies involve the production of a "negative damping" force. Vertebrates appeared more than 400 million years ago, and it is likely that force production by the stereocilia bundle was the negative damping strategy used for the relatively low frequencies of the early vertebrates. The current models for this mechanism propose motor units near the tips of the stereocilia (as reviewed in Fettiplace and Ricci, Chapter 4). Since

---

*Thomas Gold described the reasons for and the characteristics of what I am calling the piezoelectric outer hair cell in the late 1940s (Gold 1948). This chapter is dedicated to his memory.

the force required for negative damping increases with frequency, either more stereocilia must be recruited or more powerful motors (or both) are required to counteract damping at higher frequencies. When mammals appeared on the scene more than 220 million years ago, they adopted a different mechanism associated with the lateral wall of their cylindrically shaped outer hair cells (OHCs; see Furness and Hackney, Chapter 3, Fig. 3.1B, for an image of OHCs).

The spiral shaped cochlea appeared with the OHCs. The cochlea is located deep in the skull roughly at the intersection of two lines pointing inwards from the ear canal and the eye (Fig. 7.1a). Sound waves captured by the external ear pass through the ear canal and impinge on the tympanic membrane. The sound vibrations are transmitted through the bones of the middle ear and create pressure differences in the fluid-filled cochlea. The pressure differences between

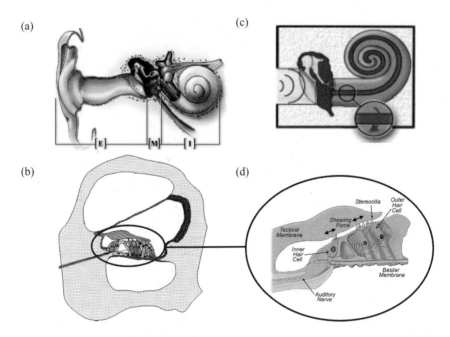

FIGURE 7.1. The mammalian cochlea. (**a**) Gross anatomy. E, External ear; middle ear; I, inner ear. (**b**) A cross section through the cochlear spiral shows the three cochlear ducts. The middle compartment, scala media, contains the hearing organ and is bounded below by the basilar membrane and above by the thin Reissner's membrane. (**c**) The pressure difference between cochlear compartments produces a traveling wave along the elastic basilar membrane. Frequency is mapped along the length of the cochlea with high frequencies mapped to the basal end closest to the middle ear. (**d**) The organ of Corti sits on the basilar membrane. (a) is Netter Plate 87, (b) is modified from http://depts.washington.edu/hearing/slice.html, (c) is copyright of Stephan Blatrix, Montpellier (France), taken from the website "Promenade round the cochlea," http://www.iurc.montp.inserm.fr/cric/audition/english/, and (d) is adapted from Brownell (1999).

cochlear compartments (Fig. 7.1b, c) produce a traveling wave along the elastic basilar membrane. Frequency mapping along the length of the cochlea occurs from systematic differences in geometry, elastic properties, and mass of the basilar membrane and cochlea partition. Sitting on the basilar membrane within the cochlear partition is the hearing organ, the organ of Corti. This sensory epithelium consists of hair cells supported by an elegant matrix of cellular and acellular structures (see Slepecky 1996 for a detailed description). There are three rows of OHCs and a single row of inner hair cells. "Inner/outer" refers to their location relative to the central axis of the cochlear spiral. The traveling wave produces a shear force on the hair bundles (Fig. 7.1d). When this occurs, hair cell stereocilia are bent, modulating the flow of ions through mechanically sensitive ion channels located near tip links connecting adjacent stereocilia (Jaramillo and Hudspeth 1991; Eatock 2000). Both inner and outer hair cells are mechanoreceptors, but only inner hair cell (IHC) receptor potentials trigger afferent nerve activity sending signals to the brain. OHCs have the specialized role of enhancing the micromechanical tuning of the basilar membrane by generating a mechanical force. The force alters the vibrations of the organ of Corti and refines the stimulation of the IHCs, thereby enhancing the perception and discrimination of the high-frequency sounds.

The need for a force generating mechanism to counteract viscous damping by cochlear fluids was first recognized in the late 1940s by the young polymath Thomas Gold (Gold 1948). His arguments and the evidence he presented to support them were largely ignored by the hearing science community and he moved on to make a number of astrophysical discoveries including pulsars. Some 30 years later direct measures of basilar membrane motion revealed that its tuning could not be explained by a passive basilar membrane and Gold's suggestion of an active mechanism began to take hold. The discovery that the living ear generated sounds (Kemp 1978) seemed to confirm the existence of what Hallowell Davis was to call the cochlear amplifier (Davis 1983). Gold was ultimately vindicated when OHC electromotility was observed, and it became the likely cellular basis for the cochlear amplifier (Brownell 1983, 1984; Brownell et al. 1985).

OHC electromotility is a length change resulting from a direct conversion of the transmembrane potential into mechanical force. A number of reviews describe the phenomenon (Brownell 1990, 1999, 2002; Dallos 1992, 1997; Holley 1996; Brownell and Popel 1998; Brownell and Oghalai 2000; Brownell et al. 2001; Dallos and Fakler 2002; Santos-Sacchi 2003; Snyder et al. 2003). Hyperpolarization leads to cell elongation while depolarization leads to cell shortening (see the voltage–displacement function in Fig. 7.2a). The displacements reach saturating plateaus for both hyperpolarizing and depolarizing potentials. The OHC radius undergoes a reciprocal change to the length change, presumably because the rapid electromotile movements preclude changes in cell volume. The mechanism underlying the force production, while unknown, is membrane based and located in the OHC lateral wall (see Figs. 7.3 and 7.4). Electromotility does not depend directly on either ATP or calcium ions but it

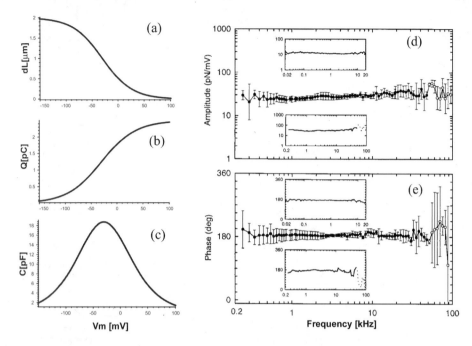

FIGURE 7.2. Electromotility—OHC response to electrical stimulation. Idealized (**a**) voltage-displacement. (**b**) voltage-charge, and (**c**) nonlinear capacitance functions. The nonlinear or voltage dependent capacitance is considered the electrical signature of electromotility. The capacitance (c) is calculated from the charge movement measured simultaneously with electromotility (b) and is the ratio of the charge to voltage. In most cells the capacitance is constant and proportional to the cell's surface area. Note that the peak of the nonlinear capacitance function (V_{pkCm}) occurs at the voltage for maximum gain (steepest slope) of the voltage–displacement function. (**d, e**) High-frequency isometric mechanical force production generated by electrically stimulated isolated OHCs. Cells were held at their basal ends with a microchamber and an AFM cantilever was pressed against their apical ends. The magnitude (d) and phase (e) of the force generated in response to electrical stimulation is plotted as a function of frequency. (a–c) from Snyder et al. 2003; (d and e) from Frank et al. (1999).

does require the cytoplasm to have a positive turgor pressure (see Section 7.2.3.4).

Electromotility is affected by drugs such as the lanthinides and ionic amphipaths. Gadolinium blocks the response while a variety of ionic amphipaths alters either the electromotile magnitude or shift its voltage–displacement function. Amphipaths are both lipid and water soluble and are therefore able to interact with cell membranes. A variety of charged (or ionic amphipaths) are ototoxic. Salicylate (an anionic amphipath and the active ingredient in aspirin) diminishes the electromotile response changing the voltage–displacement re-

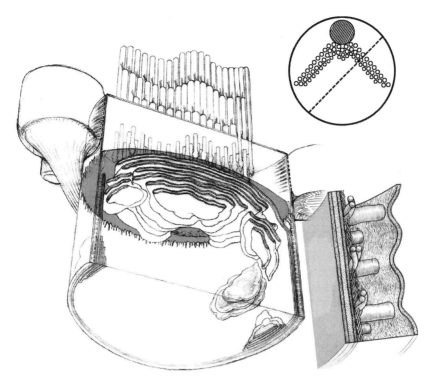

FIGURE 7.3. Cut-away of an OHC highlighting the organization of its apical pole. Inset at upper right is a top down view of the stereociliar bundle showing the plane of the cut. Stereociliar rootlets do not penetrate the cuticular plate, which is concave and extends to the edges of the cell. Tight junctional complexes are found between the OHC and the supporting cell phalangeal processes. The membranous organelle immediately below the cuticular plate is called the canalicular reticulum. It is contiguous with the lateral wall subsurface cisterna. Inset on the lower right is a high magnification rendering of the lateral wall. Modified from Brownell (2002).

sponse from a nonlinear saturating-sigmoidal function (Fig. 7.2a) to a linear function. Decreases in turgor pressure and chlorpromazine (a cationic amphipath used to treat schizophrenia) shift the voltage–displacement function in the depolarizing direction.

The electrically evoked cell displacements follow the transmembrane potential at auditory frequencies. A recent study has used atomic force microscope cantilevers to compress OHCs and monitor their force production under isometric conditions (Frank et al. 1999). An OHC was held by a large suction pipette at its basal end and was also used to electrically stimulate the cell with a constant-voltage frequency sweep. The magnitude (Fig. 7.2d) and phase (Fig. 7.2e) of the response remain essentially constant until well above 50 kHz. A stereo

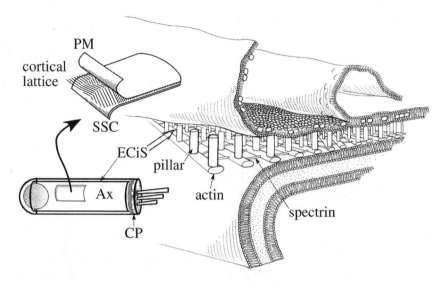

FIGURE 7.4. Diagram of OHC showing organization of lateral wall. Axial core (Ax), extracisternal space (ECiS), subsurface cisterna (SSC), cuticular plate (CP), plasma membrane (PM). From Brownell and Popel (1998).

speaker with a similar "flat" frequency response is the dream of every audiophile. Clearly the OHC has the capability to generate mechanical force at high frequencies and thereby serve as the cochlear amplifier.

The OHC is unique to the mammalian cochlea and is perhaps the most exotically specialized hair cell. Morphological and molecular features of its lateral wall (a three-layered composite of membranes and cytoskeletal proteins as shown in Figs. 7.3 and 7.4) endow it with the ability to generate mechanical force at high frequencies. The force generating mechanism in the OHCs is a direct conversion of transmembrane electrical potentials into mechanical energy. While the OHC can produce high-frequency mechanical force, the functional significance of this ability has been questioned because most biological membranes are "low-pass" and unable to sustain transmembrane receptor potentials at high frequencies (Hudspeth and Logothetis 2000). What good would it be to have high-frequency–force generating ability if the membranes do not allow high-frequency receptor potentials?

Here I develop the argument that evolution solved the low-pass problem for the OHC by strongly coupling electrical and mechanical energy in its lateral wall. The OHC has been shown to display piezoelectricity because mechanical deformation of the lateral wall plasma membrane changes its electrical polarization, and conversely a change in electrical polarization leads to membrane deformation. The ready conversion of one form of energy to the other endows the lateral wall with sensory–motor abilities, and this novel biological piezoelectricity pushes the cell membrane cutoff frequency to higher frequencies. Prestin is an integral membrane protein belonging to the Slc26A family of anion transporters that was recently isolated from OHCs, cloned, and shown to enhance the piezoelectric properties of transfected test cells (Zheng et al. 2000; Ludwig et al. 2001). Recent studies on prestin transfected cells have revealed that small intracellular anions such as chloride are also required for the optimal performance of the bidirectional OHC piezomotor (Oliver et al. 2001; Fakler and Oliver 2003).

This chapter provides an update on OHC structure, followed by the evidence for OHC piezoelectricity, its functional significance, and possible molecular basis. Specifically two plausible models for OHC electromotility involving electromechanical transduction in the plasma membrane (PM) will be discussed. One involves in-plane changes of the membrane area (Kalinec et al. 1992; Dallos et al. 1993; Iwasa 1993, 1994; Huang and Santos-Sacchi 1994; Iwasa and Adachi 1997) and the other involves out-of-plane flexoelectric bending (Raphael et al. 2000b), portrayed in Figure 7.5. The review builds on previous reviews of the OHC that have emphasized electromotility (electromechanical transduction) and its impact on cochlear function (Brownell 1990, 1999, 2002; Dallos 1992, 1997; Holley 1996; Brownell and Popel 1998; Brownell and Oghalai 2000; Brownell et al. 2001; Dallos and Fakler 2002; Santos-Sacchi 2003; Snyder et al. 2003). My overview began with a description of the evolutionary forces that led to high-frequency hearing in mammals and the appearance of the OHC. The review also attempts to raise questions about the cellular and molecular evolution underlying OHC piezoelectricity.

2. Structure of the OHC

The apical pole of the OHC, like all hair cells, is capped by a cuticular plate and stereocilia bundle (Fig. 7.3). The OHC's hemispherical basal end (Fig. 7.4) envelops a portion of the cell's nucleus along with pre- and postsynaptic specializations. The remainder of the OHC is cylindrical. The membrane-based mechanism responsible for OHC piezoelectricity is located in the lateral wall of the cylindrical region (Figs. 7.3, inset, and 7.4). OHCs in the same animal have a uniform diameter (between 7 and 9 μm depending on species). Their length varies systematically with the tonotopic organization of the cochlea, becoming shorter on moving toward the high-frequency, basal end of the cochlea near the oval window. The longest cells (> 80 μm in some species) are located in the low-frequency apical end of the cochlea near the heliocotrema (where scala tympani and scala vestibuli join at the apex of the cochlea). The longest OHCs are found in the outermost or third row and can be over twice as long as those found in the innermost (first) row. The difference in length results in a steep tilt in the reticular lamina (the top of the organ of Corti). The length difference between rows decreases gradually on moving to the high frequency base of the cochlea until all three rows are the same length (Lim 1980; Pujol et al. 1992). The reticular lamina is roughly parallel to the basilar membrane in the highest frequency regions.

The apical and synaptic regions of the OHC differ from those of other hair cells and will be described before the middle cylindrical portion, which is responsible for OHC electromotility.

2.1 Apical Pole

The OHC cuticular plate (CP) is larger than it is in other hair cells. It spans the width of the cell contacting the lateral wall plasma membrane and it is so thick that stereocilia rootlets do not penetrate it (Takasaka et al. 1983; Liberman 1987). Stereocilia rootlets penetrate the thinner CP of other hair cells. CP diameter in all other hair cells is less than that of the cell and the CP is convex, in contrast to the OHC CP, which is concave (Fig. 7.3). The thickness of the OHC CP and its contact with the OHC membrane in the area of the tight junctional contacts between the OHCs and the apical ends of the supporting cells undoubtedly contributes to the mechanical rigidity of the reticular lamina. Cytoskeletal proteins radiate away from the stereocilia rootlets toward the lateral wall in the OHC CP while the CP of inner hair cells appears morphologically isotropic (Raphael et al. 1994). The OHC stereocilia bundle contains about 130 stereocilia of average diameter 145 nm arranged in four rows in a W pattern with a mean spacing between rows of 253nm (Géléoc et al. 1997). The tallest stereocilia are embedded in the tectorial membrane and their deflection is produced by shear displacement of the tectorial membrane. Large transducer currents have been recorded in OHCs from mouse cochlear explants (Kros et al. 1992) and from developing and mature OHCs from an in situ rat organ of Corti preparation (Kennedy et al. 2003). The stereocilia are connected by lateral links (see Furness and Hackney, Chapter 3, Fig. 3.8), which become more extensive with development (Furness et al. 1989) so that deflection of the tallest stereocilia is assumed to move the entire bundle.

2.2 Basal Pole

The OHC's motor role is reflected in the afferent and efferent synapses found at the cell's basal pole. The few afferent synapses found at the base of an OHC do not appear to excite the auditory nerve fibers they contact while the efferent synapses are capable of modulating OHC force generation. Presynaptic dense bodies (synaptic ribbons) touch presynaptic thickenings of afferent synapses and are surrounded by synaptic vesicles (reviewed in Fuchs and Parsons, Chapter 6). There are fewer synaptic ribbons in OHCs than in IHCs and in some species they are missing in adult OHCs. Developmental studies on cats indicate that synaptic ribbons are abundant in OHCs of fetal cats but become difficult to find shortly after birth (Pujol et al. 1979) and are missing in adults (Spoendlin 1966; Dunn and Morest 1975). Synaptic ribbons are plastic and change their appearance in response to changes in synaptic activity (see Brownell 1982 for review). Synaptic ribbons depolymerize when cold temperatures interfere with membrane reuptake and block synaptic vesicle recycling in turtle photoreceptors (Schaeffer and Raviola 1978). Synaptic ribbons polymerize in OHCs when vesicle release is blocked (Siegel and Brownell 1981). There is no evidence that the OHC afferent synapses activate the eighth nerve fibers they contact. There is only one study in which an eighth nerve fiber was reconstructed after intracellular recording and found to terminate on an OHC. The fiber did not fire action potentials in response to acoustic stimuli (Robertson 1984). The absence of acoustically evoked activity in nerve fibers coming from the OHC is consistent with the absence of evidence for vesicle membrane recycling at the OHC afferent synapse (Siegel and Brownell 1986). Those eighth nerve fibers that innervate OHCs may communicate information to the CNS but it is unlikely to be encoded in spike train activity.

The OHC synaptic pole is dominated by efferent synapses. Postsynaptic cisternae are found in the cytoplasm adjacent to the postsynaptic thickening of efferent synapses. Most of the efferent endings originate from neurons located in the contralateral medial olivary complex. The effect of stimulating the efferent fibers is to decrease the sensitivity of eighth nerve fibers. Since the fibers innervate the OHCs and not the IHCs, they perform their function indirectly through an effect on OHC force production. The synapse is cholinergic and cell shortening was observed when acetylcholine (ACh) was iontophoretically applied to isolated OHCs (Brownell 1983, 1984; Brownell et al. 1985). It was later found that ACh modulates the magnitude of the OHC's electrically evoked movments (Sziklai et al. 1996; Dallos et al. 1997; Kalinec et al. 2000). The novel heteromeric nicotinic receptors at the efferent synapse contain both $\alpha 9$- and $\alpha 10$-subunits (Fuchs and Parsons, Chapter 6), which may activate cell signaling pathways that alter mechanical properties of cytoskeletal proteins in the OHC lateral wall. A family of small GTPases regulates the polymerization and depolymerization of cytoskeletal proteins, and studies that have biochemically dissected the action of several of these (specifically RAC, RHO, and CDC42) suggest that signaling triggered by the nicotinic receptor at the efferent synapse may regulate the mechanical properties of the OHC lateral wall (Kalinec et al. 2000; Zhang et al. 2003).

2.3 *Lateral Wall*

The organization of the cylindrical portion of the OHC is unique among hair cells and possibly among all the approximately 200 mammalian cell types. The lateral wall is an elegant, nanoscale (approximately 100 nm thick), trilaminate structure (Figs. 7.3 and 7.4) that is self assembling and self repairing. The outer and inner layers are the PM and subsurface cisternae (SSC), respectively, and sandwiched between them is a layer of cytoskeletal proteins with preferential orientations called the cortical lattice (CL). These three layers form three axially concentric cylinders at the perimeter of the cell's axial core.

The apical end of all hair cells is trilaminate and is the likely precursor for the trilaminate organization in the OHC lateral wall. The outermost layer in both locations is the PM. The innermost layer is the canalicular reticulum in the apex, which is contiguous with the SSC in the lateral wall (Fig. 7.3). In between the membranes is a cytoskeletal structure, the CP in the apex and the CL in the lateral wall. The two trilaminate structures differ in thickness, 1000 nm for the apex and 100 nm for the lateral wall.

2.3.1 Lateral Wall Subsurface Cisterna

The PM and SSC are separated by an approximately 30 nm wide extracisternal space (ECiS). While the SSC structurally resembles the endoplasmic reticulum, it is stained with both endoplasmic reticulum and Golgi body markers (Pollice and Brownell 1993; Nguyen and Brownell 1998; Oghalai et al. 1998). It shows no evidence of belonging to the PM–endoplasmic reticulum–Golgi membrane pool (Siegel and Brownell 1986). The SSC may be more closely related to the canalicular reticulum (Fig. 7.3), a structure that occurs in ion-transporting epithelia (Spicer et al. 1998). The membranes of the canalicular reticulum are contiguous with those of the SSC as well as the outer membrane of mitochondria. The molecular composition of the SSC (its phospholipid composition and complement of integral membrane proteins) and the content of the narrow approximately 20-nm lumen lying between the inner and outer SSC membranes have not been identified. The SSC partitions the cytoplasm into two domains, the axial core and the ECiS. The axial core is the larger of the compartments and it contains only a few structures, such as most of the cell's complement of mitochondria, which are located adjacent to the lateral wall. This partitioning may help to maintain the OHC's cytoskeletal organization.

2.3.2 Lateral Wall Cortical Lattice

F-actin is present in stereocilia at the OHC's apex and near the synaptic structures at the base of the cell but poorly represented in the cell's axial core (Raphael et al. 1994; Kuhn and Vater 1995; Oghalai et al. 1998; Brownell and Oghalai 2000). The cytoskeletal proteins that maintain the OHC's cylindrical shape are found in the narrow confines of the ECiS where they make up the CL with its well-defined orthotropic organization (Figs. 7.3 and 7.4). Radially oriented pillars (of unknown molecular composition) tether the plasma membrane to parallel bands of F-actin that lie adjacent to the SSC. The PM is firmly attached to the pillars, as shown by the large pressure required to separate the PM from the OHC in micropipette aspiration experiments (Sit et al. 1997; Oghalai et al. 1998; Morimoto et al. 2002) and the force required to form membrane tethers using optical tweezers (Li et al. 2002). The tether formation force for OHCs is the largest value reported for any cell. The strong attachment is functionally important as the OHC's electromechanical force generation takes place in the PM and that force must be summed across the cell and ultimately be transmitted to the other structures in the organ of Corti. Electron microscopy reveals attachment between pillars and the underlying actin filaments but the nature of the attachment to the PM is unknown. Either the pillar itself has hydrophobic domains that interact with the membrane or it binds to some as yet to be defined membrane protein that serves as an anchor in the otherwise fluid membrane environment. The actin filaments are spaced between 40 and 50 nm apart and cross-linked with molecules of spectrin. The cortical lattice is organized in microdomains defined by regions of parallel actin filaments. Actin orientation varies between the microdomains but on average is circumferential. The regular orientation and separation of CL actin has recently been imaged in lightly fixed OHCs using atomic force microscopy (Le Grimellec et al. 2002; Wada et al. 2003).

2.3.3 Lateral Wall Plasma Membrane

The appearance of the lateral wall PM differs from that of SSC membranes under transmission electron microscopy. The PM is rippled or folded (Smith 1968; Ulfendahl and Slepecky 1988; Dieler et al. 1991) while SSC membranes retain the crisp bilayer appearance seen in many biological membranes. Atomic force microscopy also reveals the presence of PM ripples with dimensions similar to that of the separation of the underlying actin filaments in the CL (Le Grimellec et al. 2002). Figure 7.5 presents a hypothetical rendering of the PM suggesting how its anchorage to the CL actin filaments might help to organize ripples. Although rippling can be produced by fixation procedures, support for folding or microscopically hidden membrane comes from micropipette aspiration experiments that demonstrate excess membrane of approximately 10% of the apparent surface area of the OHC (Morimoto et al. 2002). This value is similar to the excess membrane in ripples as measured in transmission electron microscopic images of fixed material (Ulfendahl and Slepecky 1988). Additional evidence comes from the effect of curvature altering reagents on the diffusion of lipid probes in the membrane (Oghalai et al. 2000), membrane-associated charge (Kakehata and Santos-Sacchi 1996), lateral wall and cell mechanics (Hallworth 1997; Lue and Brownell 1999), and electromotility (Oghalai et al. 2000; Lue et al. 2001). Membrane folding is a common feature of many cell membranes (Raucher and Sheetz 1999), particularly those whose function requires shape changes.

The lipid composition of OHC membranes is unknown but the constituent lipids are in the fluid and not the gel phase, allowing for free diffusion (Oghalai et al. 1999, 2000). Fluorescent labeling studies suggest that lateral wall membranes contain less cholesterol than the apical and basal plasma membrane (Nguyen and Brownell 1998). In addition, the PM does not retain exogenously applied cholesterol (Oghalai et al. 1999; Brownell and Oghalai 2000). The lipid composition of the membrane can modulate the activity of membrane proteins. Just as low membrane cholesterol is required for rhodopsin to undergo a conformational change in response to light (Boesze-Battaglia and Albert 1990), it may be that low cholesterol is required for OHC electromotility.

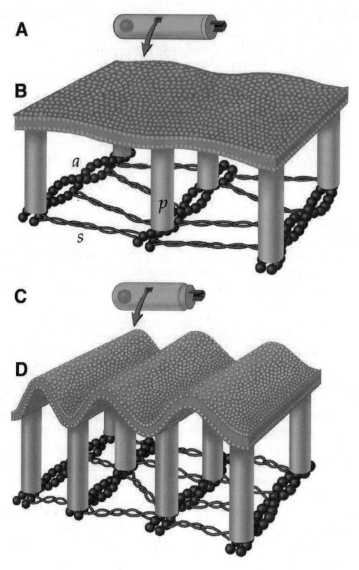

FIGURE 7.5. Illustration of the membrane bending model. The plasma membrane is tethered to a subplasmalemmal cytoskeleton by 30-nm-long "pillars" (p). The molecular composition of the pillars is unknown. The pillars bond to parallel actin filaments (a) that run circumferentially around the cell. The actin filaments are spaced about 40 nm apart and cross-linked with molecules of spectrin (s) that run longitudinally along the cell. (A, C) The OHC when hyperpolarized and depolarized, respectively. Depolarization makes the OHC shorter and wider. (B, D) Potential alterations in membrane curvature resulting from electromotile length changes. Note the increased membrane crenulations in (D). From Oghalai et al. (2000).

The different functional roles of the apex, lateral wall, and base of the OHC are associated with known differences in their integral membrane proteins. Immunohistochemical studies have shown the presence of a modified anion exchanger AE2 (Kalinec et al. 1997), a sugar transporter (Nakazawa et al. 1995; Géléoc et al. 1999; Belyantseva et al. 2000), and prestin (Belyantseva et al. 2000) in the lateral wall membranes. Stretch-activated ion channels have been demonstrated in the OHC lateral wall membrane (Ding et al. 1991; Iwasa et al. 1991; Rybalchenko and Santos-Sacchi 2003). It is not known if these channels are proteins or are lipid-based channels such as those observed in mitochondrial membranes (Siskind and Colombini 2000; Siskind et al. 2002, 2003) and predicted by molecular dynamic simulations of lipid bilayers (Marrink et al. 2001). Voltage-gated K^+ channels along the lateral wall appear to be concentrated at the base of the cell along with ACh receptor channels (Santos-Sacchi et al. 1997). Differences in protein composition between the apical and lateral wall membranes are expected because the tight junctions between the OHC and supporting cells would prevent lateral diffusion of membrane proteins. There is, however, no obvious barrier to lateral diffusion in the plasma membrane between the basal end and the lateral wall. The documented functional differences between the regions raise the possibility that the function of integral membrane proteins between the basal PM and the lateral wall PM is regulated by differences in lipid composition. It has also been proposed that integral membrane proteins prefer regions of the cell with a specific membrane curvature (Chou et al. 1997; Dan and Safran 1998; Kim et al. 1998; Ramaswamy et al. 2000). Changes in membrane curvature therefore represent a mechanism by which different functional domains may be achieved in contiguous membranes.

Freeze–fracture images of the OHC lateral wall PM reveal an orderly array of large (9–12 nm) closely packed particles in the cytoplasmic face of the bilayer (Gulley and Reese 1977; Forge 1991; Kalinec et al. 1992; Holley 1996) while the outer surface is reported as appearing flat. The particles resemble freeze fracture images of integral membrane proteins such as ion channels and it has been suggested that they represent a protein important to electromotility, possibly the motor protein itself. Alternatively, the particles may be lipidic and represent a phase transition when the membranes are frozen (the preparation by which they have been visualized). Some biologically important lipids, particularly those that support membrane curvature, can enter an inverted hexagonal phase and form lipidic particles that resemble the particles found in the OHC PM (see Mouritsen and Kinnunen 1996 for review). Lipidic particles have not been reported in biological membranes but lipids that contribute to these phase transitions, if present in OHC lateral wall membranes, would greatly alter the membrane properties. An understanding of how membrane mechanics are related to the transmembrane potential necessarily involves determination of both the lipid and protein composition of OHC membranes.

2.3.4 Lateral Wall and the Hydraulic Skeleton

A hydrostat is a mechanical structure formed from an elastic outer shell enclosing a pressurized core (Wainwright 1988). The OHC is a cellular hydrostat (Fig. 7.6) with a modest cytoplasmic turgor (approximately 1 to 2 kPa) (Brownell et al. 1985; Ratnanather et al. 1993). Other mammalian cells have no pressure while plants and bacteria are more pressurized, with trees having as much as 5 atm of internal pressure. The cytosolic turgor pressure and the tensile properties of the lateral wall together form an hydraulic skeleton that facilitates shape changes and hydraulic force transmission at high frequencies (Brownell 1990; Brownell et al. 2001). OHC electromotility diminishes and vanishes if the cell becomes flaccid (Brownell 1983, 1990; Brownell et al. 1985; Santos-Sacchi 1991; Shehata et al. 1991).

The requirement for intracellular turgor pressure places constraints on the OHC plasma membrane that are not shared by the other cells of the body. Most eukaryotic cells burst when their internal pressure is increased by even a small amount. The reinforcement provided by the trilaminate lateral wall and more specifically the cortical lattice (Oghalai et al. 1998) prevents this happening in the OHC. The maintenance of high turgor by plant and bacterial cell walls is associated with low water permeability through their membranes. It would be energetically costly if water could easily pass through their membranes; imagine trying to keep a leaky balloon full of air. The OHC plasma membrane shares the low water permeability of plant and bacterial cells (Chertoff and Brownell 1994; Ratnanather et al. 1996). The water permeability is increased when the

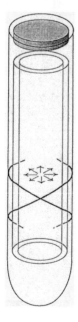

FIGURE 7.6. Schematic of OHC as a pressure vessel. The mechanical properties of cortical lattice actin filaments is represented by oppositely oriented helices. The cell is inflated by the turgor pressure of the cytoplasm indicated by the radially oriented arrows. From Brownell (1990).

membrane is mechanically deformed either with fluid jets or through contact with micropipettes. When test solutions are applied to OHCs by pressure injection through pipettes with tip diameters of between 1 and 5 μm, the membrane is deformed by the jet and the associated turbulent flow. The strain-dependent water permeability is more than an order of magnitude greater than that measured with equally rapid fluid exchange under laminar flow conditions (Brownell 1990; Brownell et al. 1994; Chertoff and Brownell 1994; Ratnanather et al. 1996; Belyantseva, 2000; Morimoto et al. 2000, 2002). The strain-dependent increase in water permeability may be attributed to mechanosensitive ion channels in the plasma membrane (Ding et al. 1991; Iwasa et al. 1991; Rybalchenko and Santos-Sacchi 2003) or to an increase of spontaneously formed meta-stable (10 to 100 ns) water defects (Marrink et al. 2001) in the plasma membrane lipid bilayer.

The micropipette aspiration technique, originally developed to study red blood cell (RBC) mechanics, has been applied to the OHC (Sit et al. 1997). The pressure required to deform the OHC lateral wall is an order of magnitude greater that that required for the RBC and is reduced by the anionic amphipath salicylate (Lue and Brownell 1999). The technique may also be used to dissect the lateral wall and observe its individual layers. Oghalai et al. (1998) selectively labeled and imaged each of the lateral wall components under confocal microscopy and evaluated the interactions among the three layers. The PM was found to be tethered to the CL and SSC until, at a critical deformation pressure, the PM separated, allowing visualization of the extracisternal space. Ultimately the PM pinched off and formed a vesicle. After detaching, the stiffness parameter of the PM (length of the aspirated tongue per unit pressure) was approximately 20% of that of the intact lateral wall. These results suggest that most of lateral wall's resistance to local bending derives from the underlying CL–SSC complex.

2.3.5 Lateral Wall and the Cell Wall of Gram-Negative Bacteria

OHCs and gliding bacteria share a surprising number of structural motifs. The mechanism underlying this type A gliding motility (like OHC electromotility) is enigmatic and a variety of models have been proposed (Burchard 1981; Siddiqui et al. 2001). Many bacteria are able to "glide" along surfaces without the aid of flagella or pseudopodia (Burchard 1981; Pate 1988; McBride 2000). Many of these are rod shaped having diameters of about 500 nm and lengths < 1000 nm (wider and shorter than most stereocilia). The cells move parallel to their long axis at speeds that can reach 50 μm/s (Hoiczyk 2000) while at the same time revolving around their long axis (Burchard 1981; Hoiczyk and Baumeister 1995). There is no apparent bending, rippling, or change in cell length under the light microscope. Microbeads attached to a gliding bacterium move along the cell surface parallel to the cell's long axis (Lapidus and Berg 1982) at speeds comparable to their locomotion.

Even though gliding bacteria are well over an order of magnitude smaller than OHCs there are a number of striking similarities (Brownell 2002). While plant and bacterial cells have pressurized fluid cores, the OHC is the only vertebrate cell known to share this feature. Bacteria, like all prokaryotes, do not have the cytoskeletal proteins of eukaryotes. The poorly developed central cytoskeleton of the OHC axial core is therefore another similarity. The elastic outer wall of both the OHC and bacteria is made of three layers. The most peripheral membrane in the bacterial cell wall is called the outer membrane and it establishes a permeability barrier in which transport molecules are located. Its innermost membrane is called the cytoplasmic membrane. The cytoplasmic membrane contains proton pumps and other integral membrane proteins crucial to bacterial energy utilization. In between the membranes is found the peptidoglycan layer serving a cytoskeletal function similar to that of the OHC's cortical lattice. Cholesterol is found only in eukaryotic membranes. The low concentration of cholesterol in mitochondrial membranes reflects their prokaryotic origins. The inferred low concentration of cholesterol in the OHC PM and SSC (Forge 1991; Nguyen and Brownell 1998) is yet another similarity between the OHC lateral wall and the bacterial cell wall. These similarities may contribute to the fact that both gram-negative bacteria and OHCs are particularly vulnerable to aminoglycosides. Rapid freezing reveals paracrystalline arrays of particles in the outer membrane of type A gliding bacteria (Hoiczyk and Baumeister 1995). Transmission electron microscopy reveals ripples in their outer membrane (Dickson et al. 1980; Adams et al. 1999) that, like those of the OHC, appear to flatten during freezing. The ripples of type A gliding bacteria spiral around the cell and the direction of the spiral is in the same direction as the rotation of the cell as it glides (Halfen and Castenholz 1970; Hoiczyk and Baumeister 1995). Membrane potential rather than ATP drives both type A gliding motility (Pate 1988) and OHC electromotility. Lanthanides block gliding motility (Pitta et al. 1997) and OHC electromotility (Kakehata and Santos-Sacchi 1996). Given the surprising number of structural and functional similarities it is possible that gliding bacteria and OHC share similar mechanisms for converting electrochemical gradients to mechanical force. The possibility that the OHC mechanism may be based, in part, on a motile mechanism found in some of the earliest life forms on the planet needs to be considered when thinking about its molecular basis.

3. Prestin Is Required for OHC Electromotility

Perhaps the most exciting single discovery in hair cell research during the past five years has been the isolation and cloning of a membrane protein that plays a central role in OHC electromotility (Zheng et al. 2000). The normal electromotile response of standard epithelial cell lines (Mosbacher et al. 1998; Zhang et al. 2001) is enhanced and strain dependent displacement currents (nonlinear capacitance) appear on transfection with prestin (Zheng et al. 2000; Ludwig et al. 2001). Prestin has provided a new tool for the dissection of OHC and cochlear function. Physiological measures of hearing in prestin knockout mice provide additional evidence for the OHC's involvement in the mammalian cochlear amplifier (Liberman et al. 2002). Mutations of Prestin have been associated with nonsyndromic hearing loss (Liu et al. 2003). In addition to its presence in OHCs, prestin has been found in vestibular hair cells (Adler et al. 2003) and prestin orthologues have been identified in the ears of fish and insects (Weber et al. 2003). Prestin's appearance and level of expression in developing OHCs matches that of the appearance and maturation of OHC electromotility (Belyantseva et al. 2000).

Prestin is a member of the Slc26A family of sulfate transporter proteins which in turn is a member of the anion transporter superfamily. The other Slc26A membrane proteins facilitate the movement of anionic substrates such as chloride, iodide, bicarbonate, oxalate, and hydroxyl across membranes. There are currently 10 members of the Slc26A family (Vincourt et al. 2003). Their amino acid sequences reveal between 11 and 13 putative transmembrane domains and the second of these hydrophobic regions contains the highly conserved sulfate transporter motif (Markovich 2001). Prestin is likely to have an even number of membrane spanning helices because both its N- and C-terminals are cytoplasmic (Ludwig et al. 2001; Zheng et al. 2001). These membrane proteins passively transport anions by allowing them to move down their electrochemical gradients with different specificity. The absence of ATP binding cassettes is consistent with the fact that Slc26A family members do not actively pump anions. Prestin is unique in the group because it is not associated with conventional anion transport capabilities. However, reports in OHCs of nonselective stretch-activated transport (Rybalchenko and Santos-Sacchi 2003) and possibly stretch-activated sugar transport (Chambard and Ashmore 2003) may reflect prestin's membership in the anion transporter family. The family member with the closest sequence similarity is pendrin. Pendrin is expressed in the tissue forming the membranous labyrinth where it may be involved in the maintenance of the unique endolymphatic fluids (Everett et al. 1997, 1999, 2001). We have no knowledge of the three-dimensional organization of any member of the Slc26A family (including prestin) and as a result we have no idea of how they function. The absence of structural information reflects the difficulty in crystallizing membrane proteins (only a handfull has been crystallized). The absence of definitive information about prestin's structure and the nature of its interaction with the PM and CL has provided fertile ground for speculation on how it alters electromechanical coupling in the OHC PM.

4. Small Intracellular Anions Are Also Required for OHC Electromotility

Prestin transfection studies have revealed that electromotility in OHCs and nonlinear charge movement in prestin-transfected cells both vary with the concentration and species of small intracellular monovalent anions (Oliver et al. 2001;

Dallos and Fakler 2002; Fakler and Oliver 2003). Early studies showed that changes in extracellular and intracellular cation concentrations have no immediate effect on the magnitude and direction of electrically evoked OHC length changes (Brownell et al. 1985; Ashmore 1987; Santos-Sacchi and Dilger 1988). Santos-Sacchi and Dilger (1988) used ion channel blockers together with alteration of transmembrane cation gradients to demonstrate that the electromotile response faithfully followed the applied voltage even under conditions where the direction of the cationic current was reversed. The concept that OHC electromotilty is not a function of cationic currents is consistent with the absence of voltage-gated cation channels in the lateral wall PM (Santos-Sacchi et al. 1997). It required transfection studies to reveal that physiological concentrations of intracellular chloride or bicarbonate are also required for electromotility (Fig. 7.2a) and its associated nonlinear capacitance (Fig. 7.2c). The nonlinear capacitance is a reliable "electrical signature" of electromotility both in OHCs and prestin transfected test cells (Snyder et al. 2003). The voltage dependences for the nonlinear capacitance in transfected cells resembles the nonlinear capacitance in OHCs, including a dependence on cytoplasmic pressure (Santos-Sacchi et al. 2001). In the OHC the magnitude of the nonlinear capacitance is about the same as the membrane capacitance (this "linear" capacitance is proportional to the cell's surface area) but in prestin-transfected cells the nonlinear capacitance is variable and usually smaller than the cell's membrane capacitance. The dependence on small univalent anions was revealed when a robust nonlinear capacitance recorded with 150 mM $[Cl^-]_{cytoplasmic}$ reversibly vanished into the noise floor in prestin-transfected cells with 2 mM $[Cl^-]_{cytoplasmic}$ (Oliver et al. 2001). Studies using inside-out patches from OHC lateral wall revealed that anions other than chloride support nonlinear capacitance but with different peak magnitudes (Oliver et al. 2001) and V_{pkCm} (Fakler and Oliver 2003). The peak magnitude for 150 mM concentration of I^- is approximately three times that of F^-, with other anions intermediate: $I^- \approx B^- > NO_3^- \approx SCN^- > Cl^- > HCO_3^- > F^-$. The V_{pkCm} value for I^- is shifted to -90 mV and value for F^- is shifted $+30$ mV respectively relative to the Cl^- value. The absolute V_{pkCm} values follow an order similar to that of the peak magnitude ($I^- > B^- \approx NO_3^- > SCN^- > Cl^- \approx HCO_3^- > F^-$). The peak magnitude and V_{pkCm} also depend on the intracellular concentration of the anion. The concentration dependence functions for the peak magnitude of Cl^- and HCO_3^- have a Hill coefficient of approximately 0.9, revealing an absence of cooperativity and suggesting that binding of the anion does not involve a structure with multiple subunits. As the peak magnitude decreases with decreases in $[Cl^-]_{cytoplasmic}$ the V_{pkCm} becomes progressively more depolarized. Other anionic amphipaths alter the nonlinear capacitance and specifically, salicylate was demonstrated to compete with chloride.

The requirement for small anions has a profound impact on our understanding of how prestin interacts with the membrane. If prestin was both the motor and the voltage sensor for the electromotility, then electromotility and its associated charge movement should continue in the absence of intracellular anions. The fact that it does not means that prestin is not the voltage sensor. It has been suggested that chloride acts as an "extrinsic voltage sensor," binding to an anion-selective site on prestin, which then moves into and out of the membrane in response to changes in the transmembrane potential (Oliver et al. 2001; Dallos and Fakler 2002). The fact that prestin is a member of the Slc26A family of anion transporters is a strong argument for such a site, even though no such binding site has been identified for any member of the family.

What are the requirements for such a binding site? The anion transporters in the Slc26A family allow specific anions to cross the membrane but only by moving down their electrochemical gradients. They do not have the machinery (e.g., ATP binding cassettes) to utilize cellular stores of energy and cannot move the ion against a gradient. In this respect, they function like chloride channels. The molecular structure of a chloride channel has been recently characterized and it possesses weak binding sites for the chloride ion (Dutzler et al. 2002, 2003). A strong binding would defeat the purpose of the channel as it would greatly slow the movement of Cl^-. Weak binding is also required for prestin as any restriction for the anion to either enter the membrane or return to the cytoplasm would reduce the charge movement that characterizes electromotility. The weak binding required for easy ion movement is at odds with the implicit requirement that the binding be strong enough that a change in membrane potential leads to a conformational change of the protein on which the binding site is located. If the binding is too weak, the protein will not move, if it is too strong, the charge movement will be restricted. For charge movement and a change in molecular conformation to occur, the binding energy would have to be greater than the force exerted by the electric field. An analysis of the kinetics of such a two-step process (binding followed by conformational change) is required to determine if it is compatible with the short time constants of electromotility. It is also difficult to explain the concentration dependent change in V_{pkCm} by means of an interaction with a binding site on prestin (Fakler and Oliver 2003). An alternative possibility is that anion interactions with phospholipid head groups on the cytoplasmic face of the plasma membrane contribute to the change in electromotility and nonlinear capacitance.

Inorganic and organic ions affect several membrane based physiological processes such as muscle twitch tension, the gating of ion channels, and the effectiveness of ion pumps (for review see Baldwin 1996; Cacace et al. 1997; Clarke and Lupfert 1999). The effectiveness of different ions follows what is called a Hofmeister series after Franz Hofmeister who, in the late nineteenth century, examined the ability of neutral salts to precipitate proteins. Hofmeister series (also called lyotropic or chaotropic series) are listings of anions or cations in order of their effect on a specific process. Anions are typically more effective than cations and the halides almost invariably follow a sequence where $I^- > B^- > Cl^- > F^-$ as was observed for the nonlinear capacitance of excised patches from the OHC (Oliver et al. 2001). While the physical basis for the difference in effectiveness is not known a popular hypothesis is that the series reflect a competition between a substrate and the ion for water. A study on unilamellar vesicles reveals that the native electric field of the pure lipid

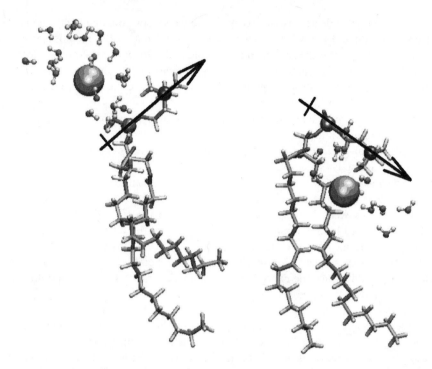

FIGURE 7.7. Molecular dynamic simulation snapshots of changes in phospholipid headgroup orientation associated with anionic penetration into the bilayer interface. The anion is the largest ball. The simulation did not include a transmembrane potential. Deep penetration occurs only in large anion simulations. Only those waters within the first minimum of the bulk anion are shown. Note the folding of the headgroup over the anion on the right. There is a change in its electrical dipole orientation. The folding may affect the apparent area of the phospholipids. The addition of an electrical field would strengthen the response and the apparent charge movement associated with the change in dipole orientation would supplement the actual entry of cytoplasmic anions into the membrane on hyperpolarization, consistent with the measured anion movement associated with OHC electromotility. The differential areal expansion of the inner and outer leaflets would contribute to changes in membrane curvature. From Sachs and Woolf (2003).

membrane (a bilayer formed by mixing a single phospholipid with water) is altered by anions following a lyotropic series (Clarke and Lupfert 1999) similar to that reported for the effectiveness of the same anions on nonlinear capacitance of the OHC lateral wall PM. The vesicles were composed of dimyristoylphophatidylcholine and electric fields were measured using voltage-sensitive dyes. The authors note that the order of effectiveness in reducing the membrane's electric field followed the order predicted from the anion's free energy of hydration. Those with the highest values were least effective, while those with the lowest value were more effective. They interpreted this as a measure of how far an anion species might enter the membrane. Anion penetration into lipid bilayers is consistent with recent molecular dynamic studies (Pandit et al. 2003; Sachs and Woolf 2003) which have investigated the interaction between ions and the head group region of model membranes (Fig. 7.7). Molecular dynamic simulations, which consider the interaction between molecules on an atom by atom basis, are often used to calculate the possible three-dimensional structure of proteins by determining how they might fold to achieve a minimum energy profile. Early molecular dynamic simulations of membranes demonstrated the self-assembly of the bilayer from a mixture of phospholipids and water and show the dynamic behavior of the bilayer. The bilayer undergoes spontaneous undulations in thickness while individual phosolipids can move in and out of the membrane by as much as half their length (Lindahl and Edholm 2000). The ability of anions to move in and out of membranes may have an as yet unexplored relation to the phospholipid composition. The factors that regulate the phospholipid profile include integral membrane proteins (such as prestin).

There is no contradiction between the early observations that electromotility is independent of ion flow and the need for intracellular anions for the expression of nonlinear capacitance. Ionic currents differ from displacement (or capacitive) currents in that the former results from charge passing through the membrane while there is no net charge movement through the membrane in the later. A simple capacitive current involves the movement of charge onto the plates of the capacitor with no charge passing through the insulator that separates the plates.

As discussed next, the nature of the membrane capacitance challenges the functional significance of OHC electromotility at high frequencies.

5. The RC Paradox and OHC Piezoelectricity

The electrical behavior of biological membranes has long been modeled as a resistor in parallel with a capacitor. The hydrophobic nature of an intact membrane does not allow strong ionic currents so that the membrane resistance is high (unless ion channels are present and open). Capacitors are electrical devices that are used to store energy in electrical circuits. They are typically fabricated from a material with strong insulating abilities (high resistance) that is sandwiched between two conductors. The membrane is a good insulator and it is surrounded on both sides by conductive salt filled solutions (the extracellular and intracellular fluids). The energy storage ability of a capacitor is measured in terms of the amount of charge that the capacitor can hold at a given voltage difference between the two conductors. The storage may be viewed as a conversion of the kinetic energy of charge flow (current) into a potential energy across the capacitor. When an alternating electric current is applied the charge is stored and discharged. In normal membranes the amount of stored charge is proportional to the voltage across the membrane as long as the field is changing with a time course that is longer than that required to charge the capacitor. The time required to charge the membrane capacitance to approximately 63% of its steady state value is determined by the product of the resistance (R) and capacitance (C) of the membrane and is known as the RC time constant. As a result, there is a time delay between current (the movement of charge) and potential. As the rate of charge movement increases the voltage decreases when the rate of change exceeds the time required to store the charge. The ions continue to be influenced by the electric field but they are no longer under the influence of the membrane (which introduces a delay as energy is exchanged between the potential field and charge storage). The peak potential difference experienced by the membrane drops off in what has been called a capacitive short circuit and, although the ions move, the membrane does not experience high frequency potential differences.

The analysis of a simple equivalent electric circuit of the OHC with their resistive and capacitive (RC) properties predicts a severe attenuation of the receptor potential (Housley and Ashmore 1992; Santos-Sacchi 1992) that would preclude the production of a physiologically significant electromotile force. There have been several proposals to resolve this "RC" paradox. It has been suggested that OHC electromotility is driven not by the OHC receptor potentials but rather by an extracellular electric field in the organ of Corti (Dallos and Evans 1995a,b) or that increases in the number of motor elements might offset decreases in the receptor potential (Santos-Sacchi et al. 1998). The possibility that multistate voltage-gated channels might increase the high-frequency response of the OHC has also been proposed (Ospeck et al. 2003). Alternatively extracellular chloride may reach an active site in the PM via nonspecific stretch activated channels, turning the mechanism into one that is current driven rather than a receptor potential driven (Rybalchenko and Santos-Sacchi 2003).

An alternative possibility is that piezoelectricity of the OHC enhances voltage changes at high frequencies. Piezoelectricity is a tight coupling between the electrical polarization and deformation of a material. It is inherently bidirectional in that deformation is associated with the production of an electric potential and the application of an electric potential deforms the material. Almost all materials are piezoelectric with some having piezoelectric coefficients of sufficient magnitude to be of use for devices like submarine sonar or clinical ultrasound. Jacques and Pierre Curie discovered piezoelectricity in the late nineteenth century. They found that the application of mechanical stress to a variety of crystalline materials produced electricity and coined the term piezoelectricity (from the Greek, *piezein*—to press). Shortly thereafter they found that the same materials showed a converse piezoelectric effect wherein the crystals changed shape in response to an applied electric field. Crystalline piezoelectric materials have been used in applications that range from marine sonar to mechanical actuators to electronic delay lines. Noncrystalline, biological piezoelectricity has been observed in bone (Fukada and Yasuda 1957) and ligament (Korostoff 1977; Fukada 1982).

OHC electromotility is a direct conversion of membrane potential to cell length change and is functionally equivalent to the converse piezoelectric effect. The OHC lateral wall also supports a mechanoelectrical conversion (the direct piezoelectric effect) (Gale and Ashmore 1994; Zhao and Santos-Sacchi 1999), which strengthens the view that the lateral wall is piezoelectric. Several theoretical treatments of the OHC and its role in hearing have assumed the OHC to be piezoelectric (Mountain and Hubbard 1994; Tolomeo and Steele 1995; Raphael et al. 2000a; Iwasa 2001; Spector et al. 2003a,b; Weitzel et al. 2003). The models all infer or assume Maxwell reciprocity (Cady 1946) in which the coefficients for the direct and converse effects are equal. There is experimental evidence providing quantitative support that Maxwell reciprocity is found in OHCs (Dong et al. 2002). Figure 7.8 shows the effect of including a piezoelectric component in the equivalent circuit of an OHC. Piezoelectricity not only pushes the low-pass corner frequency to higher values but also predicts ultrasonic resonances in the OHC lateral wall admittance (Spector et al. 2003a,b; Weitzel et al. 2003).

The extended frequency range predicted in the models is consistent with the low-pass corner frequencies measured for mechanically unloaded OHCs when stimulated in the microchamber configuration (Dallos and Evans 1995a; Frank et al. 1999). Even though a mechanical resonant peak has not been observed in such isolated OHCs, Frank et al. (1999) attributed the electrically induced displacement response, with its elevated cutoff frequency, to the presence of a resonance because only an overdamped second-order resonant system, in which the resonant frequency was found to depend on cell length, fit their data.

The OHC is mechanically loaded in the fluid filled organ of Corti. It is firmly anchored to supporting cells by tight junctions in the reticular lamina and the tectorial and basilar membranes introduce additional mechanical loads. The in vivo OHC is therefore a viscous-damped mechanically loaded resonator in contrast to the unloaded or free resonator so far considered in the models. Radiation damping of the resonance would be expected to result in a broadening and weakening of the impedance peak as a function of frequency. The observation of a broad, ultrasonic mechanical resonance in electrically evoked basilar membrane vibrations near the round window of the guinea pig inner ear (Grosh et al. 2004) is consistent with this expectation. The OHCs near the round window are all short and close to the same length. By inference they should have similar resonance behavior. Electrical stimulation around this frequency would result in a summed response that would be broader than that expected in an unloaded OHC which is what was modeled in Figure 7.8.

OHCs appeared with mammals more than 200 million years ago. The OHCs of the small rodent-like early mammals enabled them to hear frequencies above 50 kHz. They were the basis of a cochlear amplifier that counteracted viscous damping and refined the passive mechanical filtering of the organ of Corti (Brownell et al. 2001; Snyder et al. 2003). OHC piezoelectricity allows the high-frequency transmembrane potentials required for electromotility to act as the cochlear amplifier throughout the entire hearing range of mammals. In most

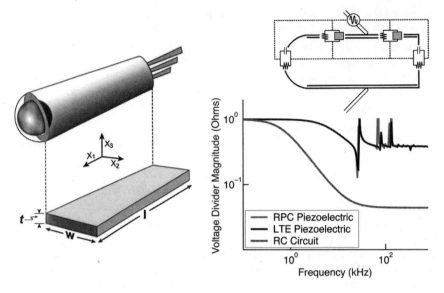

FIGURE 7.8. Model of OHC piezoelectricity. Schematic of OHC lateral wall is shown on the left. The lateral wall is represented by a radial-poled cylinder (RPC) resonator on the *top* which is cut longitudinally to form a length thickness extension (LTE) resonator on the *bottom*. Expressions for the specific electrical impedances were derived and inserted into an equivalent circuit analysis of an OHC in a microchamber as shown at the upper right. The gray box impedances in the lateral wall represent either a piezoelectric element or pure capacitance (for conventional RC analysis). The results of the RC and RPC and LTE piezoelectric representations are shown in the plot on the bottom right. The piezoelectric models reveal strong enhancement of the response at high frequencies and ultrasonic resonances. Modified from Weitzel et al. (2003).

cases somatic resonances are not required to explain the OHC contribution to mammalian hearing. It is of interest that the predicted resonance frequencies are generally beyond the frequency range that OHCs at a specific location on the basilar membrane would operate. In addition, the impedance mismatch between the OHCs and the organ of Corti is large enough (Allen and Neely 1992) that any resonance away from the best frequency would be expected to remain confined within the cell and have little impact on the vibrations of the organ of Corti.

The first bats appeared more than 60 million years ago with an elegant echolocation mechanism that allowed them to navigate in a nocturnal environment. An additional resonance has been invoked to explain their enhanced hearing sensitivity at the echolocation frequencies (Kossl and Russell 1995; Russell and Kossl 1999). Bats may have co-opted OHC piezoelectric resonance to achieve this. Some 10 million years later cetaceans adopted a remarkably similar specialization (Ketten 1997). The organ of Corti in echolocating members of the orders Chiroptera and Cetacea is greatly expanded at the region devoted to the echolocation frequencies. The OHCs in the region are short and of uniform length so that the reticular lamina is parallel to the basilar membrane. This organization would allow OHC resonance to be summed, overcoming the impedance mismatch that normally obscures the resonance. While echolocating mammals may be the only ones to utilize OHC resonances, the piezoelectric mechanism that underlies those resonances helps to counteract the RC paradox and enhance high-frequency receptor potentials in OHCs throughout the entire class Mammalia.

6. Final Considerations: Molecular Mechanisms

In the enthusiasm following its discovery, prestin has been repeatedly identified as the motor molecule responsible for electromotility. A motor is a device that converts one form of energy into another. The most common motor is the internal combustion engine. Prestin is more like a piston than a motor. Just as a car's motor will not work without pistons so too the OHC will not work without prestin (recall the prestin knockout experiments). Prestin does not operate in isolation. It requires the membrane with its transmembrane electric field in the megavolt/meter range to support electromotility (the converse piezoelectric effect). Anions are also required, not just as part of the charge movement but intimately involved in the production of mechanical force. Prestin may be viewed as increasing the piezoelectric coefficients associated with electromotility in native cells. The most dramatic increases following prestin transfection are in the magnitude of the nonlinear capacitance (charge movement resulting from OHC piezoelectricity) while the converse piezoelectric effect (electomotility) increases by less than an order of magnitude (Ludwig et al. 2001). The underlying molecular mechanism must ultimately explain OHC piezoelectricity.

The mechanism(s) by which stereocilia generate force is thought to be based on bundle motor units associated with mechanoelectrical transduction channels. The bundle's net force is directly proportional to the number of stereocilia. The characteristics of the OHC electromotility may be compared with the emerging characteristics of the bundle mechanism. The motor mechanism underlying OHC electromotility generates forces with gains of approximately 50 pN/mV at frequencies approaching 100 kHz. Two plausible molecular mechanisms for OHC electromotility have been proposed based on an electromechanical transduction process in the PM. One is driven by in-plane area changes of the membrane (Kalinec et al. 1992; Dallos et al. 1993; Iwasa 1993, 1994; Huang and Santos-Sacchi 1994; Iwasa and Adachi 1997) and the other by out-of-plane flexoelectric bending (Raphael et al. 2000b) portrayed in Fig. 7.5. The area motor model is thought to possess an associated charge density that varies between 5000 and 50,000 elements/μm^2 depending on the length of the cell (Santos-Sacchi et al. 1998), or approximately 15 to 30 million elements per cell. In the membrane-bending, flexoelectric model, the motor unit is a longitudinally oriented ripple with a width of approximately 40 nm or 25,000/μm. A 50-μm long cell will have > 1 million motor units. The number of mechanoelectrical transduction channels determines the number of bundle motor units so that the bundle's net force is proportional to the number of stereocilia ($\leq$ 300). The number of motor units in the lateral wall is between 10^3 and 10^4 times the number of stereocilia bundle motor units. The increased number of lateral wall motor units makes the relatively large OHC electromotility force production possible. A cochlear amplifier based on stereocilia appears to have reached a limit at around 10 kHz (the upper limit of hearing of most birds). Increasing the frequency range for hearing another order of magnitude requires the negative damping force of the mammalian cochlear amplifier to increase by the same amount. The apical pole cannot support 10^4 stereocilia.

Biological membranes are electrically polarized because electrical charge is not evenly distributed within them and charge separation results in electrical polarization. Membranes have an intrinsic polarization associated with charge differences between the inner and outer leaflet. There are charged moieties in phospholipid head groups as well as in those portions of integral membrane proteins outside their membrane-spanning domains. Even in the presence of equal numbers of positive and negative charges, electrical dipoles will be present. The deformation of a membrane alters membrane polarization when the "fixed" charges move and/or the orientation of the electrical dipoles change. In addition to the intrinsic polarization associated with membrane-bound charge there is an extrinsic contribution resulting from transmembrane electric fields that lead to differences in the charge distribution based on membrane permeability and electrochemical gradients. Most biological membranes have a negative charge associated with their surface and it attracts positively charged anions, forming what is known as an electrical double layer. The concentration of the counterions in the double layer can be substantially greater than their concentration in the bulk solution only a few nanometers away. The deformation of the membrane leads to changes in its polarization (Gale and Ashmore 1994; Zhao and Santos-Sacchi 1999; Dong et al. 2002). The flexoelectric mechanism associated with the out-of-plane bending of the membrane model has a converse flexoelectric effect. The flexoelectric and its converse effects satisfy Maxwell reciprocity both theoretically and experimentally (Petrov 1999). A two-state area motor model can also approach the requisite relationship between deformation and polarization (Iwasa 2001). The determination of whether the mechanism is one of these, a combination of both or entirely different awaits clarification as to the structure and function of prestin and its interaction with the membrane and anions. The molecular dynamics simulation of anions interacting with the membrane in Figure 7.7 reveals a tantalizing glimpse of how anions might be involved in both the change in membrane polarization and the differential change in bilayer leaflet area that can contribute to membrane bending. Notice that as the anion penetrates the phospholipid headgroup the resulting change in the electrical dipole orientation produces an apparent charge movement in the same direction as the anion's. There is also a change in the molecule's surface area. The summation of area change across all the phospholipids in the inner leaflet leads to differential area change between the inner and outer leaflet and results in membrane bending contributing to the converse flexoelectric effect.

Acknowledgments. Supported by research Grants RO1 DC00354 and RO1 DC 02775 from the National Institute on Deafness and Other Communicative Disorders and by the Jake and Nina Kamin Chair of Otorhinolaryngology. The concepts developed in this review benefited from the generous contribution of many colleagues.

References

Adams DG, Ashworth D, Nelmes B (1999) Fibrillar array in the cell wall of a gliding filamentous cyanobacterium. J Bacteriol 181:884–892.

Adler HJ, Belyantseva IA, Merritt RC, Frolenkov GI, Dougherty GW, Kachar B (2003) Expression of prestin, a membrane motor protein, in the mammalian auditory and vestibular periphery. Hear Res 184:27–40.

Allen JB, Neely ST (1992) Micromechanical models of the cochlea. Physics Today 45: 40–47.

Ashmore JF (1987) A fast motile response in guinea-pig outer hair cells: the cellular basis of the cochlear amplifier. J Physiol (Lond) 388:323–347.

Baldwin RL (1996) How Hofmeister ion interactions affect protein stability. Biophys J 71:2056–2063.

Belyantseva IA, Adler HJ, Curi R, Frolenkov GI, Kachar B (2000) Expression and localization of prestin and the sugar transporter GLUT-5 during development of electromotility in cochlear outer hair cells. J Neurosci 20:RC116.

Boesze-Battaglia K, Albert AD (1990) Cholesterol modulation of photoreceptor function in bovine retinal rod outer segments. J Biol Chem 265:20727–20730.

Brownell WE (1982) Cochlear transduction: an integrative model and review. Hear Res 6:335–360.

Brownell WE (1983) Observations on a motile response in isolated outer hair cells. In: Webster WR, Aitken LM (eds), Mechanisms of Hearing. Monash: Monash University Press, pp. 5–10.

Brownell WE (1984) Microscopic observation of cochlear hair cell motility. Scan Microsc Pt 3:1401–1406.

Brownell WE (1990) Outer hair cell electromotility and otoacoustic emissions. Ear Hear 11:82–92.

Brownell WE (1999) How the ear works—nature's solutions for listening. Volta Rev 99: 9–28.

Brownell WE (2002) On the origins of outer hair cell electromotility. In: Berlin CI, Hood LJ, Ricci AJ (eds), Hair Cell Micromechanics and Otoacoustic Emissions. Clifton Park, NY: Thomson Delmar Learning, pp. 25–46.

Brownell WE, Oghalai JS (2000) Structural basis of outer hair cell motility or where's the motor? In: Lim D (ed), Cell and Molecular Biology of the Ear. New York: Plenum Press, pp. 69–83.

Brownell WE, Popel AS (1998) Electrical and mechanical anatomy of the outer hair cell. In: Palmer AR, Rees A, Summerfield AQ, Meddis R (eds), Psychophysical and Physiological Advances in Hearing. London: Whurr, pp. 89–96.

Brownell WE, Bader CR, Bertrand D, de Ribaupierre Y (1985) Evoked mechanical responses of isolated cochlear outer hair cells. Science 227:194–196.

Brownell WE, Ratnanather JT, Popel AS, Zhi M, Sit PS (1994) Labyrinthine lateral walls: cochlear outer hair cell permeability and mechanics. In: Flock A, Ottoson D, Ulfendahl M (eds), Active Hearing. Amsterdam: Elsevier, pp. 167–179.

Brownell WE, Spector AA, Raphael RM, Popel AS (2001) Micro- and nanomechanics of the cochlear outer hair cell. Annu Rev Biomed Eng 3:169–94.

Burchard RP (1981) Gliding motility of prokaryotes: ultrastructure, physiology, and genetics. Annu Rev Microbiol 35:497–529.

Cacace MG, Landau EM, Ramsden JJ (1997) The Hofmeister series: salt and solvent effects on interfacial phenomena. Q Rev Biophys 30:241–277.

Cady WG (1946) Piezoelectricity. New York: McGraw-Hill.

Chambard JM, Ashmore JF (2003) Sugar transport by mammalian members of the SLC26 superfamily of anion-bicarbonate exchangers. J Physiol 550:667–677.

Chertoff ME, Brownell WE (1994) Characterization of cochlear outer hair cell turgor. Am J Physiol 266:C467–C479.

Chou T, Jaric MV, Siggia ED (1997) Electrostatics of lipid bilayer bending. Biophys J 72:2042–2055.

Clarke RJ, Lupfert C (1999) Influence of anions and cations on the dipole potential of phosphatidylcholine vesicles: a basis for the Hofmeister effect. Biophys J 76:2614–2624.

Dallos P (1992) The active cochlea. J Neurosci 12:4575–4585.

Dallos P (1997) Outer hair cells: the inside story. Ann Otol Rhinol Laryngol Suppl 168: 16–22.

Dallos P, Evans BN (1995a) High-frequency motility of outer hair cells and the cochlear amplifier. Science 267:2006–2009.

Dallos P, Evans BN (1995b) High-frequency outer hair cell motility: corrections and addendum [letter]. Science 268:1420–1421.

Dallos P, Fakler B (2002) Prestin, a new type of motor protein. Nat Rev Mol Cell Biol 3:104–111.

Dallos P, Hallworth R, Evans BN (1993) Theory of electrically driven shape changes of cochlear outer hair cells. J Neurophysiol 70:299–323.

Dallos P, He DZ, Lin X, Sziklai I, Mehta S, Evans BN (1997) Acetylcholine, outer hair cell electromotility, and the cochlear amplifier. J Neurosci 17:2212–2226.

Dan N, Safran SA (1998) Effect of lipid characteristics on the structure of transmembrane proteins. Biophys J 75:1410–1414.

Davis H (1983) An active process in cochlear mechanics. Hear Res 9:79–90.

Dickson MR, Kouprach S, Humphrey BA, Marshall KC (1980) Does gliding motility depend on undulating membranes? Micron 11:381–382.

Dieler R, Shehata-Dieler WE, Brownell WE (1991) Concomitant salicylate-induced alterations of outer hair cell subsurface cisternae and electromotility. J Neurocytol 20: 637–653.

Ding JP, Salvi RJ, Sachs F (1991) Stretch-activated ion channels in guinea pig outer hair cells. Hear Res 56:19–28.

Dong XX, Ospeck M, Iwasa KH (2002) Piezoelectric reciprocal relationship of the membrane motor in the cochlear outer hair cell. Biophys J 82:1254–1259.

Dunn RA, Morest DK (1975) Receptor synapses without synaptic ribbons in the cochlea of the cat. Proc Natl Acad Sci USA 72:3599–3603.

Dutzler R, Campbell EB, Cadene M, Chait BT, MacKinnon R (2002) X-ray structure of a ClC chloride channel at 3.0 Å reveals the molecular basis of anion selectivity. Nature 415:287–294.

Dutzler R, Campbell EB, MacKinnon R (2003) Gating the selectivity filter in ClC chloride channels. Science 300:108–112.

Eatock RA (2000) Adaptation in hair cells. Annu Rev Neurosci 23:285–314.

Everett LA, Glaser B, Beck JC, Idol JR, Buchs A, Heyman M, Adawi F, Hazani E, Nassir E, Baxevanis AD, Sheffield VC, Green ED (1997) Pendred syndrome is caused by mutations in a putative sulphate transporter gene (PDS). Nat Genet 17:411–422.

Everett LA, Morsli H, Wu DK, Green ED (1999) Expression pattern of the mouse ortholog of the Pendred's syndrome gene (Pds) suggests a key role for pendrin in the inner ear. Proc Natl Acad Sci USA 96:9727–9732.

Everett LA, Belyantseva IA, Noben-Trauth K, Cantos R, Chen A, Thakkar SI, Hoogstraten-Miller SL, Kachar B, Wu DK, Green ED (2001) Targeted disruption of mouse Pds provides insight about the inner-ear defects encountered in Pendred syndrome. Hum Mol Genet 10:153–161.

Fakler B, Oliver D (2003) Functional properties of prestin—how the motor molecule works work. In: Gummer AW (ed), Biophysics of the Cochlea from Molecules to Model. Singapore: World Scientific, pp. 110–115.

Forge A (1991) Structural features of the lateral walls in mammalian cochlear outer hair cells. Cell Tissue Res 265:473–483.

Frank G, Hemmert W, Gummer AW (1999) Limiting dynamics of high-frequency electromechanical transduction of outer hair cells. Proc Natl Acad Sci USA 96:4420–4425.

Fukada E (1982) Electrical phenomena in biorheology. Biorheology 19:15–27.

Fukada E, Yasuda I (1957) On the piezoelectric effect of bone. J Phys Soc Jpn 12:1158–1162.

Furness DN, Richardson GP, Russell IJ (1989) Stereociliary bundle morphology in organotypic cultures of the mouse cochlea. Hear Res 38:95–109.

Gale JE, Ashmore JF (1994) Charge displacement induced by rapid stretch in the basolateral membrane of the guinea-pig outer hair cell. Proc R Soc Lond B Biol Sci 255:243–249.

Géléoc GS, Lennan GW, Richardson GP, Kros CJ (1997) A quantitative comparison of mechanoelectrical transduction in vestibular and auditory hair cells of neonatal mice. Proc R Soc Lond B Biol Sci 264:611–621.

Géléoc GS, Casalotti SO, Forge A, Ashmore JF (1999) A sugar transporter as a candidate for the outer hair cell motor. Nat Neurosci 2:713–719.

Gold T (1948) Hearing. II. The physical basis of the action of the cochlea. Proc R Soc Lond B Biol Sci 135:492–498.

Grosh K, Zheng J, Zou Y, de Boer E, Nuttall AL (2004) High-frequency electromotile responses in the cochlea. J Acoust Soc Am 115:2178–2184.

Gulley RL, Reese TS (1977) Regional specialization of the hair cell plasmalemma in the organ of corti. Anat Rec 189:109–123.

Halfen LN, Castenholz RW (1970) Gliding in a blue-green alga: a possible mechanism. Nature 225:1163–1165.

Hallworth R (1997) Modulation of outer hair cell compliance and force by agents that affect hearing. Hear Res 114:204–212.

Hoiczyk E (2000) Gliding motility in cyanobacterial: observations and possible explanations. Arch Microbiol 174:11–17.
Hoiczyk E, Baumeister W (1995) Envelope structure of four gliding filamentous cyanobacteria. J Bacteriol 177:2387–2395.
Holley MC (1996) Outer hair cell motility. In: Dallos P, Popper AN, Fay RR (eds), The Cochlea. New York: Springer-Verlag, pp. 386–434.
Housley GD, Ashmore JF (1992) Ionic currents of outer hair cells isolated from the guinea-pig cochlea. J Physiol 448:73–98.
Huang G, Santos-Sacchi J (1994) Motility voltage sensor of the outer hair cell resides within the lateral plasma membrane. Proc Natl Acad Sci USA 91:12268–12272.
Hudspeth AJ, Logothetis NK (2000) Sensory systems. Curr Opin Neurobiol 10:631–641.
Iwasa KH (1993) Effect of stress on the membrane capacitance of the auditory outer hair cell. Biophys J 65:492–498.
Iwasa KH (1994) A membrane motor model for the fast motility of the outer hair cell. J Acoust Soc Am 96:2216–2224.
Iwasa KH (2001) A two-state piezoelectric model for outer hair cell motility. Biophys J 81:2495–2506.
Iwasa KH, Adachi M (1997) Force generation in the outer hair cell of the cochlea. Biophys J 73:546–555.
Iwasa KH, Li MX, Jia M, Kachar B (1991) Stretch sensitivity of the lateral wall of the auditory outer hair cell from the guinea pig. Neurosci Lett 133:171–174.
Jaramillo F, Hudspeth AJ (1991) Localization of the hair cell's transduction channels at the hair bundle's top by iontophoretic application of a channel blocker. Neuron 7:409–420.
Kakehata S, Santos-Sacchi J (1996) Effects of salicylate and lanthanides on outer hair cell motility and associated gating charge. J Neurosci 16:4881–4889.
Kalinec F, Holley MC, Iwasa KH, Lim DJ, Kachar B (1992) A membrane-based force generation mechanism in auditory sensory cells. Proc Natl Acad Sci USA 89:8671–8675.
Kalinec F, Kalinec G, Negrini C, Kachar B (1997) Immunolocalization of anion exchanger 2alpha in auditory sensory hair cells. Hear Res 110:141–146.
Kalinec F, Zhang M, Urrutia R, Kalinec G (2000) Rho GTPases mediate the regulation of cochlear outer hair cell motility by acetylcholine. J Biol Chem 275:28000–28005.
Kemp DT (1978) Stimulated acoustic emissions from within the human auditory system. J Acoust Soc Am 64:1386–1391.
Kennedy HJ, Evans MG, Crawford AC, Fettiplace R (2003) Fast adaptation of mechano-electrical transducer channels in mammalian cochlear hair cells. Nat Neurosci 6:832–836.
Ketten DR (1997) Structure and function in whale ears. Bioacoustics 8:103–135.
Kim KS, Neu J, Oster G (1998) Curvature-mediated interactions between membrane proteins. Biophys J 75:2274–2291.
Korostoff E (1977) Stress generated potentials in bone: relationship to piezoelectricity of collagen. J Biomech 10:41–44.
Kossl M, Russell IJ (1995) Basilar membrane resonance in the cochlea of the mustached bat. Proc Natl Acad Sci USA 92:276–279.
Kros CJ, Rusch A, Richardson GP (1992) Mechano-electrical transducer currents in hair cells of the cultured neonatal mouse cochlea. Proc R Soc Lond B Biol Sci 249:185–193.

Kuhn B, Vater M (1995) The arrangements of F-actin, tubulin and fodrin in the organ of Corti of the horseshoe bat (*Rhinolophus rouxi*) and the gerbil (*Meriones unguiculatus*). Hear Res 84:139–156.

Lapidus IR, Berg HC (1982) Gliding motility of Cytophaga sp. strain U67. J Bacteriol 151:384–398.

Le Grimellec C, Giocondi MC, Lenoir M, Vater M, Sposito G, Pujol R (2002) High-resolution three-dimensional imaging of the lateral plasma membrane of cochlear outer hair cells by atomic force microscopy. J Comp Neurol 451:62–69.

Li Z, Anvari B, Takashima M, Brecht P, Torres JH, Brownell WE (2002) Membrane tether formation from outer hair cells with optical tweezers. Biophys J 82:1386–1395.

Liberman MC (1987) Chronic ultrastructural changes in acoustic trauma: serial-section reconstruction of stereocilia and cuticular plates. Hear Res 26:65–88.

Liberman MC, Gao J, He DZ, Wu X, Jia S, Zuo J (2002) Prestin is required for electromotility of the outer hair cell and for the cochlear amplifier. Nature 419:300–304.

Lim DJ (1980) Cochlear anatomy related to cochlear micromechanics. A review. J Acoust Soc Am 67:1686–1695.

Lindahl E, Edholm O (2000) Mesoscopic undulations and thickness fluctuations in lipid bilayers from molecular dynamics simulations. Biophys J 79:426–433.

Liu XZ, Ouyang XM, Xia XJ, Zheng J, Pandya A, Li F, Du LL, Welch KO, Petit C, Smith RJ, Webb BT, Yan D, Arnos KS, Corey D, Dallos P, Nance WE, Chen ZY (2003) Prestin, a cochlear motor protein, is defective in non-syndromic hearing loss. Hum Mol Genet 12:1155–1162.

Ludwig J, Oliver D, Frank G, Klocker N, Gummer AW, Fakler B (2001) Reciprocal electromechanical properties of rat prestin: the motor molecule from rat outer hair cells. Proc Natl Acad Sci USA 98:4178–4183.

Lue AJ, Brownell WE (1999) Salicylate induced changes in outer hair cell lateral wall stiffness. Hear Res 135:163–168.

Lue AJ, Zhao HB, Brownell WE (2001) Chlorpromazine alters outer hair cell electromotility. Otolaryngol Head Neck Surg 125:71–76.

Markovich D (2001) Physiological roles and regulation of mammalian sulfate transporters. Physiol Rev 81:1499–1533.

Marrink SJ, Lindahl E, Edholm O, Mark AE (2001) Simulation of the spontaneous aggregation of phospholipids into bilayers. J Am Chem Soc 123:8638–8639.

McBride MJ (2000) Bacterial gliding motility: mechanisms and mysteries. ASM News 66:203–210.

Morimoto N, Nygren A, Brownell WE (2000) Quantitative assessment of drug-induced change in OHC lateral wall mechanics. In: Wada H, Takasaka T, Ikeda K, Ohyama K, Koike T (eds), Recent Developments in Auditory Mechanics. Singapore: World Scientific, pp. 261–267.

Morimoto N, Raphael RM, Nygren A, Brownell WE (2002) Excess plasma membrane and effects of ionic amphipaths on mechanics of outer hair cell lateral wall. Am J Physiol Cell Physiol 282:C1076–1086.

Mosbacher J, Langer M, Horber JK, Sachs F (1998) Voltage-dependent membrane displacements measured by atomic force microscopy. J Gen Physiol 111:65–74.

Mountain DC, Hubbard AE (1994) A piezoelectric model of outer hair cell function. J Acoust Soc Am 95:350–354.

Mouritsen OG, Kinnunen PKJ (1996) Role of lipid organization and dynamics for membrane functionality. In: Merz KM Jr, Roux B (eds), Biological Membranes: A

Molecular Perspective from Computation and Experiment. Boston: Birkhauser, pp. 463–502.

Nakazawa K, Spicer SS, Schulte BA (1995) Postnatal expression of the facilitated glucose transporter, GLUT 5, in gerbil outer hair cells. Hear Res 82:93–99.

Nguyen TV, Brownell WE (1998) Contribution of membrane cholesterol to outer hair cell lateral wall stiffness. Otolaryngol Head Neck Surg 119:14–20.

Oghalai JS, Patel AA, Nakagawa T, Brownell WE (1998) Fluorescence-imaged microdeformation of the outer hair cell lateral wall. J Neurosci 18:48–58.

Oghalai JS, Tran TD, Raphael RM, Nakagawa T, Brownell WE (1999) Transverse and lateral mobility in outer hair cell lateral wall membranes. Hear Res 135:19–28.

Oghalai JS, Zhao HB, Kutz JW, Brownell WE (2000) Voltage- and tension-dependent lipid mobility in the outer hair cell plasma membrane. Science 287:658–661.

Oliver D, He DZ, Klocker N, Ludwig J, Schulte U, Waldegger S, Ruppersberg JP, Dallos P, Fakler B (2001) Intracellular anions as the voltage sensor of prestin, the outer hair cell motor protein. Science 292:2340–2343.

Ospeck M, Dong XX, Iwasa KH (2003) Limiting frequency of the cochlear amplifier based on electromotility of outer hair cells. Biophys J 84:739–749.

Pandit SA, Bostick D, Berkowitz ML (2003) Molecular dynamics simulation of a dipalmitoylphosphatidylcholine bilayer with NaCl. Biophys J 84:3743–3750.

Pate JL (1988) Gliding motility in procaryotic cells. Can J Microbiol 34:459–465.

Petrov AG (1999) The Lyotropic State of Matter: Molecular Physics and Living Matter Physics. Amsterdam: Gordon and Breach.

Pitta TP, Sherwood EE, Kobel AM, Berg HC (1997) Calcium is required for swimming by the nonflagellated cyanobacterium *Synechococcus* strain WH8113. J Bacteriol 179: 2524–2528.

Pollice PA, Brownell WE (1993) Characterization of the outer hair cell's lateral wall membranes. Hear Res 70:187–196.

Pujol R, Carlier E, Devigne C (1979) Significance of presynaptic formations in early stages of cochlear synaptogenesis. Neurosci Lett 15:97–102.

Pujol R, Lenoir M, Ladrech S, Tribillac F, Rebillard G (1992) Correlation between the length of outer hair cells and the frequency coding in the cochlea. In: Cazals Y, Horner K, Demany L (eds), Auditory Physiology and Perception: Proceedings of the 9th International Symposium on Hearing, Carcens, France, on June 9–14, 1991, Oxford: Pergamon Press, pp. 45–52.

Ramaswamy S, Toner J, Prost J (2000) Nonequilibrium fluctuations, traveling waves, and instabilities in active membranes. Phys Rev Lett 84:3494–3497.

Raphael RM, Popel AS, Brownell WE (2000a) A membrane bending model of outer hair cell electromotility. Biophys J 78:2844–2862.

Raphael RM, Popel AS, Brownell WE (2000b) An orientational motor model of outer hair cell electromotility. In: Wada H, Takasaka T, Ikeda K, Ohyama K, Koike T (eds), Recent Developments in Auditory Mechanics. Singapore: World Scientific, pp. 344–350.

Raphael Y, Athey BD, Wang Y, Lee MK, Altschuler RA (1994) F-actin, tubulin and spectrin in the organ of Corti: comparative distribution in different cell types and mammalian species. Hear Res 76:173–187.

Ratnanather JT, Brownell WE, Popel AS (1993) Mechanical properties of the outer hair cell. In: Duifhuis H, Horst JW, van Dijk P, van Netten SM (eds), Biophysics of Hair Cell Sensory Systems. Singapore: World Scientific, pp. 199–206.

Ratnanather JT, Zhi M, Brownell WE, Popel AS (1996) Measurements and a model of the outer hair cell hydraulic conductivity. Hear Res 96:33–40.

Raucher D, Sheetz MP (1999) Characteristics of a membrane reservoir buffering membrane tension. Biophys J 77:1992–2002.

Robertson D (1984) Horseradish peroxidase injection of physiologically characterized afferent and efferent neurones in the guinea pig spiral ganglion. Hear Res 15:113–21.

Russell IJ, Kossl M (1999) Micromechanical responses to tones in the auditory fovea of the greater mustached bat's cochlea. J Neurophysiol 82:676–686.

Rybalchenko V, Santos-Sacchi J (2003) Cl^- flux through a non-selective, stretch-sensitive conductance influences the outer hair cellmotor of the guinea-pig. J Physiol 547:873–891.

Sachs JN, Woolf TB (2003) Understanding the Hofmeister effect in interactions between chaotropic anions and lipid bilayers: molecular dynamics simulations. J Am Chem Soc 125:8742–8743.

Santos-Sacchi J (1991) Reversible inhibition of voltage-dependent outer hair cell motility and capacitance. J Neurosci 11:3096–3110.

Santos-Sacchi J (1992) On the frequency limit and phase of outer hair cell motility: effects of the membrane filter. J Neurosci 12:1906–1916.

Santos-Sacchi J (2003) New tunes from Corti's organ: the outer hair cell boogie rules. Curr Opin Neurobiol 13:459–468.

Santos-Sacchi J, Dilger JP (1988) Whole cell currents and mechanical responses of isolated outer hair cells. Hear Res 35:143–150.

Santos-Sacchi J, Huang GJ, Wu M (1997) Mapping the distribution of outer hair cell voltage-dependent conductances by electrical amputation. Biophys J 73:1424–1429.

Santos-Sacchi J, Kakehata S, Kikuchi T, Katori Y, Takasaka T (1998) Density of motility-related charge in the outer hair cell of the guinea pig is inversely related to best frequency. Neurosci Lett 256:155–158.

Santos-Sacchi J, Shen W, Zheng J, Dallos P (2001) Effects of membrane potential and tension on prestin, the outer hair cell lateral membrane motor protein. J Physiol 531:661–666.

Schaeffer SF, Raviola E (1978) Membrane recycling in the cone cell endings of the turtle retina. J Cell Biol 79:802–825.

Shehata WE, Brownell WE, Dieler R (1991) Effects of salicylate on shape, electromotility and membrane characteristics of isolated outer hair cells from guinea pig cochlea. Acta Otolaryngol (Stockh) 111:707–718.

Siddiqui AM, Burchard RP, Schwarz WH (2001) An undulating surface model for the motility of bacteria gliding on a layer of non-Newtonian slime. Int J nonlinear Mech 36:743–761.

Siegel JH, Brownell WE (1981) Presynaptic bodies in outer hair cells of the chinchilla organ of Corti. Brain Res 220:188–193.

Siegel JH, Brownell WE (1986) Synaptic and golgi membrane recycling in cochlear hair cells. J Neurocytol 15:311–328.

Siskind LJ, Colombini M (2000) The lipids C2- and C16-ceramide form large stable channels. Implications for apoptosis. J Biol Chem 275:38640–38644.

Siskind LJ, Kolesnick RN, Colombini M (2002) Ceramide channels increase the permeability of the mitochondrial outer membrane to small proteins. J Biol Chem 277:26796–26803.

Siskind LJ, Davoody A, Lewin N, Marshall S, Colombini M (2003) Enlargement and contracture of C2-ceramide channels. Biophys J85:1560–1575.

Sit PS, Spector AA, Lue AJ, Popel AS, Brownell WE (1997) Micropipette aspiration on the outer hair cell lateral wall. Biophys J 72:2812–2819.

Slepecky NB (1996) Structure of the mammalian cochlea. In: Dallos P, Popper AN, Fay RR (eds), The Cochlea. New York: Springer-Verlag, pp. 44–129.

Smith CA (1968) Ultrastructure of the organ of Corti. Adv Sci 122:419–433.

Snyder KV, Sachs F, Brownell WE (2003) The outer hair cell: a mechanoelectrical and electromechanical sensor/actuator. In: Barth FG, Humphrey JAC, Secomb TW (eds), Sensors and Sensing in Biology and Engineering. Vienna: Springer-Verlag, pp. 71–95.

Spector AA, Brownell WE, Popel AS (2003a) Effect of outer hair cell piezoelectricity on high-frequency receptor potentials. J Acoust Soc Am 113:453–461.

Spector AA, Popel AS, Brownell WE (2003b) Piezoelectric properties enhance outer hair cell high-frequency response. In: Gummer AW, Dalhoff E, Scherer MP (eds), Biophysics of the Cochlea: From Molecule to Model. Singapore: World Scientific, pp. 152–160.

Spicer SS, Thomopoulos GN, Schulte BA (1998) Cytologic evidence for mechanisms of K^+ transport and genesis of Hensen bodies and subsurface cisternae in outer hair cells. Anat Rec 251:97–113.

Spoendlin H (1966) The organization of the cochlear receptor. Fortschr Hals Nasen Ohrenheilkd 13:1–227.

Sziklai I, He DZ, Dallos P (1996) Effect of acetylcholine and GABA on the transfer function of electromotility in isolated outer hair cells. Hear Res 95:87–99.

Takasaka T, Shinkawa H, Hashimoto S, Watanuki K, Kawamoto K(1983) High-voltage electron microscopic study of the inner ear. Technique and preliminary results. Ann Otol Rhinol Laryngol Suppl 101:1–12.

Tolomeo JA, Steele CR (1995) Orthotropic piezoelectric properties of the cochlear outer hair cell wall. J Acoust Soc Am 97:3006–3011.

Ulfendahl M, Slepecky N (1988) Ultrastructural correlates of inner ear sensory cell shortening. J Submicrosc Cytol Pathol 20:47–51.

Vincourt JB, Jullien D, Amalric F, Girard JP (2003) Molecular and functional characterization of SLC26A11, a sodium-independent sulfate transporter from high endothelial venules. FASEB J 17:890–892.

Wada H, Usukura H, Sugawara M, Katori Y, Kakehata S, Ikeda K, Kobayashi T (2003) Relationship between the local stiffness of the outer hair cell along the cell axis and its ultrastructure observed by atomic force microscopy. Hear Res 177:61–70.

Wainwright SA (1988) Axis and Circumference the Cylindrical Shape of Plants and Animals. Cambridge, MA: Harvard University Press.

Weber T, Gopfert MC, Winter H, Zimmermann U, Kohler H, Meier A, Hendrich O, Rohbock K, Robert D, Knipper M (2003) Expression of prestin-homologous solute carrier (SLC26) in auditory organs of nonmammalian vertebrates and insects. Proc Natl Acad Sci U S A 100:7690–7695.

Weitzel EK, Tasker R, Brownell WE (2003) Outer hair cell piezoelectricity: frequency response enhancement and resonance behavior. J Acoust Soc Am 114:1462–1466.

Zhang M, Kalinec GM, Urrutia R, Billadeau DD, Kalinec F(2003) ROCK-dependent and ROCK-independent control of cochlear outer hair cell electromotility. J Biol Chem 278:35644–35650.

Zhang PC, Keleshian AM, Sachs F (2001) Voltage-induced membrane movement. Nature 413:428–432.

Zhao HB, Santos-Sacchi J (1999) Auditory collusion and a coupled couple of outer hair cells. Nature 399:359–362.
Zheng J, Shen W, He DZ, Long KB, Madison LD, Dallos P (2000) Prestin is the motor protein of cochlear outer hair cells. Nature 405:149–155.
Zheng J, Long KB, Shen W, Madison LD, Dallos P (2001) Prestin topology: localization of protein epitopes in relation to the plasma membrane. NeuroReport 12:1929–1935.

8
Mammalian Vestibular Hair Cells

RUTH ANNE EATOCK AND ANNA LYSAKOWSKI

1. Introduction

Vestibular hair cells provide signals about head movements and head tilt to motor reflexes controlling eye, head and body position. Chronic bilateral loss of vestibular signals can produce oscillopsia (inability to maintain gaze during head movements), blurry vision, distorted perception of self-orientation, imbalance, gait ataxia, and abnormal posture (Hess 1996).

Vestibular hair cells resemble their auditory counterparts in many fundamental features: The stimulus is transduced into a current flowing into the hair bundle by similar mechanisms. Both auditory and vestibular hair cells have diverse potassium (K^+)-selective ion channels in their basolateral membranes to shape the receptor potential. Both have the synaptic machinery for glutamate release onto the dendrites of primary afferent neurons. Both auditory and vestibular epithelia are innervated by cholinergic nerve fibers originating in the brainstem, which terminate on hair cells and the peripheral endings of primary afferent neurons.

But auditory and vestibular epithelia also differ significantly. The vestibular system works at subacoustic frequencies, from tens of Hertz down to DC (head tilt). In auditory epithelia, finely graded mapping of hair cell biophysical properties contributes to the spectral analysis of sound (Fettiplace and Ricci, Chapter 4; Art and Fettiplace, Chapter 5). In contrast, vestibular epithelia are organized in relatively coarse maps, typically comprising several zones with distinct afferent physiology. We shall review regional differences in hair cell properties that may contribute to the regional differences in afferent properties.

Most sensory epithelia have morphologically and physiologically distinct classes of receptor cell, such as the rods and cones of retinas and the inner and outer hair cells of mammalian cochleas. Wersäll (1956) described two classes of mammalian vestibular hair cell, distinguished principally by the kind of postsynaptic contact made by vestibular afferent fibers. On type II hair cells, the primary afferents form small rounded ("bouton") contacts similar to the afferent contacts formed on all auditory hair cells and on all vestibular hair cells in fish and amphibians. Type I hair cells, in contrast, bear a cup-shaped afferent ending, or "calyx." Although large afferent endings occur in the vestibular epithelia of fish and amphibians, full calyces occur only in amniotes. The unusual morphology of the type I–calyx complex and its relatively late appearance in evolution suggest a distinctive contribution to vestibular processing in amniotes. We shall review several ideas at various junctures in the chapter, especially Sections 2.2 and 2.3 of the Overview and 6.2 of Concluding Remarks.

We begin with background on the anatomy of the vestibular periphery and the physiology of primary vestibular afferents in mammals (Section 2). Reviews by Goldberg (2000) and Lysakowski and Goldberg (2004) discuss these subjects in more detail and cover other vertebrate groups as well. We then describe mammalian vestibular hair cells by following the afferent signal from the accessory structures through the stages of mechanoelectrical transduction (Section 3), post-transduction processing by basolateral ion channels (Section 4), afferent transmission, and efferent feedback (Section 5). The coverage of mammalian cristae and maculae within the literature can be uneven; for example, there is more known about transduction and bundle morphology for otolith hair cells than for crista hair cells. We draw comparisons with cochlear hair cells and refer to results from non-mammalian vestibular hair cells where they provide context, throughout the chapter and specifically in Section 6.1.

2. Overview of the Vestibular Periphery

2.1 Organization and Anatomy

The mammalian inner ear (Fig. 8.1A) has five vestibular organs: three semicircular canal organs (horizontal or lateral, anterior, and posterior), which detect angular head movements, and two otolith organs (the utriculus and sacculus), which detect linear head movements and head tilt. Some mammalian saccular afferents also have modest sensitivity to acoustic stimuli between 200 and 2000 Hz (McCue and Guinan 1995). Within each organ is a specialized patch of epithelium, the sensory epithelium (Fig. 8.1B), comprising hair cells, supporting cells, and afferent and efferent nerve endings. The hair bundles of the hair cells project apically into specialized extracellular matrices (*accessory structures*). In the otolith organs, the sensory epithelia (*maculae*; Fig. 8.1C) are overlain by a multilayered gel supporting calcareous crystals (*otoconia*, "ear dust"; Fig. 8.1B). In the semicircular canals, the sensory epithelia are called *cristae ampullares* because of their crest shape in transverse section (Fig. 8.1D) and are located in swellings of the canals called *ampullae* (Fig. 8.1A). The accessory structure is a delicate *cupula* that surmounts the hair bundles and spans the ampulla parallel to the long axis of the crista (Fig. 8.1C,D).

Head movements deflect the accessory structures and the underlying hair bundles. The hair cells transduce the bundle deflections into receptor potentials (Section 3.3), which in turn modulate the release of excitatory transmitter onto the peripheral endings of primary vestibular neurons (Section 5.1). The primary

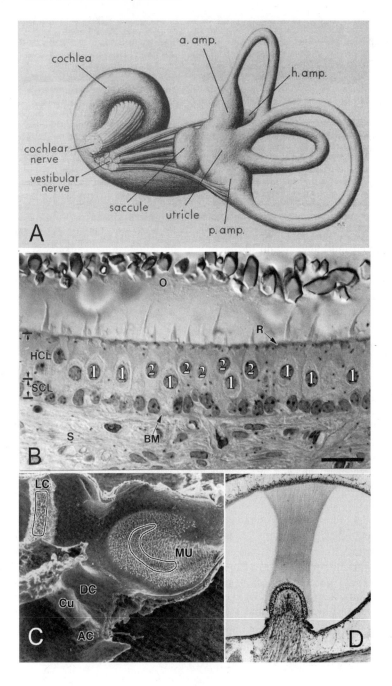

afferent transmitter of both hair cells and afferent fibers is the excitatory amino acid, glutamate. In hair cells synaptic vesicles cluster around electron-dense structures called synaptic ribbons.

The maculae of mammalian otolith organs are organized into *striolar* and *extrastriolar* zones (Figs. 8.2A and 8.3A) and mammalian cristae have distinct *central* and *peripheral* zones. These zones are differentiated by numerous morphological criteria as well as afferent nerve physiology (Section 2.2) and hair cell properties (Section 4.1.4). In the utricular maculae of reptiles (Jørgensen 1988) and birds (Jørgensen 1989; Si et al. 2003), type I hair cells (i.e., calyx terminals) are concentrated in the striolar zone, but in mammalian maculae, type I cells are also found in significant numbers in the extrastriolar zone (Lindeman 1969; Fernández et al. 1988, 1990). Some cristae, including the anterior and posterior cristae of rats and mice, are bisected by a nonsensory structure called an *eminentia cruciatum* or *septum cruciatum*. It is not known whether the septum interrupts a single map similar to that in cristae lacking a septum or whether each hemicrista has its own central and peripheral zones.

In each vestibular epithelium, thousands of hair cells are contacted by thousands of afferent neurons. The cell bodies of the afferent nerve fibers reside in the two divisions of the vestibular (Scarpa's) ganglion. The superior division of the vestibular ganglion innervates the utricular macula, the anterior part of the saccular macula, and the anterior and horizontal cristae. The inferior division innervates the posterior saccular macula and the posterior crista. The cen-

◄─────────────────────

FIGURE 8.1. Vestibular organs. (**A**) Membranous labyrinth. a. amp., h. amp., p. amp: ampullae of the anterior vertical, horizontal (lateral) and posterior vertical canals, respectively. Reproduced with permission from Figure 1.1, p. 4, Highstein SM (1996) How does the vestibular part of the inner ear work? In: Disorders of the Vestibular System (Baloh RW, Halmagyi GM, eds), pp. 3–19. New York: Oxford University Press. (**B**) Cross section through adult rat utricular macula. BM, Basement membrane; HCL and SCL, hair cell and supporting cell nuclear layers; O, otoconia, R, reticular lamina; S, stroma; 1, 2, type I and II hair cells, respectively. Reproduced with permission from Figure 1, p. 180, Oesterle EC, Cunningham DE, Westrum LE, Rubel EW (2003) Ultrastructural analysis of [^{3}H]thymidine-labeled cells in the rat utricular macula. J Comp Neurol 463: 177–195. (**C**) Scanning electron micrograph from the hamster inner ear. The accessory structures have been removed from the horizontal (lateral) crista (LC) and utricular macula (MU), revealing the hair bundles. The cupula (Cu) over the anterior crista (AC) has been retained, although it has shrunk from dehydration. Black on white lines indicate the boundaries of the central zone on the LC and the striolar region on the MU, respectively. DC, dark cell epithelium. Modified with permission from Figure 9.33 in Hunter-Duvar IM, Hinojosa R (1984) Vestibule: Sensory epithelia. In: Ultrastructural Atlas of the Inner Ear (Friedmann I, Ballantyne J, eds), pp. 211–244. London: Butterworths, with permission from Elsevier. (**D**) Cross-section through squirrel monkey lateral crista and cupula. Reproduced with permission from Figure 6, p. 952, Igarashi M (1966) Architecture of the otolith end organ: some functional considerations. Ann Otol Rhinol Laryngol 75:945–955.

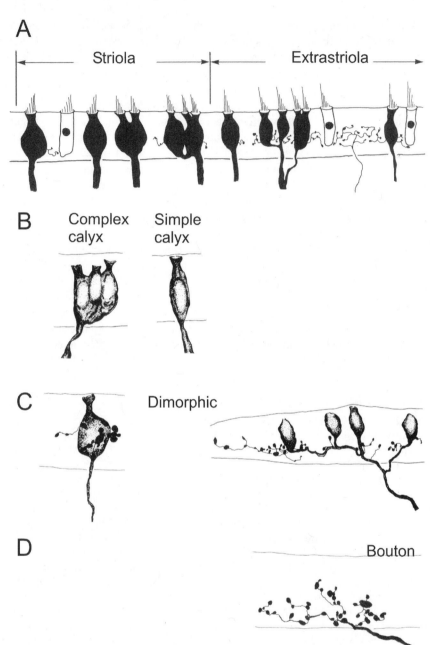

tral processes of the ganglion cells coalesce to form the vestibular nerve, which projects centrally to brainstem vestibular nuclei and parts of the cerebellum (reviewed in Barmack 2003; Newlands and Perachio 2003).

Fernández and colleagues have divided afferents into three morphological classes (Fig. 8.2). *Bouton afferents* contact type II cells in the peripheral/extrastriolar zones, arise from small neuronal somata in the ganglion, and express peripherin (Lysakowski et al. 1999; Kevetter and Leonard 2002; Leonard and Kevetter 2002). *Calyx afferents* contact type I cells in the central/striolar zones, arise from large cell bodies, and express calretinin (Section 4.2.2). *Dimorphic afferents* contact both type I and type II hair cells in both regions of the epithelium, and arise from medium-sized ganglion cells, and have no known specific marker. Each afferent samples from between one and thirty hair cells while each hair cell is contacted by three to four afferents (Fernández et al. 1988, 1990, 1995).

The cell bodies of several hundred efferent neurons form loose clusters near the genu of the facial nerve in the brainstem. The axons of the efferent neurons terminate on the basolateral membranes of type II hair cells and on bouton and calyx terminals of primary afferent fibers, where they release acetylcholine and probably other transmitters (Section 5.2.4).

The vestibular sensory epithelium is surrounded by a *transitional epithelium* and a *dark cell epithelium* (DC in Fig. 8.1C). Tight junctions segregate the apical surface of the epithelium from the basolateral compartment. Within the sensory epithelium, the basolateral compartment comprises the cell bodies of hair cells and supporting cells plus afferent and efferent nerve terminals. The basolateral compartment is bathed in perilymph, a conventional extracellular saline, while the apical compartment contains endolymph, an unusual saline with a high concentration of potassium ions (K^+) and low concentrations of sodium ions (Na^+) and calcium ions (Ca^{2+}). The endolymph is produced by the dark cell epithelium, which is analogous to the stria vascularis of the cochlea (Wangemann 1995). The endolymph has a potential of about 0 mV (0 mV in the ampulla and +3 mV in the utricle; see analysis and discussion in Marcus et al.

◄─────────────────────────────────

FIGURE 8.2. Morphology and regional localization of vestibular afferents. (**A**) Schematic cross-section of the utricular macula, showing striolar and extrastriolar zones. Note the reversal of bundle polarity within the striola. Reproduced with permission from Fig. 1 in Goldberg JM, Lysakowski A, Fernández C (1992) Structure and function of vestibular nerve fibers in the chinchilla and squirrel monkey. Ann NY Acad Sci 656:92–107, Copyright 1992 New York Academy of Sciences. (**B–D**) Horseradish peroxidase fills of chinchilla utricular afferent terminals. Calyx afferents (B) innervate principally the striola, dimorphic afferents (C) innervate both regions, and bouton afferents (D) innervate the extrastriola. Modified with permission from Fig. 6 (B, D) and Figs. 7 and 8 (C) in Fernández C, Goldberg JM, Baird RA (1990) The vestibular nerve of the chinchilla. III. Peripheral innervation patterns in the utricular macula. J Neurophysiol 63:767–780.

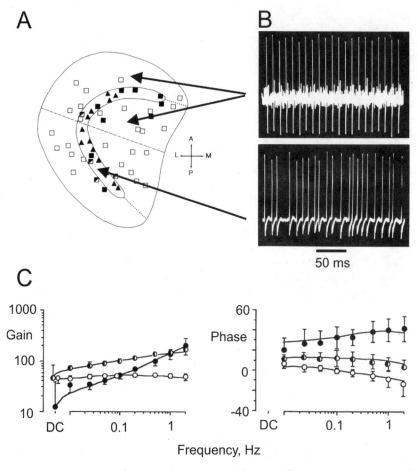

FIGURE 8.3. The discharge properties of mammalian vestibular afferents covary with location in the sensory epithelium. (**A**) Map of terminal locations of labeled, physiologically characterized afferents in the chinchilla utricular macula. The U-shaped central strip is the striola. In A, C: *Filled symbols:* irregular units; *open symbols,* regular units; *mixed symbols,* intermediate units. Reproduced with permission from Fig. 1 in Goldberg JM, Desmadryl G, Baird RA, Fernández C (1990b) The vestibular nerve of the chinchilla. V. Relation between afferent discharge properties and peripheral innervation patterns in the utricular macula. J Neurophysiol 63:791–804. (**B**) Spike data from squirrel monkey crista afferents: *top,* a regular afferent; *bottom,* an irregular afferent. Modified with permission from Fig. 8 in Goldberg JM, Fernández C (1971) Physiology of peripheral neurons innervating semicircular canals of the squirrel monkey. I. Resting discharge and response to constant angular accelerations. Striolar afferents tend to be irregular and extrastriolar afferents tend to be regular (*arrows*; Goldberg et al. 1990b). (**C**) Gain (spike/ s/g) and phase re: linear acceleration, averaged for samples of chinchilla utricular afferents. Taken together, results in this figure show that striolar afferents (*irregular, filled symbols*) are more phasic (adapting) and extrastriolar afferents (*regular, open symbols*) are more tonic (nonadapting). Reproduced with permission from Fig. 6 in Goldberg JM, Desmadryl G, Baird RA, Fernández C (1990a) The vestibular nerve of the chinchilla. IV. Discharge properties of utricular afferents. J Neurophysiol 63:781–790.

1994), in contrast to the large endolymphatic potential in the mammalian cochlea (approximately +80 mV; Wangemann and Schacht 1996).

2.2 Afferent Nerve Physiology

Two striking features of mammalian vestibular afferent physiology are the high firing rates (background rates of 60 to 120 spikes/s in anesthetized mammals) and the regular interspike intervals of many afferents. In squirrel monkeys, the coefficient of variation (CV) for rates of 50 to 100 spikes/s ranges from a highly irregular value of 0.6, comparable to values in cochlear and retinal afferents, to a highly regular value of 0.025 (reviewed in Goldberg 2000; see his Table 2). Figure 8.3B shows samples of spiking in two squirrel monkey semicircular canal afferents, one regular and the other irregular. Covariation of spike regularity with other properties (next section) has led naturally to suggestions that irregular and regular afferents represent parallel channels of vestibular information (see Peterson, 1998, and Goldberg, 2000, for discussions of this question).

2.2.1 Morphological Correlates of Afferent Discharge Properties

Afferents from mammalian canal and otolith organs show a bimodal distribution of spike regularity, measured as the CV at a particular spike rate. For chinchilla utricular afferents, for example, the modes occur at 0.03 and 0.3 for a spike rate of 67 spikes/s (Goldberg et al. 1990a). Spike regularity covaries with the following properties: (1) sensitivity to electrical stimulation: current injected into the perilymphatic space triggers spikes at lower current levels in irregular afferents than in regular afferents; (2) sensitivity to efferent activation: shocks applied to efferent fibers elicit larger changes in discharge rate at lower shock levels in irregular afferents than in regular afferents; (3) response dynamics to stimulus trapezoids (in which stimulus levels are separated by ramps) or sinusoids (Fig. 8.3C): regular afferents respond tonically to stimulus trapezoids and respond to low-frequency sinusoidal head movements with spike rate changes that are nearly in phase and relatively constant in gain. In irregular afferents, in contrast, responses to trapezoids follow a mixed, "phasic-tonic" firing pattern, and responses to low-frequency sinusoids lead the stimulus in phase and increase in gain with frequency.

Goldberg et al. (1984) proposed that the regularity of afferent spiking is determined by the complement of voltage-gated ion channels in the afferent nerve ending. The model assumes similar neurotransmitter input from the hair cell to regular and irregular afferents, in the form of excitatory quanta. It proposes that afferents with the regular firing pattern have an after-hyperpolarizing potential (AHP) conductance that sets interspike interval according to its deactivation kinetics. In this way, hair cell output drives the postsynaptic firing in both regular and irregular afferents but is not responsible for the difference in spike regularity. The AHP conductance could be a Ca^{2+}-activated K^+ conductance, as in hippocampal neurons with regular ("pacemaker") firing (Storm 1990; Faber and Sah 2003). Differential expression of this conductance across fibers might explain other differences: all else being equal, the reduced expression of the AHP conductance in irregular afferents would increase their input resistance and therefore their sensitivity to the effects of current from any source, whether a perilymphatic electrode or postsynaptic currents evoked by neurotransmitter quanta from the hair cell or from the efferent terminal. Thus, this model can account for the correlation between a vestibular afferent's spike regularity and its sensitivity to electrical stimuli and efferent activation. It further suggests that regional variation in regularity arises from regional variation in AHP channel expression, a highly testable prediction. The model does not explain the regional variation in response dynamics. Simple macromechanical models of the accessory structures (Section 3.1) predict tonic responses, suggesting that the phasic component of the irregular afferents' responses arises at a subsequent stage, from hair cell micromechanics to afferent synaptic transmission (Highstein et al. 1996).

In a series of papers from the late 1980s through to the mid 1990s, Goldberg, Fernández, and colleagues investigated whether afferent discharge properties correlate with terminal morphology (calyx vs. bouton) or with location within the sensory epithelium. Spontaneous firing and responses to low-frequency head movements ($\leq$ 4 Hz in semicircular canals; $\leq$ 2 Hz in utricles) were characterized in large samples of semicircular canal and utricular afferents from a rodent, the chinchilla, and a primate, the squirrel monkey. A subset of physiologically characterized afferents was labeled for morphological analysis of their peripheral inputs; some of the fill data are shown in Figure 8.2, and the results are described next.

Fiber diameter was found to vary across afferent groups, with calyx afferents the largest and bouton afferents the smallest (Fernández et al. 1988, 1990). In contrast to illustrations showing separate afferent endings on type I and type II cells, most afferents contact both cell types (dimorphic afferents), with relatively small percentages contacting just type I cells (calyx afferents) or just type II cells (bouton afferents) (Fig. 8.2). In chinchilla cristae, 70% of afferents are dimorphic, 10% are calyx afferents, and 20% are bouton afferents (Fernández et al. 1988). In squirrel monkey cristae, Fernández et al. (1995) estimated proportionally more calyx afferents (30%) and fewer dimorphic afferents (60%) and bouton afferents (10%). The higher proportion of calyx afferents in squirrel monkey relative to the chinchilla is consistent with the larger average ratio of type I to type II hair cells in primate vestibular epithelia (3:1; Fernández et al. 1995) relative to rodent epithelia (1:1 to 2:1; Lindeman 1969; Desai et al. 2005a,b). Indeed, in the *central* crista of the squirrel monkey, both the type I: type II ratio (5:1) and the percentage of calyx afferents (70%) are very high. For chinchilla utricular maculae (Fig. 8.2), Fernández et al. (1990) estimated more dimorphic afferents (86%), fewer bouton afferents (12%), and fewer calyx afferents (2%) than for the cristae. More recent data on vestibular epithelia for various rodents show greater similarity between the percentages for maculae and cristae: according to calretinin immunoreactivity (Section 4.2.2), calyx afferents constitute 10% to 20% of afferents from rodent maculae (Desai et al. 2005b) and about 10% for rodent cristae (Desai et al. 2005a).

Physiological properties were found to correlate strongly with the afferent terminals' epithelial region and weakly with the type of hair cell providing input. Distinguishing the separate contributions of hair cell type and epithelial location to afferent diversity was complicated by two factors. First, most afferents are dimorphic, so that it is difficult to separate the contributions of type I and type II hair cells. Second, even when comparing calyx and bouton afferents, region is a confounding factor. Calyx afferents are restricted to the central/striolar zones and bouton afferents to the peripheral/extrastriolar zones (Fig. 8.2). When calyx and bouton afferent data are compared to dimorphic afferent data, afferent properties appear to correlate more strongly with epithelial zone than with hair cell type (Fig. 8.3A). That is, the properties of dimorphic afferents vary with region of innervation, and the properties of calyx and bouton afferents tend to match those of dimorphic afferents in the same region. Central/striolar afferents, which comprise calyx and dimorphic afferents, have more irregular firing and higher response gains and phase leads at 2 Hz than do peripheral/extrastriolar afferents, which comprise dimorphic and bouton afferents (Fig. 8.3C).

Baird et al. (1988) did, nevertheless, suggest a difference between type I and type II input to afferents: specifically, that transmission at each type I ribbon synapse is weaker than at each type II ribbon synapse. This suggestion derives from their observation that calyx afferents of the chinchilla crista have lower response gains at 2 Hz than do dimorphic afferents of similar irregularity. In the chinchilla crista, type I and type II hair cells have similar average numbers of synaptic ribbons (Lysakowski and Goldberg 1997). A simple calyx ending, enclosing one type I cell, receives synaptic input from an average of 20 ribbons, while a bouton ending on a type II hair cell receives input from just one or two ribbons (but the type II hair cell contacts 10 to 20 boutons). All else being equal, then, synaptic transmission should be 10- or 20-fold more powerful at the hair cell-calyx contact than at the hair cell-bouton contact. Baird et al. (1988), however, calculated that the hair cell input to a calyx is only threefold more potent than the hair cell input to a single bouton, that is, that each type I ribbon synapse is substantially weaker than each type II ribbon synapse. While it may seem counterintuitive to suggest that the calyx synapse serves to reduce gain, such a reduction may expand the operating range of stimulus levels for calyx afferents such that they will "not become saturated by the large velocities characterizing active head movements in terrestrial vertebrates" (Baird et al. 1988, p. 201).

These arguments were based on the low gain of crista calyx afferents in response to 2-Hz angular head movements. By extending the stimulus frequency range of angular head movements, Hullar et al. (2004) have recently shown that the relative gains of the different afferent classes depend on the stimulus frequency. The response–frequency relations of regular afferents (which tend to be peripheral), of irregular afferents with high gains at 2 Hz (which tend to be central and dimorphic), and of irregular afferents with low gains at 2 Hz (which tend to be central and calyx), all show some high-pass filtering, but the frequency dependence is steepest for calyx afferents. Thus, as frequency increases above 2 Hz, the gain of calyx afferents surpasses that of regular afferents, and comes to match that of other irregular afferents by 15 Hz. Thus, calyx afferents are better characterized as high-pass than as low-gain.

2.3 Why Type I Hair Cells?

Efforts to determine the relative influences of zone and hair cell type on afferent diversity may fall short by treating the two as independent factors. As reviewed in later sections, central type I and II hair cells and synaptic contacts differ from their peripheral counterparts in numerous ways (e.g., Lysakowski and Goldberg 1997; Sans et al. 2001). Furthermore, the organization of avian and reptilian epithelia suggests that type I cells evolved to process signals particular to central/striolar zones, with the spread of type I cells into peripheral and extrastriolar zones in mammalian epithelia being a later development. Because response dynamics is a salient difference between regions, we speculate that type I cells evolved to process high stimulus frequencies (phasic signals) represented by central/striolar zones. The relatively steep high-pass filtering of stimuli up to approximately 20 Hz by calyx afferents to the chinchilla crista (Hullar et al. 2004) is consistent with this suggestion. Data from the bird crista also support this suggestion. By applying a mechanical indenter to the pigeon ampulla, Dickman and Correia (1989) were able to explore very high frequencies and determine the upper corner frequencies of the afferents' responses. Corner frequencies ranged from 30 to 80 Hz; irregular afferents had higher corner frequencies than did regular afferents. If central afferents in birds tend to be irregular, like central afferents in mammals, then again we have a correlation between high-frequency signaling, central zones, and type I hair cells (restricted to central zones in birds).

This is not to suggest that high-frequency signaling in central zones is peculiar to amniotes. On the contrary, evidence from fish and frogs suggests that this association predates the evolution of type I cells. For example, in central afferents to the toadfish crista, gains increase with stimulus frequency up to at least 10 Hz (Boyle and Highstein 1990; Boyle et al. 1991). In the frog utriculus, Baird (1994a,b) found that both transduction and basolateral membrane properties are tuned to higher frequencies in striolar hair cells than in extrastriolar hair cells. Afferents from the extrastriola of the frog utriculus have tonic response properties while those from the borders of the striola have phasic-tonic response properties, consistent with a high-frequency component (Baird and Lewis 1986).

If anamniotes represent high stimulus frequencies in central/striolar zones, and if type I cells are specialized to represent high frequencies, why don't anamniotes have type I cells? In fact, modern anamniotes have vestibular afferents that resemble calyx afferents in some ways: large afferents to the toadfish crista (Boyle et al. 1991) and the frog utricular striola (Baird and Schuff 1994) have thick axons and make relatively large club- or cup-shaped synaptic contacts on hair cells. Thus, the type I hair cell and the calyx ending may have evolved from precursors present in anamniote ancestors. The more extreme morphology of the amniote calyx may reflect more extreme, or different, evolutionary pressures on amniote vestibular signals. Amniotes tend to have necks, which grant more freedom of head movement (and more involuntary wobble), and some have highly acute, foveate vision. Both of these features may apply strong evolutionary pressure on vestibular processing by taxing the ability of vestibuloocular and vestibulocollic reflexes to maintain visual fixation.

We will return to these ideas at various places in the remainder of the chapter and especially at the end (Section 7.2). In the next section, we begin our review of the afferent cascade of processing in mammalian vestibular hair cells, with mechanoelectrical transduction.

3. Mechanical Stimulation and Transduction

3.1 Accessory Structures

For a recent review of the structure and mechanics of vestibular accessory structures, see Rabbitt et al. (2004).

3.1.1 The Cupula

The cupula extends from the apical surface of the crista to the roof of the ampulla, filling the entire longitudinal section of the ampulla and tethered at the edges like a diaphragm (Fig. 8.1D). (In textbooks, the cupula is sometimes incorrectly depicted as detached from the wall of the ampulla.) In the toadfish, the cupula has three distinct components, a central "pillar" flanked by two "wings," all of which extend from the epithelium to the ampulla roof (Silver et al. 1998). The rodent cupula appears more uniform (Hunter-Duvar and Hinojosa 1984). The matrix comprises a loose array of randomly oriented filaments and an amorphous substance (Goodyear and Richardson 2002). Associated with the filaments are secretory droplets, presumably secreted by the supporting cells of the epithelium and forming the substance of the cupula. The long stereocilia of crista hair cells project up through a subcupular space, filled with a uniform refractile material, into the cupula proper; in the toadfish there is no indication that the bundles project into canals (Silver et al. 1998), as was suggested by earlier scanning electron microscopy on guinea pig cupulae (Lim 1971, 1976). Because older microscopic methods were harsher and introduced more artifacts, the question of whether mammalian cupulae have canals should be reexamined.

Angular head movements cause endolymph in the canals to lag the motion of the canal walls, producing volume displacement of the cupula and deflection of the hair bundles (Rabbitt et al. 1999, 2004). The cupula moves maximally in the center of the ampulla, with motion tapering off to zero at the edges (McLaren and Hillman 1979; Yamauchi et al. 2002). Models of the cupular response show it to be in phase with angular velocity over most of the relevant stimulus frequency range, from about 0.01 Hz to 10 Hz (Wilson and Melvill Jones 1979; Rabbitt et al. 2004). Although the lower corner frequency has not been experimentally measured, data from regular afferents are reasonably consistent with the predicted macroscopic motion of the cupula, consistent with angular head velocity causing hair bundle deflection and afferent discharge with little additional filtering. (The reality may be more complex; because in vitro studies show filtering by transducer adaptation and basolateral hair cell conductances [Sections 3.3.1 and 4.1.5] it may be that such filtering processes cancel each other out to produce the flat in vivo frequency response.) The high-pass filtering of irregular afferents, evident as phase leads and increasing gain with frequency, could arise through either fluid coupling of central-zone bundles to cupular movements or such hair cell mechanisms as transducer adaptation (Section 3.3.1), voltage-dependent channel activation (Section 4.1.5), or synaptic adaptation (Rabbitt et al. 2005).

3.1.2 The Otolithic Membrane

The accessory structures of otolith organs comprise several distinct layers: (1) a fine filamentous layer surrounding the hair bundles (guinea pig and chinchilla: Lindeman 1973; "column filament layer" in frog saccule: Jacobs and Hudspeth 1990; Kachar et al. 1990); (2) a denser gelatinous layer (otoconial membrane or "mesh" layer, Jaeger and Haslwanter 2004) with an isotropic filamentous structure resembling that of the tectorial membrane in the cochlea (Lins et al. 2000); and (3) the otoconial layer, comprising calcium carbonate crystals (otoconia) in an organic matrix (Fig. 8.1B). Models of otolith organ mechanics treat the otoconial mass as a rigid unit, the gel layer as a uniform viscoelastic substance, and the endolymph as a Newtonian fluid with uniform viscosity (reviewed in Wilson and Melvill Jones 1979 and Rabbitt et al. 2004; see also De Vries 1950; Grant et al. 1994; Jaeger et al. 2002; Kondrachuk 2002; Jaeger and Haslwanter 2004). Translational head movements or head tilts that apply linear forces parallel to the plane of the epithelium displace the relatively dense otoconia less than the epithelium. The hair bundles, coupled to the apical surface of the epithelium at their bases and to the otolithic membrane at their tips, are deflected. There is, however, little empirical information about the mechanical properties of these structures, and none reported in mammals. In the frog sacculus, the point stiffness of the gel layer and the stiffness of a hair bundle are of similar magnitude (Benser et al. 1993), implying that the gel layer does not move the bundles as a rigid plate would, but as a softer structure.

How otoconia are made is a mystery. Otoconia are calcite in mammals, birds, and sharks; aragonite and vaterite in other fishes; and aragonite and calcite in amphibians and reptiles (reviewed in Fermin et al. 1998). In mammals, each otoconium has a dense filamentous core, the principal protein of which is otoconin-90 (Y. Wang et al. 1998), with an outer cortex of microcrystals linked by a filamentous organic matrix (Lins et al. 2000). A central core is not seen in fractured frog otoconia (Kurc et al. 1999). Thalmann et al. (2001) suggest that otoconia form when otoconin-90, which is secreted by cells of the nonsensory epithelium, interacts with microvesicles secreted by supporting cells within the sensory epithelium (macula). Other proteins are implicated by the analysis of mice with genetic balance disorders not associated with hearing loss. Three such mutants have defective or missing otoconia and evidently present similar phenotypes: the tilted-head mouse, *thd* (Lim and Erway 1974); the tilted mouse, *tlt/mlh* (Ornitz et al. 1998; Y. Wang et al. 1998); and the head tilt mouse, *het* (Bergstrom et al. 1998). The *het* mice completely lack vestibular evoked potential responses to pulsed linear acceleration (Jones et al. 1999). The *het* mice have a mutant NADPH oxidase 3 (Paffenholz et al. 2004). The mutant protein in *tlt* mice is "otopetrin 1," a novel transmembrane domain protein of the macula (Hurle et al. 2003).

Richardson's group has evaluated expression of various proteins associated with accessory structures in both auditory and vestibular epithelia in mice and chicks. With in situ hybridization experiments on mouse saccular and utricular maculae, Rau et al. (1999) detected mRNA for α-tectorin between E12.5 and P15 and mRNA for β-tectorin between E14.5 and adult. Goodyear and Richardson (2002) used antibodies to compare expression of various extracellular matrix proteins across mouse and chick inner ear accessory structures. Mouse cupulae and otolithic membranes are both immunoreactive for otogelin and not immunoreactive for type II and IX collagens. Mouse otolithic membranes, but not cupulae, are immunoreactive for α-tectorin (throughout the otolithic membrane) and for β-tectorin in the meshwork around the bundles and in the gel. In the α-tectorin knockout mouse (Legan et al. 2000), the saccular and utricular otolithic membranes and otoconia were severely reduced, while the cupulae appeared normal.

The accessory structures show marked regional variations that may contribute to regional variations in afferent properties. In fact, the name "striola" was first used by Werner (1933) to describe the stripe of comparatively small crystals overlying the striolar zone of the macula (first noted by Lorente de Nó, 1926). In the turtle utriculus (fixed without dehydration to avoid disrupting the accessory matrices), Xue and Peterson (2003) found that relative to the extrastriola, the striola had a thinner otoconial layer, smaller crystals, a thicker gel layer, and a thinner column filament layer with fewer filaments. Furthermore, in the striola but not the extrastriola, channels ran through the gel and column filament layers; at the bottom of each was an eccentrically located hair bundle. Such channels are reported in other otolith organs (Jacobs and Hudspeth 1990) and may arise because the material of the accessory matrices is secreted by supporting cells, leaving holes over the hair cells. In conventionally fixed material from rodents, the gel layer has larger channels in the striola than in the extrastriola (Lim 1984). While such channels may be large because of dehydration artifact, the results nevertheless indicate regional variation in gel structure.

In chicks, the striolar zones of otolith maculae are, in common with the auditory epithelium, strongly immunoreactive for β-tectorin (Goodyear and Richardson 2002), provoking the investigators to make the intriguing suggestion that the auditory organs of higher vertebrates evolved from the striolar zone of an otolith organ.

3.2 Hair Bundles

Like other vertebrate hair bundles, vestibular bundles have rows of elongate microvilli (stereocilia) ascending stepwise in height (Fig. 8.4). At the tall edge of the bundle is a single true cilium, the kinocilium. Unlike the kinocilia of mammalian and avian cochlear bundles, which are an immature feature, the kinocilia of vestibular bundles persist throughout life. The stereocilia are filled with filamentous actin, cross-linked by fimbrin and espin (Furness and Hackney, Chapter 3). Extensions of the central set of actin filaments anchor each stereocilium in the cuticular plate, a specialized version of the terminal web. The kinocilium arises from a basal body located outside the cuticular plate and has a 9 + 2 array of microtubules. The kinocilia of amniotes (arrows, Fig. 8.4A) do not have the large distal bulb found in some bundles in amphibian otolith organs (Lewis et al. 1985; Jacobs and Hudspeth 1990).

As in all hair bundles, multiple extracellular linkages connect stereocilia within and between rows (Goodyear and Richardson 1999; Goodyear et al., Chapter 2). These include the tip links (Osborne et al. 1984) that run from the

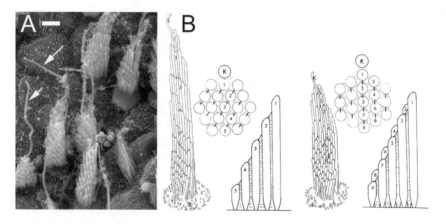

FIGURE 8.4. Hair bundles in mammalian otolith organs. (**A**) Adult mouse sacculus. *Arrows* point to long kinocilia found on some hair bundles. Most of the bundles in this field are oriented toward the upper right. *Scale bar*: 2 μm. Reproduced with permission from Figure 2 in Denman-Johnson K, Forge A (1999) Establishment of hair bundle polarity and orientation in the developing vestibular system of the mouse. J Neurocytol 28:821–835. (**B**) Schematic showing (*left*) tight and (*right*) loose arrangements of stereocilia in mammalian utricular bundles. Reproduced with permission from Figure 11 in Bagger-Sjöbäck and Takumida (1988) Geometrical array of the vestibular sensory hair bundle. Acta Otolaryngol (Stockh) 106:393–403.

apical tips of stereocilia to the backs of adjacent stereocilia in the neighboring, taller row (Bagger-Sjöbäck and Takumida 1988). Tip links have a central two or three helically entwined filaments and branch at their ends to several insertion sites on the stereociliary membrane (Kachar et al. 2000; Tsuprun and Santi 2000). One of the constituent proteins is cadherin 23 (Siemens et al. 2004; Söllner et al. 2004); the links between the kinocilium and tallest stereocilia are also immunoreactive for cadherin 23. Although tip links were postulated to be the gating springs that apply force to the transduction channels (Pickles and Corey 1992), cadherin 23 may not be elastic enough to act as the gating spring (Corey and Sotomayor 2004). Nevertheless, a critical role for tip links in transduction is suggested by multiple observations, including their proximity to the transduction channels, known to be near the tips of the bundles (Jaramillo and Hudspeth 1991); their alignment with the sensitive axis of the transduction apparatus (Shotwell et al. 1981); the sensitivity of both tip links and transduction to exogenous Ca^{2+} chelators (Assad et al. 1991; Zhao et al. 1996); and the sudden appearance of both tip links and transduction, at embryonic day 16 in developing mouse utricular hair bundles (Géléoc and Holt 2003), or a day after tip link breakage in recovering chick hair bundles (Zhao et al. 1996). Tip links might help position the transduction apparatus in its narrow operating range, while the force is applied by another route, possibly through ankyrin domains on the intracellular side of the channel protein (Howard and Bechstedt 2004).

3.2.1 Variations in Bundle Morphology with Zone and Cell Type

Variation in bundle morphology, commonplace in hair cell epithelia, presumably leads to diversity in transduction properties by affecting bundle mechanics and interactions with accessory structures. In the maculae of amniote otolith organs, bundles appear to differ between type I and type II hair cells and between zones, although there are few quantitative supporting data in mammals. One difficulty has been that scanning electron microscopy, ideal for examining bundle structure, does not reveal whether a cell is type I or type II, a classification based on postsynaptic structures, except on the rare occasions where the epithelium has been fractured to expose the hair cells' synaptic poles (Lapeyre et al. 1992). Most vestibular bundle data are from maculae, possibly reflecting the difficulty of preserving the shape of the long stereocilia in cristae. Features of stereocilia with potential functional sequelae include their number and arrangement within the bundle (the stereociliary array), stereociliary dimensions and linkages:

Stereociliary array. Variations have been noted in the shape of the bundle's footprint in the apical cell surface and in the stereociliary packing. In mammals, striolar bundles tend to be more wing shaped, with more stereocilia along the axis orthogonal to the sensitivity axis, while peripheral bundles have a rounder and more compact footprint (Lindeman 1973).

The hexagonal packing of stereocilia takes one of two orientations in mammalian vestibular hair bundles. In hair bundles with the "loose" arrangement (Fig. 8.4B, right), stereocilia in adjacent rows are aligned so that tip links run parallel to the bundle's axis of bilateral symmetry (the "orientation axis," running through the kinocilium). This is the arrangement typically reported in other hair cell epithelia. In the "tight" hair bundles of mammalian vestibular organs (Fig. 8.4B, left), in contrast, the stereocilia in adjacent rows are staggered, reportedly resulting in tip links oriented 30° off the line of bilateral symmetry. One third of the bundles in guinea pig utricles have tight packing (Bagger-Sjöbäck and Takumida 1988); these are distributed throughout the epithelium, may be either short or tall bundles, and may belong to type I and type II hair cells (Lapeyre et al. 1992). Semicircular canal cristae also have both tight and loose bundles (Bagger-Sjöbäck and Takumida 1988). If tip-link orientation determines response directionality (either as gating springs or as accessories to gating), then the intermingling of tight and loose bundles in vestibular epithelia appears to add a random element to the directionality. In cristae, as in auditory epithelia, stimuli are uniformly directed perpendicular to the long axis of the epithelium. Here the inclusion of tight bundles, which are maximally sensitive to stimuli 30° from the stimulus axis, would reduce an afferent's sensitivity proportional to its fractional input from such bundles. In otolith maculae, the deflection of the otoconial gel can occur in any direction, and all directions within the plane of the epithelium are mapped systematically on the epithelium (Jacobs and Hudspeth 1990). Here the random interjection of tight bundles may broaden the orientation sensitivity of afferents sampling from multiple hair cells.

Bundle stiffness and stereocilia dimensions. By Hooke's law, the stiffness of the hair bundle determines its deflection by an applied force. Stereociliary number, height and thickness will affect bundle stiffness. There are no estimates of stiffness in crista hair bundles. The static stiffness of hair bundles in cultured neonatal mouse utriculi and sacculi has been estimated, from the deflections evoked by fluid jets, as 1 mN/m (Géléoc et al. 1997). This value is 10-fold lower than estimates made the same way for the very short bundles of cultured outer hair cells (Géléoc et al. 1997), and is comparable to flexible probe measurements from frog saccular bundles (Howard and Hudspeth 1988). Experiments on frog saccular and turtle cochlear hair cells show that stiffness plummets by as much as 80% in the midpoint of the bundle's operating range, reflecting the "gating compliance" of channel opening and closing (Howard and Hudspeth 1988; Ricci et al. 2002).

In mammalian maculae, the kinocilium is sometimes significantly longer than the tallest stereocilia (arrows, Fig. 8.4A) and sometimes of similar height (Lim 1976). For both maculae and cristae, the kinocilia and the tallest row of stereocilia are reportedly shorter in striolar and central zones than in extrastriolar and peripheral zones (Lindeman 1969, 1973; Lim 1979; Hunter-Duvar and Hinojosa 1984; Lapeyre et al. 1992). This is partly true in the turtle utricular macula, where Peterson and colleagues are quantifying bundle morphology: kinocilia are shorter but stereocilia are taller in the striola relative to the extrastriola (Fontilla and Peterson 2000; Xue and Peterson 2004). This combination may make the turtle striolar bundles stiffer than the extrastriolar bundles. Qualitatively similar regional variations in bundle profiles have been described in the bullfrog utriculus (Baird 1992). Lapeyre et al. (1992) fractured guinea pig saccular epithelia to permit simultaneous visualization of cell somata and bundles in the scanning electron microscope. In all regions, the tallest stereocilia of type II bundles were shorter than those of type I bundles, with overall mean heights of about 4 and 9 µm, respectively. The heights of the tallest stereocilia were similar for type II cells across regions, but tended to be shorter in striolar type I bundles (median 7 µm) than in peripheral extrastriolar type I bundles (median 11 µm). Cristae are noted for the presence of very tall hair bundles; kinocilia and stereocilia are as tall as 80 µm and 60 µm, respectively, in the peripheral zone of the guinea pig crista (Lindeman 1969). According to Lim (1976), stereocilia in the periphery can also be much shorter, while the shortest stereocilia of all are found centrally.

Based on images from fixed tissue, investigators have suggested that bundles in the striolar zone are not directly coupled to the overlying gel, so that they are stimulated by fluid movements rather than motion of an attached structure (Lim 1979). At vestibular frequencies (approximately 10 Hz and below), vestibular bundles are shorter than the boundary layer, and viscous forces will dominate over inertial forces in determining fluid-evoked motions of the hair bundles (Freeman and Weiss 1988). Such movements will have a component determined by gel velocity (in addition to gel displacement), which could account for the phase lead of striolar afferent responses relative to extrastriolar afferent responses (Fig. 8.2C). Images from the turtle utriculus after fixation in nondehydrating conditions, however, show coupling in both striolar and extrastriolar zones (Xue and Peterson 2003). The coupling between bundles and extracellular matrices in mammalian vestibular epithelia should be reexamined in nondehydrating conditions.

Stereocilia diameter varies with both hair cell type and location, being larger for type I bundles than for neighboring type II bundles (Lindeman 1969; Hunter-Duvar and Hinojosa 1984; Morita et al. 1997; Rüsch et al. 1998) and larger for central/striolar bundles than for peripheral/extrastriolar bundles (Rüsch et al. 1998; Denman-Johnson and Forge 1999). Thus, the largest diameters belong to central/striolar type I bundles and the smallest to peripheral/extrastriolar type II bundles. During development, stereocilia begin as elongate microvilli (for immature mouse macular bundles, see Rüsch et al. 1998; Denman-Johnson and Forge 1999; Géléoc and Holt 2003; reviewed in Eatock and Hurley 2003).

Stereociliary numbers have not been counted in mammalian vestibular bundles. In the turtle utriculus, the bundles of type I cells have significantly more stereocilia than do those of type II cells (Moravec and Peterson 2004). Assuming that transduction channels are evenly distributed across stereocilia, the difference in stereociliary number can explain the larger transduction currents reported for type I cells relative to type II cells in the turtle utricle (Rennie et al. 2004; Rennie and Ricci 2004).

In summary, bundles in mammalian otolith organs have varied profiles (heights of stereocilia and kinocilia), stereociliary arrays (tight vs. loose, wide vs. compact), tip-link orientations, and stereociliary diameter. Cell type, region, and developmental stage affect these parameters, but we have few systematic observations of these relationships and no tests of their functional significance. Pulling together the scattered observations, however, the bundles of striolar type I cells appear to differ from their extrastriolar counterparts in ways that should increase their stiffness and move their frequency range up: they are shorter, with a larger footprint, reflecting thicker and probably more stereocilia.

3.3 Mechanoelectrical Transduction

Our direct information on mechanoelectrical transduction in mammalian vestibular organs comes from excised sensory epithelia of embryonic and early postnatal mouse utricles (E15 to P10). The otoconia and otolithic membrane are removed and the hair bundles are deflected, either with fluid jets of external solution delivered by wide-bore pipettes positioned near the bundle, or with rigid glass probes brought into contact with the bundle. In most experiments, bundles are deflected with 10- to 500-ms steps (Fig. 8.5A, C). Transduction currents have been recorded with the whole-cell ruptured patch method, with either K^+ or Cs^+ internal solutions, at room temperature and in a perilymph-like medium (with high Na^+, low K^+, and 1 to 2 mM Ca^{2+}). As described below, these experiments show fundamental similarities and a few differences between transduction in mammalian and nonmammalian hair cells.

The largest transduction currents reported in mouse utricular hair cells are 450 pA, corresponding to about two channels per stereociliary pair (or tip link; Géléoc et al. 1997; Vollrath and Eatock 2003). Peak transduction currents are plotted against bundle displacement, measured looking down on the bundle. As in other hair cells, the peak current–displacement relationships are fit with either first- or second-order Boltzmann functions (Fig. 8.5B, D). A second-order Boltzmann is consistent with a three-state model of transduction channel gating, with two closed states, one open state and mechanosensitive transitions (C↔C↔O; Corey and Hudspeth 1983b). The instantaneous operating range (the range of bundle displacements corresponding to 10% to 90% of the maximal response) is broader for mouse utricular hair cells than for frog saccular hair cells (Fig. 8.5B) or mouse cochlear hair cells (Géléoc et al. 1997; Holt et al. 1997; J.R. Holt et al. 2002; Vollrath and Eatock 2003). Correcting for the taller bundles of mouse utricular maculae, the transduction channels appear similarly sensitive to angular deflection (Géléoc et al. 1997; Holt et al. 1997). In mouse otolithic organs, the measured angular operating ranges vary from 2° (Géléoc et al. 1997) to 7° (Vollrath and Eatock 2003). These numbers agree with in vivo experiments on the transparent semicircular canals of young eels (Rüsch and Thurm 1989), which showed operating ranges of 5° of bundle deflection for extracellular summed responses from hair cells (the transepithelial voltage) and from afferent fibers (the compound nerve action potential).

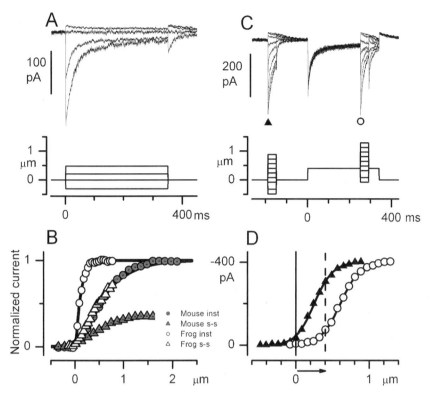

FIGURE 8.5. Transduction and adaptation in immature mouse utricular hair cells. (**A**) Transduction currents evoked by a family of step displacements, measured as probe displacement parallel to the plane of the epithelium. (**B**) Instantaneous and steady-state I–X relationships from a mouse utricular hair cell (*filled symbols*) and a frog saccular hair cell (*open symbols*). In terms of displacement at probe height, the instantaneous relation of the frog hair cell (*open circles*) is the most sensitive, while that of the mouse hair cell (*filled circles*) overlaps the steady-state relation of the frog hair cell (*open triangles*). The steady-state relation of the mouse hair cell is compressed relative to all of the other curves (see text). (**C**) Test steps superimposed on an adapting step (*bottom*) evoke large responses even after adaptation has occurred, showing that the adapted hair cell is sensitive to novel stimuli. (**D**) Peak, or instantaneous, stimulus-response (I–X) relationships from the cell in C, made by plotting the peak responses to the test steps before (*filled triangles*) and during (*open circles*) the adapting step. The step shifts the peak I–X relationship in the direction of the adapting step by 89% of the step amplitude (*arrow*). Reproduced with permission from (A, C, D) Fig. 1 and (B) Fig. 9 in Vollrath MA, Eatock RA (2003) Time course and extent of mechanotransducer adaptation in mouse utricular hair cells: comparison with frog saccular hair cells. J Neurophysiol 90: 2676–2689.

3.3.1 Adaptation

3.3.1.1 In Vitro Experiments

The transduction current of a mouse utricular hair cell in response to a positive bundle displacement decays rapidly and by a large amount (Figs. 8.5A, B and 8.6; Holt et al. 1997; J.R. Holt et al. 2002; Géléoc and Holt 2003; Vollrath and Eatock 2003; but see Géléoc et al. 1997). The adaptation produces a shift in the instantaneous $I(X)$ relation along the displacement axis in the direction of the adapting step (Fig. 8.5C, D; Corey and Hudspeth 1983a; Eatock et al. 1987; Holt et al. 1997; Vollrath and Eatock 2003). When a fast rigid probe is used, the adaptation has two kinetic components (Fig. 8.6; J.R. Holt et al. 2002; Géléoc and Holt 2003; Vollrath and Eatock 2003), as also found in frog saccular and turtle auditory hair cells (Howard and Hudspeth 1987; Wu et al. 1999; Vollrath and Eatock 2003). In development, the two adaptation processes appear as soon as the hair cells begin to transduce, at about E16 in the mouse utricular macula, coincident with the appearance of tip links (Fig. 8.6A; Géléoc and Holt 2003). The faster component dominates for stimuli in the lower half of the operating range while the slower component dominates for larger stimuli (Vollrath and Eatock 2003), as also described in frog saccular and turtle auditory hair cells (Shepherd and Corey 1994; Wu et al. 1999). At a displacement corresponding to half-maximal activation of the transduction current, the mean time constants are 5 and 45 ms (J.R. Holt et al. 2002; Vollrath and Eatock 2003), slower than corresponding estimates for frog saccular hair cells (Vollrath and Eatock 2003) and turtle cochlear hair cells (Wu et al. 1999). In the frequency domain, the adaptation processes act as high-pass filters with mean corner frequencies in mouse utricular hair cells near 30 and 3.5 Hz. Figure 8.7 shows an example in which slow adaptation caused the transduction current to fall off at frequencies below 2.5 Hz (filled triangles in Fig. 8.7B). As discussed in the next section, these processes might have different kinetics in vivo.

In a study of the early postnatal utricular epithelium of the mouse, adaptation to rigid-probe steps did not vary systematically between type I and type II hair cells or between striolar and extrastriolar hair cells (Vollrath and Eatock 2003). Thus, these experiments found no evidence that properties of the hair cell transduction apparatus contribute to the regional variation in response dynamics of rodent utricular afferents (see afferent data plotted as lines in Fig. 8.7B; Section 2.2.1). More heterogeneity may emerge as the vestibular epithelium matures; other caveats associated with this comparison are discussed in the next section. In contrast, in the mature turtle utriculus, where type I hair cells are confined to part of the striolar zone, transduction currents are threefold larger in type I cells than in type II cells (Rennie et al. 2004; Rennie and Ricci 2004), consistent with differences in stereocilia numbers between type I cells and extrastriolar type II cells (Moravec and Peterson 2004).

Work on frog and turtle hair cells has shown that both fast and slow adaptation processes are Ca^{2+}-dependent (reviewed in Eatock 2000; Fettiplace and Ricci,

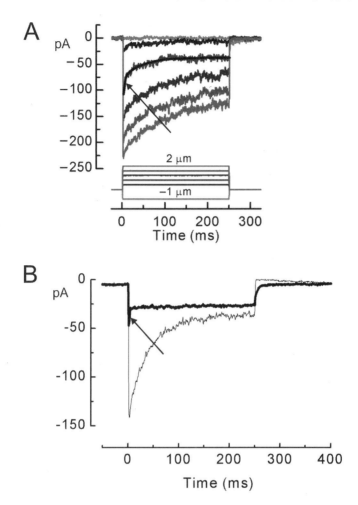

FIGURE 8.6. Development and analysis of two components of adaptation in mouse utricular hair cells. (**A**) Both fast and slow adaptation components are evident early in development, as soon as transduction can be detected (E16 to 17 in the mouse utriculus). The current evoked by an intermediate step clearly has a double exponential decay (*arrow*). Reproduced with permission from Fig. 3b in Géléoc GS, Holt JR (2003) Developmental acquisition of sensory transduction in hair cells of the mouse inner ear. Nat Neurosci 6:1019–1020. (**B**) Slow adaptation requires myosin 1c. To test whether myosin 1c drives slow adaptation, transgenic mice were created with mutant myosin 1c proteins that would freeze in the presence of certain modified ADP analogues, introduced via patch recording pipettes. Whole-cell recordings are from a single utricular hair cell, showing transduction currents evoked by a step bundle displacement, early in the recording (*fine trace*) and late in the recording (*heavy trace*), after the modified ADP molecules reached the site of adaptation. As predicted, the slow adaptation is blocked late in the recording, leaving behind a fast component (*arrow*). Modified from Fig. 4G in Holt JR, Gillespie SK, Provance DW, Shah K, Shokat KM, Corey DP, Mercer JA, Gillespie PG (2002) A chemical-genetic strategy implicates myosin-1c in adaptation by hair cells. Cell 108: 371–381, with permission from Elsevier.

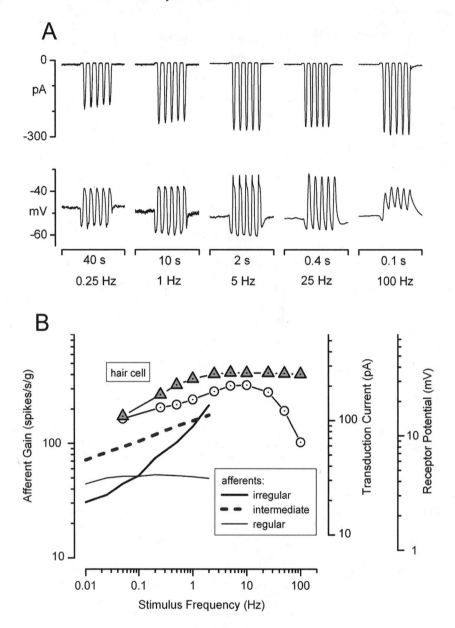

Chapter 4). The fast component is proposed to be an amplifying mechanism that tunes the transduction process to the natural stimulus frequency of the hair cell in vivo (the "stereociliary amplifier"; Hudspeth et al. 2000; Fettiplace et al. 2001). In frog and turtle hair cells, when the Ca^{2+} conditions are changed to better resemble in vivo conditions (by reducing apical Ca^{2+} and using the hair cell's endogenous Ca^{2+} buffers), the fast decay becomes a damped series of oscillations at lower frequency. One model of how this process works is that Ca^{2+} enters the transduction channel and binds to the channel, promoting its closure (Howard and Hudspeth 1987; Crawford et al. 1991). According to the gating spring hypothesis, channel closure will increase tension in the attached gating spring, feeding back positively on the transduction process (Hudspeth et al. 2000). This process will tune the transduction current at a frequency determined by its kinetics.

Based on results from the frog sacculus, the slow adaptation process has been modeled as a movement of the transduction channel and attached gating spring along the stereocilium in the direction to restore resting levels of tension (Howard and Hudspeth 1987; Assad and Corey 1992; Shepherd and Corey 1994). The movement is envisioned as driven by ensembles of myosin molecules interacting with the actin core of the stereocilium. For positive bundle deflections (toward the tall edge of the bundle), slipping of the myosin relative to the actin filaments relieves stretch on the gating spring and makes the bundle substantially less stiff. Under a positive force step, as applied by a flexible probe or a fluid jet, this mechanical change produces a forward creep in the direction of the force step (Howard and Hudspeth 1987). There is a great deal of evidence for

◀―――――――――――――――――――――――――――

FIGURE 8.7. Comparison of the frequency dependence of hair cell and afferent responses. (A) Whole-cell transduction currents (*top*) and receptor potentials (*bottom*) recorded in voltage-clamp and current-clamp modes, respectively, from a mouse utricular hair cell (neonatal, no $g_{K,L}$) in response to sinusoidal bursts of five cycles each at frequencies ranging from 100 Hz to 0.25 Hz, delivered by a fluid jet. Note the different time scales for each burst. (B) Peak-peak transduction currents (*triangles*), receptor potentials (*circles*) and afferent gains (*lines*) as functions of stimulus frequency; hair cell data are from the traces in A and others. The transduction current is highly asymmetric and decays below 2.5 Hz, reflecting slow transducer adaptation, which dominates for these relatively large stimuli (Vollrath and Eatock 2003). The receptor potential decays more than the transduction current does between 5 and 2.5 Hz and is increasingly symmetric as frequency is lowered. Both effects may reflect K^+ current activation at low stimulus frequencies (see text). The low-frequency slope of the receptor potential falls between the slopes of the intermediate and irregular groups of chinchilla utricular afferents. Modified with permission from (A) Figure 2 and (B) Figure 3 in Holt JR, Vollrath MA, Eatock RA, Stimulus processing by type II hair cells in the mouse utricle. Ann NY Acad Sci 871:15–26; Copyright 1999 New York Academy of Sciences. The chinchilla afferent data in (B) are replotted from Figure 6 in Goldberg JM, Desmadryl G, Baird RA, Fernández C (1990a) The vestibular nerve of the chinchilla. IV. Discharge properties of utricular afferents. J Neurophysiol 63:781–790.

this model, much of it reviewed previously (e.g., Hudspeth and Gillespie 1994; Eatock 2000). More recently, the model received strong support from experiments on transgenic mice with a mutant form of myosin 1c that can be blocked by an ADP analogue (J.R. Holt et al. 2002). This manipulation blocks the slow component, but not the fast component, of transduction current decay in mouse utricular hair cells (Fig. 8.6B).

3.3.1.2 Implications of Hair Cell Transducer Adaptation for Vestibular Function In Vivo

The fast adaptation in mouse utricular hair cells, which dominates for small stimuli, has an average time constant of 5 ms, corresponding to a corner frequency of 30 Hz. This is likely to be within the frequency range of mouse utricular afferents. In preliminary data from mouse otolith afferents (Lasker et al. 2004), gain re: linear acceleration peaks at 2 Hz for regular afferents and 4 Hz for irregular afferents, but decreases less than twofold as frequency is increased to 10 Hz. Chinchilla semicircular canal afferents are sensitive to head movement frequencies up to at least 20 Hz (Hullar and Minor 1999) and in fact, gain increases with frequency over this range (Hullar et al. 2004). Otolith-driven reflexes operate up to at least 25 Hz in squirrel monkeys (Angelaki 1998). It is possible that high-pass filtering by the fast adaptation process of hair cells in otolith organs helps compensate for low-pass filtering of linear forces by the mass of the head.

The slow component of hair cell adaptation, which dominates for large stimuli and has a corner frequency of about 5 Hz, could contribute to the high-pass filtering of some vestibular afferents in this low frequency range (Goldberg et al. 1990a; Hullar et al. 2004). Figure 8.7B compares the frequency dependence of receptor potential data from young mouse utricular hair cells (Holt et al. 1999) and from mature chinchilla utricular afferents (Goldberg et al. 1990a). Frequency filtering therefore shifts to lower frequencies as stimulus level increases. Such a shift might correlate with a level-dependent transition in head movements, from small-amplitude, high-frequency involuntary vibrations to larger, slower postural motions.

The slow component of adaptation also shifts the instantaneous $I(X)$ relationship along the displacement axis. As illustrated by Figure 8.5C, this restores transducer sensitivity in the face of large steady stimuli that saturate the instantaneous operating range, as might occur, for example, during head tilts. In frog saccular hair cells, the shift saturates at about 80% of the step size no matter what the step size. In many mouse utricular hair cells, however, the percent shift increases for large bundle deflections (Vollrath and Eatock 2003). This compresses the steady-state operating range (Fig. 8.5B; see explanation in Vollrath and Eatock 2003) but expands the range of background displacements over which the mouse utricular hair cells can adjust their instantaneous operating range. Together, the relatively large operating range for adaptation and instantaneous operating range for transduction (Fig. 8.5B) suggest that mouse utricular bundles handle relatively large deflections in vivo.

Although it is reasonable to ask how the measured properties of hair cell transduction and adaptation manifest themselves in afferent responses to head movements, we must recognize the substantial differences between in vitro and in vivo recording configurations. Of many factors that may affect adaptation, three obvious examples are intrastereociliary Ca^{2+}, temperature, and the nature of the stimulus. Ricci et al. (1998) studied how adaptation in turtle auditory hair cells changed when they went from a perilymph-like solution and artificial Ca^{2+} buffering to an endolymph-like solution and endogenous Ca^{2+} buffering— manipulations that should produce an intrastereociliary Ca^{2+} level close to the in vivo value. The step-evoked adaptation changed from a fast decay to a slower resonance; in other words, from high-pass filtering at a particular corner frequency to tuning with a lower peak frequency. Whether similar effects would be seen in mammalian vestibular hair cells is not known. To get more physiological data from mammalian hair cells, we should both manipulate Ca^{2+} and move from room to body temperature; increasing the temperature might offset the frequency-lowering effects of the Ca^{2+} manipulations.

The stimulus delivered by the otolithic membrane or cupula to the bundle during a head movement may resemble a force stimulus more than a displacement stimulus. The relative stiffnesses of mammalian accessory structures and hair bundles are not known. Although the otoconial mass may move as a rigid body, the gel layer probably does not, so that bundles are normally moved by a more compliant structure than the rigid probes used in many experiments. This will affect the transduction current decay because the adaptation process can feed back on bundle position. Indeed, when the slow adaptation process is activated by a force step rather than a displacement step, the decay of the transduction current is considerably slowed (Vollrath and Eatock 2003). This effect may have contributed to the lack of apparent adaptation in cultured mouse vestibular hair cells when deflected by short fluid-jet steps (Géléoc et al. 1997; see discussion in Vollrath and Eatock 2003) and in toadfish crista hair cells when stimulated by cupular deflections in vivo (Rabbitt et al. 2005). It is important to note that although use of a more compliant stimulus reduces the rate of transduction current decay, it does not eliminate the underlying adaptation processes, which will continue to affect the sensitivity of the transduction process.

In addition to changing the average behavior, the stimulation and recording conditions may obliterate naturally occurring heterogeneity in transduction and adaptation. The rigid probe will override mechanical differences between bundles of different geometry. The ruptured-patch whole-cell recording configuration, which washes out the native intracellular milieu with a simple pipette solution, should reduce the impact of natural variations in Ca^{2+} binding proteins on Ca^{2+}-dependent processes.

3.3.2 Ca^{2+} Regulation in Vestibular Hair Bundles

Intrastereociliary Ca^{2+} controls multiple aspects of transduction and adaptation (Fettiplace and Ricci, Chapter 4). Factors affecting intrastereociliary Ca^{2+} levels include extracellular Ca^{2+} levels; transduction channel conductance; Ca^{2+} pumps and Ca^{2+} binding proteins.

The free Ca^{2+} in vestibular endolymph has been estimated at 200 to 250 μM, about 10-fold higher than estimates for mammalian cochlear endolymph (reviewed in Sterkers et al. 1988). The higher value in the vestibular endolymphatic compartment may reflect the presence there of calcium carbonate crystals. New data from a lizard's auditory organ, which shares an otic compartment with an otolith organ (the lagena), suggest that endolymphatic Ca^{2+} is an order of magnitude higher, making it similar to perilymph levels (Manley et al. 2004). This question needs to be resolved before we can know the in situ behavior of the many processes that depend on endolymphatic Ca^{2+} levels.

The principal entry route for Ca^{2+} into stereocilia is transduction channels, which are significantly Ca^{2+} permeable (Jørgensen and Kroese 1995). It is estimated that in endolymph, 15% to 20% of the transduction current is carried by Ca^{2+} ions despite their relatively low concentration (Ricci and Fettiplace 1998). These estimates have not been done for mammalian vestibular hair bundles. In the mammalian cochlea, Ca^{2+} may also enter stereocilia via $P2X_2$ ATP receptor channels, which are present at high density and flux significant amounts of Ca^{2+} (Housley et al. 1999), but there is no evidence for such receptors in vestibular bundles (G.D. Housley, personal communication).

The Ca^{2+} gradient across the stereociliary membrane is maintained by plasma membrane Ca^{2+} ATPases (PMCA) (Yamoah et al. 1998). Hair bundles lack intracellular Ca^{2+} storage organelles. Sans and colleagues have used functional and immunocytochemical methods to show that hair cells of the guinea pig utricular macula and cristae express PMCAs in both bundles and basolateral membranes (Boyer et al. 2001) and K^+-dependent Na^+–Ca^{2+} exchangers in the basolateral membrane (Chabbert et al. 1995; Boyer et al. 1999). (Note that Na^+–

FIGURE 8.8. Calcium regulation. (**A**) Expression of PMCA isoforms and calretinin in rat vestibular hair cells. *Top row*, rat utricular macula, viewed from above. *Bottom row*, Side view of rat crista. PMCA3 labels hair cell basolateral membranes (*top, left*) and PMCA2 labels hair bundles (*bottom, left*). Calretinin antibody (*right*) stains numerous cell bodies in both the utricular macula and the crista. Hair cells with low PMCA3 immunoreactivity have high calretinin levels (*arrows, top row*). Modified with permission from Figure 7 in Dumont RA, Lins U, Filoteo AG, Penniston JT, Kachar B, Gillespie PG (2001) Plasma membrane Ca^{2+}-ATPase isoform 2a is the PMCA of hair bundles. J Neurosci 21:5066–5078. (**B**) Calretinin antibody staining in mature rodent vestibular epithelia. Calyx afferent terminals in the central zone of the crista and striolar zone of the macula are strongly immunoreactive. In the mouse, numerous extrastriolar/peripheral type II cells are also calretinin-positive, as also seen in the rat vestibular epithelia in A (*right column*). *Scale bars*, 100 µm; A, anterior, M, medial. Modified with permission from Fig. 4 in Desai SS, Ali H, Lysakowski A (2005b) Comparative morphology of the rodent vestibular periphery: II. Cristae ampullares. J Neurophysiol 93(1):267–280; and Fig. 4, Desai SS, Zeh C, Lysakowski A (2005a) Comparative morphology of the rodent vestibular periphery: I. Saccular and utricular maculae. J Neurophysiol 93(1):251–266.

8. Mammalian Vestibular Hair Cells 375

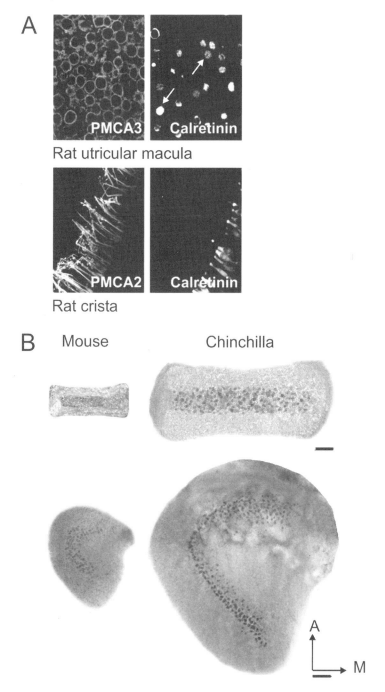

Ca^{2+} exchangers would not work well to extrude Ca^{2+} across the bundle membrane because of the low Na^+ gradient in endolymph.)

Different PMCA isoforms occur in the apical and basolateral membranes of hair cells. In frog saccular and rat vestibular and auditory hair cells, hair bundles express PMCA2a, a splice variant of PMCA isoform 2, and basolateral membranes express other isoforms, specifically PMCA3 and PMCA1 in rat vestibular hair cells (Fig. 8.8A; Dumont et al. 2001). Defects or null mutations in PMCA2 lead to vestibular dysfunction as well as hearing problems (Kozel et al. 1998; Street et al. 1998). The deafwaddler mouse mutant ($dfw^{-/-}$) has a single-nucleotide substitution for Gly^{238}, a nucleotide that is conserved across PMCA isoforms. Both $dfw^{-/-}$ mice and $Pmca2^{-/-}$ mice eventually lose otoconia, indicating that PMCA2-mediated extrusion of Ca^{2+} into endolymph is necessary to maintain these calcium carbonate crystals.

Calmodulin-like immunoreactivity is ubiquitous and intense in hair bundles and cell bodies of adult rodent vestibular hair cells (Ogata and Slepecky 1998; Ogata et al. 1999). Calmodulin can be both a Ca^{2+} buffer and a Ca^{2+}-activated facilitator of various processes. But calmodulin is estimated at just 70 μM in frog saccular hair bundles (Walker et al. 1993) and therefore is not likely to be a significant source of Ca^{2+} buffering (Heller et al. 2002). Calmodulin-facilitated processes in vestibular hair cells may include PMCA activity, myosin-based transducer adaptation (Walker and Hudspeth 1996; Cyr et al. 2002), and nitric oxide synthase activity (see Section 5.2.4.2). Most bundle calmodulin may be bound to PMCA (Yamoah et al. 1998). In frog utricular hair cells, there are correlations between epithelial location, bundle type, calcium-binding protein expression, and adaptation rate: certain striolar bundles are fast adapting and strongly immunoreactive for the high-affinity calcium-binding proteins, calmodulin and parvalbumin, while peripheral bundles are slowly adapting and more immunoreactive for the lower affinity calbindin-D28k (Baird et al. 1997).

3.4 Summary of Mechanical Stimulation and Transduction

Multiple regional variations in accessory structures and bundle morphology may help tune the distinctive response dynamics of central/striolar versus peripheral/extrastriolar vestibular afferents. In the striola, otoconia are smaller and the gel layer is more porous. Models of otolith and gel movement have not yet taken these variations into account, and generally treat the input to the hair bundles as uniform across the epithelium. Bundles in the striola tend to be have morphological features likely to increase their stiffness and, consequently, sensitivity at high frequencies. Possible effects of regional and cell type variations in calcium pumps and binding proteins on transduction remain to be explored.

Transduction and adaptation in mouse utricular hair cells resemble reports from nonvestibular hair cells in several fundamental ways. They have similar sensitivity to angular bundle deflection and similar adaptation mechanisms. Because their bundles are taller than those of cochlear and frog saccular hair cells, they are less sensitive in terms of displacement near the tips of the bundles, for example, displacement of the overlying gel. For step deflections effected by a rigid glass probe, the transduction current exhibits both fast and slow phases of adaptation, with time constants an order of magnitude apart. The two components dominate over different ranges of stimulus level such that the hair cell's corner frequency for high-pass filtering moves to lower frequencies as level increases. One difference relative to other hair cells is that the steady-state extent of adaptation increases for large displacements, rather than being a constant proportion of the displacement, allowing the mouse utricular hair cells to shift their instantaneous operating ranges over a very broad range of displacements. Even after adaptation is complete, significant steady-state transduction currents (5% to 30% of the peak current) remain.

4. Post-Transduction Processing

In this section we will review voltage-gated conductances (see Section 4.1) and Ca^{2+} buffering in the basolateral, rather than bundle, compartment (see Section 4.2). A basolateral localization for most voltage-gated ion channels is supported by immunocytochemical evidence and whole-cell recordings from isolated hair cells, which are frequently shorn of their hair bundles. Thus, the hair cell appears neatly segregated into an apical compartment that performs mechano-electrical transduction and a basolateral compartment that shapes the receptor potential and transmits the signal to the afferent nerve terminals.

4.1 Voltage-Gated Conductances

The transduction current initiates a receptor potential that is shaped by currents flowing through voltage-dependent channels in the basolateral membrane. Each morphologically distinct hair cell type is endowed with a distinct complement of ionic conductances. Presumably the driving force behind these tightly regulated differences is the need for differential stimulus processing at the sensory periphery, although voltage-gated conductances may participate in development of the epithelium as well (reviewed in Eatock and Hurley 2003, and Goodyear et al., Chapter 2). The conductances of rodent type I and type II hair cells are illustrated schematically in Figure 8.9. We use g_x and I_x to refer to the whole-cell conductance and current, respectively, corresponding to conductance type X.

In this section, we review how voltage-gated conductances vary with hair cell type, developmental stage, and epithelial zone in mammalian vestibular organs, with reference to analogous and sometimes more thorough studies in avian and turtle vestibular organs. These physiologically characterized conductances may reflect the activity of one or more channel types (e.g., Selyanko et al. 1999) and we are making progress in learning the molecular identities of the responsible channels (Table 8.1).

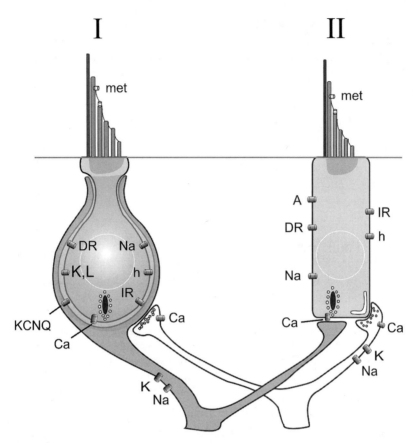

FIGURE 8.9. Ionic conductances in rodent utricular hair cells. Both type I and type II hair cells have mechanoelectrical transducer channels (met), voltage-gated Ca^{2+} channels (Ca), a mixed K^+/Na^+ conductance that activates with hyperpolarization (h), and a fast, steep inward rectifier (IR), which may be carried by Kir4.1 (Acuna et al. 2004). The largest conductances comprise outwardly rectifying K^+ channels. In type II cells these include A-current (A) and delayed rectifier (DR) channels, which may be differently distributed across the epithelium. In type I cells the K^+ conductances that activate with depolarization include $g_{K,L}$, with a very negative activation range, and one or more others with a more positive activation range (DR). Afferent (*gray*) and efferent (*white*) fibers presumably use Na^+ and K^+ channels to make action potentials. The latter may include an AHP conductance in regular afferents (see Section 2.2.3) and other outward rectifiers. Several outwardly rectifying K^+ channels have been described in immature rat vestibular ganglion neurons (Chabbert et al. 2001). The afferent neuron shown is dimorphic: it innervates a type I hair cell and a type II hair cell. The calyx inner face is strongly immunoreactive for KCNQ4, hence KCNQ channels are shown. Efferent terminals presumably have Ca^{2+} channels to mediate neurotransmitter release and K^+ and Na^+ channels to mediate spikes; these have not been studied.

TABLE 8.1. Molecular candidates for voltage-gated conductances in amniote vestibular hair cells.

Conductance	Type I	Type II	Molecular candidates
K,L (KI)	X		KCNQ subunits (Kharkovets et al. 2000; J.R. Holt et al. 2004)
IR (K1)	X	X	Kir2.1 (Correia et al. 2004); Kir4.1 (Acuna et al. 2004)
h	X	X	HCN isoforms (Cho et al. 2003)
L-type Ca^{2+}	X	X	$Ca_V1.3$ (Bao et al. 2003; Dou et al. 2004)
Non-L-type Ca^{2+}		X	$Ca_V2.x$? (Bao et al. 2003; Dou et al. 2004)
Na^+, TTX-sensitive		X	$Na_V1.2$, 1.6 (Chabbert et al. 2003)
Na^+, TTX-insensitive	X	X	$Na_V1.5$ (Wooltorton et al. 2005)

4.1.1 Differences Between Type I and Type II Hair Cells

The most striking known physiological difference between type I and type II hair cells is in the outwardly rectifying conductances. In most hair cells, including type II cells, such conductances have voltage ranges of activation that are positive of resting potential. The outwardly rectifying conductances of type II cells comprise delayed rectifiers and, in some cases, A-type conductances and Ca^{2+}-activated K conductances; some of the variation between type II cells correlates with location in the epithelium (see Section 4.1.3). Type I hair cells have a conventional outwardly rectifying current as well, but are dominated by a large and very negatively activating conductance, called g_{KI} (for "type I specific K conductance," Rennie and Correia 1994) or $g_{K,L}$ (for "low-voltage-activating potassium conductance," Eatock et al. 1994). This conductance was first reported in pigeon type I hair cells by Correia and Lang (1990) and has since been documented in type I hair cells from the maculae and cristae of diverse amniotes, including turtles (Brichta et al. 2002), birds (Ricci et al. 1996; Masetto et al. 2000), and rodents (Eatock and Hutzler 1992; Rennie and Correia 1994; Rüsch and Eatock 1996a; Chen and Eatock 2000). Unlike the outward rectifiers in type II cells, $g_{K,L}$ is substantially activated at resting potential. In comparison to the outward rectifiers in type II cells, $g_{K,L}$ has relatively slow activation kinetics and inactivates little. It confers distinctive whole-cell current and voltage responses to a standard series of voltage and current steps (Fig. 8.10). As described in Section 4.1.5.2, $g_{K,L}$ is sensitive to modulation by a variety of agents.

The pore-forming part of a K^+ channel protein comprises four α subunits, which may be the same (forming homomultimeric proteins) or dissimilar but from the same family (forming heteromultimeric proteins). Based on antibody staining and mRNA expression, Kharkovets et al. (2000) suggested that $g_{K,L}$ comprises subunits from the KCNQ superfamily of K channels, specifically KCNQ4, possibly in combination with another subunit. Consistent with this prediction, transfection with a dominant negative variant of KCNQ4 knocked out the expression of $g_{K,L}$ in young mouse utricular hair cells (preliminary report from J.R. Holt et al. 2004). Because the mutation is in the pore, the presence of just one mutant subunit per channel protein suffices to block K^+ flux. Together, the results of Kubisch et al. (1999), Kharkovets et al. (2000) and J.R. Holt et al. (2004) all strongly suggest that $g_{K,L}$ includes KCNQ4 subunits. This would make it a sister conductance to $g_{K,n}$, a negatively activating conductance that includes KCNQ4 subunits and is expressed in cochlear hair cells (Marcotti and Kros 1999; Oliver et al. 2003; Marcotti et al. 2003a).

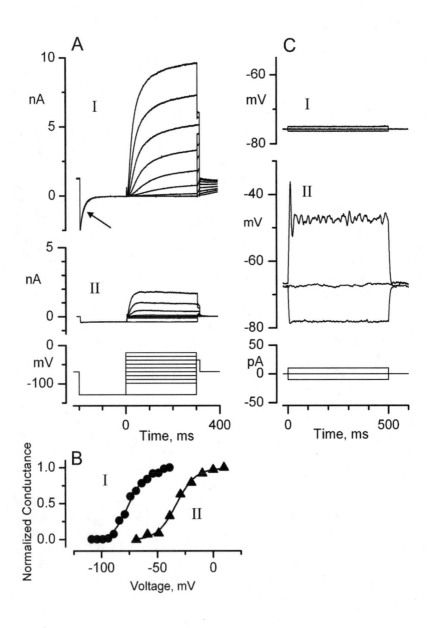

The full story may be more complicated, however. Quantitative electron microscopic analysis of the KCNQ4 immunoreactivity in adult rat vestibular epithelia reveals it to be principally localized on the postsynaptic inner face of the calyx, with surprisingly light staining of the hair cells (Lysakowski and Price 2003; see Fig. 8.14C,D). Furthermore, a comparison of $g_{K,L}$ and $g_{K,n}$ in 2 to 4-week-old rat utricular and cochlear hair cells revealed major qualitative differences in biophysical and pharmacological properties (Wong et al. 2004). Relative to $g_{K,n}$, $g_{K,L}$ has a slower activation time course and different deactivation kinetics. It also has a less negative and more labile activation range, which tends to shift positively during ruptured-patch recordings (Rüsch and Eatock 1996a; Chen and Eatock 2000). $g_{K,n}$ channels are much more sensitive to the KCNQ ion channel blocker linopirdine than either KCNQ4 homomultimeric channels (Kubisch et al. 1999) or $g_{K,L}$ channels; indeed, the outwardly rectifying current of vestibular type II hair cells is more sensitive to linopirdine (Rennie et al. 2001). There are several possible explanations for the biophysical and pharmacological differences. One is that $g_{K,n}$ and $g_{K,L}$ are both KCNQ channels but have different primary and/or accessory subunit composition, for example, they could be different heteromultimers of KCNQ3, KCNQ4 and/or KCNQ5. Another possibility is that $g_{K,n}$ and $g_{K,L}$ have similar subunit composition but are post-translationally modified in different ways. Finally, the negatively activating conductance of mature rat utricular type I hair cells, which is substantially larger than that at younger ages (Eatock and Hurley 2003), may comprise KCNQ

FIGURE 8.10. The voltage-gated conductances of type I and type II hair cells differ. (**A**) Whole-cell currents evoked by voltage steps (*bottom panel*), from a type I hair cell (*top panel*) and a type II hair cell (*middle panel*) isolated from the rat utricular macula (K.M. Hurley and R.A. Eatock, unpublished observations). Standard external (perilymph-like) solution and perforated patch recording, which by preserving internal messengers stabilizes $g_{K,L}$ and other labile conductances. At the holding potential (−69 mV) the type II hair cell has very little current, while the type I cell has an outward current through open $g_{K,L}$ channels. A prepulse to −129 mV activates g_{K1} in the type II cell but deactivates $g_{K,L}$ in the type I cell (*arrow in A*). Steps positive to −69 mV activate outwardly rectifying current in the type II cell, which may include g_{KCNQ}. Steps positive to −109 mV activate $g_{K,L}$ in the type I cell. At more positive potentials, the $g_{K,L}$ channels may be joined by outwardly rectifying K⁺ channels with a more positive activation range. (**B**) Activation curves showing how the outwardly rectifying K⁺ conductance varies with voltage for a type I hair cell (*circles*) and a type II hair cell (*triangles*), both isolated from the rat utricular macula in the second postnatal week. Single-Boltzmann fits of the data yielded voltages of half-maximal activation of −76 mV for the type I cell and −33 mV for the type II cell. (**C**) Voltage responses to current steps (shown in *bottom panel*) differ markedly in the type I hair cell (*top*) and the type II hair cell (*middle*). Same cells as in A. Because of $g_{K,L}$, type I cells have a very low input resistance and therefore, by Ohm's law, generate small voltage responses to 10-pA current steps. Type II cells have a high input resistance because very few channels are activated at the resting potential, and therefore generate large receptor potentials in response to small currents.

channels plus other negatively activating channels. This last possibility could help explain why the KCNQ4 immunoreactivity of mature rat type I cells is light while the conductance is so large.

Why do type I hair cells have $g_{K,L}$? $g_{K,n}$ may be important in clearing K^+ from cochlear hair cells; K^+ enters through transduction channels apically and exits through K^+ channels basolaterally (Kennedy et al. 2003). Rennie and Ricci (2004) put forth a similar argument for $g_{K,L}$, based on their observations that in the turtle utriculus, type I cells have larger transduction currents than do type II hair cells. $g_{K,L}$ is well suited to this role because it is activated at all physiological voltages and does not inactivate even in the face of steady depolarization (Goldberg and Brichta 2002). In the remainder of this chapter we will discuss such other possibilities as a role in nonconventional afferent transmission (see Section 5.1.1) and the potential value of $g_{K,L}$ as a target of modulators such as efferent transmitters (see Section 5.2.1). It should be noted that the density of the negatively activating conductance in rat utricular type I hair cells is as high as 50 nS/pF, compared to just 500 pS/pF in rat cochlear outer hair cells (Wong et al. 2004). The extraordinarily large size of $g_{K,L}$, together with its very negative activation range, gives type I cells unusually low input resistances. The expected impact on receptor potentials is described in Section 4.1.4.

There may be other differences in the voltage-gated ion channels of type I and type II hair cells, but these are less dramatic and less certain. As described in the next section, voltage-gated Ca^{2+} currents differ only slightly between type I and type II cells isolated from the cristae of young rats.

4.1.2 Voltage-Gated Ca^{2+} Channels

The voltage-gated Ca^{2+} channels of hair cells are of particular interest because they mediate release of afferent neurotransmitter. In cochlear hair cells from diverse vertebrates, it appears that most of the voltage-gated Ca^{2+} current flows through one type of channel. In vestibular hair cells, however, significant voltage-gated Ca^{2+} current may flow through other channel classes.

Voltage-gated Ca^{2+} channels fall into six pharmacologically and biophysically defined classes: L, N, P, Q, R, and T. The channel comprises a pore-forming α subunit plus additional subunits. In cochlear hair cells, unlike neurons, transmitter release is principally mediated by L-type Ca^{2+} channels (Fuchs and Parsons, Chapter 6). Relative to other L-type channels, L-type channels in hair cells have an unusually negative voltage range of activation, little inactivation, and low sensitivity to dihydropyridine antagonists. The channels activate close to the cells' resting potential; whether they activate negative to resting potential has been hard to establish and is critical to the question of whether all transmitter release, including that at background stimulus levels, is mediated by L-type Ca^{2+} channels.

In chick and mouse cochleae, there appears to be no contribution of non-L-type channels either to whole-cell Ca^{2+} currents or to transmitter release (Zidanic and Fuchs 1995; Spassova et al. 2001; Brandt et al. 2003). The predominant L-type α subunit in these hair cells is $Ca_V1.3$ (formerly $α_{1D}$; Kollmar et al. 1997;

Platzer et al. 2000; Brandt et al. 2003). In mice that are null for $Ca_V1.3$, inner hair cells have a small residual Ca^{2+} current (Platzer et al. 2000; Dou et al. 2004) carried by another L-type α subunit, $Ca_V1.2$ ($α_{1C}$; Brandt et al. 2003) but virtually no synaptic exocytosis, measured as depolarization-evoked increases in cell surface area (Brandt et al. 2003).

In vestibular hair cells, there may be both L-type channels and non-L-type Ca^{2+} channels. In frog saccular hair cells, single-channel studies reveal two channel types (Su et al. 1995; Rodriguez-Contreras and Yamoah 2001), one L-type and the other possibly N-type: fluorescently labeled ω-conotoxin, an N-type channel blocker, labels the hair cells at sites that are likely to be active zones (Rodriguez-Contreras and Yamoah 2001). In frog crista hair cells, pharmacological analysis of whole-cell Ca^{2+} currents suggests that there are R-type channels as well as L-type channels (Martini et al. 2000). Both type I and type II mammalian hair cells have L-type channels (Boyer et al. 2001; Almanza et al. 2003; Bao et al. 2003; Dou et al. 2004), some of which are $Ca_V1.3$ channels (Bao et al. 2003: Dou et al. 2004). Not all, however: in $Ca_V1.3$-null mice, the Ca^{2+} current of vestibular hair cells is 50% to 60% that of wild-type currents (Fig. 8.11; Dou et al. 2004) versus just 10% for cochlear cells (Platzer et al. 2000). The greater functionality of the residual current in vestibular hair cells is suggested by the fact that $Ca_V1.3$-null mice, while deaf, have no obvious balance problems (Platzer et al. 2000; Dou et al. 2004).

In hair cells isolated from the rat crista in the first postnatal month, there is little difference in the Ca^{2+} currents of type I and type II hair cells (Bao et al. 2003). The recorded channel densities are slightly smaller in type I cells and in both cell types are smaller than those of frog saccular and turtle cochlear hair cells. Ca^{2+} channels are believed to be predominantly located in the plasma membrane adjacent to synaptic ribbons, that is, at active zones (Roberts et al. 1990; Martinez-Dunst et al. 1997; Sidi et al. 2004). Ca^{2+} current amplitudes in rodent vestibular hair cells meet expectations based on the total plasma membrane area devoted to active zones, which is smaller than that in turtle cochlear and frog saccular hair cells (see Discussion in Bao et al. 2003).

Boyer et al. (2001) found that fluorescently tagged dihydropyridine labeled not just basolateral membranes, as expected, but also the kinocilia and cuticular plate regions of type I hair cells. They suggest that L-type channels in the kinocilia contribute to the two-fold higher basal Ca^{2+} levels that they measured in type I cells relative to type II cells. The lack of labeling of type II kinocilia argues for the specificity of the labeling on type I kinocilia. Other possible reasons for the higher basal Ca^{2+} levels of type I cells include larger resting transduction conductance (see Section 4.1.1) and different Ca^{2+} buffering (see Section 4.2).

4.1.3 Developmental Changes in Voltage-Gated Conductances

Many experiments involving excised rodent inner ear organs have used relatively young preparations, from P0 to P20, often for technical reasons. For experiments with isolated hair cells, the cells are easier to dissociate from younger

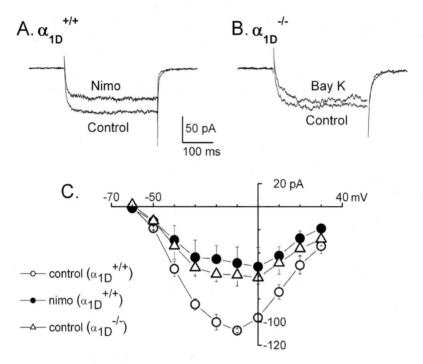

FIGURE 8.11. Mouse utricular hair cells have two Ca^{2+} conductances, one L-type and one non-L-type. (**A, B**) Whole-cell Ba^{2+} currents evoked by voltage steps to -10 mV. Wild-type currents (A) have an L-type component, as shown by partial block by 20 μM of the dihydropyridine (DHP) antagonist, nimodipine. Currents from the Ca_V 1.3-null mutant (B) do not have a DHP-sensitive component, as shown by the lack of enhancement by the DHP agonist, Bay K 8644 (10 μM). Thus, the current in the null mutants is not through L-type channels, despite its overall resemblance to wild-type currents (see kinetics in B and voltage dependence in C). (**C**) The Ba^{2+} current-voltage relation from wild type mouse utricular hair cells in control conditions is compared to the relation in 20 μM nimodipine and in Ca_V 1.3-null mutants. The nimodipine-resistant current (*filled circles*) is similar to the current in the null mutants (*open triangles*) and about half that in wild-type control conditions (*open circles*). Reproduced with permission from Fig. 8 in Dou H, Vazquez AE, Namkung Y, Chu H, Cardell EL, Nie L, Parson S, Shin HS, Yamoah EN (2004) Null mutation of $α_{1D}$ Ca^{2+} channel gene results in deafness but no vestibular defect in mice. J Assoc Res Otolaryngol 5:215–226.

epithelia. For experiments on semi-intact epithelia, young epithelia are thinner and flatter, and therefore optically superior and mechanically more stable. The first postnatal month, however, is a time of intense developmental change within the rodent inner ear. The onset of hearing is in the second postnatal week, the membranous labyrinth does not reach full size until P30 (Curthoys 1983), and both afferent and efferent synapses form on hair cells throughout this period. It is also a period of hair cell differentiation. Here we briefly describe the acquisition of conductances by differentiating vestibular hair cells. The reader is referred to Chapter 2 by Goodyear and colleagues and to Eatock and Hurley (2003) for more comprehensive reviews of developmental changes in ion channels in hair cells.

At birth, mouse utricular hair cells have functional transduction channels (Holt et al. 1997) and several voltage-gated channels (Rüsch et al. 1998): outwardly rectifying K^+ channels, fast inwardly rectifying channels and voltage-gated Na^+ and Ca^+ channels. Recent data show that these channels are acquired and functional between embryonic day (E) 14 and birth at E19 to E21 (Géléoc and Holt 2003; Géléoc et al. 2004). Over the first postnatal week, the complements of ion channels change in both type I and type II hair cells. The incidence of $g_{K,L}$ (percentage of cells with $g_{K,L}$) is low until midway through the first postnatal week in mice (Rüsch et al. 1998), although it can be recorded as early as E18 (Géléoc et al. 2004). In rats, development is slightly delayed: $g_{K,L}$ first appears in the rat utricle at postnatal day (P) 7 and increases in incidence and size over the second postnatal week (Eatock and Hurley 2003, and unpublished observations). $g_{K,L}$'s incidence grows even in organotypic cultures that lack innervation, indicating that although expansion of calyces and acquisition of $g_{K,L}$ are normally contemporaneous, calyx formation does not itself trigger $g_{K,L}$ expression (Rüsch et al. 1998). The incidence of the hyperpolarization-activated conductance, g_h, also increases postnatally (Rüsch et al. 1998). In rat utricular hair cells, Na^+ currents decrease in incidence and size between P3 and P21 (Chabbert et al. 2003). In mouse utricular hair cells, Na^+ current amplitude may peak before birth (Géléoc et al. 2004).

Some of these changes resemble those described for cochlear hair cells. Rodent cochlear hair cells acquire negatively activating conductances (Marcotti and Kros 1999; Marcotti et al. 2003a) and lose some Ca^{2+} channels (Beutner and Moser 2001; Marcotti et al. 2003b) and Na^+ channels (Oliver et al. 1997; Marcotti et al. 2003b), all changes that help transform the hair cells from spiking to nonspiking cells (Marcotti et al. 2003b). Spike activity in developing hair cell epithelia may help development both locally and at higher levels in the pathway. Chabbert et al. (2003) suggested a possible mechanism for such effects. They found that tetrodotoxin, a specific blocker of Na^+ channels, partly blocks the release of brain-derived neurotrophic factor (BDNF) from cultured immature rat utricles. The implication is that Na^+ channel activity, which affects spike shape and frequency, enhances neurotrophin release, promoting growth and maintenance of neural contacts.

Hair cells from chick cristae acquire ion channels in a fairly similar progression but at a faster rate (Masetto et al. 2000, 2003), as is true in general for chick development vs. rodent development. At E10, all chick hair cells have short, immature hair bundles and express a voltage-dependent Ca^{2+} current and a delayed rectifier with a relatively positive activation range. The comparable stage in the mouse utricle appears to be E14 to E15 (Denman-Johnson and Forge 1999). By E14, afferent synaptic contacts appear mature and some hair cells have acquired additional outwardly rectifying K^+ currents (A currents and Ca^{2+}-

dependent K^+ currents, probably of the BK type); the comparable stage in the mouse utricle may be P10 or later (Rüsch et al. 1998). By the time of hatching (E21), chick crista hair cells have mature complements of ion channels; the comparable stage in the rat utricle may be about P14 (Eatock and Hurley 2003). There are some differences in the sequence of expression between chick crista and mouse utricle: for example, g_h appears before g_{K1} in the chick crista (Masetto et al. 2000) but after g_{K1} in the mouse utricle (Rüsch et al. 1998). In the chick crista, Na^+ currents do not go away as the hair cells mature (Masetto et al. 2003). Thus, in some vestibular hair cells, Na^+ currents may play a sensory signaling role in addition to the postulated developmental role.

In adult pigeon cristae, hair cells regenerate following treatment with ototoxic aminoglycoside antibiotics. During the regeneration period, the hair cells recapitulate the developmental sequence of acquisition of basolateral ion channels, although $g_{K,L}$ is very slow to return (Masetto and Correia 1997a,b; Correia et al. 2001). Afferent calyces and other morphological attributes of type I cells recover more rapidly, showing that, just as calyces do not induce $g_{K,L}$ expression, $g_{K,L}$ expression is not required for calyx formation.

4.1.4 Variation in Ion Channels with Epithelial Location

Relatively little is known about how voltage-gated currents vary with location in mammalian vestibular epithelia. In frog and chick cristae (Masetto et al. 1994, 2003), pigeon utricle and crista (Weng and Correia 1999), and turtle crista (Brichta et al. 2002), K^+ currents of type II hair cells tend to be slower and larger in central and striolar zones than in peripheral and extrastriolar zones. Similar trends are evident in a preliminary report from type II hair cells in the young mouse utricle (Holt et al. 1999). Thus, a distinctive feature of the striola and central zones is the expression of relatively large, slow K^+ currents by both type I and type II hair cells.

The voltage-gated currents of type II cells also vary with position on the long axis of the cristae of frogs (Masetto et al. 1994), turtles (Brichta et al. 2002) and birds (Masetto and Correia 1997a; Weng and Correia 1999; Masetto et al. 2000). The hair cells at the outside edge of each hemicrista (toward the planum in turtle, "zone I" in bird) tend to have larger, more rapidly inactivating outward rectifiers and smaller fast inward rectifiers than the hair cells located mid-way along the crista (toward the torus in turtle, "zone III" in bird). Whether such longitudinal variation in hair cell properties occurs in mammalian cristae is not known; the afferent properties have a concentric, rather than longitudinal, organization (Baird et al. 1988).

In contrast to data from frogs, turtles, and birds, most data on ionic currents in rodent vestibular epithelia are from immature preparations (see Section 4.1.3). Because the striolar and central zones appear to be developmentally ahead of the extrastriolar and peripheral zones (Sans and Chat 1982; Rüsch et al. 1998), differences between zones may reflect differences in their developmental status rather than characteristics of the mature epithelium. For example, immature hair cells express Na$^+$ conductances with inactivation ranges separated by 20 to 30 mV and different sensitivities to tetrodotoxin (TTX) (Fig. 8.12A; Wooltorton et al., 2005). Between P0 and P8, most hair cells in the striola express the more negatively inactivating, TTX-insensitive conductance, while more than one half of the hair cells in the extrastriola express the less negatively inactivating, TTX-sensitive conductance. The less negatively inactivating form goes away by P20 (Chabbert et al. 2003). Thus, the extrastriolar expression of the less negatively inactivating conductance in young rats may simply reflect the more immature state of that zone. Recall that Chabbert et al. (2003) found evidence for Na$^+$ current involvement in neurotrophin release in the immature rat utricle.

In the same study, Chabbert and colleagues used single-cell reverse-transcriptase polymerase chain reaction (RT-PCR) to identify molecular candidates for the α subunits of the less negatively inactivating Na$^+$ current (Fig. 8.12B). The prevalence of Na$_V$ 1.2 and Na$_V$ 1.6 subunits suggests that these are the most important components. Channels formed from these subunits share many properties, including voltage dependence and kinetics of fast inactivation, so that channel heterogeneity may have been too subtle to detect with whole-cell recordings. The biophysical and pharmacological properties of the more negatively inactivating conductance, in contrast, are distinctive, and most closely resemble those of the Na$_V$1.5 α subunit (Wooltorton et al., 2005).

4.1.5 Impact of Basolateral Currents on the Receptor Potential

How far have we progressed in analyzing the significance of the distinctive complements of ion channels borne by different vestibular hair cells? Most efforts in this direction have examined voltage responses of isolated hair cells to injected current steps. This approach reveals the impact of the voltage-dependent properties of the cell on the response to a square pulse of current. To fully understand the basolateral channels' roles in vestibular processing, we need to compare the actual receptor potentials evoked by bundle deflections in hair cells with different channels. There are just a few examples of such recordings.

4.1.5.1 Impact of Basolateral Currents on Receptor Potentials in Type II Hair Cells

Because relatively few channels of any type are open at the resting bundle position and resting potential, type II cells have a high input resistance (R_{in}; hundreds of MΩ to GΩ), and relatively small transduction currents evoke large peak receptor potentials. Depolarization rapidly activates g_{Ca} and g_{Na} and deactivates g_{K1}, effects that tend to boost the depolarization. Depolarization more slowly activates g_A and g_{DR} and deactivates g_h, actions that tend to repolarize the cell membrane. Together these effects produce an initial peak in the voltage response to a current step injected through the recording pipette, as illustrated for a rat utricular type II hair cell in Figure 8.10B. In hair cells with sharp electrical tuning, notably frog saccular hair cells and turtle cochlear hair cells,

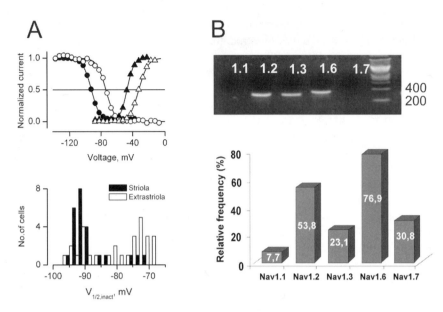

FIGURE 8.12. Na$^+$ currents in the immature rodent utricular macula. (**A**) Two kinds of Na$^+$ current are differently distributed in the two major zones of the immature rat utricular epithelium (Wooltorton et al. 2005). *Top*, Voltage dependence of inactivation and activation of whole-cell Na$^+$ currents in a striolar hair cell (*filled symbols*) and an extrastriolar cell (*open symbols*). *Triangles*: Activation curves: Peak Na$^+$ currents as functions of voltage. The curves are fit with single Boltzmann functions, with similar slopes and voltages of half-maximal activation of -32 mV (striolar cell) and -47 mV (extrastriolar cell). *Circles*: Inactivation curves: Peak Na$^+$ currents evoked by steps to -7 mV from various pre-pulse voltages, as functions of the prepulse voltage. As the prepulse voltage becomes less negative, the peak current is more inactivated. Data were fit by Boltzmann functions, yielding half-maximal inactivation voltages ($V_{1/2,inact}$) values of -92 mV (striolar cell) and -74 mV (extrastriolar cell). *Bottom*, Distributions of $V_{1/2,inact}$ for hair cells in the striola and extrastriola. Striolar cells preferentially express the Na$^+$ current with the more negative inactivation range. About one-third of hair cells in the extrastriola express the same current, while about two thirds express the less negatively inactivating current. Because the development of the striola is advanced over that of the extrastriola, this regional variation may change with maturation. (**B**) Na$^+$ channel α-subunits that are candidates for the less negatively inactivating Na$^+$ current in rat utricular hair cells (Chabbert et al. 2003). *Top panel*, RT-PCR products corresponding to Na$_v$1.2, 1.3 and 1.6 were detected for a single utricular hair cell. *Bottom panel*, Frequency of expression of Na$^+$ channel subunits encountered in single rat utricular hair cells. Modified with permission from Fig. 2 in Chabbert C, Mechaly I, Sieso V, Giraud P, Brugeaud A, Lehouelleur J, Couraud F, Valmier J, Sans A (2003) Voltage-gated Na$^+$ channel activation induces both action potentials in utricular hair cells and brain-derived neurotrophic factor release in the rat utricle during a restricted period of development. J Physiol 553:113–123.

depolarizing current steps evoke a strongly oscillating voltage response at the frequency of the electrical resonance (Crawford and Fettiplace 1981; Lewis and Hudspeth 1983; reviewed in Art and Fettiplace, Chapter 5). In vestibular epithelia, where afferents are not sharply tuned, the hair cells also are not sharply tuned—the initial peak typically decays with few if any secondary oscillations (Correia et al. 1989, 1996; Rüsch and Eatock 1996b). A step bundle deflection also produces a receptor potential with an initial peak that decays without further oscillation. It is possible to suggest separate contributions of transducer adaptation and K^+ channels to the repolarization of the receptor potential (e.g., Fig. 4 in Eatock 2000), but these remain to be demonstrated by selectively blocking the K^+ channels.

To see the effect of voltage-gated conductances in the frequency domain, we can compare how transduction current and receptor potential vary with stimulus frequency (Fig. 8.7A,B; Holt et al. 1997). The transduction current in the mouse utricular hair cell in Figure 8.7 showed high-pass filtering, attributable to adaptation (see Section 3.3.1.2), while the receptor potential was broadly tuned to about 10 Hz (Holt et al. 1999). The peak–peak receptor potential fell off at high frequencies with a corner frequency of 30 Hz and a limiting slope of approximately 20 dB/decade, consistent with low-pass filtering by a membrane with a time constant of 5 ms. On the low-frequency side, the peak–peak receptor potential began to fall off at higher frequencies (<5 Hz) than did the transduction current (<2 Hz), suggesting filtering by voltage-gated conductances in addition to the high-pass filtering by transducer adaptation. At low frequencies, outward K^+ conductances have enough time to activate appreciably during depolarizing half-cycles, attenuating the voltage response. Their selective attenuation of the depolarizing part of the response may account for the increasingly symmetric (linear) appearance of the receptor potential as frequency is lowered (Fig. 8.7A; Holt et al. 1999). In this way, the asymmetric voltage-current relation of the hair cell membrane may compensate for the asymmetric current–displacement relation (Fig. 8.5B) and contribute to the high linearity of vestibular afferent responses at low stimulus frequencies, where nonlinear distortion is $<20\%$ (Goldberg et al. 1990a).

K^+ conductances may enhance the linearity of the afferent signal in other ways as well. By attenuating the receptor potential, K^+ conductances may keep membrane potential in the range over which Ca^{2+} current is a monotonic function of voltage. Positive to -20 mV, the Ca^{2+} current decreases because the driving force for Ca^{2+} entry decreases (see Fig. 8.11C). Such nonmonotonic behavior would be an additional source of nonlinearity in the afferent spike rate, since Ca^{2+} current size controls transmitter release. In addition, without the counteracting effect of large K^+ conductances, the influx of Ca^{2+} and Na^+ through voltage-gated basolateral channels during positive bundle deflections may lead to spiking, which in turn might degrade representations of stimulus timing and size. Indeed, early in cochlear hair cell development, the combination of relatively large Na^+ and Ca^{2+} currents and relatively small K^+ conductances results in spiking (Goodyear et al., Chapter 2).

Figure 8.7B also compares the frequency dependence of the peak-peak receptor potential and the gain of chinchilla utricular afferents, replotted from Goldberg et al. (1990a) (see original data in Fig. 8.3C). The low-frequency slope of the receptor potential data resembles that of intermediate afferents, although, as discussed earlier, similarities between the datasets may be fortuitous given the large differences in how they were collected. The large, relatively slow outward K^+ currents of striolar hair cells might contribute to the low gain of striolar afferents at stimulus frequencies low enough to permit significant modulation of these voltage-gated currents. The prominent inactivation of K^+ currents in some extrastriolar hair cells (see Section 4.1.4) might help produce the flat frequency response of extrastriolar afferents below 2 Hz, by compensating for the decline in depolarizing transducer current during slow adaptation with a concurrent decline in K^+ currents.

Na^+ currents and spiking. When hair cells are found to have voltage-gated Na^+ currents, it is assumed that such currents contribute to spiking. But this has been demonstrated in very few instances, and the very negative inactivation range of many hair cell Na^+ currents casts doubt on their functionality. The most thorough analysis has been done in immature mouse cochlear hair cells, where the Na^+ current has a less negative inactivation range and does contribute to spikes from resting potential (Marcotti et al. 2003b). The spikes have a larger Ca^{2+} component; the effect of the Na^+ current is to hasten the upstroke and thereby increase the frequency of spontaneous spiking. Na^+ current will also speed up depolarizations that are subthreshold for spiking. A similar Na^+ current is found in many hair cells in the immature rat utriculus (see Section 4.1.4; Chabbert et al. 2003), and it may also contribute to spiking from resting potential. The very negatively inactivating Na^+ current in other rat utricular hair cells (Section 4.1.4, Wooltorton et al. 2005) may contribute to spikes only at the offset of hyperpolarizations that relieve the inactivation, as Masetto et al. (2003) found for chick crista hair cells with Na^+ currents with a similarly negative inactivation range. Alternatively, its inactivation range may be less negative at mammalian body temperature (Oliver et al. 1997), such that it is not fully inactivated at resting potential.

4.1.5.2 Impact of Basolateral Currents on Receptor Potentials in Type I Hair Cells

Type I hair cells in rodents have small membrane capacitances (C_m) because they are small cells and have remarkably low R_{in} values because they express $g_{K,L}$. By Ohm's law, the voltage change per given current is proportional to R_{in}; thus, rodent type I cells are expected to produce unusually small receptor potentials for a given transduction current. This effect is illustrated in Figure 8.10C, which compares the voltage responses to small injected currents of a type I cell and a type II cell. The membrane time constant, which determines the speed at which the membrane voltage responds to a given current, is the product of R_{in} and C_m and therefore is also very small in type I cells. As a consequence, they can respond to high-frequency currents (Rennie et al. 1996;

Rüsch and Eatock 1996b). In the rat utricular macula, median R_{in} at -64 mV was 37 MΩ for 55 type I cells and 1.2 GΩ for 39 type II cells (ruptured-patch recordings, M. Saeki and R.A. Eatock, unpublished results). Thus, at -64 mV, a transduction current of 10 pA would generate just 370 µV ($R_{in} \times I_{met}$) in an average type I cell and 12 mV in an average type II cell. Assuming membrane capacitances of 4 pF, our mean value, the median membrane time constants ($R_{in} \times C_m$) of the rat utricular type I and type II hair cells were about 160 ms and 5 ms, respectively, corresponding to low-pass corner frequencies of about 1000 and 30 Hz. Similarly, the membranes of type I and type II hair cells from gerbil semicircular canals have low-pass corner frequencies of 400 to 5000 Hz and 40 Hz, respectively (Rennie et al. 1996).

The functional significance of the high bandwidth of type I cells is not clear. It fits qualitatively with the notion that type I hair cells are a specialization of the central/striolar zones, which seem to represent higher stimulus frequencies than do peripheral/extrastriolar zones (see Section 2.2.1). But the highest head movement frequencies known to be transduced in mammals are in the range of 20 Hz (Grossman et al. 1988; Angelaki 1998; Hullar and Minor 1999). At such frequencies, even the receptor potentials of hair cells lacking $g_{K,L}$ are only slightly diminished by low-pass filtering (Fig. 8.7B, open circles). Moreover, given that type I hair cells acquire bandwidth at the cost of receptor potential gain, it is not clear whether type I cells or type II cells generate larger receptor potentials at high stimulus frequencies.

$g_{K,L}$ also confers upon the cells relatively negative resting potentials (-70 to -80 mV in standard solutions, vs. -60 to -70 mV in type II hair cells). This poses the following conundrum: For a type I cell with R_{in} of 40 MΩ and a resting potential of -80 mV, a very large transduction current (400 pA) would depolarize the cell to -64 mV, where Ca^{2+} channels are just 1% activated (Bao et al. 2003; see Fig. 8.11C). It therefore seems likely that in vivo, either R_{in} is not so low, or resting potential is not so negative, or transmission is not exclusively mediated by Ca^{2+}-regulated exocytosis.

The first possibility, that R_{in} is not so low, would be the case if $g_{K,L}$ is smaller at resting potential in vivo than in our experimental conditions. $g_{K,L}$ is resistant to inactivation (Goldberg and Brichta 2002), but its size at resting potential could be reduced by shifting its activation range positive. Several lines of evidence support this possibility. The activation range of $g_{K,L}$ is more labile than that of the K^+ conductances in type II cells. In the ruptured-patch mode of whole-cell recording, the activation range of $g_{K,L}$ can slip positively with time by tens of millivolts, presumably reflecting loss of an intracellular factor (Rüsch and Eatock 1996a; Chen and Eatock 2000). Indeed, in the perforated-patch mode of recording, which largely preserves the normal intracellular milieu, the activation range holds stable (Wong et al. 2004). A broad-spectrum protein kinase inhibitor can shift the activation range positively (Eatock et al. 2002), consistent with results obtained in the absence of intracellular ATP (Lennan et al. 1999; see discussion in Chen and Eatock 2000). The size of $g_{K,L}$ at resting potential can be reduced by nitric oxide, a possible neurotransmitter in vestibular epithelia (see Section 5.2.4.2). In sum, there is reason to speculate that $g_{K,L}$'s activation range is modulated in vivo; even a modest positive shift would have a potent impact on the size of the receptor potential.

Another factor influencing the activation range for $g_{K,L}$ is temperature. In mouse utricular type I cells, the activation range shifted by 0.8 mV per °C as temperature was raised to 36°C (Rüsch and Eatock 1996a). But resting potential shifted by 0.6 mV per °C, so that raising the temperature only modestly increased R_{in} at resting potential. This shift could, however, bring resting potential into line with the activation range of voltage-gated Ca^{2+} channels (provided that their activation ranges do not shift similarly), permitting modulation of transmitter release.

Factors other than $g_{K,L}$ that affect resting potential include the transduction current and the extracellular K^+ concentration. In isolated hair cells, transduction channels are often not functional. Resting potential may be hyperpolarized in such cells by the absence of the resting inward transduction current. With this in mind, Goldberg and Brichta (2002) investigated the effect of a constant +50-pA current on the voltage responses of isolated turtle crista hair cells to steps and sinusoidal currents. $g_{K,L}$ in turtle type I cells has a somewhat more positive activation range than in rodent hair cells (Brichta et al. 2002). Depolarizing the type I cell increased the activation of $g_{K,L}$, which did not inactivate. Thus, the 50-pA current made the type I cells' voltage responses to current steps and sinusoids smaller and more linear (more like those shown for rodent type I cells in Fig. 8.10C). In type II cells, in contrast, the steady +50-pA current inactivated the outwardly rectifying K^+ current, increasing R_{in} and the voltage gain. Thus, the background current increased the electrophysiological difference between type I and type II cells.

As described in Section 4.1.4, outwardly rectifying currents in type II cells vary with location in the vestibular epithelium. Inactivating the outwardly rectifying currents with a backround current made the type II population more electrophysiologically uniform. It seems unlikely that the differential expression of outward K^+ conductances by type II hair cells in different regions of the epithelium is non-functional. Rather, a steady positive current may not adequately mimic the situation in vivo. Fluctuations in bundle position (and therefore transduction current) and efferent inputs may relieve the inactivation of type II outward rectifiers so that they do not spend most of their time inactivated. Nevertheless, the study by Goldberg and Brichta (2002) identified a key difference between the outwardly rectifying currents of type I and type II hair cells, which may in some situations strongly differentiate the voltage responses of the two cell types.

It has also been suggested that the average membrane potential of type I hair cells is elevated by K^+ accumulation in the thin shell of extracellular solution in the synaptic cleft (Goldberg 1996). This may depolarize the type I hair cell into the activation range of voltage-gated Ca^{2+} channels. Because K^+ accumulation would depolarize the postsynaptic calyx membrane as well, K^+ may be thought of as a nonconventional transmitter (see Section 5.1.3.1).

In summary, we cannot be certain of the activation range of $g_{K,L}$ in vivo. In hair cells of the turtle crista and utriculus recorded with the ruptured-patch method at room temperature, $g_{K,L}$ is not as negatively activating and R_{in} is appreciable, such that the large currents through turtle type I hair cells evoke sizeable receptor potentials (Brichta et al. 2002; Rennie and Ricci 2004). It is plausible that the same is true for mammalian type I hair cells in native conditions. Or $g_{K,L}$'s activation range may vary in vivo; changing the range would be a potent way to manipulate the electrical properties of the type I hair cell and therefore the gain and filtering of calyx-bearing afferents (Chen and Eatock 2000).

4.2 Calcium Buffering

Ca^{2+} regulation in hair bundles was reviewed in Section 3.3.2. Basolateral routes for Ca^{2+} entry include voltage-gated Ca^{2+} channels (see Section 4.1.2) and ligand-gated ion channels, including nicotinic acetylcholine receptors and possibly $P2X_2$ receptors at the sites of efferent synapses (see Section 5.2.4.4). Plasma membrane Ca^{2+}-ATPases are expressed not just in hair bundles (see Section 3.2.1) but also in the basolateral membranes of both type I and type II hair cells (Fig. 8.8A; Dumont et al. 2001). Here we review the literature suggesting regional, cell type and developmental variation in Ca^{2+} buffering proteins and organelles in vestibular epithelia.

4.2.1 Ca^{2+} Sequestering Organelles

Hair cells are richly endowed with mitochondria, which in addition to being energy sources are important Ca^{2+} sinks. Kennedy and Meech (2002) showed that a mitochondrial inhibitor slowed the rate of recovery from Ca^{2+} loads in inner hair cells. Ca^{2+} ions within mitochondria are believed to be mostly bound (immobile), although little is known about Ca^{2+} buffering properties therein (Ganitkevich 2003). Ca^{2+} overload in mitochondria can initiate cell death by necrotic or apoptotic processes. Mitochondria in type I hair cells are larger than those of type II hair cells (Rüsch et al. 1998), possibly reflecting the greater metabolic and buffering needs of the former as a consequence of their larger transduction and voltage-gated currents.

Smooth endoplasmic reticulum (SER) accumulates Ca^{2+} by means of a Ca^{2+}-ATPase (SERCA) in the SER membrane. This accumulated Ca^{2+} is released upon binding of cytoplasmic Ca^{2+} to ryanodine receptors in the SER membrane (Ca^{2+}-induced Ca^{2+} release). There is evidence for both ryanodine receptors and SERCA in the synaptic cisterns of outer hair cells (Fuchs and Parsons, Chapter 6) and preliminary evidence for ryanodine and inositol triphosphate (IP_3) receptors in the smooth endoplasmic reticulum of rat vestibular hair cells (Cameron et al. 2004). Thus, Ca^{2+} entry through nicotinic acetylcholine receptors postsynaptic to efferent boutons may trigger Ca^{2+}-induced Ca^{2+} release from the synaptic cistern in vestibular hair cells, as proposed for cochlear hair cells (Sridhar et al. 1997; Fuchs and Parsons, Chapter 6; and Section 5.2.4.1, this chapter).

4.2.2 Ca^{2+} Binding Proteins

Because Ca^{2+} binding proteins (CBPs) have different affinities and kinetics, differences in CBP expression between hair cell types will affect such functions as transducer adaptation, K(Ca) channel activation and transmitter release. CBPs in hair cells and eighth-nerve afferents include calretinin, calbindin-D28K, parvalbumin- and S-100, all of which are members of the EF-hand family of CBPs (Baimbridge et al. 1992). All of our information on Ca^{2+} binding proteins in mammalian vestibular epithelia comes from immunohistochemical studies. There are species variations either in the expression of particular proteins or in immunoreactivity against particular antibodies, but some general observations can be made for the vestibular hair cells of rodents.

Calretinin expression is an early marker of rodent hair cells (Zheng and Gao 1997). Calbindin-D28k may be an early marker for supporting cells (Dechesne and Thomasset 1988). In mature epithelia, CBP immunoreactivity varies with epithelial zone and cell type. In the peripheral/extrastriolar zones of rats and mice, type II cells show calretinin-like immunoreactivity (Dechesne et al. 1991; Desai et al. 2005a,b) and type I and type II cells show S-100-like immunoreactivity (Foster et al. 1994). Type I hair cells in central and striolar zones are strongly immunoreactive to antibodies against conventional parvalbumin (Demêmes et al. 1993; Raymond et al. 1993; Sans et al. 2001) and calbindin-D28k (Dechesne et al. 1988; Dechesne and Thomasset 1988; Baird et al. 1997; Sage et al. 2000; Kevetter and Leonard 2002, Leonard and Kevetter 2002). Calyx afferents, which are restricted to the central/striolar zones, stain brilliantly for calretinin (Fig. 8.8B; Desmadryl and Dechesne 1992; Kevetter and Leonard 2002; Desai et al. 2005a,b) and calbindin-D28k (Dechesne and Thomasset 1988; Leonard and Kevetter 2002); the calbindin-like immunoreactivity extends over a wider zone than the calretinin-like immunoreactivity.

In mammalian vestibular epithelia, CBP staining patterns have been used principally as markers: of hair cells versus supporting cells, of epithelial zone, or of hair cell type or developmental stage. Experiments on frog saccular and turtle cochlear hair cells have addressed the functional significance of different Ca^{2+} buffers. Compact frog saccular hair cells and turtle cochlear hair cells have fast, mobile buffers equivalent to 8 mM Ca^{2+} binding sites (Roberts 1994; Tucker and Fettiplace 1996). CBPs responsible for this buffering include parvalbumin 3 in compact frog saccular hair cells (Heller et al. 2002) and calretinin in tall frog saccular hair cells (Edmonds et al. 2000) and a parvalbumin 3 homologue (oncomodulin or parvalbumin β) and calbindin-D28k in turtle cochlear hair cells (Hackney et al. 2003). In their sharp electrical tuning and large K(Ca) conductances, frog saccular and turtle cochlear hair cells differ from mammalian vestibular hair cells; thus, further experiments will be needed to understand the functional significance of CBP variation with developmental stage, cell type and location in mammalian vestibular epithelia. A promising place to look is the vestibular nerve, where calretinin and calbindin-D28k expression correlate with irregular spike patterns. What CBPs do regular afferents express, and might they control spike timing by regulating Ca^{2+}-gated K^+ channels?

4.3 Summary of Post-Transduction Processing in Hair Cells

As in most hair cells, the dominant conductances in vestibular hair cells are outwardly rectifying K^+ currents. These tend to be particularly large and slow in type I cells and in type II cells in the central/striolar zones, where, by reducing the input resistance once activated, they may contribute to the low gain of afferents at low frequencies. The expression of Ca^{2+}-binding proteins also differs between zones in both hair cells and afferents, possibly allowing for differential regulation of transducer adaptation, Ca^{2+}-gated K^+ channels, and afferent and efferent transmission.

Both type I and type II hair cells express voltage-gated Ca^{2+} channels of the particular noninactivating L-type variety found in cochlear hair cells ($Ca_V1.3$), plus an additional type with grossly similar biophysical characteristics. The currents are of a size consistent with the amount of membrane devoted to presynaptic active zones. Voltage-gated Na^+ currents, present in immature vestibular hair cells in at least two varieties, may play a role in synaptogenesis.

5. Synaptic Transmission

Much of the diversity in vestibular epithelia concerns synapses. In Section 5.1, we shall discuss how afferent endings vary with both hair cell type and with location. Such variation may play an important role in setting such physiological functions as regularity of spike timing and sensitivity to efferent inputs. In Section 5.2, we will discuss variation in the morphology and neurotransmitters of efferent endings with both hair cell type and location. Such regional differences in feedback from the brain add to evidence that the regions fulfill different functional roles.

5.1 Afferent Transmission

Having sketched the kinds of afferents and afferent terminals in vestibular epithelia (Figs. 8.2 and 8.3), we review here detailed morphology of afferent synapses and discuss possible physiological correlates.

5.1.1 Morphology of Synaptic Contacts

5.1.1.1 Synaptic Ribbons

Except in birds and mammals, synaptic bodies of vestibular hair cells are spherical in shape (Wersäll and Bagger-Sjöbäck 1974; Jacobs and Hudspeth 1990;

Lysakowski 1996). In mammalian vestibular hair cells, synaptic bodies vary in number, shape and size. Differences have been reported between species (cat vs. chinchilla), epithelium (crista vs. utricular macula), epithelial zone (central/striolar vs. peripheral/extrastriolar), hair cell type (I vs. II) and developmental stage. In hair cells of the adult chinchilla crista, synaptic bodies may be elongate, spherical, barrel-like or plate-like, with diameters ranging from 90 nm to 400 nm (Fig. 8.13; Lysakowski and Goldberg 1997). We shall refer to all of these structures as ribbons from here on.

In type II cells, there is an average of one synaptic ribbon facing each afferent bouton (see Table 5 in Lysakowski and Goldberg 1997). Central type II cells receive fewer afferent boutons than do peripheral type II cells, but have similar numbers of ribbons because: (1) they tend to make more than one ribbon synapse per bouton and (2) they make more ribbon synapses on the outer faces of neighboring calyces (Lysakowski and Goldberg 1997). Via the outer-face ribbons, even calyx afferents often receive a fraction of their input from type II cells. The fraction is minor, however: just one-eighth of the synaptic ribbons on calyx afferents in the chinchilla crista are made by type II cells (Lysakowski and Goldberg 1997). Moreover, the type II input left no clear mark on the physiological responses of calyx afferents in the chinchilla (Baird et al. 1988), which closely resemble those of the squirrel monkey (Lysakowski et al. 1995), where outer-face ribbons are very rare (Lysakowski 1996).

The number of ribbons in mature type I hair cells may vary with species. In a comparison of synaptic ribbons in hair cells in the vestibular epithelia of newborn and adult cats, Favre and Sans (1979) observed a steep decline (93%) in the incidence of ribbons in type I cells between newborn and adult, with no change in type II ribbon numbers. The timing of this decrease differs from that of the decline in the number of ribbons in mouse cochlear inner hair cells; the latter occurs in the early postnatal period and may reflect a pruning of afferent contacts (Sobkowicz et al. 1982, 1986). The equivalent period in the cat occurs in utero, long before the postnatal period studied by Favre and Sans (1979). Favre and Sans (1979) suggested that the decline in type I ribbons reflects a decline in the importance of vesicular transmission. In the mature chinchilla crista, however, both type I and type II hair cells have, on average, between 15 and 20 ribbons (Fig. 8.13, Lysakowski and Goldberg 1997). Preliminary data from mature mouse utricular maculae suggest 10 to 13 ribbons per hair cell (Ahmad and Lysakowski 2004). Thus, in rodents, synaptic ribbons in type I cells are not likely to be vestigial (see Section 5.1.3).

The different sizes and shapes affect the number of ribbon-associated vesicles and may also affect the number of voltage-gated Ca^{2+} channels. Recall that the size of voltage-gated Ca^{2+} current in tall hair cells of the chick cochlea correlates with the projected area of the synaptic body on the presynaptic membrane (Martinez-Dunst et al. 1997). The large synaptic bodies of frog saccular hair cells bear 400 to 500 vesicles (Parsons et al. 1994; Lenzi et al. 1999). In chinchilla crista, the number ranges from 20 to 30 vesicles around small spherules to 150 to 300 around the larger synaptic "plates"; synaptic ribbons in central

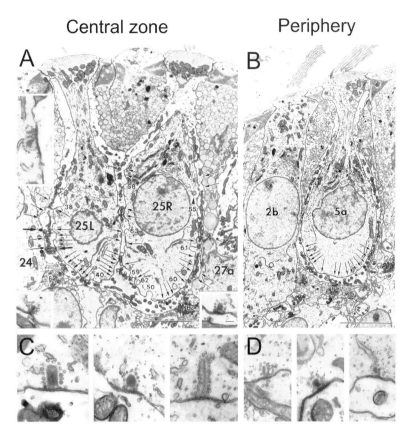

FIGURE 8.13. Afferent synapses in the (**A**) central zone and (**B**) peripheral zone of the adult chinchilla crista. Note that the central type I cells (labeled 25L and 25R in their nuclei) are larger than the peripheral type I cell (labeled 5a). In B, a peripheral type II cell is shown (labeled 2b). Note the larger-diameter stereocilia and bulging apical surface of the neighboring type I cell (5a). *Arrows* in A and B point to projected locations of synaptic ribbons in type I hair cells, as reconstructed from serial sections. Note that the ribbons in the central type I cells are more numerous and extend from the basal pole further up the side of the hair cell, relative to the ribbons in the peripheral type I cell. Multiple calyceal invaginations can be seen in both hair cells in A. *Scale bars:* A: 5 µm; B: 5 µm. (**C, D**) Ribbons in type II cells from the central and peripheral zones, respectively. Ribbons from central type II cells tend to be larger than ribbons from peripheral type II cells. In addition to spherical ribbons, examples of barrels (C, *left*), plates (C, *right*), and rods (D, *left*) are shown. Modified with permission from Figs. 9, 10, and 16 in Lysakowski A, Goldberg JM (1997) A regional ultrastructural analysis of the cellular and synaptic architecture in the chinchilla cristae ampullares. J Comp Neurol 389:419–443, by permission of Wiley-Liss, Inc., a subsidiary of John Wiley & Sons, Inc.

type II hair cells are particularly large and complex (Fig. 8.13C; Lysakowski and Goldberg 1997). The number, shape and size of synaptic bodies also vary in avian vestibular hair cells (Lysakowski 1996). Such differences may contribute to regional differences in afferent gain. All else being equal, a given receptor potential should activate a larger Ca^{2+} current and more vesicle fusion at larger ribbons.

5.1.1.2 Calyceal Specializations

In development, the calyx grows from the base of the type I hair cell apically, with finger-like projections, stopping short of the tight junctions and eventually filling in (Desmadryl and Sans 1990; Sans and Scarfone 1996). The calyx is usually continuous and not penetrated by other structures such as bouton afferent endings and efferent endings (Fig. 8.13A, B); such contacts may be made early in development but are pushed off by the growing calyx (Pujol and Sans 1986). Because there are no obvious specializations closing off the synaptic cleft at its apical extent (Fig. 8.13A, B), there appears to be exchange between the synaptic cleft, a thin shell of liquid between the hair cell and the calyx, and the perilymph bathing the sublumenal compartment of the neuroepithelium. In complex calyces, the calyx ending expands to engulf multiple type I cells (Figs. 8.2, 8.13A). In rodents, complex calyces are common in central and striolar zones and rare elsewhere (Fig. 8.2; Fernández et al. 1988, 1990; Desai et al. 2005a,b).

The calyx develops concurrently with the neck of the type I hair cell, in the first few postnatal weeks in rodents. Development of the neck is not secondary to development of the calyx, however; type I hair cells raised in denervated organotypic cultures acquire necks (Rüsch et al. 1998). In the type I cell, many mitochondria cluster below the cuticular plate, apical to and at the level of the top of the calyx. Extending from the mitochondrial zone through the neck proper are many microtubules (Favre and Sans 1983; Kikuchi et al. 1991) that are recognized by antibodies to βI and βIV tubulin (Perry et al. 2003). The numbers of microtubules in type I cells are not that much higher than those in type II cells (100 to 200 per mature guinea pig vestibular hair cell; Kikuchi et al. 1991), but they are more concentrated in type I cells because they occur where the soma is constricted at the neck. Vestibular ganglion neurons and their peripheral terminals, including calyces, are strongly immunoreactive for βIII tubulin, which is commonly found in neurons (Perry et al. 2003).

No gap junctions occur between type I hair cells and calyces (Gulley and Bagger-Sjöbäck 1979; Yamashita and Ohmori 1991). The calyx ending pushes into the hair cell at multiple sites, the so-called *calyceal invaginations* (Fig. 8.14A). In the chinchilla crista, Lysakowski and Goldberg (1997) found about 10 invaginations and 10 ribbons per peripheral type I hair cell and about 50 invaginations and 20 ribbons per central type I hair cell. Thus, in the central zone, invaginations outnumber ribbons. Preliminary data suggest the same is true in type I hair cells in the mouse utriculus (Ahmad and Lysakowski 2004). Both ribbons and invaginations are concentrated toward the basal pole of the type I hair cell, just as ribbons are in type II hair cells. That central type I cells

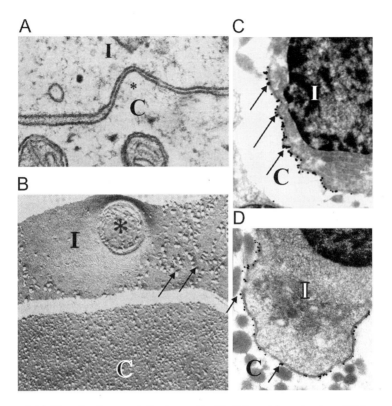

FIGURE 8.14. Type I-calyx specializations. (**A, B**) Calyceal invaginations (C) into type I hair cells (I). (A) Invaginations (*) have narrow intercellular clefts (6-7 nm) and are not necessarily associated with synaptic ribbons. (B) Freeze–fracture image of type I and calyx ending; * shows invaginating calyx; note the particles (*arrows*) in the hair cell membrane around the invagination. (Modified with permission from (A) Fig. 3 and (B) Fig. 7 in Gulley RL, Bagger-Sjöbäck D [1979] Freeze-fracture studies on the synapse between the type I hair cell and the calyceal terminal in the guinea-pig vestibular system. J Neurocytol 8:591–603.) (**C, D**) Strong KCNQ4-like immunoreactivity in the inner face of the calyx membrane enveloping a striolar hair cell (C) and an extrastriolar hair cell (D), as shown with silver enhancement of the immunogold method. *Arrows* point to several silver-enhanced gold particles, which are especially dense in the striolar hair cell. Lysakowski and Price (2003); antibody provided by T. Jentsch.

have more ribbons and invaginations than do peripheral type I cells may reflect both the larger size of central type I cells and that ribbons and invaginations occur along a larger extent of the basolateral membrane in central type I cells (Fig. 8.13; Lysakowski and Goldberg 1997). Ribbons and invaginations are often close to each other but do not appear to have an obligatory close association (Lysakowski and Goldberg 1997).

At calyceal invaginations, the synaptic cleft narrows from 25 to 30 nm to 6 to 7 nm and the electron-dense cleft material disappears (Fig. 8.14A; Spoendlin 1965; Hamilton 1968; Favre and Sans 1979; Gulley and Bagger-Sjöbäck 1979). In transmission electron micrographs, the invaginations have no membrane-associated densities or other specializations typical of either gap junctions or synaptic active zones. With the freeze-fracture method, however, Gulley and Bagger-Sjöbäck (1979) saw annuli of particles in the hair cell membrane around the invagination (Fig. 8.14B). It is possible that the particles correspond to ion channel proteins such as those corresponding to $g_{K,L}$, which are present at high density (see Discussion in Chen and Eatock 2000).

It has been suggested that calyceal invaginations help hold the pre- and post-synaptic membranes together (Goldberg 1996), like the morphologically-unrelated puncta adherens of the calyx of Held and other large synapses in the brain (Satzler et al. 2002). By increasing membrane surface areas, the invaginations might also allow proportional increases in channels and pumps for handling ion transport out of the cleft. (But KCNQ4 channels, at least, appear to be excluded from the invaginations, while densely present in the calyx membrane outside the invaginations; A. Lysakowski and S.D. Price, unpublished observations.)

The defining feature of the type I hair cell is its calyx ending. How unique is this structure? As far as we know it is the only postsynaptic calyx in the nervous system; the calyx of Held in the auditory brainstem and the calyx ending in the ciliary ganglion are presynaptic structures. But there are specialized large afferent endings in fish and amphibians that may be related evolutionarily or by analogy to the amniote calyx: candelabra endings in the lamprey (Lowenstein et al. 1968), club endings in the toadfish (Boyle et al. 1991) and cup-shaped or claw endings in the frog and goldfish (Honrubia et al. 1989; Baird and Schuff 1994; Lanford and Popper 1996; Lanford et al. 2000). In the goldfish crista, large afferent nerve terminals may cup the basal parts of as many as five hair cells (Lanford and Popper 1996). The endings do not cover as much of the basolateral membrane as the calyx ending does, nor do the hair cells within have the classic type I amphora shape. As Popper and colleagues have noted (Yan et al. 1991; Chang et al. 1992), however, there are other parallels between striolar hair cells in certain fish otolith organs and type I hair cells of amniotes. (Note that although rodent vestibular epithelia have quite a few type II cells in the striola, in many other amniotes, from turtles to squirrel monkeys, type I cells dominate in the striola.) Type I hair cells (Wersäll 1981) and striolar hair cells in a fish utricle and lagena (Yan et al. 1991) are both more sensitive to gentamicin treatment than are type II and extrastriolar hair cells. Striolar hair cells of some fish are immunoreactive for the Ca^{2+} binding protein, S-100, while extrastriolar hair cells are not (Saidel et al. 1990). Although S-100 is not specific to type I cells, there are regional and cell-type-specific patterns of Ca^{2+} binding protein expression in amniote epithelia (see Section 4.2.2). Striolar hair cells of both amniotes and fish are larger than extrastriolar hair cells and are contacted by relatively large afferent fibers. Striolar hair cells have large mitochondria, like type I cells (Chang et al. 1992). Thus, it is reasonable to wonder if the fish striolar cells that are large, gentamicin-sensitive and S-100–positive are evolutionarily related to the type I hair cells of amniotes (Yan et al. 1991). Finally, many physiological features of amniote vestibular afferents are shared by anamniote vestibular afferents (reviewed in Lysakowski and Goldberg 2004), suggesting that some of the functions of the amniote type I/calyx complex can be achieved without a full calyx.

5.1.2 Transmitters and Receptors in Vestibular Hair Cells and Afferent Neurons

The hair cell's principal excitatory transmitter activates postsynaptic glutamate receptors, so is either glutamate or a substance with glutamate-like action (see Sewell 1996, for a review, and Kataoka and Ohmori 1994). The pharmacological and physiological evidence in vestibular epithelia for glutamatergic transmission is principally from non-mammals (e.g., Annoni et al. 1984; Prigioni et al. 1994), but there is immunocytochemical evidence from mammals for glutamate in the hair cells and for various kinds of glutamate receptors on the afferent terminals.

Demêmes et al. (1990) and Usami and Ottersen (1995) observed glutamate-like immunoreactivity in vestibular ganglion cells and in both type I and type II hair cells, but not in supporting cells. Mammalian vestibular ganglion cells have ionotropic glutamate receptors of the α-amino-3-hydroxy-5-methyl-4-isoxazole propionic acid (AMPA), kainate, and N-methyl-D-aspartate (NMDA) classes, as shown by multiple approaches. There is evidence for both protein (Fig. 8.15A; Demêmes et al. 1995; Matsubara et al. 1999; Puyal et al. 2002) and mRNA (Niedzielski and Wenthold 1995) corresponding to the AMPA-type glutamate receptor subunits GluR2, R3 and R4, but not GluR1. With in situ hybridization and RT-PCR, Niedzielski and Wenthold (1995) also detected mRNA for the kainate (GluR5-6, KA1-2) and NMDA classes (NR1, NR2A, NR2C) of glutamate receptors. The strongest expression levels were for GluR2-5 and NR1, but only GluR1 (AMPA) and GluR7 (kainate) were absent. Ishiyama et al. (2002) found that rat calyx terminals, but not bouton terminals, are strongly immunoreactive for the NMDA receptor subunits NR1, NR2A, and NR2B. NMDA receptors require prior depolarization to relieve Mg^{2+} block of the pore (reviewed in Hille 2001), but the high transmission rates at hair cell synapses plus the possibility that calyces are depolarized by elevated cleft K^+ (see Section 5.1.3.1) make it likely that the NMDA receptors of calyces are activatable most of the time.

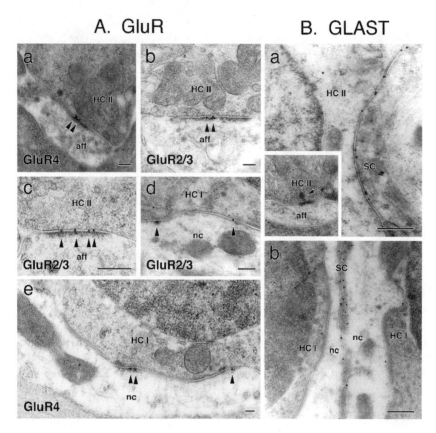

FIGURE 8.15. Glutamatergic transmission. (**A**) Glutamate receptor immunoreactivity. Adult rat. *a–c,* Type II contacts with boutons; *d,e,* type I contacts with calyces; *arrows* point to postsynaptic particles. There are also several presynaptic particles. Modified from (a–c) Fig. 1 and (d, e) Fig. 2 in Matsubara A, Takumi Y, Nakagawa T, Usami S, Shinkawa H, Ottersen OP (1999) Immunoelectron microscopy of AMPA receptor subunits reveals three types of putative glutamatergic synapse in the rat vestibular end organs. Brain Res 819:58–64, with permission from Elsevier. (**B**) Distribution of the glutamate/aspartate transporter, GLAST (EAAT-1), in vestibular epithelia. GLAST is concentrated in supporting cell membranes adjacent to type II basolateral membranes (*a*) and adjacent to the basal pole of the calyx outer face (*b*). Modified from Fig. 2 in Takumi Y, Matsubara A, Danbolt NC, Laake JH, Storm-Mathisen J, Usami S, Shinkawa H, Ottersen OP (1997) Discrete cellular and subcellular localization of glutamine synthetase and the glutamate transporter GLAST in the rat vestibular end organ. Neurosci 79:1137–1144, with permission from Elsevier.

How glutamate is cleared from the calyceal synaptic cleft is mysterious. The calyces and hair cells do not label with antibodies to known glutamate transporters, including the glial glutamate transporters GLAST (EAAT1) and GLT1 (EAAT2) and a neuronal form, EAAC1 (EAAT3) (Takumi et al. 1997). GLAST is a key glutamate transporter in the cochlea, where it is concentrated in supporting cells (Furness and Lawton 2003). In rodent vestibular epithelia, GLAST-like immunoreactivity is highest in supporting cell membranes facing the basal outer face of the calyx (Fig. 8.15Bb; Takumi et al. 1997). How does glutamate released into the synaptic cleft get taken up by supporting cells on the other side of the calyx? Glutamate might diffuse out at the apical tip of the calyx, or the calyx inner face might express a neuronal glutamate transporter other than EAAC1.

5.1.3 Transmission at the Type I-Calyx Synapse

Transmission at the type I-calyx synapse includes conventional vesicular transmission, as indicated by the following observations in diverse species. Excitatory postsynaptic potentials (EPSPs) have been recorded in calyx afferents in a lizard vestibular organ (Schessel et al. 1991). The occurrence of numerous presynaptic ribbons and associated synaptic vesicles in mature rodent type I hair cells (see Section 5.1.1.1) provides a priori evidence for vesicular release. This is buttressed by evidence for glutamate in type I hair cells and for various glutamate receptors on calyx endings (see Section 5.1.2) and for $Ca_V1.3$ Ca^{2+} channels in type I hair cells (see Section 4.1.2). Also, the irregular firing of calyx afferents in squirrel monkey vestibular epithelia has been attributed to quantal input (Smith and Goldberg 1986).

A case can also be made for nonvesicular afferent transmission at this synapse. In an excised preparation of the chick crista, Yamashita and Ohmori (1990) recorded afferent responses to vibrations of a ball in the medium above the epithelium (Fig. 8.16A). Intracellular recordings from bouton afferents exhibited the noisy EPSPs that are typical of eighth-nerve afferents (e.g., Furukawa et al. 1972; Fuchs and Parsons, Chapter 6) and indicative of quantal release of excitatory transmitter. Recordings from calyx afferents, in contrast, had no identifiable EPSPs, looking instead like a low-pass-filtered version of the hair cell receptor potential. The lack of EPSPs in the chick crista calyx afferents is not consistent with recordings of EPSPs made in vivo from lizard calyx afferents (Schessel et al. 1991) nor expectations of vesicular release based on the presence of ribbons in avian type I cells (Lysakowski 1996). Possibly vesicular release was not functional in the in vitro chick preparation. Nevertheless, the responses that were recorded from chick calyx afferents suggest some form of nonvesicular transmission. Yamashita and Ohmori looked for gap junctions that would mediate electrical transmission, but found none (Yamashita and Ohmori 1991). Two other kinds of nonvesicular transmission have been considered. In one form, postsynaptic voltage change occurs via stimulus-evoked modulation of K^+ concentration in the synaptic cleft (Goldberg 1996). In the other, postsynaptic

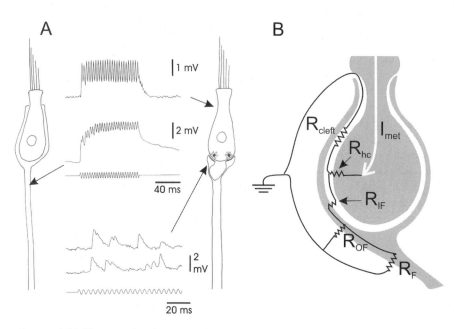

FIGURE 8.16. Unconventional transmission at the type I-calyx synapse? (**A**) Intracellular voltage recordings from a hair cell and from labeled afferents in a semi-intact chick crista. Sinusoidal vibrations at 250 Hz were delivered by a glass ball above the epithelium. The hair cell's receptor potential (*upper trace*) included an oscillating (AC) component at 250 Hz and an average (DC) depolarization that reflects low-pass filtering of the transduction current by the hair cell's membrane capacitance. The bouton afferent responded with highly variable EPSPs (*bottom traces*), as expected for quantal transmission. The calyx afferent, however, produced a response (*middle trace*) similar to the hair cell's but with additional low-pass filtering (slower rise and proportionally smaller AC component). Reproduced from Fig. 1 in Yamashita M, Ohmori H (1991) Synaptic bodies and vesicles in the calix type synapse of chicken semicircular canal ampullae. Neurosci Lett 129:43–46, with permission from Elsevier. (**B**) Schematic showing routes for current flow out of the type I hair cell to ground, either through the cleft or across the calyx inner-face and outer-face membranes. The amount traveling by each route depends on the relative resistances of the two paths: (1) R_{cleft} and (2) the summed resistances of the calyx inner face, R_{IF}, and the parallel combination of the outer face, R_{OF}, and the fiber, R_F. Because of the large K^+ conductances on both sides of the cleft, most of the ionic current is carried by K^+. I_{met}: transduction current; R_{hc}: resistance of the hair cell's basolateral membrane.

voltage change occurs via the flow of current from the hair cell into the calyx. We will consider these in the next two sections.

5.1.3.1 K^+ Accumulation in the Cleft

During transduction in vivo, K^+ enters the hair cell through apical transduction channels. In a typical hair cell, including vestibular type II cells, the influx of K^+ depolarizes the membrane and activates outwardly rectifying K^+ channels in the basolateral membrane. As they open, K^+ exits the basolateral membrane along its concentration gradient. In type I hair cells at rest, the channels are already open, allowing immediate K^+ efflux into the synaptic cleft: a thin shell of liquid with a maximal thickness of 25 nm (as low as 5 to 10 nm at invaginations) connected to the extracellular space of the basolateral epithelium by a thin collar at the neck of the hair cell. It seems likely that K^+ concentration in the cleft rises and falls with the transduction current with little lag because of the large standing K^+ conductance in the type I hair cell's membrane. This stimulus-modulated change in the driving force for K^+ will modulate K^+ flow and voltage across both the hair-cell and calyx inner-face membranes, which are both richly endowed with K^+ channels (Fig. 8.14C,D). The increased K^+ during a positive bundle deflection would depolarize the calyx inner face by decreasing the electrochemical gradient for K^+.

Such a scheme assumes that the K^+ transport pathways out of the cleft (channels and pumps) do not fully compensate for K^+ efflux from the hair cell. The synaptic cleft is well equipped with Na^+,K^+ ATPases and $Na^+-K^+-Cl^-$ cotransporters, however. Staining for Na^+,K^+-ATPase is very strong in mammalian vestibular epithelia, where it appears to localize to calyces (Spicer et al. 1990; Ichimiya et al. 1994) and possibly to hair cell and supporting cell membranes (ten Cate et al. 1994), although the low-resolution of the latter images would seem to preclude localization. Of the subunit isoforms tested (α_1, α_2, α_3, β_1, β_2), α_3 and β_1 subunits predominate in the sensory epithelium and α_1 and β_2 subunits predominate in the dark cells (ten Cate et al. 1994). The α-subunit distribution pattern fits with observations in the brain, where α_3 and α_1 are associated with neurons and glia, respectively (Blanco and Mercer 1998). Injection of high K^+ into perilymph stimulates strong Na^+,K^+-ATPase activity in the vestibular epithelium, principally at the type I-calyx synaptic cleft (Hozawa et al. 1991). In addition, Rennie et al. (1997) obtained physiological and pharmacological evidence for $Na^+-K^+-Cl^-$ cotransporter molecules in the type I hair cell membrane. High external K^+ causes slow shortening of the type I hair cell (Lapeyre and Cazals 1991). Rennie et al. (1997) blocked the shortening with bumetanide, a blocker of the $Na^+-K^+-Cl^-$ cotransporter, and not by ouabain, which blocks ATPase activity. The high K^+ presumably stimulates cotransporter activity to cause an influx of salt and water that increases cell volume, which manifests itself in isolated type I hair cells as a shortening of the neck.

It remains to be seen whether the combined activities of Na^+,K^+-ATPases and $Na^+-K^+-Cl^-$ cotransporters prevent changes in cleft K^+ levels at vestibular frequencies. Goldberg (1996) did not explore the effect of pump and transporter activities in his model but separately calculated that the fastest K^+ transport rates known for Na^+, K^+ ATPase would not prevent K^+ accumulation in the synaptic cleft.

5.1.3.2 Ephaptic Transmission

Even without changes in cleft K^+, there might be nonvesicular *ephaptic* transmission in which hair cell currents influence the calyx membrane potential.

Current entering through the transduction channels reaches ground by passing across the basolateral hair cell membrane and then taking one of two paths: an extracellular path through the synaptic cleft and out at the top (neck), and a path into the postsynaptic calyx membrane (calyx inner face) and out across the outer face and the nerve fiber (Fig. 8.16B; see also Goldberg 1996). The ratio of current taking the calyx route rather than the cleft route (I_{calyx}/I_{cleft}) is (R_{cleft}/R_{calyx}), where R_{calyx} is the summed resistance of the inner face (R_{IF}) and the parallel combination of the outer face (R_{OF}) and fiber (R_F): $R_{calyx} = R_{IF} + (R_{OF}*R_F)/(R_{OF} + R_F)$. The question of interest is how much voltage drop is produced at the spike-initiating zone in the fiber, probably somewhere below the cup. At a conventional synapse, most of the current will follow the low-resistance extracellular route (R_{cleft}) rather than traverse the high-resistance postsynaptic membrane, so that ephaptic transmission should be small. But the calyx may enhance ephaptic transmission by creating a high R_{cleft} and by having a large inner-face K^+ conductance, such that R_{IF} is low. In a model that took into account morphological features of the type I-calyx synapse but treated calyx inner and outer face membranes as electrically similar, Goldberg (1996) calculated that ephaptic transmission from the hair cell to the calyx would produce a voltage drop across the outer calyx face that was < 10% of the voltage across the hair cell membrane. Recent KCNQ4 immunoreactivity data (Fig. 8.14C,D; Kharkovets et al. 2000; Lysakowski and Price 2003), however, strongly suggest that the calyx inner face has a low membrane resistance. Lowering R_{IF} decreases not just the total calyx resistance (R_{calyx}), but also R_{IF} relative to R_{OF+F} (which lacks KCNQ4 immunoreactivity), thereby increasing the proportion of the voltage drop across R_{OF+F}, where spike initiation must occur. Moreover, Goldberg (1996) assumed that K^+ channels are evenly distributed along the hair cell and calyceal membranes. However, the pattern of KCNQ4 immunoreactivity on the calyx inner face membrane (Fig. 8.14C,D) and the particles around invaginations on freeze-fractured hair cell membranes (Fig. 8.14B) together suggest that K^+ channels may be concentrated toward the basal pole of the hair cell and synaptic cleft. This would increase R_{cleft} by increasing the average path length to ground.

5.1.3.3 Retrograde Transmission at the Calyx Synapse

Another plausible function for the calyx terminal is to provide feedback, by chemical or ephaptic mechanisms, to the type I hair cell. Data that suggest retrograde chemical transmission from calyces to hair cells include reports of vesicles in calyces (Scarfone et al. 1988, 1991) that are immunoreactive for glutamate (Raymond et al. 1984) and of such presynaptic markers as synaptophysin (Scarfone et al. 1988, 1991) and Rab3A (Dechesne et al. 1997). Moreover, isolated type I hair cells produce intracellular Ca^{2+} signals in response to exogenously supplied glutamate (Devau et al. 1993). In vivo, such responses might be evoked by retrograde glutamate release from the calyx and/or by glutamate released from the hair cell into the cleft. The pharmacology of the response is consistent with a mixture of NMDA receptors, AMPA receptors, and metabotropic receptors. Demêmes et al. (1995) found immunoreactivity in both type I and II hair cells with an antibody that recognized both GluR2 and GluR3 subunits; staining was most intense in type I cells. The mechanism linking glutamate receptors on the hair cells to the measured intracellular Ca^{2+} signals has not been elucidated. In addition, there might be retrograde transmission of nitric oxide from the calyx to the hair cell (see Section 5.2.4.2).

Ephaptic transmission may also work in the retrograde direction. If there is a low-resistance pathway for current flow from the hair cell to the afferent ending, there may also be back-propagation of current flow from afferent postsynaptic potentials and spikes into the hair cell.

5.2 Efferent Transmission

As in auditory epithelia (Fuchs and Parsons, Chapter 6), brainstem neurons project onto hair cells and primary afferent endings in vestibular epithelia. As we shall see, the form and neurotransmitters of the efferent endings vary with region, as do the effects of shocking the efferents on afferent activity. The effects of efferent shocks are more diverse in vestibular than auditory epithelia, although some molecular mediators are shared between the two systems. In the mammalian vestibular periphery, as in the cochlea (see Discussion in Maison et al. 2002), it is proving challenging to figure out the function of efferents at the level of behavior. For more detailed reviews of efferent transmission in vestibular organs, including those of non-mammals, see Goldberg et al. (2000) and Lysakowski and Goldberg (2004).

5.2.1 Anatomical Location of Vestibular Efferents

The efferent vestibular innervation to one ear originates from a group of 200 to 300 neurons on each side of the brain stem, adjacent to the genu of the facial nerve (Gacek and Rasmussen 1961; Gacek and Lyon 1974; Warr 1975; Goldberg and Fernández 1980; Marco et al. 1993). These neurons send compact bundles of axons via the vestibular nerve to the periphery. There, the axons branch profusely to form a rich plexus of fibers and terminals innervating hair cells, calyces and other afferent processes (Fig. 8.17; Purcell and Perachio 1997), frequently forming single boutons on each of multiple targets (Fig. 8.18; Dohlman et al. 1958; Hilding and Wersall 1962; Lysakowski and Goldberg 1997). The divergent nature of the projection might suggest that efferents act nonspecifically, but recent anatomical and neurochemical studies show them to be systematically heterogeneous. Such heterogeneity may explain the broad range of effects that vestibular efferents have on the firing patterns and dynamic responses of afferents.

Vestibular efferents are subdivided into two or three groups on the basis of soma location. The main vestibular efferent nucleus in mammals, group e, has been described in primates, cats, and rodents. In some species, group e is located lateral to the genu of the facial nerve and the abducens nucleus; in other species, group e has two parts, one lateral and one medial to the facial genu.

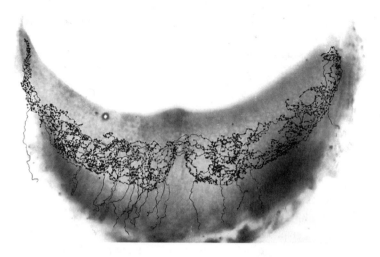

FIGURE 8.17. Efferent innervation to a gerbil crista. Terminal reconstructions of seven efferent neurons to a posterior crista, shown in side view (central zone at the *top*), following extracellular biocytin labeling of the contralateral e group. All of the efferents project to the peripheral zone, leaving the central zone (at the top) free of labeled endings. Reproduced with permission from Fig. 4A in Purcell IM, Perachio AA (1997) Three-dimensional analysis of vestibular efferent neurons innervating semicircular canals of the gerbil. J Neurophysiol 78:3234–3248.

The group e neurons project contralaterally or bilaterally, with considerable variation between species.

Vestibular efferents can also be subdivided according to their postsynaptic targets (see Section 5.2.2), whether hair cell types, afferents types or specific epithelial regions. In gerbil cristae, Purcell and Perachio (1997) found that group e efferent fibers terminate preferentially in the central zones of ipsilateral cristae and the peripheral zones of contralateral cristae (Fig. 8.17). Based on Figure 8.17, one would expect efferents from the contralateral brainstem to have little effect on irregular afferents, which principally represent central zones. But in the chinchilla crista, both ipsilateral and contralateral efferents can have large effects on irregular afferents (Marlinski et al. 2004). This discrepancy raises the unattractive possibility that innervation patterns differ significantly among rodent species.

5.2.2 Efferent Synaptic Contacts

Efferent fibers form bouton terminals on several kinds of targets in vestibular epithelia (Fig. 8.18A). "Presynaptic" efferent boutons, so-called because they act before the afferent synapse (Flock 1971), contact the base of type II hair cells. "Postsynaptic" efferent boutons contact the outside surfaces of calyces, afferent boutons on type II hair cells, and occasionally the distal processes or even parent axons of afferent nerve fibers (Smith and Rasmussen 1968; Iurato

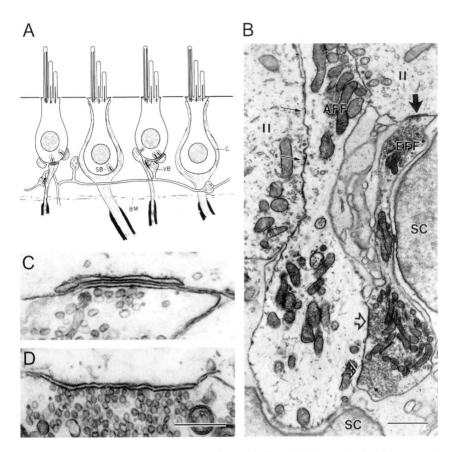

FIGURE 8.18. Efferent synapses in vestibular epithelia. (**A**) Schematic showing types of efferent contacts. (Reproduced from Fig. 11 in Smith CA, Rasmussen GL (1968) Nerve endings in the maculae and cristae of the chinchilla vestibule with a special reference to the efferents. In: Graybiel A, ed, Third Symposium on the Role of Vestibular Organs in Space Exploration, NASA SP-152 pp. 183–201. Washington, DC.: US Government Printing Office.) (**B**) Efferent (EFF) contacting both a type II hair cell (presynaptic contact, *thick arrow*; see cistern in the type II cell shown at higher magnification in **C**) and an afferent (AFF; postsynaptic contact; *open arrow*; expanded scale in **D**). From the central zone of an adult chinchilla posterior crista. *Scale bars*: B: 2.5 μm. D: 500nm, also applies to C. Modified from Fig. 14 in Lysakowski A, Goldberg JM (1997) A regional ultrastructural analysis of the cellular and synaptic architecture in the chinchilla cristae ampullares. J Comp Neurol 389:419–443, by permission of Wiley-Liss, Inc., a subsidiary of John Wiley & Sons, Inc.

et al. 1972; Lysakowski and Goldberg 1997). Single efferent fibers (Iurato and Taidelli 1964), even single efferent boutons (Lysakowski and Goldberg 1997), can contact multiple postsynaptic targets (Fig. 8.18B). Efferent boutons have plentiful synaptic vesicles (Fig. 8.18B-D). Adjacent to efferent contacts, type II hair cells have subsurface cisternae which may be Ca^{2+} stores (Fig. 8.18C; Fuchs and Parsons, Chapter 6). At postsynaptic contacts on afferents, including calyces, the afferent membranes have typical asymmetric postsynaptic densities (Fig. 8.18D).

5.2.3 Physiological Effects of Vestibular Efferents

The background firing of vestibular afferents in anesthetized mammals increases when vestibular efferents are activated, whether by electrical shocks (Goldberg and Fernández 1980; McCue and Guinan 1994; Marlinski et al. 2004) or by strong vestibular stimulation (Plotnik et al. 2002). The excitatory effect on background discharge has two components, a fast one on the order of tens of milliseconds and a slower one on the order of tens of seconds (Goldberg and Fernández 1980; Marlinski et al. 2004). For irregular afferents, the excitatory effect is relatively large and the fast component dominates. For regular afferents, the total excitatory effect is smaller and the fast component is proportionally smaller than in irregular afferents. Thus, maximal efferent effects on background rates tend to be largest and fastest in the central/striolar zones, where firing tends to be most irregular (see Section 2.2.2). The more modest fast effects in the regular afferents, which tend to be peripheral, may be related to the different nature of the rich efferent innervation in this zone (Fig. 8.17; Purcell and Perachio 1997). Peripheral efferent fibers tend to be thin and probably unmyelinated (Wackym et al. 1991) and stain for the peptide neurotransmitter, calcitonin gene-related peptide (CGRP) (see Section 5.2.4.3) and the muscarinic acetylcholine receptor (see Section 5.2.4.1), both of which could exert slow modulatory effects.

Based on their observations that efferent activation produces inhibitory postsynaptic potentials (IPSPs) in fish lateral line hair cells, Russell and colleagues proposed that the efferents provide an "efference copy" of motor signals, which suppresses incoming information about the fish's own movements (Russell 1971; Roberts and Russell 1972). By firing in anticipation of motion, vestibular efferents might modulate afferent sensitivity during active head movements. In toadfish, generalized arousal augments the background discharge rates of vestibular afferents and reduces their sensitivity (gain; reviewed in Highstein 1992); the arousal presumably acts through efferent neurons to the vestibular periphery. The dual action might reflect functional differences at presynaptic and postsynaptic efferent endings, with efferents mediating excitatory effects on afferents and inhibitory effects on hair cells. In the turtle crista, afferents that are inhibited by efferent activation are high-gain (sensitive), while those that are excited by efferent activation are low-gain (Brichta and Goldberg 2000). This correlation led Brichta and Goldberg to suggest that the efferent activity switches vestibular organs from a "postural" mode, in which small postural movements are sensed with high gain, to a "volitional" mode, in which low-gain afferents provide useful signals about large voluntary head movements.

Cullen and Minor (2002) tested the general idea that efferents modulate sensitivity to voluntary head movements by comparing, in awake rhesus monkeys, vestibular afferent activity during passive and active head movements of similar size and frequency (peak velocity ± 100–400 °/s, predominate frequency 0.5 to 1.5 Hz). The activity of semicircular canal afferents was similar in the two conditions. Thus, for primates and the kinds of head movements tested, there was no indication of efferent feedback influencing semicircular canal responses to voluntary, angular head movements. Cullen and Minor suggest that efferents may instead be important in the long-term balancing of activity between vestibular organs on the two sides of the head. They comment that this kind of action might be mediated by the slow component evident in the responses of regular afferents to efferent shocks, which we have speculated is a peptidergic effect (above and Section 5.2.4.3), and which would not be detectable in the short time course of their experiments.

Another possibility is that fast efferent effects, which the experiments were designed to detect, are strongest at higher head movement frequencies than were investigated (i.e., above 1.5 Hz). Fast effects on background discharge dominate in the central zone, where afferents appear to be tuned to higher frequencies than in the periphery (Fig. 8.3 and Hullar et al. 2004). The effects of efferents on auditory afferents are most prominent at best frequency (Brown and Nuttall 1984; Art et al. 1985). This hypothesis might also explain the weakness and variability of efferent effects on afferent gain (sensitivity to head movements) seen by Goldberg and Fernández (1980), who used just low-frequency head movement stimuli. It would be valuable to investigate efferent effects at higher head movement frequencies.

5.2.4 Efferent Transmitters and Receptors

As in the cochlea, there is evidence for multiple efferent transmitters in the vestibular periphery: acetylcholine (ACh) acting through neuronal nicotinic receptors (nAChRs) or muscarinic receptors (mAChRs); ATP acting on purinergic receptors, either ionotropic (P2X) or metabotropic (P2Y); calcitonin gene-related peptide (CGRP) acting on CGRP receptors or ACh receptors; enkephalin acting on opiate receptors and possibly ACh receptors; nitric oxide (NO) acting on K^+ channels or other targets; and γ-aminobutyric acid (GABA). Much of the work on vestibular efferent transmission has been done in non-mammals, and in the following descriptions we will borrow from that literature as needed.

5.2.4.1 Acetylcholine

Most vestibular efferents contain the synthetic enzyme for ACh, choline acetyltransferase (Perachio and Kevetter 1989; Ishiyama et al. 1994), and their terminals have round clear vesicles (Fig. 8.18). Both nicotinic and muscarinic ACh receptor subunits are found in vestibular epithelia. Thus, most vestibular efferent neurons are considered to be, like cochlear efferents, cholinergic.

As described in Chapter 6 (Fuchs and Parsons), the neuronal nicotinic receptors on avian and mammalian cochlear hair cells are heteromultimers of α9 and α10 nAChR subunits. These ionotropic receptors pass both monovalent cations and significant amounts of Ca^{2+} (Weisstaub et al. 2002). Cochlear efferents release ACh, which opens nAChR channels in the hair cell membrane; the ensuing Ca^{2+} influx activates SK channels (small Ca^{2+}-activated K^+ channels), through which K^+ exits the cell. The hair cell's efferent response comprises a small fast depolarization, reflecting cation influx through the nAChR channels, and a slower, larger hyperpolarization, reflecting K^+ efflux through SK channels, which accounts for the efferent-evoked IPSP and the predominantly inhibitory effect of cochlear efferents.

The most detailed information about the efferent cascade in amniote vestibular epithelia is emerging from data on the turtle crista, where efferent activation can have excitatory, inhibitory or mixed effects (Brichta and Goldberg 2000). α-Bungarotoxin, which specifically binds to neuronal nicotinic acetylcholine receptors, labels afferent calyces and type II hair cells in the turtle crista, consistent with expression of neuronal nicotinic receptors at both sites (Dailey et al. 2000). The inhibitory mechanism in turtle crista is thought to resemble that in the mammalian and avian cochleas, where presynaptic α9/10 receptors are coupled to SK channels (J.C. Holt et al. 2002). SK channel blockers convert the inhibition to excitation, which in turn is blocked by α9/10 antagonists.

There is evidence for portions of this cascade in the vestibular periphery of mammals. Again, α-bungarotoxin binding patterns are consistent with expression of neuronal nicotinic receptors on afferent calyces and type II hair cells (Wackym et al. 1995; Ishiyama et al. 1995). α9 and α10 subunit mRNA and protein have been reported in both vestibular hair cells and ganglia (Hiel et al. 1996; Anderson et al. 1997; Maroni et al. 1998; Guth et al. 1999), although some of the results are contradictory. With in situ hybridization on adult rat tissue, Hiel et al. (1996) reported mRNA for α9 receptors in type I and type II hair cells but not in ganglion cells. Expression in mature type I cells is a curiosity, given the lack of access of efferent terminals to the type I soma. Moreover, antibody to the α9 subunit does not label type I hair cells but does label type II cells, calyces and the ganglion cell bodies of calyx and dimorphic afferents (Maroni et al. 1998; Guth et al. 1999). It is not yet known whether mammalian vestibular hair cells have SK channels. Mammalian type II hair cells do have cisternae postsynaptic to efferent boutons (Fig. 8.18C). In cochlear and other hair cells, it is speculated that cisternae influence the activation of SK channels either as sources of Ca^{2+}-induced Ca^{2+} release or as Ca^{2+} buffers (Fuchs and Parsons, Chapter 6).

Other nicotinic subunits may also be involved in vestibular efferent transmission. Preliminary pharmacological data on efferent effects in turtle crista support the existence of non-α9/10 nicotinic receptors in addition to the α9/10 receptors already described (J.C. Holt et al. 2004). Vestibular ganglion neurons express mRNA for all α subunits from α4 to α7 (α8 is specific to chicks) (Wackym et al. 1995; Hiel et al. 1996; Anderson et al. 1997), as well as the auxiliary neuronal nicotinic subunits β_2, β_3 and β_4 (Anderson et al. 1997). RT-PCR experiments (Anderson et al. 1997) also showed expression of α_2, α_3, and β_4 subunits, whereas in situ hybridization experiments (Hiel et al. 1996) did not. The discrepancy may mean that the adult ganglion neurons express relatively small amounts of message for these subunits, below the detection level of in situ studies.

Muscarinic receptors are also a candidate target for acetylcholine. These metabotropic receptors activate different G proteins to modulate different enzymes, for example, to stimulate (via G_s) or inhibit (via G_i) adenylyl cyclase. RT-PCR studies show that mammalian vestibular organs and ganglia express several mAChR subunits, including m1, m2, and m5 in humans and m1 to m5 in the rat; the G proteins, G_s and G_i; and adenylyl cyclase (Wackym et al. 1996, 2000a,b). The binding pattern of a muscarinic receptor antagonist in rat vestibular organs is consistent with the sites of efferent terminals, particularly in the peripheral zones of cristae (Drescher et al. 1999). This localization is particularly interesting given the importance of slow efferent effects in the periphery (see Section 5.2.3). Drescher et al. (1995) also found histochemical evidence for adenylyl cyclase in efferent boutons of the trout saccule.

Parallels with the M current suggest intriguing possibilities for efferent action at the vestibular periphery. M current is so named because it is inhibited by ACh acting via mAChR's, and some M current channels are heteromultimers of KCNQ subunits (H.-S. Wang et al. 1998; Selyanko et al. 1999; Shah et al. 2002). This raises the possibility that mAChRs interact with KCNQ channels in calyx afferent endings and type II hair cells (see Section 4.1.1).

5.2.4.2 Nitric Oxide

NO, a gaseous neurochemical, can operate either as a transmitter or as a second messenger (reviewed in Garthwaite and Boulton, 1995). NO is synthesized from nitric oxide synthase (NOS), a calmodulin-dependent enzyme that is activated by Ca^{2+}, and can diffuse through membranes to activate soluble guanylyl cyclase, either in neighboring target cells or within the cell of origin, to produce cGMP and activate cGMP-dependent processes. It can also act directly upon ion channels (e.g., Ciorba et al. 1999). About 20% of group e efferent neurons, all confined to the lateral group, and about 20% of efferent boutons are NADPH diaphorase-positive (Lysakowski and Singer 2000). Calyces in the central zone of the rodent crista are immunoreactive for neuronal NOS (Fig. 8.19B) and soluble guanylate cyclase (Lysakowski et al. 2000; Desai et al. 2004).

How might NO act in the vestibular periphery? In rat crista hair cells, the large type-I-specific conductance, $g_{K,L}$, can be turned off by NO (Fig. 8.19C), possibly acting through cGMP-dependent protein kinase (Behrend et al. 1997; Chen and Eatock 2000; Rennie 2002). According to the diaphorase and NOS immunoreactivity results, NO may be synthesized within hair cells, calyx endings, and/or efferent terminals when Ca^{2+} enters as a consequence of excitation.

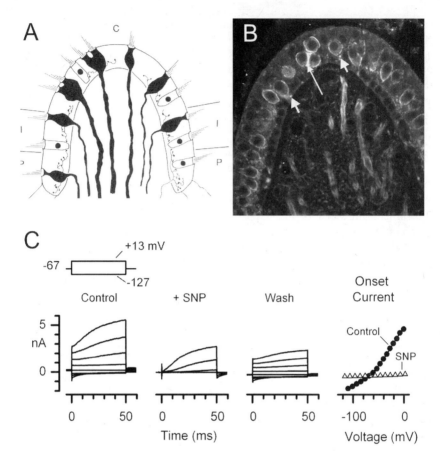

FIGURE 8.19. Nitric oxide (NO) is a possible neurotransmitter in vestibular epithelia. (**A**) Schematic cross-section of a crista. C, I, P: central, intermediate, peripheral zones. Reproduced with permission from Fig. 1 in Goldberg JM, Lysakowski A, Fernández C, Structure and function of vestibular nerve fibers in the chinchilla and squirrel monkey. Ann NY Acad Sci 656:92–107, Copyright 1992 New York Academy of Sciences. (**B**) Immunoreactivity for NOS I, the neuronal form of nitric oxide synthase, in calyx endings and fibers, adult rat crista (Desai et al. 2004). *Long arrow* points to labeled complex calyx. *Short arrows* point to labeled efferent boutons, which appear as light dots. (**C**) NO donor blocks $g_{K,L}$ in a type I hair cell from the rat crista. Whole-cell currents through $g_{K,L}$ channels were activated by voltage steps iterated between −127 and +13 mV, from a holding potential of −67 mV. In control and wash conditions, current flowed instantly at the voltage step onset, through open $g_{K,L}$ channels. The NO donor, sodium nitroprusside (SNP), reversibly blocked this onset current. Reproduced with permission from Fig. 9 in Chen JWY, Eatock RA (2000) Major potassium conductance in type I hair cells from rat semicircular canals: Characterization and modulation by nitric oxide. J Neurophysiol 84:139–151.

For example, when efferents are activated by vestibular or other forms of stimulation, Ca^{2+} entry into efferent terminals through voltage-gated Ca^{2+} channels will trigger both vesicular release of efferent transmitter and NO synthesis via calmodulin activation of NOS I (also called neuronal NOS). The resulting NO will turn off $g_{K,L}$ in any type I hair cells within broadcast range from the source (approximately 300 μm; Garthwaite and Bolton 1995), an effect likely mediated in part by cGMP-dependent protein kinase (Behrend et al. 1997; Chen and Eatock 2000; Rennie 2002). Both NO and 8-bromo-cGMP, a CGMP analogue, turn off $g_{K,L}$ recorded in the whole-cell mode or in cell-attached patches, but not in excised patches, indicating that the effects involve cellular intermediaries. The effect of 8-bromo-cGMP can be blocked by an inhibitor of cGMP-dependent protein kinase (Rennie 2002).

NO-mediated block of $g_{K,L}$ will depolarize the type I cell and increase its input resistance and receptor potential, enhancing vesicular transmitter release and possibly blocking ephaptic transmission. NO can act on targets other than $g_{K,L}$, so its net effect should be evaluated in an intact system. NO production in the axolotl inner ear enhances vestibular afferent firing (Flores et al. 1996).

5.2.4.3 Peptides

CGRP has different actions on different neurons (for review, see Van Rossum et al. 1997). CGRP exists in two forms in rats and humans (Amara et al. 1982; Steenbergh et al. 1986) and there are two types of CGRP receptor, CGRP1 and CGRP2 (Quirion et al. 1992). CGRP is commonly found in efferents to hair cell epithelia, where it typically co-localizes with ACh (reviewed in Eybalin 1993).

In the mammalian vestibular efferent system, both the cell bodies of group e neurons in the brainstem and efferent boutons in the periphery are immunoreactive for CGRP (Tanaka et al. 1988, 1989; Perachio and Kevetter 1989; Wackym et al. 1991; Scarfone et al. 1996). The CGRP-immunoreactive fibers are unmyelinated (Tanaka et al. 1989; Lysakowski 1999) and predominate in the periphery; in chinchilla vestibular epithelia, 25% of central/striolar efferent boutons and 90% of peripheral/extrastriolar efferent boutons are CGRP-immunoreactive (Lysakowski 1999). These results together suggest that the extensive efferent innervation of the peripheral zones of vestibular epithelia may have slow modulatory effects mediated by CGRP. Efferent effects lasting more than 20 s have been observed in chinchilla vestibular afferents in vivo (Plotnik et al. 2002). CGRP application to lateral line afferents can evoke long-lasting increases in background discharge and attenuation of the mechanically evoked response (Bailey and Sewell 2000).

Opioid peptides may also have a role in efferent transmission. There are no pharmacological studies in mammalian vestibular epithelia, but opioid peptides inhibit basal and glutamate-induced activity of frog vestibular afferents (Andrianov and Ryzhova 1999). In larval axolotls, opioids may regulate the activity of vestibular afferent neurons in a complex fashion, with μ opioid receptors mediating an excitatory, postsynaptic modulatory input to afferent neurons and κ receptors mediating an inhibitory presynaptic input to hair cells (Vega and Soto 2003). Alternatively, opioids, CGRP and other peptides released by efferent terminals may block α9/α10 nicotinic ACh receptors on hair cells or afferent nerve endings (Lioudyno et al. 2000, 2002).

In the gerbil, lateral group e neurons are immunoreactive for Met-enkephalin (Perachio and Kevetter 1989) and all efferents were positive for preproenkephalin in an in situ hybridization study (Ryan et al. 1991). Although Ylikoski et al. (1989) found no enkephalinergic fibers within the vestibular nerve, ganglion, or epithelia of rats and guinea pigs, Popper et al. (2002) obtained preliminary evidence for μ opioid receptors in hair cells and μ opiate receptors in nerve processes, including calyx outer membranes.

5.2.4.4 Purinergic Transmission

ATP is coreleased with ACh at certain cholinergic synapses (e.g., Schweitzer 1987), and there is some evidence that this occurs at efferent synapses in the inner ear. Both ionotropic (P2X) and metabotropic (P2Y) ATP receptors are reported in hair cells and ganglion cells in auditory and vestibular compartments of the inner ear (Housley and Ryan 1997; Troyanovskaya and Wackym 1998; Kreindler et al. 2001; Syeda and Lysakowski 2001). RT-PCR revealed several splice variants of $P2X_2$ mRNA in the vestibular organs but not the vestibular ganglia of rats (Troyanovskaya and Wackym 1998). Recent Western blot and immunocytochemical data, however, provide evidence for $P2X_2$ protein in both places (Syeda and Lysakowski 2001).

In cochlear outer hair cells, ATP opens $P2X_2$ cation-selective channels on the apical surface (Nakagawa et al. 1990; Housley et al. 1992; Jarlebark et al. 2002). ATP evokes a rapid inward current in guinea pig crista hair cells (Rennie and Ashmore 1993), probably through $P2X_2$ channels in the basolateral membrane (Housley and Ryan 1997; Brändle et al. 1999). Rossi et al. (1994) found that adding exogenous ATP to the perilymphatic compartment increased the background rate of miniature excitatory postsynaptic potentials recorded from frog crista afferents, consistent with depolarization of the hair cells.

5.2.4.5 GABA

GABA, the principal inhibitory transmitter in the central nervous system, may be a neurotransmitter in the mammalian vestibular periphery. GABA acts on ionotropic $GABA_A$ receptors, which form chloride channels, and metabotropic $GABA_B$ receptors, which activate G protein cascades that have inhibitory effects, for example, by opening K^+ channels. Diverse reports suggest that it is released by hair cells, afferent endings, efferent endings, or some combination. Immunoreactivity for GABA and GAD (L-glutamic acid decarboxylase, the synthetic enzyme) has been reported in rodent vestibular hair cells (Lopez et al. 1990; 1992), in calyx endings and/or vestibular ganglion cells (Didier et al. 1990; Lopez et al. 1990), and in presumptive efferent fibers and terminals (Usami et al. 1987). K⁺ currents in isolated type I hair cells are inhibited by exogenous GABA acting via $GABA_B$ receptors (Lapeyre et al. 1993). For this pathway to be functional in vivo, GABA from hair cells might feed back onto $GABA_B$ autoreceptors or calyces might release GABA onto hair cells. Immunoreactivity of rat calyx endings for $β_2$ and $β_3$ subunits of the $GABA_A$ receptor (Foster et al. 1995) is similarly consistent with GABA release from hair cells and/or efferent endings onto the calyx or with GABA release from the calyx onto itself. The cell bodies of group e neurons are not immunoreactive (Usami et al. 1987; Perachio and Kevetter 1989), so that for GABA to be an efferent transmitter, it must be either synthesized in the efferent terminals of group e neurons or released from unknown efferent neurons.

5.3 Summary of Transmission

In rodents, both type I and type II hair cells have the standard machinery serving chemical transmission, although the larger, more complex shapes of type II ribbons may support more vesicles and thereby provide more physiological impact per ribbon. Numerous immunohistochemical experiments indicate glutamatergic transmission, mediated postsynaptically by AMPA, kainate and NMDA receptors (Fig. 8.20). EPSPs have been recorded in vestibular afferents, including afferents that are likely to be calyx afferents (the latter not in rodents).

In type I cells, all synaptic structures are enhanced in the central/striolar zones: there are more ribbons, more calyceal invaginations, more complex calyces, even more type II ribbons on calyx outer faces. Some of the difference reflects the larger size of central/striolar type I cells, but in addition ribbons, invaginations and K⁺ channels extend further up the basolateral surface from the depth of the calyx in central/striolar cells. There are substantial regional differences proximal to the synaptic endings as well: calyx afferents and other central afferents have larger axonal and somatic diameters than do peripheral/extrastriolar afferents (both dimorphic and bouton). And calretinin is a useful marker for central/striolar zones because it strongly and selectively labels the calyx afferents.

We have speculated that type I hair cells also transmit via current flow, with or without changes in cleft K⁺. How would vesicular and nonvesicular transmission work together? Conditions that enhance one form of transmission may work against the other—for example, the high concentration of KCNQ4 channels in the inner face membrane might enhance ephaptic transmission but shunt glutamate-evoked EPSPs. The K⁺ conductances of the hair cell and calyx may be highly modulatable by efferent transmitters such as NO and ACh (Fig. 8.20), such that efferent feedback can adjust the relative importance of vesicular and ephaptic transmission.

There are far more candidate efferent transmitter systems in mammalian vestibular epithelia than there are known efferent functions. Stimulating mammalian vestibular efferents increases the background firing rate of afferent neurons

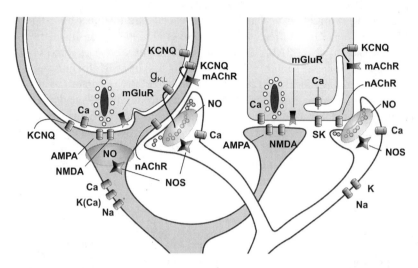

FIGURE 8.20. Afferent and efferent transmission in mammalian vestibular epithelia. In hair cells, there are Ca^{2+} channels (possibly two classes), K^+ channels, possibly glutamate receptors and NOS I. In the vestibular afferents, there is evidence for glutamate receptors of the AMPA, Kainate and NMDA classes, muscarinic and nicotinic AChRs, KCNQ channels, and NOS I. We speculate that type II cells have the Ach-nAChR-SK efferent cascade of cochlear hair cells, but SK channels and IPSPs have not been demonstrated. The diagram suggests two mechanisms by which efferent transmitters may inhibit K^+ channels in the afferent pathway: Arrival of action potentials at efferent terminals leads to Ca^{2+} entry, stimulating (1) synthesis of NO by NOS; NO turns off $g_{K,L}$, probably through a cGMP-dependent protein kinase; or (2) vesicular release of ACh, activating muscarinic receptors that are coupled negatively by a G protein to KCNQ channels, possibly on both calyx endings and type II hair cells.

with at least two time courses: the fast component dominates in irregular afferents, which tend to come from the central/striolar zones, and the slow component is proportionally larger in regular afferents, which tend to come from peripheral/extrastriolar zones. Efferents acting on hair cells in other systems produce IPSPs via a cascade involving ACh release from the efferent bouton and neuronal nicotinic receptors and SK channels in the hair cells. In mammals, vestibular efferents release ACh, vestibular epithelia express neuronal nicotinic receptors, and type II hair cells have postsynaptic specializations (cisternae) similar to those of other hair cells, but many parts of the cascade remain to be tested—for example, whether type II hair cells have SK channels and respond to efferent activation with IPSPs. In addition, a test of the efference copy hypothesis of efferent function failed to find any efference copy effect on vestibular afferent responses to active head movements. Thus, substantial questions remain on multiple levels, from the molecular cascade of efferent transmitter action to effects on the animal's behavior.

6. Concluding Remarks

6.1. Comparison with Cochlear Hair Cells

At a superficial level, hair cells share many transduction properties, kinds of basolateral conductance, synaptic transmission mechanisms, and the underlying morphological substrates. Yet each type of hair cell organ is specialized to handle a particular kind of mechanical input, and while we have known of some of the morphological specializations for years, our characterization of the biophysical specializations is still at an early stage.

Transduction properties seem to be grossly conserved across vertebrate hair cells. Transduction channels are of approximately the same conductance and selectivity, and there is evidence for two transducer adaptation mechanisms in all hair cells that have been examined closely. The adaptation processes in certain conditions act like tuning mechanisms: stereociliary amplifiers that increase the transducer sensitivity to mechanical input over a narrow frequency range. The significance of these mechanisms in vestibular hair cells awaits clarification of their properties in more in vivo-like conditions.

In mammals, both vestibular and cochlear hair cells have low-quality electrical resonances, unlike the auditory hair cells of turtles and chicks and the saccular hair cells of frogs. In vestibular hair cells, low-quality tuning is not surprising, given that vestibular afferents are only weakly tuned. Indeed, sharp electrical resonances would serve no clear benefit and would interfere with representation of timing (Lewis 1990), which may be more critical to the major function of the vestibular system in providing input to motor reflexes. In mammalian cochlear hair cells, there is also a lack of sharp electrical tuning (Art and Fettiplace, Chapter 5), presumably because electrical tuning mechanisms do not work at the high frequencies that occupy much of the mammalian audiogram. Instead, the basolateral membranes of mammalian cochlear hair cells show graded low-pass filtering, with hair cells that are acoustically tuned to higher sound frequencies having higher cut-off frequencies. A critical factor in setting the quality of electrical tuning appears to be the number of BK Ca^{2+}-dependent K^+ channels, which are hyper-sensitive to voltage via their dependence on Ca^{2+} influx through voltage-gated Ca^{2+} channels (Ashmore and Attwell 1985). While such channels are present in many vestibular hair cells, the dominant conductances tend to be purely voltage dependent.

Mammalian type I vestibular and cochlear hair cells both acquire negatively activating voltage-dependent K^+ conductances as they mature. These conductances lower their input resistance and extend their frequency bandwidth to higher frequencies. In the cochlea, it is striking that the conductances appear at the same time as hearing onset, i.e., when the system becomes functional. This timing has led to suggestions that the negatively activating conductances are required to clear the large K^+ load that enters via transduction channels and/ or to eliminate electrical tuning or spiking. Type I hair cells do have larger bundles, especially in the central/striolar zones, so it is possible that their larger

K⁺ conductances reflect larger transduction currents. In cochlear hair cells, the increased bandwidth provided by the channels may enhance signaling at acoustic frequencies, while lowering the input resistance of outer hair cells may reduce voltage-dependent nonlinearities that could interfere with somatic electromotility. Neither of these possible benefits of a large K^+ conductance clearly applies to vestibular hair cells, which operate at subacoustic frequencies and without somatic electromotility (although they do, curiously, express the motor protein of outer hair cells; Adler et al. 2003). Instead, a tempting possibility for type I hair cells is that their large K^+ conductance facilitates ephaptic transmission.

Variation in hair cell properties with hair cell type and location is common to auditory and vestibular hair cell epithelia. Such properties include hair bundle dimensions and transduction currents (Fettiplace and Ricci, Chapter 4) and voltage-dependent ion channels (Art and Fettiplace, Chapter 5). Regional variations in these properties are presumably coordinated with regional variations in accessory and synaptic structures to produce maps of sensory information. In the auditory organs of mammals and birds, zones tend to be oriented along the characteristic frequency axis—for example, zones of inner and outer hair cells. In the auditory organs of lizards, there are also abrupt transitions along this axis, with quite different zones representing different frequency bands (reviewed in Manley 2002). This arrangement is reminiscent of vestibular epithelia, where two or three morphological zones are served by afferents with distinctive physiological properties.

Key properties of mammalian vestibular afferents, including the regularity of interspike intervals, efferent actions and response dynamics, vary substantially between the central/striolar and peripheral/extrastriolar zones. The regularity of intervals may be determined postsynaptically by electrical properties of the afferent, which vary between zones in diameter (central/striolar afferents are larger), Ca^{2+}-binding protein expression (calyx afferents, which are exclusively central, express calretinin), and K^+ channel expression (central/striolar calyces show more intense immunoreactivity for KCNQ4 channels). There appear to be regional differences in expression of efferent transmitters, in efferent fiber diameter and myelination, and in types and abundance of postsynaptic targets (e.g., the proportion of type II hair cells is low in the central zones of some species). Regional differences in response dynamics may reflect variation in accessory structures (e.g., striolar otoconia are smaller, the striolar gel has a different composition and dimensions), hair bundles (striolar bundles have a larger footprint, fatter stereocilia and shorter kinocilia) and, we have speculated, the impact of type I hair cells and their calyx endings.

6.2 Why Type I Cells?

Mammalian vestibular epithelia may have diverged from the amniote archetype in that type I cells and their calyx endings have spread extensively into peripheral and extrastriolar zones. The central/striolar zones still have a higher "calyx index" in mammals, however, as follows: the proportion of type I cells is often larger in central and striolar zones; afferents that make exclusively calyx endings are restricted to central/striolar zones; and complex calyx endings are much more prevalent in central/striolar zones than in the surrounding territory.

It seems fair to suggest, then, that type I cells and their calyx endings contribute to the special physiology of the central and striolar zones in amniotes. These zones appear specialized for higher frequency signaling than are peripheral zones, although the extent of that specialization will not be clear until further experiments are done at frequencies above a few Hertz. Some of the properties of central/striolar type I cells and calyx afferents may facilitate high-frequency signaling: short bundles may be stiff; large transduction and K^+ conductances reduce membrane time constants both pre- and postsynaptically and might enhance ephaptic transmission; large synaptic contacts and large-diameter afferents seem designed for secure and fast signaling; high Ca^{2+} buffering in calyx afferents may speed up gating of Ca^{2+}-activated K^+ conductances; and fast efferent responses would allow for rapid modulation.

Data from toadfish (Boyle et al. 1991) and frogs (Baird 1994b) suggest that high-frequency signaling and large afferents with large terminals were features of central/striolar zones before the divergence of anamniotes and amniotes. Indeed, the sharing of properties between amniote and anamniote vestibular afferents has made it hard to relate a particular function to the calyx. But calyces may be an extreme form of large afferent endings present in anamniotes (see Section 5.1.1.2), so that shared properties are expected.

After amniotes and anamniotes diverged, factors that may have selectively affected the evolution of amniote vestibular epithelia include changes in the size and frequency content of head-on-body movements, as necks elongated and heads grew; and the development of foveate vision, requiring more precise and faster vestibular reflexes to stabilize gaze. Thus, the calyx may be especially important at high stimulus frequencies, a range that investigators have only begun to explore.

6.3 Future Directions

From the 1970s to the 1990s, investigators characterized vestibular afferent responses to simple head movement stimuli and the corresponding innervation patterns and morphology of afferent endings within the epithelia. In mammals, the physiological data are most complete for the squirrel monkey, and the morphophysiological correlations are most complete for the chinchilla. Ongoing research is extending the stimulus frequency range at the high end (Hullar et al. 2004), and characterizing vestibular afferent activity in mice, for comparison with hair cell and genetic data (Lasker et al. 2004).

From 1990 to the present, a number of studies have characterized different aspects of the hair cells' ion channels, both the apical mechanosensitive channels and the basolateral voltage-, Ca^{2+}-, and ligand-gated ion channels. We know a fair amount about how basolateral channels vary with cell type, epithelium type, location in the epithelium, developmental stage and even, in birds, regeneration stage. We can make educated guesses about their molecular identities, and with present techniques (e.g., J.R. Holt et al. 2002; Chabbert et al. 2003) will be able to test those guesses and localize the ion channels at the subcellular level (e.g., Kharkovets et al. 2000). The characterization of such specialized hair cell properties is incomplete, and we are even further from understanding their functional significance. Few studies bridge the gap between the molecular, anatomical and biophysical level of much hair cell research and the systems level of much vestibular research.

For example, to understand how accessory structures and bundles work together to shape the stimulus to the transduction apparatus, we need molecular and genetic investigations to reveal the key molecules determining the material properties of gels and bundles; experiments to examine how bundles move and transduction currents vary when coupled to a gel; and a systems-level approach to put the gel-bundle interactions together with other neuroepithelial properties to create a model of peripheral vestibular processing.

The puzzles of afferent and efferent transmission in mammalian vestibular epithelia also have both reductionist and systems level aspects. How transmission occurs and is modulated at the type I hair cell/calyx will require a synthesis of molecular, ultrastructural and biophysical data. Figuring out what drove the evolution of this special synapse will require a systems-level analysis of its functional impact. Similarly, neurochemical and neuroanatomical experiments have revealed a plethora of efferent transmitter candidates, but we neither understand the reductionist molecular details of the efferent actions on afferent activity, nor the systems-level effects on movement, balance and posture.

Acknowledgements. Research in the authors' laboratories is supported by grants from the National Institute on Deafness and Communicative Disorders (to R.A.E. and A.L.) and by the Karim Al-Fayed Hearing Reseach fund (to R.A.E.) and by NASA (to A.L.). We thank Drs. Sophie Gaboyard, Chris Holt, Karen Hurley, and Ellengene Peterson, and Mrs. Jasmine Garcia, for helpful comments on the manuscript.

References

Acuna D, Ishiyama A, Lopez IA (2004) Kir4.1 potassium channel subunit expression in the vestibular sensory epithelia of mice. Abstr Assoc Res Otolaryngol 27:1313.

Adler HJ, Belyantseva IA, Merritt RC Jr, Frolenkov GI, Dougherty GW, Kachar B (2003) Expression of prestin, a membrane motor protein, in the mammalian auditory and vestibular periphery. Hear Res 184:27–40.

Ahmad M, Lysakowski A (2004) The synaptic ultrastructure of the adult mouse utricular macula. Abstr Assoc Res Otolaryngol 27:918.

Almanza A, Vega R, Soto E (2003) Calcium current in type I hair cells isolated from the semicircular canal crista ampullaris of the rat. Brain Res 994:175–180.

Amara SG, Jonas V, Rosenfeld MG, Ong ES, Evans RM (1982) Alternative RNA processing in calcitonin gene expression generates mRNAs encoding different polypeptide products. Nature 298:240–244.

Anderson AD, Troyanovskaya M, Wackym PA (1997) Differential expression of alpha2-7, alpha9 and beta2-4 nicotinic acetylcholine receptor subunit mRNA in the vestibular end-organs and Scarpa's ganglia of the rat. Brain Res 778:409–413.

Andrianov GN, Ryzhova IV (1999) Opioid peptides as possible neuromodulators of the afferent synaptic transmission in the frog semicircular canal. Neuroscience 93:801–806.

Angelaki DE (1998) Three-dimensional organization of otolith-ocular reflexes in rhesus monkeys. III. Responses to translation. J Neurophysiol 80:680–695.

Annoni J-M, Cochran SL, Precht W (1984) Pharmacology of the vestibular hair cell-afferent fiber synapse in the frog. J Neurosci 4:2106–2116.

Art JJ, Crawford AC, Fettiplace R, Fuchs PA (1985) Efferent modulation of hair cell tuning in the cochlea of the turtle. J Physiol 360:397–421.

Ashmore JF, Attwell D (1985) Models for electrical tuning in hair cells. Proc R Soc Lond B 226:325–344.

Assad JA, Corey DP (1992) An active motor model for adaptation by vertebrate hair cells. J Neurosci 12:3291–3309.

Assad JA, Shepherd GMG, Corey DP (1991) Tip-link integrity and mechanical transduction in vertebrate hair cells. Neuron 7:985–994.

Bagger-Sjöbäck D, Takumida M (1988) Geometrical array of the vestibular sensory hair bundle. Acta Otolaryngol (Stockh) 106:393–403.

Bailey GP, Sewell WF (2000) Calcitonin gene-related peptide suppresses hair cell responses to mechanical stimulation in the *Xenopus* lateral line organ. J Neurosci 20:5163–5169.

Baimbridge KG, Celio MR, Rogers JH (1992) Calcium-binding proteins in the nervous system. Trends Neurosci 15:303–308.

Baird RA (1992) Morphological and electrophysiological properties of hair cells in the bullfrog utriculus. Ann N Y Acad Sci 656:12–26.

Baird RA (1994a) Comparative transduction mechanisms of hair cells in the bullfrog utriculus. I. Responses to intracellular current. J Neurophysiol 71:685–705.

Baird RA (1994b) Comparative transduction mechanisms of hair cells in the bullfrog utriculus. II. Sensitivity and response dynamics to hair bundle displacement. J Neurophysiol 71:685–705.

Baird RA, Lewis ER (1986) Correspondences between afferent innervation patterns and response dynamics in the bullfrog utricle and lagena. Brain Res 369:48–64.

Baird RA, Schuff NR (1994) Peripheral innervation patterns of vestibular nerve afferents in the bullfrog utriculus. J Comp Neurol 342:279–298.

Baird RA, Desmadryl G, Fernández C, Goldberg JM (1988) The vestibular nerve of the chinchilla. II. Relation between afferent response properties and peripheral innervation patterns in the semicircular canals. J Neurophysiol 60:182–203.

Baird RA, Steyger PS, Schuff NR (1997) Intracellular distributions and putative functions of calcium-binding proteins in the bullfrog vestibular otolith organs. Hearing Res 103:85–100.

Bao H, Wong WH, Goldberg JM, Eatock RA (2003) Voltage-gated calcium channel currents in type I and type II hair cells isolated from the rat crista. J Neurophysiol 90:155–164.

Barmack NH (2003) Central vestibular system: vestibular nuclei and posterior cerebellum. Brain Res Bull 60:511–541.

Behrend O, Schwark C, Kunihiro T, Strupp M (1997) Cyclic GMP inhibits and shifts the activation curve of the delayed-rectifier (I_{K1}) of type I mammalian vestibular hair cells. NeuroReport 8:2687–2690.

Benser ME, Issa NP, Hudspeth AJ (1993) Hair-bundle stiffness dominates the elastic reactance to otolithic-membrane shear. Hear Res 68:243–252.

Bergstrom RA, You Y, Erway LC, Lyon MF, Schimenti JC (1998) Deletion mapping of the head tilt (het) gene in mice: a vestibular mutation causing specific absence of otoliths. Genetics 150:815–822.

Beutner D, Moser T (2001) The presynaptic function of mouse cochlear inner hair cells during development of hearing. J Neurosci 21:4593–4599.

Blanco G, Mercer RW (1998) Isozymes of the Na-K-ATPase: heterogeneity in structure, diversity in function. Am J Physiol. 275:F633–650.

Boyer C, Sans A, Vautrin J, Chabbert C, Lehouelleur J (1999) K^+-dependence on Na^+-Ca^{2+} exchange in type I vestibular sensory cells of guinea-pig. Eur J Neurosci 11: 1955–1959.

Boyer C, Art JJ, Dechesne CJ, Lehouelleur J, Vautrin J, Sans A (2001) Contribution of the plasmalemma to Ca^{2+} homeostasis in hair cells. J Neurosci 21:2640–2650.

Boyle R, Highstein SM (1990) Resting discharge and response dynamics of horizontal semicircular canal afferents of the toadfish, *Opsanus tau*. J Neurosci 10:1557–1569.

Boyle R, Carey JP, Highstein SM (1991) Morphological correlates of response dynamics and efferent stimulation in horizontal semicircular canal afferents of the toadfish, *Opsanus tau*. J Neurophysiol 66:1504–1521.

Brändle U, Zenner HP, Ruppersberg JP (1999) Gene expression of P2X-receptors in the developing inner ear of the rat. Neurosci Lett 273:105–108.

Brandt A, Striessnig J, Moser T (2003) $Ca_V1.3$ channels are essential for development and presynaptic activity of cochlear inner hair cells. J Neurosci 23:10832–10840.

Brichta AM, Goldberg JM (2000) Responses to efferent activation and excitatory response-intensity relations of turtle posterior-crista afferents. J Neurophysiol 83: 1224–1242.

Brichta AM, Aubert A, Eatock RA, Goldberg JM (2002) Regional analysis of whole cell currents from hair cells of the turtle posterior crista. J Neurophysiol 88:3259–3278.

Brown MC, Nuttall AL (1984) Efferent control of cochlear inner hair cell responses in the guinea-pig. J Physiol 354:625–46.

Cameron P, Modi D, Price SD, Lysakowski A (2004) Multiple isoforms of the ryanodine and IP3 receptors are expressed in rat inner ear ganglia and endorgans. Abstr Assoc Res Otolaryngol 27:633.

Chabbert C, Canitrot Y, Sans A, Lehouelleur J (1995) Calcium homeostasis in guinea pig type-I vestibular hair cell: possible involvement of an Na^+-Ca^{2+} exchanger. Hear Res 89:101–108.

Chabbert C, Chambard JM, Sans A, Desmadryl G (2001) Three types of depolarization-activated potassium currents in acutely isolated mouse vestibular neurons. J Neurophysiol 85:1017–1026.

Chabbert C, Mechaly I, Sieso V, Giraud P, Brugeaud A, Lehouelleur J, Couraud F, Valmier J, Sans A (2003) Voltage-gated Na^+ channel activation induces both action potentials in utricular hair cells and brain-derived neurotrophic factor release in the rat utricle during a restricted period of development. J Physiol 553:113–123.

Chang JS, Popper AN, Saidel WM (1992) Heterogeneity of sensory hair cells in a fish ear. J Comp Neurol 324:621–640.

Chen JWY, Eatock RA (2000) Major potassium conductance in type I hair cells from rat semicircular canals: characterization and modulation by nitric oxide. J Neurophysiol 84:139–151.

Cho WJ, Drescher MJ, Hatfield JS, Bessert DA, Skoff RP, Drescher DG (2003) Hyperpolarization-activated, cyclic AMP-gated, HCN1-like cation channel: the primary, full-length HCN isoform expressed in a saccular hair-cell layer. Neuroscience 118:525–534.

Ciorba MA, Heinemann SH, Weissbach H, Brot N, Hoshi T (1999) Regulation of voltage-dependent K^+ channels by methionine oxidation: effect of nitric oxide and vitamin C. FEBS Lett 442:48–52.

Corey DP, Hudspeth AJ (1983a) Analysis of the microphonic potential of the bullfrog's sacculus. J Neurosci 3:942–961.

Corey DP, Hudspeth AJ (1983b) Kinetics of the receptor current in bullfrog saccular hair cells. J Neurosci 3:962–976.

Corey DP, Sotomayor M (2004) Hearing: tightrope act. Nature 428:901–903.

Correia MJ, Lang DG (1990) An electrophysiological comparison of solitary type I and type II vestibular hair cells. Neurosci Lett 116:106–111.

Correia MJ, Christensen BN, Moore LE, Lang DG (1989) Studies of solitary semicircular canal hair cells in the adult pigeon. I. Frequency- and time-domain analysis of active and passive membrane properties. J Neurophysiol 62:924–945.

Correia MJ, Ricci AJ, Rennie KJ (1996) Filtering properties of vestibular hair cells: an update. Ann NY Acad Sci 781:138–149.

Correia MJ, Rennie KJ, Koo P (2001) Return of potassium ion channels in regenerated hair cells: possible pathways and the role of intracellular calcium signaling. Ann N Y Acad Sci 942:228–240.

Correia MJ, Wood TG, Prusak D, Weng T, Rennie KJ, Wang HQ (2004) Molecular characterization of an inward rectifier channel (IKir) found in avian vestibular hair cells: cloning and expression of pKir2.1. Physiol Genom 19:155–169.

Crawford AC, Fettiplace R (1981) An electrical tuning mechanism in turtle cochlear hair cells. J Physiol 312:377–412.

Crawford AC, Evans MG, Fettiplace R (1991) The actions of calcium on the mechano-electrical transducer current of turtle hair cells. J Physiol 434:369–398.

Cullen KE, Minor LB (2002) Semicircular canal afferents similarly encode active and passive head-on-body rotations: implications for the role of vestibular efference. J Neurosci 22:RC226.

Curthoys IS (1983) The development of function of primary vestibular neurons. In: Romand R (ed), Development of Auditory and Vestibular Systems New York: Academic Press, pp. 425–461.

Cyr JL, Dumont RA, Gillespie PG (2002) Myosin-1c interacts with hair-cell receptors through its calmodulin-binding IQ domains. J Neurosci 22:2487–2495.

Dailey SH, Wackym PA, Brichta AM, Gannon PJ, Popper P (2000) Topographic distribution of nicotinic acetylcholine receptors in the cristae of a turtle. Hear Res 141:51–56.

Dechesne CJ, Thomasset M (1988) Calbindin (CaBP 28 kDa) appearance and distribution during development of the mouse inner ear. Brain Res 468:233–242.

Dechesne CJ, Thomasset M, Brehier A, Sans A (1988) Calbindin (CaBP 28 kDa) localization in the peripheral vestibular system of various vertebrates. Hear Res 33:273–278.

Dechesne CJ, Winsky L, Kim HN, Goping G, Vu TD, Wenthold RJ, Jacobowitz DM

(1991) Identification and ultrastructural localization of a calretinin-like calcium-binding protein (protein 10) in the guinea pig and rat inner ear. Brain Res 560:139–148.
Dechesne CJ, Kauff C, Stettler O, Tavitian B (1997) Rab3A immunolocalization in the mammalian vestibular end-organs during development and comparison with synaptophysin expression. Brain Res Dev Brain Res 99:103–111.
Demêmes D, Wenthold RJ, Moniot B, Sans A (1990) Glutamate-like immunoreactivity in the peripheral vestibular system of mammals. Hear Res 46:261–269.
Demêmes D, Eybalin M, Renard N (1993) Cellular distribution of parvalbumin immunoreactivity in the peripheral vestibular system of three rodents. Cell Tissue Res 274:487–492.
Demêmes D, Lleixa A, Dechesne CJ (1995) Cellular and subcellular localization of AMPA-selective glutamate receptors in the mammalian peripheral vestibular system. Brain Res 671:83–94.
Denman-Johnson K, Forge A (1999) Establishment of hair bundle polarity and orientation in the developing vestibular system of the mouse. J Neurocytol 28:821–835.
Desai SS, Dhaliwal J, Lysakowski A (2004) NOS immunochemical staining in calyces in chinchilla vestibular endorgans. Abstr Assoc Res Otolaryngol 27:853.
Desai SS, Ali H, Lysakowski A (2005a) Comparative morphology of the rodent vestibular periphery: II. Cristae ampullares. J Neurophysiol 93:267–280.
Desai SS, Zeh C, Lysakowski A (2005b) Comparative morphology of the rodent vestibular periphery: I. Saccular and utricular maculae. J Neurophysiol 93:251–266.
Desmadryl G, Dechesne CJ (1992) Calretinin immunoreactivity in chinchilla and guinea pig vestibular end organs characterizes the calyx unit subpopulation. Exp Brain Res 89:105–108.
Desmadryl G, Sans A (1990) Afferent innervation patterns in crista ampullaris of the mouse during ontogenesis. Brain Res Dev Brain Res 52:183–189.
Devau G, Lehouelleur J, Sans A (1993) Glutamate receptors on type I vestibular hair cells of guinea-pig. Eur J Neurosci 5:1210–1217.
De Vries HL (1950) Mechanics of the labyrinth organs. Acta Otolaryngol 38:262–273.
Dickman JD, Correia MJ (1989) Responses of pigeon horizontal semicircular canal afferent fibers. II. High-frequency mechanical stimulation. J Neurophysiol 62:1102–1112.
Didier A, Dupont J, Cazals Y (1990) GABA immunoreactivity of calyceal nerve endings in the vestibular system of the guinea pig. Cell Tissue Res 260:415–419.
Dohlman G, Farkashidy J, Salonna F (1958) Centrifugal nerve-fibres to the sensory epithelium of the vestibular labyrinth. J Laryngol Otol 72:984–991.
Dou H, Vazquez AE, Namkung Y, Chu H, Cardell EL, Nie L, Parson S, Shin HS, Yamoah EN (2004) Null mutation of α_{1D} Ca^{2+} channel gene results in deafness but no vestibular defect in mice. J Assoc Res Otolaryngol 5:215–226.
Drescher MJ, Kern RC, Hatfield JS, Drescher DG (1995) Cytochemical localization of adenylyl cyclase activity within the sensory epithelium of the trout saccule. Neurosci Lett 196:145–148.
Drescher DG, Kerr TP, Drescher MJ (1999) Autoradiographic demonstration of quinuclidinyl benzilate binding sites in the vestibular organs of the gerbil. Brain Res 845:199–207.
Dumont RA, Lins U, Filoteo AG, Penniston JT, Kachar B, Gillespie PG (2001) Plasma membrane Ca^{2+}-ATPase isoform 2a is the PMCA of hair bundles. J Neurosci 21:5066–5078.
Eatock RA (2000) Adaptation in hair cells. Annu Rev Neurosci 23:285–314.

Eatock RA, Hurley KM (2003) Functional development of hair cells. In: Romand R, Varela-Nieto I (eds), Development of the Auditory and Vestibular Systems 3: Molecular Development of the Inner Ear. San Diego: Academic Press, pp. 389–448.

Eatock RA, Hutzler MJ (1992) Ionic currents in mammalian vestibular hair cells. Ann NY Acad Sci 656:58–74.

Eatock RA, Corey DP, Hudspeth AJ (1987) Adaptation of mechanoelectrical transduction in hair cells of the bullfrog's sacculus. J Neurosci 7:2821–2836.

Eatock RA, Chen W-Y, Saeki M (1994) Potassium currents in mammalian vestibular hair cells. Sensory Systems 8:21–28.

Eatock RA, Hurley KM, Vollrath MA (2002) Mechanoelectrical and voltage-gated ion channels in mammalian vestibular hair cells. Audiol Neurootol 7:31–35.

Edmonds B, Reyes R, Schwaller B, Roberts WM (2000) Calretinin modifies presynaptic calcium signaling in frog saccular hair cells. Nat Neurosci 3:786–790.

Eybalin M (1993) Neurotransmitters and neuromodulators of the mammalian cochlea. Physiol Rev 73:309–373.

Faber ES, Sah P (2003) Calcium-activated potassium channels: multiple contributions to neuronal function. Neuroscientist 9:181–194.

Favre D, Sans A (1979) Morphological changes in afferent vestibular hair cell synapses during the postnatal development of the cat. J Neurocytol 8:765–775.

Favre D, Sans A (1983) Organization and density of microtubules in the vestibular sensory cells in the cat. Acta Otolaryngol 96:15–20.

Fermin CD, Lychakov D, Campos A, Hara H, Sondag E, Jones T, Jones S, Taylor M, Meza-Ruiz G, Martin DS (1998) Otoconia biogenesis, phylogeny, composition and functional attributes. Histol Histopathol 13:1103–1154.

Fernández C, Baird RA, Goldberg JM (1988) The vestibular nerve of the chinchilla. I. Peripheral innervation patterns in the horizontal and superior semicircular canals. J Neurophysiol 60:167–181.

Fernández C, Goldberg JM, Baird RA (1990) The vestibular nerve of the chinchilla. III. Peripheral innervation patterns in the utricular macula. J Neurophysiol 63:767–780.

Fernández C, Lysakowski A, Goldberg JM (1995) Hair-cell counts and afferent innervation patterns in the cristae ampullares of the squirrel monkey with a comparison to the chinchilla. J Neurophysiol 73:1253–1281.

Fettiplace R, Ricci AJ, Hackney CM (2001) Clues to the cochlear amplifier from the turtle ear. Trends Neurosci 24:169–175.

Flock Å (1971) Sensory transduction in hair cells. In: Loewenstein WR (ed), Handbook of Sensory Physiology, Vol. 1. Berlin: Springer-Verlag, pp. 396–441.

Flores A, León-Olea M, Vega R, Soto E (1996) Histochemistry and role of nitric oxide synthase in the amphibian (*Ambystoma tigrinum*) inner ear. Neurosci Lett 205:131–134.

Fontilla MF, Peterson EH (2000) Kinocilia heights on utricular hair cells. Hear Res 145:8–16.

Foster JD, Drescher MJ, Hatfield JS, Drescher DG (1994) Immunohistochemical localization of S-100 protein in auditory and vestibular end organs of the mouse and hamster. Hear Res 74:67–76.

Foster JD, Drescher MJ, Drescher DG (1995) Immunohistochemical localization of GABAA receptors in the mammalian crista ampullaris. Hear Res 83:203–208.

Freeman DM, Weiss TF (1988) The role of fluid inertia in mechanical stimulation of hair cells. Hear Res 35:201–207.

Furness DN, Lawton DM (2003) Comparative distribution of glutamate transporters and receptors in relation to afferent innervation density in the mammalian cochlea. J Neurosci 23:11296–11304.

Furukawa T, Ishii Y, Matsuura S (1972) Synaptic delay and time course of postsynaptic potentials at the junction between hair cells and eighth nerve fibers in the goldfish. Jpn J Physiol 22:617–635.

Gacek RR, Lyon M (1974) The localization of vestibular efferent neurons in the kitten with horseradish peroxidase. Acta Otolaryngol 77:92–101.

Gacek RR, Rasmussen GL (1961) Fiber analysis of the statoacoustic nerve of guinea pig, cat, and monkey. Anat Rec 139:455–463.

Ganitkevich VY (2003) The role of mitochondria in cytoplasmic Ca^{2+} cycling. Exp Physiol 88:91–97.

Garthwaite J, Boulton CL (1995) Nitric oxide signaling in the central nervous system. Annu Rev Physiol 57:683–706.

Géléoc GS, Holt JR (2003) Developmental acquisition of sensory transduction in hair cells of the mouse inner ear. Nat Neurosci 6:1019–1020.

Géléoc GS, Lennan GWT, Richardson GP, Kros CJ (1997) A quantitative comparison of mechanoelectrical transduction in vestibular and auditory hair cells of neonatal mice. Proc R Soc Lond B 264:611–621.

Géléoc GS, Risner JR, Holt JR (2004) Developmental acquisition of voltage-dependent conductances and sensory signaling in hair cells of the embryonic mouse inner ear. J Neurosci 24:11148–11159.

Goldberg JM (1996) A theoretical analysis of intercellular communication between the vestibular type I hair cell and its calyx ending. J Neurophysiol 76:1942–1957.

Goldberg JM (2000) Afferent diversity and the organization of central vestibular pathways. Exp Brain Res 130:277–297.

Goldberg JM, Brichta AM (2002) Functional analysis of whole cell currents from hair cells of the turtle posterior crista. J Neurophysiol 88:3279–3292.

Goldberg JM, Fernández C (1971) Physiology of peripheral neurons innervating semicircular canals of the squirrel monkey. I. Resting discharge and response to constant angular accelerations. J Neurophysiol 34:635–660.

Goldberg JM, Fernández C (1980) Efferent vestibular system in the squirrel monkey: anatomical location and influence on afferent activity. J Neurophysiol 43:986–1025.

Goldberg JM, Smith CE, Fernández C (1984) Relation between discharge regularity and responses to externally applied galvanic currents in vestibular nerve afferents of the squirrel monkey. J Neurophysiol 51:1236–1256.

Goldberg JM, Desmadryl G, Baird RA, Fernández C (1990a) The vestibular nerve of the chinchilla. IV. Discharge properties of utricular afferents. J Neurophysiol 63:781–790.

Goldberg JM, Desmadryl G, Baird RA, Fernández C (1990b) The vestibular nerve of the chinchilla. V. Relation between afferent discharge properties and peripheral innervation patterns in the utricular macula. J Neurophysiol 63:791–804.

Goldberg JM, Lysakowski A, Fernández C (1992) Structure and function of vestibular nerve fibers in the chinchilla and squirrel monkey. Ann NY Acad Sci 656:92–107.

Goldberg JM, Brichta AM, Wackym PA (2000) Efferent vestibular system: Anatomy, physiology and neurochemistry. In: Beitz AJ, Anderson JH (eds), Neurochemistry of the Vestibular System. Boca Raton: CRC Press, pp. 61–94.

Goodyear R, Richardson G (1999) The ankle-link antigen: an epitope sensitive to calcium chelation associated with the hair-cell surface and the calycal processes of photoreceptors. J Neurosci 19:3761–3772.

Goodyear RJ, Richardson GP (2002) Extracellular matrices associated with the apical surfaces of sensory epithelia in the inner ear: molecular and structural diversity. J Neurobiol 53:212–227.

Grant JW, Huang CC, Cotton JR (1994) Theoretical mechanical frequency response of the otolithic organs. J Vestib Res 4:137–151.

Grossman GE, Leigh RJ, Abel LA, Lanska DJ, Thurston SE (1988) Frequency and velocity of rotational head perturbations during locomotion. Exp Brain Res 70:470–476.

Gulley RL, Bagger-Sjöbäck D (1979) Freeze-fracture studies on the synapse between the type I hair cell and the calyceal terminal in the guinea-pig vestibular system. J Neurocytol 8:591–603.

Guth SM, Price SD, Luebke A, Lysakowski A (1999) Alpha-9 nicotinic acetylcholine receptor is not found on the peripherin-labelled (bouton) class of afferents in the vestibular ganglion of the chinchilla. Abstr Assoc Res Otolaryngol 22: 186–187.

Hackney CM, Mahendrasingam S, Jones EM, Fettiplace R (2003) The distribution of calcium buffering proteins in the turtle cochlea. J Neurosci 23:4577–4589.

Hamilton DW (1968) The calyceal synapse of type I vestibular hair cells. J Ultrastruct Res 23:98–114.

Heller S, Bell AM, Denis CS, Choe Y, Hudspeth AJ (2002) Parvalbumin 3 is an abundant Ca^{2+} buffer in hair cells. J Assoc Res Otolaryngol 3:488–498.

Hess K (1996) Vestibulotoxic drugs and other causes of acquired bilateral peripheral vestibulopathy. In: Baloh RW, Halmagyi GM (eds), Disorders of the Vestibular System. New York: Oxford University Press, pp. 360–373.

Hiel H, Elgoyhen AB, Drescher DG, Morley BJ (1996) Expression of nicotinic acetylcholine receptor mRNA in the adult rat peripheral vestibular system. Brain Res 738: 347–352.

Highstein SM (1992) The efferent control of the organs of balance and equilibrium in the toadfish, *Opsanus tau*. Ann NY Acad Sci 656:108–123.

Highstein SM (1996) How does the vestibular part of the inner ear work? In: Baloh RW, Halmagyi GM (eds), Disorders of the Vestibular System. New York: Oxford University Press, pp. 3–19.

Highstein SM, Rabbitt RD, Boyle R (1996) Determinants of semicircular canal afferent response dynamics in the toadfish, *Opsanus tau*. J Neurophysiol 75:575–596.

Hilding D, Wersäll J (1962) Cholinesterase and its relation to the nerve endings in the inner ear. Acta Otolaryngol 55:205–217.

Hille B (2001) Ion Channels of Excitable Membranes. Sunderland, MA: Sinauer.

Holt JC, Xue J-T, Goldberg JM (2002) A cellular and pharmacological analysis of the responses of turtle posterior crista afferents to efferent activation. Abstr Assoc Res Otolaryngol 25:480.

Holt JC, Xue J-T, Goldberg JM (2004) Afferent responses to efferent activation in the turtle posterior canal involve pharmacologically-distinct nicotinic receptors. Abstr Assoc Res Otolaryngol 27:629.

Holt JR, Corey DP, Eatock RA (1997) Mechanoelectrical transduction and adaptation in hair cells of the mouse utricle, a low-frequency vestibular organ. J Neurosci 17:8739–8748.

Holt JR, Vollrath MA, Eatock RA (1999) Stimulus processing by type II hair cells in the mouse utricle. Ann NY Acad Sci 871:15–26.

Holt JR, Gillespie SK, Provance DW, Shah K, Shokat KM, Corey DP, Mercer JA, Gillespie PG (2002) A chemical-genetic strategy implicates myosin-1c in adaptation by hair cells. Cell 108:371–381.

Holt JR, Abraham D, Géléoc GS (2004) Adenoviral-mediated dominant-negative suppression of a low-voltage-activated potassium conductance in type I vestibular hair cells. Abstr Assoc Res Otolaryngol 27:153.

Honrubia V, Hoffman LF, Sitko S, Schwartz IR (1989) Anatomic and physiological correlates in bullfrog vestibular nerve. J Neurophysiol 61:688–701.

Housley GD, Ryan AF (1997) Cholinergic and purinergic neurohumoral signalling in the inner ear: a molecular physiological analysis. Audiol Neurootol 2:92–110.

Housley GD, Greenwood D, Ashmore JF (1992) Localization of cholinergic and purinergic receptors on outer hair cells isolated from the guinea-pig cochlea. Proc R Soc Lond B Biol Sci 249:265–273.

Housley GD, Kanjhan R, Raybould NP, Greenwood D, Salih SG, Jarlebark L, Burton LD, Setz VC, Cannell MB, Soeller C, Christie DL, Usami S, Matsubara A, Yoshie H, Ryan AF, Thorne PR (1999) Expression of the P2X(2) receptor subunit of the ATP-gated ion channel in the cochlea: implications for sound transduction and auditory neurotransmission. J Neurosci 19:8377–8388.

Howard J, Bechstedt S (2004) Hypothesis: a helix of ankyrin repeats of the NOMPC-TRP ion channel is the gating spring of mechanoreceptors. Curr Biol 14: R224–R226.

Howard J, Hudspeth AJ (1987) Mechanical relaxation of the hair bundle mediates adaptation in mechanoelectrical transduction by the bullfrog's saccular hair cell. Proc Natl Acad Sci USA 84:3064–3068.

Howard J, Hudspeth AJ (1988) Compliance of the hair bundle associated with gating of mechanoelectrical transduction channels in the bullfrog's saccular hair cell. Neuron 1:189–199.

Hozawa J, Fukuoka K, Usami S, Ikeno K, Fukushi E, Shinkawa H, Hozawa K (1991) The mechanism of irritative nystagmus and paralytic nystagmus. A histochemical study of the guinea pig's vestibular organ and an autoradiographic study of the vestibular nuclei. Acta Otolaryngol Suppl 481:73–76.

Hudspeth AJ, Gillespie PG (1994) Pulling springs to tune transduction: adaptation by hair cells. Neuron 12:1–9.

Hudspeth AJ, Choe Y, Mehta AD, Martin P (2000) Putting ion channels to work: mechanoelectrical transduction, adaptation and amplification by hair cells. Proc Natl Acad Sci USA 97:11765–11772.

Hullar TE, Minor LB (1999) High-frequency dynamics of regularly discharging canal afferents provide a linear signal for angular vestibuloocular reflexes. J Neurophysiol 82:2000–2005.

Hullar TE, Della Santina CC, Hirvonen TP, Lasker DM, Carey JP, Minor LB (2004) Responses of irregularly discharging chinchilla semicircular canal vestibular-nerve afferents during high-frequency head rotations. J Neurophysiol [Epub ahead of print] doi:10.1152/jn.01002.2004

Hunter-Duvar IM, Hinojosa R (1984) Vestibule: Sensory epithelia. In: Friedmann I, Ballantyne J (eds), Ultrastructural Atlas of the Inner Ear. London: Butterworths, pp. 211–244.

Hurle B, Ignatova E, Massironi SM, Mashimo T, Rios X, Thalmann I, Thalmann R, Ornitz DM (2003) Non-syndromic vestibular disorder with otoconial agenesis in tilted/mergulhador mice caused by mutations in otopetrin 1. Hum Mol Genet 12:777–789.

Ichimiya I, Adams JC, Kimura RS (1994) Immunolocalization of Na$^+$, K$^+$-ATPase, Ca^{++}-ATPase, calcium-binding proteins, and carbonic anhydrase in the guinea pig inner ear. Acta Otolaryngol 114:167–176.

Igarashi M (1966) Architecture of the otolith end organ: some functional considerations. Ann Otol Rhinol Laryngol 75:945–955.

Ishiyama A, Lopez I, Wackym PA (1994) Choline acetyltransferase immunoreactivity in the human vestibular end- organs. Cell Biol Int 18:979–984.

Ishiyama A, Lopez I, Wackym PA (1995) Distribution of efferent cholinergic terminals and alpha-bungarotoxin binding to putative nicotinic acetylcholine receptors in the human vestibular end-organs. Laryngoscope 105:1167–1172.

Ishiyama G, Lopez I, Williamson R, Acuna DL, Ishiyama A (2002) Subcellular immunolocalization of NMDA receptor subunit NR1, 2A, 2B in the rat vestibular periphery. Brain Res 935:16–23.

Iurato S, Taidelli G (1964) Relationships and structure of the so-called "much granulated endings" in the crista ampullaris (as studied by means of serial sections). In: Third European Regional Conference on Electron Microscopy, Publishing House of the Czechoslovak Academy of Sciences, Prague, Czechoslovakia, pp. 325–326.

Iurato S, Luciano L, Pannese E, Reale E (1972) Efferent vestibular fibers in mammals: morphological and histochemical aspects. Prog Brain Res 37:429–443.

Jacobs RA, Hudspeth AJ (1990) Ultrastructural correlates of mechanoelectrical transduction in hair cells of the bullfrog's internal ear. CSH Symp Quant Biol LV:547–561.

Jaeger R, Takagi A, Haslwanter T (2002) Modeling the relation between head orientations and otolith responses in humans. Hear Res 173:29–42.

Jaramillo F, Hudspeth AJ (1991) Localization of the hair cell's transduction channels at the hair bundle's top by iontophoretic application of a channel blocker. Neuron 7: 409–420.

Jarlebark LE, Housley GD, Raybould NP, Vlajkovic S, Thorne PR (2002) ATP-gated ion channels assembled from P2X2 receptor subunits in the mouse cochlea. NeuroReport 13:1979–1984.

Jones SM, Erway LC, Bergstrom RA, Schimenti JC, Jones TA (1999) Vestibular responses to linear acceleration are absent in otoconia-deficient C57BL/6JEi-het mice. Hear Res 135:56–60.

Jørgensen F, Kroese AB (1995) Ca selectivity of the transduction channels in the hair cells of the frog sacculus. Acta Physiol Scand 155:363–376.

Jørgensen JM (1988) The number and distribution of calyceal hair cells in the inner ear utricular macula of some reptiles. Acta Zool (Stockh) 69:169–175.

Jørgensen JM (1989) Number and distribution of hair cells in the utricular macula of some avian species. J Morphol 201:187–204.

Kachar B, Parakkal M, Fex J (1990) Structural basis for mechanical transduction in the frog vestibular sensory apparatus: I. The otolithic membrane. Hear Res 45:179–190.

Kachar B, Parakkal M, Kurc M, Zhao Y, Gillespie PG (2000) High-resolution structure of hair-cell tip links. Proc Natl Acad Sci USA 97:13336–13341.

Kataoka Y, Ohmori H (1994) Activation of glutamate receptors in response to membrane depolarization of hair cells isolated from chick cochlea. J Physiol 477:403–414.

Kennedy HJ, Meech RW (2002) Fast Ca^{2+} signals at mouse inner hair cell synapse: a role for Ca^{2+}-induced Ca^{2+} release. J Physiol 539:15–23.

Kennedy HJ, Evans MG, Crawford AC, Fettiplace R (2003) Fast adaptation of mechano-

electrical transducer channels in mammalian cochlear hair cells. Nat Neurosci 6:832–836.
Kevetter GA, Leonard RB (2002) Molecular probes of the vestibular nerve. II. Characterization of neurons in Scarpa's ganglion to determine separate populations within the nerve. Brain Res 928:18–29.
Kharkovets T, Hardelin JP, Safieddine S, Schweizer M, El Amraoui A, Petit C, Jentsch TJ (2000) KCNQ4, a K^+ channel mutated in a form of dominant deafness, is expressed in the inner ear and the central auditory pathway. Proc Natl Acad Sci USA 97:4333–4338.
Kikuchi T, Takasaka T, Tonosaki A, Watanabe H, Hozawa K, Shinkawa H, Wada H. (1991) Microtubule subunits of guinea pig vestibular epithelial cells. Acta Otolaryngol Suppl 481:107–111.
Kollmar R, Montgomery LG, Fak J, Henry LJ, Hudspeth AJ (1997) Predominance of the α_{1D} subunit in L-type voltage-gated Ca^{2+} channels of hair cells in the chicken's cochlea. Proc Natl Acad Sci USA 94:14883–14888.
Kondrachuk AV (2002) Models of otolithic membrane-hair cell bundle interaction. Hear Res 166:96–112.
Kozel PJ, Friedman RA, Erway LC, Yamoah EN, Liu LH, Riddle T, Duffy JJ, Doetschman T, Miller ML, Cardell EL, Shull GE (1998) Balance and hearing deficits in mice with a null mutation in the gene encoding plasma membrane Ca^{2+}-ATPase isoform 2. J Biol Chem 273:18693–18696.
Kreindler JL, Troyanovskaya M, Wackym PA (2001) Ligand-gated purinergic receptors are differentially expressed in the adult rat vestibular periphery. Ann Otol Rhinol Laryngol 110:277–282.
Kubisch C, Schroeder BC, Friedrich T, Lütjohann B, El-Amraoui A, Marlin S, Petit C, Jentsch TJ (1999) KCNQ4, a novel potassium channel expressed in sensory outer hair cells, is mutated in dominant deafness. Cell 96:437–446.
Kurc M, Farina M, Lins U, Kachar B (1999) Structural basis for mechanical transduction in the frog vestibular sensory apparatus: III. The organization of the otoconial mass. Hear Res 131:11–21.
Lanford PJ, Popper AN (1996) Novel afferent terminal structure in the crista ampullaris of the goldfish, carassius auratus. J Comp Neurol 366:572–579.
Lanford PJ, Platt C, Popper AN (2000) Structure and function in the saccule of the goldfish (*Carassius auratus*): a model of diversity in the non-amniote ear. Hear Res 143:1–13.
Lapeyre PNM, Cazals Y (1991) Guinea pig vestibular type I hair cells can show reversible shortening. J Vestib Res 1:241–250.
Lapeyre PNM, Guilhaume A, Cazals Y (1992) Differences in hair bundles associated with type I and type II vestibular hair cells of the guinea pig saccule. Acta Otolaryngol (Stockh) 112:635–642.
Lapeyre PNM, Kolston PJ, Ashmore JF (1993) $GABA_B$-mediated modulation of ionic conductances in type I hair cells isolated from guinea-pig semicircular canals. Brain Res 609:269–276.
Lasker D, Park HJ, Minor LB (2004) Extracellular recordings from vestibular-nerve afferents in the normal C5BL/6 mouse. Abstr Assoc Res Otolaryngol 27:936.
Legan PK, Lukashkina VA, Goodyear RJ, Kossi M, Russell IJ, Richardson GP (2000) A targeted deletion in alpha-tectorin reveals that the tectorial membrane is required for the gain and timing of cochlear feedback. Neuron 28:273–285.
Lennan GWT, Steinacker A, Lehouelleur J (1999) Ionic currents and current-clamp de-

polarisations of type I and type II hair cells from the developing rat utricle. Pflügers Arch 438:40–46.

Lenzi D, Runyeon JW, Crum J, Ellisman MH, Roberts WM (1999) Synaptic vesicle populations in saccular hair cells reconstructed by electron tomography. J Neurosci 19:119–132.

Leonard RB, Kevetter GA (2002) Molecular probes of the vestibular nerve. I. Peripheral termination patterns of calretinin, calbindin and peripherin containing fibers. Brain Res 928:8–17.

Lewis ER (1990) Electrical tuning in the ear. Comm Theor Biol 1:253–273.

Lewis ER, Leverenz EL, Bialek WS (1985) The Vertebrate Inner Ear. Boca Raton: CRC Press.

Lewis RS, Hudspeth AJ (1983) Voltage- and ion-dependent conductances in solitary vertebrate hair cells. Nature 304:538–541.

Lim DJ (1971) Vestibular sensory organs. A scanning electron microscopic investigation. Arch Otolaryngol 94:69–76.

Lim DJ (1976) Morphological and physiological correlates in cochlear and vestibular sensory epithelia. Scan Electron Microsc 1:269–276.

Lim DJ (1979) Fine morphology of the otoconial membrane and its relationship to the sensory epithelium. Scan Electron Microsc 3:929–938.

Lim DJ (1984) Otoconia in health and disease. A review. Ann Otol Rhinol Laryngol Suppl 112:17–24.

Lim DJ, Erway LC (1974) Influence of manganese on genetically defective otolith. A behavioral and morphological study. Ann Otol Rhinol Laryngol 83:565–581.

Lindeman HH (1969) Studies on the morphology of the sensory regions of the vestibular apparatus. Ergeb Anat Entwicklungsgesch 42:1–113.

Lindeman HH (1973) Anatomy of the otolith organs. Adv Otorhinolaryngol 20:405–433.

Lins U, Farina M, Kurc M, Riordan G, Thalmann R, Thalmann I, Kachar B (2000) The otoconia of the guinea pig utricle: internal structure, surface exposure, and interactions with the filament matrix. J Struct Biol 131:67–78.

Lioudyno MI, Verbitsky M, Holt JC, Elgoyhen AB, Guth PS (2000) Morphine inhibits an alpha9-acetylcholine nicotinic receptor-mediated response by a mechanism which does not involve opioid receptors. Hear Res 149:167–177.

Lioudyno MI, Verbitsky M, Glowatzki E, Holt JC, Boulter J, Zadina JE, Elgoyhen AB, Guth PS (2002) The alpha9/alpha10-containing nicotinic ACh receptor is directly modulated by opioid peptides, endomorphin-1, and dynorphin B, proposed efferent cotransmitters in the inner ear. Mol Cell Neurosci 20:695–711.

Lopez I, Wu JY, Meza G (1992) Immunocytochemical evidence for an afferent GABAergic neurotransmission in the guinea pig vestibular system. Brain Res 589:341–348.

Lopez I, Juiz JM, Altschuler RA, Meza G (1990) Distribution of GABA-like immunoreactivity in guinea pig vestibular cristae ampullaris. Brain Res 530:170–175.

Lorente de Nó R (1926) Etudes sur l'anatomie et la physiologie du labyrinth de l'oreille et du VIIIe nerf: II. Quelque données au sujet de l'anatomie des organes sensoriels du labyrinthe. Trav Lab Rech Biol Univ Madrid 24:53–153.

Lowenstein O, Osborne MP, Thornhill RA (1968) The anatomy and ultrastructure of the labyrinth of the lamprey (*Lampetra fluviatilis* L.). Proc R Soc Lond B Biol Sci 170:113–134.

Lysakowski A (1996) Synaptic organization of the crista ampullares in vertebrates. Ann NY Acad Sci 781:164–182.

Lysakowski, A (1999) CGRP shows regional variation in efferent innervation of chinchilla vestibular periphery. Soc Neurosci Abstr 25:1670.

Lysakowski A, Goldberg JM (1997) A regional ultrastructural analysis of the cellular and synaptic architecture in the chinchilla cristae ampullares. J Comp Neurol 389:419–443.

Lysakowski A, Goldberg JM (2004) Morphophysiology of the vestibular sensory periphery. In: Highstein SM, Popper AN, Fay RR (eds), Anatomy and Physiology of the Central and Peripheral Vestibular System. New York: Springer-Verlag, pp. 57–152.

Lysakowski A, Price SD (2003) Potassium channel localization in sensory epithelia of the rat inner ear. Abstr Assoc Res Otolaryngol 26:1534.

Lysakowski A, Singer M (2000) Nitric oxide synthase localized in a subpopulation of vestibular efferents with NADPH diaphorase histochemistry and nitric oxide synthase immunohistochemistry. J Comp Neurol 427:508–521.

Lysakowski A, Minor LB, Fernandez C, Goldberg JM (1995) Physiological identification of morphologically distinct afferent classes innervating the cristae ampullares of the squirrel monkey. J Neurophysiol 73:1270–1281.

Lysakowski A, Alonto A, Jacobson L (1999) Peripherin immunoreactivity labels small diameter vestibular 'bouton' afferents in rodents. Hear Res 133:149–154.

Lysakowski A, Dhaliwal J, Singer M (2000) NOS and elements of the nitric oxide cascade in vestibular efferents and hair cells. Abstr Assoc Res Otolaryngol 23:6683.

Maison SF, Luebke AE, Liberman MC, Zuo J (2002) Efferent protection from acoustic injury is mediated via alpha9 nicotinic acetylcholine receptors on outer hair cells. J Neurosci 22:10838–10846.

Manley GA (2002) Evolution of structure and function of the hearing organ of lizards. J Neurobiol 53:202–211.

Manley GA, Sienknecht UJ, Koeppl C (2004) Calcium modulates the frequency and amplitude of spontaneous otoacoustic emissions in the bobtail skink. J Neurophysiol 92:2685–2693.

Marco J, Lee W, Suarez C, Hoffman L, Honrubia V (1993) Morphologic and quantitative study of the efferent vestibular system in the chinchilla: 3-D reconstruction. Acta Otolaryngol 113:229–234.

Marcotti W, Kros CJ (1999) Developmental expression of the potassium current $I_{K,n}$ contributes to maturation of mouse outer hair cells. J Physiol 520:653–660.

Marcotti W, Johnson SL, Holley MC, Kros CJ (2003a) Developmental changes in the expression of potassium currents of embryonic, neonatal and mature mouse inner hair cells. J Physiol 548:383–400.

Marcotti W, Johnson SL, Rüsch A, Kros CJ (2003b) Sodium and calcium currents shape action potentials in immature mouse inner hair cells. J Physiol 552:743–761.

Marcus DC, Liu J, Wangemann P (1994) Transepithelial voltage and resistance of vestibular dark cell epithelium from the gerbil ampulla. Hear Res 73:101–108.

Marlinski V, Plotnik M, Goldberg JM (2004) Efferent actions in the chinchilla vestibular labyrinth. J Assoc Res Otolaryngol 5:126–143.

Maroni P, Lysakowski A, Luebke AE (1998) Alpha-9 nicotinic acetylcholine receptor subunit immunoreactivity in the rodent inner ear. Abstr Assoc Res Otolaryngol 21:65.

Martinez-Dunst C, Michaels RL, Fuchs PA (1997) Release sites and calcium channels in hair cells of the chick's cochlea. J Neurosci 17:9133–9144.

Martini M, Rossi ML, Rubbini G, Rispoli G (2000) Calcium currents in hair cells isolated from semicircular canals of the frog. Biophys J 78:1240–1254.

Masetto S, Correia MJ (1997a) Electrophysiological properties of vestibular sensory and supporting cells in the labyrinth slice before and during regeneration. J Neurophysiol 78:1913–1927.

Masetto S, Correia MJ (1997b) Ionic currents in regenerating avian vestibular hair cells. Int J Dev Neurosci 15:387–399.

Masetto S, Russo G, Prigioni I (1994) Differential expression of potassium currents by hair cells in thin slices of frog crista ampullaris. J Neurophysiol 72:443–455.

Masetto S, Perin P, Malusa A, Zucca G, Valli P (2000) Membrane properties of chick semicircular canal hair cells in situ during embryonic development. J Neurophysiol 83:2740–2756.

Masetto S, Bosica M, Correia MJ, Ottersen OP, Zucca G, Perin P, Valli P (2003) Na^+ currents in vestibular type I and type II hair cells of the embryo and adult chicken. J Neurophysiol 90:1266–1278.

Matsubara A, Takumi Y, Nakagawa T, Usami S, Shinkawa H, Ottersen OP (1999) Immunoelectron microscopy of AMPA receptor subunits reveals three types of putative glutamatergic synapse in the rat vestibular end organs. Brain Res 819:58–64.

McCue M, Guinan JJ, Jr (1994) Influence of efferent stimulation on acoustically responsive vestibular afferents in the cat. J Neurosci 14:6071–6083.

McCue M, Guinan JJ, Jr (1995) Spontaneous activity and frequency selectivity of acoustically responsive vestibular afferents in the cat. J Neurophysiol 74:1563–1572.

McLaren JW, Hillman DE (1979) Displacement of the semicircular canal cupula during sinusoidal rotation. Neuroscience 4:2001–2008.

Moravec WJ, Peterson EH (2004) Differences between stereocilia numbers on type I and type II vestibular hair cells. J Neurophysiol 92:3153–3160.

Morita I, Komatsuzaki A, Tatsuoka H (1997) The morphological differences of stereocilia and cuticular plates between type-I and type-II hair cells of human vestibular sensory epithelia. ORL J Otorhinolaryngol Relat Spec 59:193–197.

Nakagawa T, Akaike N, Kimitsuki T, Komune S, Arima T (1990) ATP-induced current in isolated outer hair cells of guinea pig cochlea. J Neurophysiol 63:1068–1074.

Newlands SD, Perachio AA (2003) Central projections of the vestibular nerve: a review and single fiber study in the Mongolian gerbil. Brain Res Bull 60:475–495.

Niedzielski AS, Wenthold RJ (1995) Expression of AMPA, kainate, and NMDA receptor subunits in cochlear and vestibular ganglia. J Neurosci 15:2338–2353.

Oesterle EC, Cunningham DE, Westrum LE, Rubel EW (2003) Ultrastructural analysis of [^{3}H] thymidine-labeled cells in the rat utricular macula. J Comp Neurol 463:177–195.

Ogata Y, Slepecky NB (1998) Immunocytochemical localization of calmodulin in the vestibular end- organs of the gerbil. J Vestib Res 8:209–216.

Ogata Y, Slepecky NB, Takahashi M (1999) Study of the gerbil utricular macula following treatment with gentamicin, by use of bromodeoxyuridine and calmodulin immunohistochemical labelling. Hear Res 133:53–60.

Oliver D, Plinkert P, Zenner HP, Ruppersberg JP (1997) Sodium current expression during postnatal development of rat outer hair cells. Pflugers Arch 434:772–778.

Oliver D, Knipper M, Derst C, Fakler B (2003) Resting potential and submembrane calcium concentration of inner hair cells in the isolated mouse cochlea are set by KCNQ-type potassium channels. J Neurosci 23:2141–2149.

Ornitz DM, Bohne BA, Thalmann I, Harding GW, Thalmann R (1998) Otoconial agenesis in tilted mutant mice. Hear Res 122:60–70.

Osborne MP, Comis SD, Pickles JO (1984) Morphology and cross-linkage of stereocilia in the guinea-pig labyrinth examined without the use of osmium as a fixative. Cell Tissue Res 237:43–48.

Paffenholz R, Bergstrom RA, Pasutto F, Wabnitz P, Munroe RJ, Jagla W, Heinzmann U, Marquardt A, Bareiss A, Laufs J, Russ A, Stumm G, Schimenti JC, Bergstrom DE (2004) Vestibular defects in head-tilt mice result from mutations in Nox3, encoding an NADPH oxidase. Genes Dev 18:486–491.

Parsons TD, Lenzi D, Almers W, Roberts WM (1994) Calcium-triggered exocytosis and endocytosis in an isolated presynaptic cell: capacitance measurements in saccular hair cells. Neuron 13:875–883.

Perachio AA, Kevetter GA (1989) Identification of vestibular efferent neurons in the gerbil: histochemical and retrograde labelling. Exp Brain Res 78:315–326.

Perry B, Jensen-Smith HC, Luduena RF, Hallworth R (2003) Selective expression of beta tubulin isotypes in gerbil vestibular sensory epithelia and neurons. J Assoc Res Otolaryngol 4:329–338.

Peterson EH (1998) Are there parallel channels in the vestibular nerve? News Physiol Sci 13:194–201.

Pickles JO, Corey DP (1992) Mechanoelectrical transduction by hair cells. Trends Neurosci 15:254–259.

Platzer J, Engel J, Schrott-Fischer A, Stephan K, Bova S, Chen H, Zheng H, Striessnig J (2000) Congenital deafness and sinoatrial node dysfunction in mice lacking class D L-type Ca^{2+} channels. Cell 102:89–97.

Plotnik M, Marlinski V, Goldberg JM (2002) Reflections of efferent activity in rotational responses of chinchilla vestibular afferents. J Neurophysiol 88: 1234–1244.

Popper P, Wackym PA, Siebenreich W, Cristobal R (2002) Distribution of opioid receptors in the vestibular periphery. Abstr Assoc Res Otolaryngol 25:111.

Prigioni I, Russo G, Masetto S (1994) Non-NMDA receptors mediate glutamate-induced depolarization in frog crista ampullaris. NeuroReport 5:516–518.

Pujol R and Sans A (1986) Synaptogenesis in the mammalian inner ear. In: Norwood NJ, Ablex, *Advances in Neural and Behavioral Development*, Vol. 2 (Aslin RN (ed), pp. 1–18.

Purcell IM, Perachio AA (1997) Three-dimensional analysis of vestibular efferent neurons innervating semicircular canals of the gerbil. J Neurophysiol 78:3234–3248.

Puyal J, Sage C, Dememes D, Dechesne CJ (2002) Distribution of alpha-amino-3-hydroxy-5-methyl-4 isoazolepropionic acid and *N*-methyl-*D*-aspartate receptor subunits in the vestibular and spiral ganglia of the mouse during early development. Brain Res Dev Brain Res 139:51–57.

Quirion R, Van Rossum D, Dumont Y, St Pierre S, Fournier A (1992) Characterization of CGRP1 and CGRP2 receptor subtypes. Ann NY Acad Sci 657:88–105.

Rabbitt RD, Boyle R, Highstein SM (1999) Influence of surgical plugging on horizontal semicircular canal mechanics and afferent response dynamics. J Neurophysiol 82:1033–1053.

Rabbitt RD, Damiano ER, Grant JW (2004) Biomechanics of the semicircular canals and otolith organs. In: (Highstein SM, Popper AN, Fay RR, (eds), Anatomy and Physiology of the Central and Peripheral Vestibular System. New York: Springer-Verlag, pp. 153–201.

Rabbitt RD, Boyle R, Holstein GR, Highstein SM (2005) Hair-cell vs. afferent adaptation in the semicircular canals. J Neurophysiol 93(1):424–436.

Rau A, Legan PK, Richardson GP (1999) Tectorin mRNA expression is spatially and temporally restricted during mouse inner ear development. J Comp Neurol 405:271–280.

Raymond J, Nieoullon A, Dememes D, Sans A (1984) Evidence for glutamate as a neurotransmitter in the cat vestibular nerve: radioautographic and biochemical studies. Exp Brain Res 56:523–531.

Raymond J, Dechesne CJ, Desmadryl G, Dememes D (1993) Different calcium-binding proteins identify subpopulations of vestibular ganglion neurons in the rat. Acta Otolaryngol Suppl 503:114–118.

Rennie KJ (2002) Modulation of the resting potassium current in type I vestibular hair cells by cGMP. In: Berlin CI, Hood LJ, Ricci AJ, (eds), Hair Cell Micromechanics and Otoacoustic Emissions. San Diego: Singular Press, pp. 79–89.

Rennie KJ, Ashmore JF (1993) Effects of extracellular ATP on hair cells isolated from the guinea-pig semicircular canals. Neurosci Lett 160:185–189.

Rennie KJ, Correia MJ (1994) Potassium currents in mammalian and avian isolated type I semicircular canal hair cells. J Neurophysiol 71:317–329.

Rennie KJ, Ricci AJ (2004) Mechanoelectrical transduction (met) and basolateral currents in hair cells of the turtle utricle. Abstr Assoc Res Otolaryngol 27:1204.

Rennie KJ, Ricci AJ, Correia MJ (1996) Electrical filtering in gerbil isolated type I semicircular canal hair cells. J Neurophysiol 75:2117–2123.

Rennie KJ, Ashmore JF, Correia MJ (1997) Evidence for an Na^+-K^+-Cl^- cotransporter in mammalian type I vestibular hair cells. Am J Physiol 273:C1972–C1980.

Rennie KJ, Weng T, Correia MJ (2001) Effects of KCNQ channel blockers on K^+ currents in vestibular hair cells. Am J Physiol Cell Physiol 280:C473-C480.

Rennie KJ, Manning KC, Ricci AJ (2004) Mechano-electrical transduction in the turtle utricle. Biomed Sci Instrum 40:441–446.

Ricci AJ, Fettiplace R (1998) Calcium permeation of the turtle hair cell mechanotransducer channel and its relation to the composition of endolymph. J Physiol 506: 159–173.

Ricci AJ, Rennie KJ, Correia MJ (1996) The delayed rectifier, I_{KI}, is the major conductance in type I vestibular hair cells across vestibular end organs. Pflügers Arch 432:34–42.

Ricci AJ, Wu Y-C, Fettiplace R (1998) The endogenous calcium buffer and the time course of transducer adaptation in auditory hair cells. J Neurosci 18:8261–8277.

Ricci AJ, Crawford AC, Fettiplace R (2002) Mechanisms of active hair bundle motion in auditory hair cells. J Neurosci 22:44–52.

Roberts BL, Russell IJ (1972) The activity of lateral-line efferent neurones in stationary and swimming dogfish. J Exp Biol 57:435–448.

Roberts WM (1994) Localization of calcium signals by a mobile calcium buffer in frog saccular hair cells. J Neurosci 14:3246–3262.

Roberts WM, Jacobs RA, Hudspeth AJ (1990) Colocalization of ion channels involved in frequency selectivity and synaptic transmission at presynaptic active zones of hair cells. J Neurosci 10:3664–3684.

Rodriguez-Contreras A, Yamoah EN (2001) Direct measurement of single-channel Ca^{2+} currents in bullfrog hair cells reveals two distinct channel subtypes. J Physiol 534:669–689.

Rossi ML, Martini M, Pelucchi B, Fesce R (1994) Quantal nature of synaptic transmission at the cytoneural junction in the frog labyrinth. J Physiol 478:17–35.

Rüsch A, Eatock RA (1996a) A delayed rectifier conductance in type I hair cells of the mouse utricle. J Neurophysiol 76:995–1004.

Rüsch A, Eatock RA (1996b) Voltage responses of mouse utricular hair cells to injected currents. Ann NY Acad Sci 781:71–84.

Rüsch A, Thurm U (1989) Cupula displacement, hair bundle deflection, and physiological responses in the transparent semicircular canal of young eel. Pflügers Arch 413:533–545.

Rüsch A, Lysakowski A, Eatock RA (1998) Postnatal development of type I and type II hair cells in the mouse utricle: Acquisition of voltage-gated conductances and differentiated morphology. J Neurosci 18:7487–7501.

Russell IJ (1971) The role of the lateral-line efferent system in *Xenopus laevis*. J Exp Biol 54:621–641.

Ryan AF, Simmons DM, Watts AG, Swanson LW (1991) Enkephalin mRNA production by cochlear and vestibular efferent neurons in the gerbil brainstem. Exp Brain Res 87:259–267.

Sage C, Venteo S, Jeromin A, Roder J, Dechesne CJ (2000) Distribution of frequenin in the mouse inner ear during development, comparison with other calcium-binding proteins and synaptophysin. Hear Res 150:70–82.

Saidel WM, Presson JC, Chang JS (1990) S-100 immunoreactivity identifies a subset of hair cells in the utricle and saccule of a fish. Hear Res 47:139–146.

Sans A, Chat M (1982) Analysis of temporal and spatial patterns of rat vestibular hair cell differentiation by tritiated thymidine radioautography. J Comp Neurol 206:1–8.

Sans A, Scarfone E (1996) Afferent calyces and type I hair cells during development. A new morphofunctional hypothesis. Ann NY Acad Sci 781:1–12.

Sans A, Dechesne CJ, Demêmes D (2001) The mammalian otolithic receptors: a complex morphological and biochemical organization. Adv Otorhinolaryngol 58:1–14.

Satzler K, Sohl LF, Bollmann JH, Borst JG, Frotscher M, Sakmann B, Lubke JH (2002) Three-dimensional reconstruction of a calyx of Held and its postsynaptic principal neuron in the medial nucleus of the trapezoid body. J Neurosci. 22:10567–10579.

Scarfone E, Demêmes D, Jahn R, De Camilli P, Sans A (1988) Secretory function of the vestibular nerve calyx suggested by the presence of vesicles, synapsin I, and synaptophysin. J Neurosci 8:4640–4645.

Scarfone E, Demêmes D, Sans A (1991) Synapsin I and synaptophysin expression during ontogenesis of the mouse peripheral vestibular system. J Neurosci 11:1173–1181.

Scarfone E, Ulfendahl M, Lundeberg T (1996) The cellular localization of the neuropeptides substance P, neurokinin A, calcitonin gene-related peptide and neuropeptide Y in guinea-pig vestibular sensory organs: a high-resolution confocal microscopy study. Neuroscience 75:587–600.

Schessel DA, Ginzberg R, Highstein SM (1991) Morphophysiology of synaptic transmission between type I hair cells and vestibular primary afferents. An intracellular study employing horseradish peroxidase in the lizard, *Calotes versicolor*. Brain Res 544:1–16.

Schweitzer E (1987) Coordinated release of ATP and ACh from cholinergic synaptosomes and its inhibition by calmodulin antagonists. J Neurosci 7:2948–2956.

Selyanko AA, Hadley JK, Wood IC, Abogadie FC, Delmas P, Buckley NJ, London B, Brown DA (1999) Two types of K^+ channel subunit, Erg1 and KCNQ2/3, contribute to the M-like current in a mammalian neuronal cell. J Neurosci 19:7742–7756.

Sewell WF (1996) Neurotransmitters and synaptic transmission. In: Dallos P, Popper AN, Fay RR (eds), The Cochlea. New York: Springer-Verlag, pp. 503–533.

Shah MM, Mistry M, Marsh SJ, Brown DA, Delmas P (2002) Molecular correlates of the M-current in cultured rat hippocampal neurons. J Physiol 544:29–37.

Shepherd GM, Corey DP (1994) The extent of adaptation in bullfrog saccular hair cells. J Neurosci 14:6217–6229.

Shotwell SL, Jacobs R, Hudspeth AJ (1981) Directional sensitivity of individual vertebrate hair cells to controlled deflection of their hair bundles. Ann NY Acad Sci 374:1–10.

Si X, Zakir MM, Dickman JD (2003) Afferent innervation of the utricular macula in pigeons. J Neurophysiol 89:1660–1677.

Sidi S, Busch-Nentwich E, Friedrich R, Schoenberger U, Nicolson T (2004) gemini encodes a zebrafish L-type calcium channel that localizes at sensory hair cell ribbon synapses. J Neurosci 24:4213–4223.

Siemens J, Lillo C, Dumont RA, Reynolds A, Williams DS, Gillespie PG, Müller U (2004) Cadherin 23 is a component of the tip link in hair-cell stereocilia. Nature 428:950–955.

Silver RB, Reeves AP, Steinacker A, Highstein SM (1998) Examination of the cupula and stereocilia of the horizontal semicircular canal in the toadfish *Opsanus tau*. J Comp Neurol 402:48–61.

Smith CA, Rasmussen GL (1968) Nerve endings in the maculae and cristae of the chinchilla vestibule with a special reference to the efferents. In: Graybiel A (ed), Third Symposium on the Role of Vestibular Organs in Space Exploration, NASA SP-152. Washington, DC: US. Government Printing Office, pp. 183–201.

Smith CE, Goldberg JM (1986) A stochastic afterhyperpolarization model of repetitive activity in vestibular afferents. Biol Cybern 54:41–51.

Sobkowicz HM, Rose JE, Scott GE, Slapnick SM (1982) Ribbon synapses in the developing intact and cultured organ of Corti in the mouse. J Neurosci 2:942–957.

Sobkowicz HM, Rose JE, Scott GL, Levenick CV (1986) Distribution of synaptic ribbons in the developing organ of Corti. J Neurocytol 15:693–714.

Söllner C, Rauch GJ, Siemens J, Geisler R, Schuster SC, Müller U, Nicolson T (2004) Mutations in cadherin 23 affect tip links in zebrafish sensory hair cells. Nature 428:955–959.

Spassova M, Eisen MD, Saunders JC, Parsons TD (2001) Chick cochlear hair cell exocytosis mediated by dihydropyridine- sensitive calcium channels. J Physiol 535:689–696.

Spicer SS, Schulte BA, Adams JC (1990) Immunolocalization of Na^+,K^+-ATPase and carbonic anhydrase in the gerbil's vestibular system. Hear Res 43:205–217.

Spoendlin H (1965) Ultrastructural studies of the labyrinth in squirrel monkeys. In: Graybel A, (ed), First Symposium on the Role of Vestibular Organs in Space Exploration, NASA SP-77. Washington, DC: US Government Printing Office, pp. 7–22.

Sridhar TS, Brown MC, Sewell WF (1997) Unique postsynaptic signaling at the hair cell efferent synapse permits calcium to evoke changes on two time scales. J Neurosci 17:428–437.

Steenbergh PH, Hoppener JW, Zandberg J, Visser A, Lips CJ, Jansz HS (1986) Structure and expression of the human calcitonin/CGRP genes. FEBS Lett 209:97–103.

Sterkers O, Ferrary E, Amiel C (1988) Production of inner ear fluids. Physiol Rev 68:1083–1128.

Storm JF (1990) Potassium currents in hippocampal pyramidal cells. Prog Brain Res 83:161–187.

Street VA, McKee-Johnson JW, Fonseca RC, Tempel BL, Noben-Trauth K (1998) Mutations in a plasma membrane Ca^{2+}-ATPase gene cause deafness in deafwaddler mice. Nat Genet 19:390–394.

Su Z-L, Jiang SC, Gu R, Yang WP (1995) Two types of calcium channels in bullfrog saccular hair cells. Hear Res 87:62–68.

Syeda SN, Lysakowski A (2001) P2X$_2$ purinergic receptor localized in the inner ear. Abstr Assoc Res Otolaryngol 24:68.

Takumi Y, Matsubara A, Danbolt NC, Laake JH, Storm-Mathisen J, Usami S, Shinkawa H, Ottersen OP (1997) Discrete cellular and subcellular localization of glutamine synthetase and the glutamate transporter GLAST in the rat vestibular end organ. Neuroscience 79: 1137–1144.

Tanaka M, Takeda N, Senba E, Tohyama M, Kubo T, Matsunaga T (1988) Localization of calcitonin gene-related peptide in the vestibular end-organs in the rat: an immunohistochemical study. Brain Res 447:175–177.

Tanaka M, Takeda N, Senba E, Tohyama M, Kubo T, Matsunaga T (1989) Localization, origin and fine structure of calcitonin gene-related peptide-containing fibers in the vestibular end-organs of the rat. Brain Res 504:31–35.

ten Cate W-JF, Curtis LM, Rarey KE (1994) Na,K-ATPase α and β subunit isoform distribution in the rat cochlear and vestibular tissues. Hear Res 75:151–160.

Thalmann R, Ignatova E, Kachar B, Ornitz DM, Thalmann I (2001) Development and maintenance of otoconia: biochemical considerations. Ann NY Acad Sci 942:162–178.

Troyanovskaya M, Wackym PA (1998) Evidence for three additional P2X2 purinoceptor isoforms produced by alternative splicing in the adult rat vestibular end-organs. Hear Res 126:201–209.

Tsuprun V, Santi P (2000) Helical structure of hair cell stereocilia tip links in the chinchilla cochlea. J Assoc Res Otolaryngol 1:224–231.

Tucker TR, Fettiplace R (1996) Monitoring calcium in turtle hair cells with a calcium-activated potassium channel. J Physiol 494:613–626.

Usami S, Ottersen OP (1995) Differential cellular distribution of glutamate and glutamine in the rat vestibular endorgans: an immunocytochemical study. Brain Res 676:285–292.

Usami S, Igarashi M, Thompson GC (1987) GABA-like immunoreactivity in the squirrel monkey vestibular endorgans. Brain Res 417:367–370.

Van Rossum D, Hanisch UK, Quirion R (1997) Neuroanatomical localization, pharmacological characterization and functions of CGRP, related peptides and their receptors. Neurosci Biobehav Rev 21:649–678.

Vega R, Soto E (2003) Opioid receptors mediate a postsynaptic facilitation and a presynaptic inhibition at the afferent synapse of axolotl vestibular hair cells. Neuroscience 118: 75–85.

Vollrath MA, Eatock RA (2003) Time course and extent of mechanotransducer adaptation in mouse utricular hair cells: comparison with frog saccular hair cells. J Neurophysiol 90:2676–2689.

Wackym PA, Chen CT, Ishiyama A, Pettis RM, Lopez IA, Hoffman L (1996) Muscarinic acetylcholine receptor subtype mRNAs in the human and rat vestibular periphery. Cell Biol Int 20:187–192.

Wackym PA, Popper P, Lopez I, Ishiyama A, Micevych PE (1995) Expression of alpha 4 and beta 2 nicotinic acetylcholine receptor subunit mRNA and localization of alpha-bungarotoxin binding proteins in the rat vestibular periphery. Cell Biol Int 19:291–300.

Wackym PA, Popper P, Ward PH, Micevych PE (1991) Cell and molecular anatomy of nicotinic acetylcholine receptor subunits and calcitonin gene-related peptide in the rat vestibular system. Otolaryngol Head Neck Surg 105:493–510.

Wackym PA, Troyanovskaya M, Popper P (2000a) Adenylyl cyclase isoforms in the vestibular periphery of the rat. Brain Res 859:378–380.

Wackym PA, Troyanovskaya M, Popper P (2000b) Partial cDNAs encoding G-protein alpha subunits in the rat vestibular periphery. Neurosci Lett 280:159–162.

Walker RG, Hudspeth AJ (1996) Calmodulin controls adaptation of mechanoelectrical transduction by hair cells of the bullfrog's sacculus. Proc Natl Acad Sci USA 93:2203–2207.

Walker RG, Hudspeth AJ, Gillespie PG (1993) Calmodulin and calmodulin-binding proteins in hair bundles. Proc Natl Acad Sci USA 90:2807–2811.

Wang H-S, Pan Z, Shi W, Brown BS, Wymore RS, Cohen IS, Dixon JE, McKinnon D (1998) KCNQ2 and KCNQ3 potassium channel subunits: Molecular correlates of the M-channel. Science 282:1890–1893.

Wang Y, Kowalski PE, Thalmann I, Ornitz DM, Mager DL, Thalmann R (1998) Otoconin-90, the mammalian otoconial matrix protein, contains two domains of homology to secretory phospholipase A2. Proc Natl Acad Sci USA 95:15345–15350.

Wangemann P (1995) Comparison of ion transport mechanisms between vestibular dark cells and strial marginal cells. Hear Res 90:149–157.

Wangemann P, Schacht J (1996) Homeostatic mechanisms in the cochlea. In: Dallos P, Popper AN, Fay RR (eds), The Cochlea. New York: Springer-Verlag, pp. 130–185.

Warr WB (1975) Olivocochlear and vestibular efferent neurons of the feline brain stem: their location, morphology and number determined by retrograde axonal transport and acetylcholinesterase histochemistry. J Comp Neurol 161:159–181.

Weisstaub N, Vetter DE, Elgoyhen AB, Katz E (2002) The alpha9alpha10 nicotinic acetylcholine receptor is permeable to and is modulated by divalent cations. Hear Res 167: 122–135.

Weng T, Correia MJ (1999) Regional distribution of ionic currents and membrane voltage responses of type II hair cells in the vestibular neuroepithelium. J Neurophysiol 82: 2451–2461.

Werner CF (1933) The differentiation of the maculae in the labyrinth, particularly in mammals. Z Anat Entwicklungsgesch 99:696–709.

Wersäll J (1956) Studies on the structure and innervation of the sensory epithelium of the cristae ampulares in the guinea pig; a light and electron microscopic investigation. Acta Otolaryngol Suppl 126:1–85.

Wersäll J (1981) Structural damage to the organ of Corti and the vestibular epithelia caused by aminoglycoside antibiotics in the guinea pig. In: Lerner SA, Matz GJ, Hawkins JE Jr. (eds), Aminoglycoside Ototoxicity. Boston: Little, Brown, pp. 197–214.

Wersäll J, Bagger-Sjöbäck D (1974) Morphology of the vestibular sense organ. In: Kornhuber HH (ed), Handbook of Sensory Physiology. Vestibular System. Basic Mechanisms. New York: Springer-Verlag, pp. 123–170.

Wilson VJ, Melvill Jones G (1979) Mammalian Vestibular Physiology. New York: Plenum Press.

Wong WH, Hurley KM, Eatock RA (2004) Differences between the negatively activating potassium conductances of mammalian cochlear and vestibular hair cells. J Assoc Res Otolaryngol 5:270–284.

Wooltorton JRA, Hurley KM, Garcia J, Eatock RA (2005) Voltage-dependent sodium channels in rat utricular hair cells. Abstr Assoc Res Otolaryngol 28:526.

Wu YC, Ricci AJ, Fettiplace R (1999) Two components of transducer adaptation in auditory hair cells. J Neurophysiol 82:2171–2181.

Xue J, Peterson EH (2003) Spatial patterns in the structure of otolithic membranes. Abstr Assoc Res Otolaryngol 26:126.

Xue J, Peterson EH (2004) Organization of the utricular striola in *Trachemys scripta*: bundle heights. Abstr Assoc Res Otolaryngol 27:1114.

Yamashita M, Ohmori H (1990) Synaptic responses to mechanical stimulation in calyceal and bouton type vestibular afferents studied in an isolated preparation of semicircular canal ampullae of chicken. Exp Brain Res 80:475–488.

Yamashita M, Ohmori H (1991) Synaptic bodies and vesicles in the calix type synapse of chicken semicircular canal ampullae. Neurosci Lett 129:43–46.

Yamauchi A, Rabbitt RD, Boyle R, Highstein SM (2002) Relationship between inner-ear fluid pressure and semicircular canal afferent nerve discharge. J Assoc Res Otolaryngol 3:26–44.

Yamoah EN, Lumpkin EA, Dumont RA, Smith PJS, Hudspeth AJ, Gillespie PG (1998) Plasma membrane Ca^{2+}-ATPase extrudes Ca^{2+} from hair cell stereocilia. J Neurosci 18: 610–624.

Yan HY, Saidel WM, Chang JS, Presson JC, Popper AN (1991) Sensory hair cells of a fish ear: evidence of multiple types based on ototoxicity sensitivity. Proc R Soc Lond B Biol Sci 245:133–138.

Ylikoski J, Pirvola U, Happola O, Panula P, Virtanen I (1989) Immunohistochemical demonstration of neuroactive substances in the inner ear of rat and guinea pig. Acta Otolaryngol 107:417–423.

Zhao Y, Yamoah EN, Gillespie PG (1996) Regeneration of broken tip links and restoration of mechanical transduction in hair cells. Proc Natl Acad Sci USA 93:15469–15474.

Zheng JL, Gao W-Q (1997) Analysis of rat vestibular hair cell development and regeneration using calretinin as an early marker. J Neurosci 17:8270–8282.

Zidanic M, Fuchs PA (1995) Kinetic analysis of barium currents in chick cochlear hair cells. Biophys J 68:1323–1336.

Index

ACh, 73
  calcium stores, 293ff
  effects on electromotility, 287
  effects on isolated hair cells, 12, 281ff
  effects on K+ currents, 282, 291
  effects on outer hair cells, 287
  effects on vestibular hair cells, 411–413
  molecular components of ACh-mediated efferent effects, 289–291
ACh receptors, 249–250
  genes, 289–290
  on outer hair cells, 325
  on vestibular hair cells and afferent endings, 411–413
Actin, circumferential ring, 119
  cuticular plate, 118–119
  hair bundles, 114ff
  stereocilia and rootlets, 116–117
Action potentials, slow, 67
  in vestibular hair cells, 390
Active cochlear processes, see also Cochlear amplifier
  evolution, 313–314
Adaptation, in auditory hair cells, 267–268
  calcium, 176ff
  fast and slow mechanisms, 169ff, 176ff
  functions, 173ff, 373
  and unconventional myosins, 170, 174
  in vestibular hair cells, 367ff
Adhesion proteins, and hair bundles, 125ff

Afferent nerve, auditory response properties, 252
  discharge properties, vestibular, 354ff
  morphological correlates to discharge, vestibular, 355ff
  regional variation in response properties, vestibular, 354ff
  responses compared to hair cell responses, 370–371
Afferent transmission, ontogeny, 272–273
Alligator lizard, electrical tuning of hair cells, 207
α9, ACh receptor genes, 289–290
α10, ACh receptor genes, 290–291
α actinin, hair bundles, 124
α subunits, BK_{ca} channel, 235–236
Ames waltzer mouse, 60
Aminoglycosides, fimbrin, 123
  hair bundles, 110–111
  hair cells and bacteria, 328
Amphipaths, electromotility, 316–317
Ankle links, 54, 108
4-AP, inhibition of K+ currents, 221ff
ATP, efferent transmission, 416
Auditory organ development, summary, 22–23
Auditory vs. vestibular function, 348

Bacteria, mechanosensing channel, MscL, 7
  type A gliding motility and outer hair cell, 327–328
Bandwidth, type I hair cells, 391
Basilar papilla, chick, effects of ACh on hair cells, 281–282

443

Basolateral currents, impact on receptor potentials, 380ff
Basolateral membrane properties, development, 66ff
β subunits, BK$_{ca}$ channel, 234
β-tectorin, otolithic membrane, 361
BK$_{ca}$ channels, 214, 216, 221, 225, 227ff
   α subunits, 235–236
   alternative splicing, 233–234
   β subunits, 234
   Ca^{2+} sensitivity, 233–234
   chick, 230–231, 233–234
   cloned, 231ff
   hearing loss, 238–239
   mammalian hair cells, 237–239
   property variations, 231ff
   turtle, 230–231, 233–234
Bone morphogenic protein 4 (BMP4), and ear development, 25–26
Brn3.1, hair-cell maintenance, 34–35

Ca^{2+}, see Calcium
Ca^{2+} binding proteins, vestibular hair cells, 394–395, 400–401
Ca^{2+} buffering, see Calcium buffering
Ca^{2+} channels, hair cells, 212–214, 260, 382–383
   roles of different types, 253–255
Ca^{2+} currents, 204ff
Ca^{2+} feedback on transduction, 215–216
Ca^{2+} regulation, vestibular hair cells, 374–375
Ca^{2+} sequestering organelles, 393–394
Ca^{2+}-activated K conductances, 379
Ca^{2+}-activated K^+ channels, see BK$_{ca}$ channels
Cadherin 23, 9, 59, 112, 363
Caenorhabditis elegans, mechanotransduction, 154
Calcitonin gene-related peptide (CGRP), 410, 411, 415–416
Calcium, activation of nitric oxide synthase, 413
   and adaptation, 169ff
   cholinergic inhibition, 281–282, 293ff
   exocytosis, 264
   in hair bundle, 176ff
   in hair cells, 253ff
   microdomains, 253ff
   in stereocilia, 178
   vestibular transduction, 367, 373ff
Calcium buffering, 131ff, 261–263
   vestibular hair cells, 393–395
Calcium channels, see Ca^{2+} channels
Calcium currents, see Ca^{2+} currents
Calcium pump, 178
   and fast adaptation, 130–131
Calcium receptors, hair cells, 263–265
Calcium release, calcium-induced, 294–296
Calmodulin, and adaptation, 135ff
   vestibular hair cells, 376
Calyx afferents, 351ff, 403ff
Calyx synapses, anterograde transmission, 403ff
   morphological specializations, 398ff
   retrograde transmission, 406–407
   synaptic ribbons, 396ff
   type I hair cells, 348–349, 400
C-Delta1, ear development, 36
Cell-surface proteins, development, 55ff
CF, see Characteristic frequency
CGRP, see Calcitonin gene-related peptide
Channel kinetics, hair cell tuning, 227ff
Characteristic frequency (CF), relationship to K_{IR}, 217–218
   variation with Ca^{2+}-dependent K^+ current, 229–230
Cholesterol, outer hair cell membrane, 323
Cholinergic, see ACh, ACh receptors
Ciliary bundles, see Hair bundles
Cloned BK$_{ca}$ channels, properties, 231ff
Cochlea, see also Ear
   BK$_{ca}$ channels, 237–239
   effects of efferent inhibition, 288–289
   efferent innervation, 275–276
   inhibition of response to sound, 286–288
   K^+ channels, 237–239
   synaptic inhibition in mammals, 284–286
   viscous damping in mammals, 325
Cochlear amplifier, 11, see also Electromotility of outer hair cells, Stereociliary amplifier
   hair bundle movements, 185ff

outer hair cell frequency response, 335–336
Cochlear development, tonotopic axis, 48
Cochlear vs. vestibular hair cells, 419–420
Corti, first descriptions of hair cells, 5
Cortical lattice, of lateral outer hair cell wall, 322–323
Crista, afferent synapses, 397
   cupula role, 359–360
   ion channel development, 385–386
   structure, 351ff
Cross-links, hair bundle, 108
Cupula, fish, 359
   function and structure, 359–360
   toadfish, 359
Current injection, receptor potentials, 207–209
Current vs. voltage clamp, hair cells, 219–221
Currents, see also Ca^{2+} currents, K^+ currents
   inhibition by 4-AP, 221ff
   location on hair cells, 226
   underlying electrical tuning, 221ff
Current–voltage relationship, hair cells, 212–214
Cuticular plate, development, 52–53
   outer hair cell, 320
Cyclic nucleotide-gated (CNG) channel, and the inner ear, 168
Cytoskeletal proteins, role in development, 55ff
Cytoskeleton, hair bundle, 114ff

Danio rerio, see Zebrafish
Dark cell epithelium, 353
Deaf waddler mutation, Ca^{2+} pump, 178
Delayed rectifier current, inner hair cell development, 67
   vestibular hair cells, 378ff
Delta, see Notch signaling pathway
Depolarization, positive feedback, 214ff
Development, afferent transmission, 272–273
   effects of efferents in cochlea, 288–289
   of hair cells, 20ff
   otoconial, 360–361

   voltage-gated conductances in vestibular hair cells, 383ff
DFN4, and kaptin, 125
Dihydrostreptomycin, 156
Drosophila sp., mechanotransduction, 154

Ear, see also Cochlea
   comparative efferent innervation, 275
   efferent innervation, 275
   innervation, 351ff
   membranous labyrinth structure, 350–351
   vestibular portion anatomy, 349ff
Ear organ development, genes, 26ff
Efferent control, outer hair cell, 70
Efferent inhibition, calcium stores, 293ff
   fast and slow, 288
   vestibular system, 291–293
Efferent innervation, comparative, 275
   mammalian, 275
   turtle, 275–276
Efferent neurons, vestibular, origin in brain, 407–408
Efferent synapse, hair cells, 274ff
   outer hair cells, 321
   ultrastructure, 275–276
   vestibular hair cells, 408–410
Efferent system, cholinergic inhibition, 293ff
Efferent transmission, vestibular hair cells, 407ff
Efferent transmitters and receptors, 411ff
Efferents, hair cells, 12
   physiological effects, 410–411
Eighth nerve, see Afferent nerve, Vestibular nerve
Electrical resonance, see Electrical tuning of hair cells
Electrical tuning of hair cells, 10–11
   BK_{ca} subunits, 235–236
   comparative, 207–209
   frogs, 207
   goldfish, 207
   mammalian cochlea, 238
   vs. micromechanical tuning, 204
   turtle hair cells, 174
Electromotility of outer hair cells, 11, 70ff, 286–287, 313ff

development, 70–71
effects of ACh, 287
Endocytosis, 270–272
Ephaptic transmission, vestibular hair cells, 405–406
Epithelial Na+ channel, and the cochlea, 167
EPSCs, see Excitatory postsynaptic currents
EPSPs, see Excitatory postsynaptic potentials
Escherichia coli, see Bacteria
Espin, 58–59, 122–123
Evolution, active cochlear processes, 313–314
  hair cells, 358–359
Excitatory postsynaptic currents (EPSCs), hair cells, 269–270
Excitatory postsynaptic potentials (EPSPs), vestibular afferents, 403ff
Exocytosis, calcium dependence, 264
  coupling to endocytosis, hair cells, 271
  fast and slow components in hair cells, 267–268
  from hair cells, 264ff
Extrastriola, 361
Extrinsic voltage sensor, chloride channels, 331
Ezrin, 125

F-actin, outer hair cell, 322
Fast adaptation, and calcium, 170
Feedback, efferent, 274ff
Fibroblast growth factor (FGF), otocyst development, 28
Fibronectin, 125
Fimbrin, and hair bundles, 117, 122–123
Fish, cupula, 359
  type I hair cells, 400
Focal adhesion kinase (FAK), 125
Frequency dependence, hair cell vs. afferent response, 370–371
Frequency selectivity, see Frequency tuning
Frequency tuning, and adaptation, 173–174
  bundle versus somatic amplifiers, 191–192
  electrical tuning of hair cells, 10, 207, 219
  K^+ conductances, 217–221
  mechanotransducer channels, 236–237
Friedmann's body, 118
Frog, electrical tuning of hair cells, 207
  hair cell resonance frequency, 219
  saccular hair cell studies, 7

GABA, efferent transmission, 416–417
G-actin, and F-actin, 116
Gadolinium, electromotility, 316–317
Gating compliance, hair bundles, 182–183
Gating spring, 9, 106–107
  transduction, 159ff
Gentamicin, sensitivity of type I hair cells, 400–401
Gfi1, hair cell maintenance, 34–35
Glutamate, receptors, vestibular hair cells, 401–403
  transporters, vestibular epithelium, 401, 403ff
  vestibular hair cells, 250, 257, 263, 270, 401–403
Gold, Thomas, 11, 315
Goldfish, electrical tuning of hair cells, 207
  retinal bipolar cell, 266

Hair bundle
  active movements, 190ff
  carbohydrates, 110
  compliance, see Stiffness
  contact region between stereocilia, 109
  cross-links, 53ff, 108–109, 111–112
  cytoskeleton, 114ff
  height and tonotopy, 106
  heights, 104–105, 365
  kinocilium, 9, 49, 363–365
  labeling techniques, 97ff
  mechanical properties, 7–8, 181ff
  orientation, *Vangl2*, 51
  polarity, development, 49ff
  shape, 52, 101ff
  stereocilia, 95ff, 104–105, 109, 363–366
  stereociliary amplifier, 11, 191–192
  stereovilli, see Stereocilia
  stiffness, and transducer channels, 181–182
  structure and composition, 95ff

summary of functions and properties, 137ff
surface coat, 108
tip links, 3ff, 9, 155, 362–363
vestibular, 362–366
Hair bundle development, 22ff, 52, 119
chick, 44ff
gradient, 49
mammal, 47ff
Hair cell adaptation, see Adaptation
Hair cell frequency response, models, 209ff
Hair cell physiological properties, development, 63ff
Hair cell resonance, see Electrical tuning of hair cells
Hair cell response, current vs. voltage clamp, 219–221
Hair cell spiking, roles in development, 75
Hair cell stimulation, 359ff
Hair cell studies, history, 5ff
Hair cell transducer channels, see Transducer channels
Hair cell transduction, see Transduction
Hair cells, see also Inner hair cells, Outer hair cells, Type I hair cells, Type II hair cells, Vestibular hair cells
ACh, 12, 411–413
ACh-sensitive potassium channels, 291
adaptation, 267–268, 367ff
afferent synapses, 251ff
BK_{ca} channels, 237–239
Ca^{2+} binding proteins, 400–401
Ca^{2+} channels, 382–383
and calcium, 96–97
calcium buffering, 393–395
calcium channels, 260ff
calcium role, 253ff
calyceal specializations, 398ff
calyx, 348–349
chemical neurotransmission, 253ff
ciliary bundle stiffness, 364
ciliary bundles, 362ff
cochlear amplifier, 11
cochlear vs. vestibular, 419–420
currents underlying resonance, 221ff
current–voltage relationship, 212–214
differentiation, 32ff

effect of ACh, 281ff
efferent synapses, 274ff
electrical tuning in turtle, 207–209
electromotility, 286–287
EPSCs, 269–270
evolution, 358–359
exocytosis, 271
exo–endocytosis coupling, 271
fast and slow exocytosis, 267–268
functional significance of inhibition, 280–281
history of studies, 5ff
inhibition in turtles, 278–280
inhibition of response to sound, 286–288
innervation, 352–353
IPSPs, 276–277
K^+ and Ca^{2+} channels, 212–214
K^+ channels, 237–239
kinetic variation, 227ff
location of ionic currents, 226
L-type channels, 382–384
models for mechanotransduction, 6ff
models for synaptic transmission, 11–12
myosin VII, 10
negative feedback, 219ff
PCMA, 376
postsynaptic response, 257
receptor potential, 277ff
receptors, 401–403
resonance frequency, 219–221
responses to efferent inhibition, 287
ribbon synapse function, 254
stereocilia, 362ff
supporting cell lineage, 31–32
synapses, 249–250
synaptic body, 258–259
synaptic calcium domains, 295
synaptic mechanisms, 258ff
synaptic morphology in vestibular hair cells, 395ff
synaptic physiology, 249ff
synaptic ribbons, 258–260, 396
synaptic vesicles, 256–257
transduction channels, 8–9
transmitters, 401–403
trichannel complexes, 229–230
tuning, 10–11, 204ff, 278–280

Hair cells (*continued*)
  tuning and calcium kinetics, 227ff
  tuning by mechanotransducer channels, 236–237
  tuning in turtle, 278
  type I, 348–349
  type I function, 358–359
  type II, 348–349
  variation in hair bundle morphology, 363ff
  vesicle cycle, 256–257
  vestibular, 292–293, 348ff
  voltage-gated conductances, 377ff
Hair-cell development, patterns, 30–31
  timing, 29–30
  tonotopy, 30
Hair-cell differentiation, theories and models, 40ff
Harmonin, 58
Head ectoderm, 22
Head movements, vestibular system, 349ff
Head tilt, otolithic organ, 360
Hes1, hair-cell development, 39–40
Hes5, hair-cell development, 39–40
Hofmeister series, 331
Hydraulic skeleton, outer hair cell, 326
Hydrostat, and outer hair cell, 326

IHCs, see Hair cells, Inner hair cells
Immunocytochemistry of hair bundles, controls, 100–101
Inhibition, calcium dependence, 282–283
  cholinergic mechanisms, 289–291
  efferent, 274ff
  fast and slow, 288
  functional significance, 280–281
  inner hair cells, 287
  intact vs. isolated hair cells, 284–286
  mammalian cochlea, 284–286
  outer hair cells, 286
  receptor potential in hair cells, 277ff
  turtle hair cell, 278–280
Inhibitory postsynaptic potentials (IPSPs), hair cells, 276–277
Inner ear, anatomy, 350ff
Inner hair cells (IHCs), see also Hair cells
  birthdates, 29–30

  currents, development, 67ff
  responses during efferent inhibition, 287
  spiking and development, 67
Innervation, comparative efferent, 275
  ear, 351ff
  efferent, 275, 407ff, 410ff
  type I and type II hair cells, 356–357
  type I hair cells, 403ff
  vestibular hair cells and afferents, 352–353, 410ff
Integrin, 125
  hair-cell development, 61
Ion channel development, cristae, 385–386
Ion channels, calcium, 162
  variation with epithelial location, 386–387
Ionic conductances, utricular hair cells, 378
Ionic currents, and capacitive currents, 333
  development of, 63ff
  locations on hair cells, 226
  models in hair cells, 209ff
  tuning in hair cells, 204ff
IPSPs, see Inhibitory postsynaptic potentials

Jagged2, ear development, 36–37, 40ff
  hair-cell orientation, 51–52
Jerker mouse, espin, 59, 123–124

K^+ channels, see also BK_{Ca} channels, K_{IR}
  comparative, 221
  hair cells, 212–214
  mammalian cochlea, 237–239
  outer hair cells, 325
  vestibular hair cells, 379–380, 389–390
K^+ conductances, repolarizing membranes, 219ff
K^+ currents, 204ff, 214
  ACh evoked, 282
  block by 4-AP, 221ff
  block by TEA, 224–225
  development, 67
  different types, 225–227
  inhibition of hair cells, 282
  mammalian hair cells, 190

outer hair cells, 71–72
vestibular hair cells, 292–293
K^+ in neurotransmission, 404–405
Kaptin, and actin, 124–125
KCNQ superfamily of K channels, 379–380
Keap 1, adhesion protein, 125
Kinocilium, function, 9
   length, 364–365
   migration, 49
   tip links, 363
K_{IR}, CF, 217–218
   conductance, 216–218
Knockout mice, BK_{Ca} and hearing loss, 238–239

Labeling methods, post-embedding, 100
   pre-embedding, 97ff
Lamprey, hair bundles, 116
Lateral inhibition, ear development, 40
Lateral links, hair bundle, 108
Lipids, outer hair cell membrane, 325
Loose hair bundles, 103
Low-affinity nerve growth factor receptor (p75NGFR), ear development, 25–26
L-type Ca^{2+} channels in hair cells, 382–384
Lunatic fringe, ear development, 37

Macula communis, ear development, 26
Maculae, otolith organs, 351ff
Mammals, cochlea, see Cochlea
   vestibular hair cells, 348ff
Math1, 3
   and atonal, 33
   hair-cell differentiation, 33–34, 41–42
Maxwell reciprocity, and outer hair cell, 335, 338
Mechanical properties, hair bundles, 7–8
   tuning, 204
Mechanoelectrical transduction in hair cells, 6ff, 359ff, see also Transduction
   channels (MET), see Transducer channels
   earliest studies, 6ff
   hair cell models, 6ff
   mammalian cochlea, 188ff
   tuning, 236–237

Mechanosensing channel in bacteria, MscL, 7
Medial olivary complex, efferent system, 321
Medial olivocochlear efferents, outer hair cells, 286
Membrane retrieval, neurotransmission, see Endocytosis
Membranous labyrinth, anatomy, 350–351
Micromechanical tuning vs. electrical tuning, 204
Microphonic potentials (Wever-Bray effect), first recordings, 6
Microtubules, hair bundles, 127ff
Mobile Ca^{2+} buffers, neurotransmission, 261–263
Modeling, role of BK_{Ca} subunits, 235–236
Models, hair cell frequency response, 209ff
   ionic currents in hair cells, 209ff
   mechanotransduction, 6ff
   synaptic transmission, 11–12
   voltage-dependent signaling, 10–11
Monovalent anions, and outer hair cell electromotility, 329–330
Mouse cochlea, in vitro, 189
Mouse utricle, in vitro, 366ff
MscL, mechanosensing channel, 7
Mutations affecting development of hair bundles, 28, 32, 38, 39, 51, 55ff, 71
   affecting otolith organ accessory structures, 361
   in KCNQ4 subunit, 379
   in mechanosensitive channels, 8–9
   in myosin VII, 10
   in PMCA isoforms, 376
   tailchaser, 62
   waltzer, 9–10
Myosin, 125
   hair bundles, 119ff
Myosin 1c, and adaptation motor, 174–175
   hair bundles, 120–121
Myosin IIIa, hair bundle, 121
Myosin motor, and adaptation, 156
Myosin VI, cuticular plate, 56
   hair bundle, 121
Myosin VII, hair cells, 10

Myosin VIIa, Usher's syndrome type 1, 121–122
Myosin XVa, hair bundles, 122

Na^+ currents, feedback, 216
　immature rodent utricle, 388–390
　spiking, 390
Nanchung (NAN) mutations, *Drosophila*, 168–169
Negative damping, mammalian cochlea, 313–314
Negative feedback, hair cell transduction, 219ff
Neomycin, and hair bundles, 110
Neurotransmission, see Synaptic transmission
Neurotransmitter release, see also Exocytosis, Synaptic transmission, Transmitters
　development, 66–67
　and neurotransmitter receptors, 401–403
　postsynaptic potentials, 268–270
Nitric oxide, efferent transmitter, 413–415
Noggin, BMP4 block, 28
NOMPC, *Drosophila*, 168–169
　and zebrafish hair cells, 169
Nonlinear capacitance, outer hair cell membrane, 330
Notch signaling pathway, 2, 32
　hair-cell orientation, 51–52
　inner ear development, 35ff
　number of hair cell rows, 37ff
　otocyst development, 28–29

OHCs, see Hair cells, Outer hair cells
Ontogeny, see also Development
　afferent transmission, 272–273, see also Development
　otoconial, 360–361
　voltage-gated channels, 383ff
Opioid peptides, efferent transmission, 415–416
Otic pit, 22
Otic placode, 22
Otic vesicle, 22
Otoacoustic emissions, 315
　and hair bundle movement, 187–188
Otoconia, 360–361

Otocyst, 22
Otolith organs, maculae, 351ff
Otolithic maculae, chick, 361–362
Otolithic membrane, 360–362
Otolithic organ, ciliary bundles, 362ff
　head tilt, 360
　transduction mechanisms, 366ff
Outer hair cell development, basolateral currents and electromotility, 70ff
　currents, 71ff
Outer hair cell structure, 319ff
Outer hair cells (OHCs), see also Hair cells
　birthdates, 29–30
　effects of ACh, 287
　efferent synapses, 321–322
　electromotility, 313ff
　enhancing micromechanical tuning, 315
　inhibition, 286
　lack of afferent activation, 321
　length variation, 319
　mammalian specialization, 318
　medical olivocochlear efferents, 286
　and RC circuits, 334
　structure, 319ff
Outward currents, identification, 221ff, 377ff
　inhibition by TEA, 221ff
Outwardly rectifying conductances, differences in type I and type II hair cells, 379ff

Patch clamp, first in hair cells, 10–11
Peptides, efferent transmitter, 415–416
Phase-locking, synaptic requirements, 253
Photoreceptors, neurotransmission, 271–272
　transduction channels, 8
Piezoelectricity, and frequency response, 337–338
　and outer hair cells, 318–319, 334ff
Plasma membrane, outer hair cell lateral wall, 323–324
　rippled appearance, 323–324
Plasma membrane Ca^{2+} ATPases (PMCA), vestibular hair cells, 376
Positive feedback, enhancing depolarization, 214ff

Postsynaptic potentials, see Excitatory postsynaptic currents, Inhibitory postsynaptic potentials
Post-transduction processing, vestibular system, 377ff
Potassium currents, see K⁺ currents
Potassium inward rectifiers, see K_{IR}
Prestin, 11, 70
  development of expression, 71
  knockout mouse, 71
  motors, 337–338
  outer hair cell electromotility, 188, 328–329
  piezoelectric outer hair cell, 319
  sulfate transporters, 329
  voltage sensor, 330–331
Profilin, 125–126
Ptprq, hair bundle, 61
Purinergic transmission, 416

Radixin, 125
"RC" paradox, outer hair cell receptor potentials, 334
Receptor potentials, characterization, 205ff
  evoked by currents, 207–209
  evoked by sounds, 205ff
  impact of basolateral currents in vestibular hair cells, 387ff
  inhibition, 277ff
  turtle, 205–207
  type I hair cells, 390ff
  type II hair cells, 387ff
Receptors, see Neurotransmitter receptors
Red-eared turtle, receptor potential, 205–207, see also Turtle
Repolarization of membranes, K⁺ conductances, 219ff
Resonance frequency, hair cells, 219–221
  outer hair cell, 336–337
Response properties, afferent neurons, 252
Retinal bipolar cell, goldfish, 266
Retinoic acid, hair-cell development, 42ff
Retrograde transmission, calyx synapses, 406–407
Ribbon synapse, function, 254
  kinetics, 266–268
  vestibular hair cells, 395ff

Saccular hair cell, bullfrog, 261–263
Saccule, 350–351
  electrical tuning of hair cells, 207
  hair cell studies, 7
Salicylate, electromotility, 316–317
Semicircular canals, 350ff
Serrate1, ear development, 37, 40ff
Shaker 1 mouse, myosin VIIa, 56, 175
Shaker 2 mouse, Myo15, 122
  myosin XVa, 57
Shear gain, 106
Signaling, calcium channels, 261–263
Signaling requirements, afferent synapse, 251–253
Single-channel analysis, basolateral channels, 227ff
  MET channels, 155–156, 158ff
Slow adaptation, 367ff
  myosin 1c, 120–121
  and tip-link mechanics, 170
Smooth endoplasmic reticulum, Ca^{2+} sequestering, 393
Snell's waltzer, Myo6, 121
  and myosin VI, 56
Sodium currents, see NA⁺ currents
Sound response, inhibition in cochlea, 286–288
Spectrin, and actin, 323
  and cell shape, 125
Spike rate, vestibular nerve, 355ff
Spiking, Na⁺ currents, 390
Spinner mutant, hair bundle, 62
Spontaneous motion of hair bundles, 184
Stereocilia, Ca^{2+} diameter, 365
  first descriptions, 5
  gradient across, 374–376
  growth, 117
  number, 365
  outer hair cell, 320
  structure, 362ff
  variation, 363–364
Stereocilia patterns, development, 46
Stereocilial staircase, 45
Stereociliary amplifier, 154ff
Stiffness, hair bundle, 364–365
Stimulation methods, hair cells, 359ff
Strain-dependent water permeability, 327
Stretch-activated membrane channels, 325

Striola, 358
  hair bundles, 365
  origin of term, 361
Subsurface cisterna, of lateral outer hair cell wall, 322
Subsynaptic cistern, 275–276
Supporting cells, hair cell lineage, 31–32
Synapses, calcium domains, 295
  efferent, 275–276
  efferent in vestibular system, 409–410
  kinetics, 266–268
  morphology in vestibular hair cells, 395ff
  phase-locking, 253
  timing, 252–253
  type I-calyx, 403ff
Synaptic body, 259–260
  morphology, 395ff
Synaptic inhibition, cellular, 276–277
  mammalian cochlea, 284–286
Synaptic mechanisms, hair cells, 258ff
  membrane retrieval, 270–272
Synaptic physiology, hair cells, 249ff
Synaptic potentials, vestibular hair cells, 292
Synaptic ribbons, 258–259
  hair cells, 258–260
  outer hair cell, 320–321
  parameters, 266
  vestibular hair cells, 396
Synaptic transmission, calcium buffering, 261–263
  calcium channels, 260ff
  control of duration, 261–262
  endocytosis, 270–272
  exocytosis, 264ff
  hair cell vesicles, 256–257
  hair cells, 11–12
  kinetics, 266–268
  phase-locking, 253
  photoreceptors, 271–272
  postsynaptic response in hair cells, 257
  timing, 252–253
  type I-calyx, 403ff
  vestibular hair cells, 395ff
Synaptic vesicles, functional pools, 265ff
  hair cells, 256–257

Tailchaser mutant, outer hair cells, 62
Tasmanian devil mutant, stereocilia, 62
Tetraethylamonium (TEA), effects on K^+, 224–225
  inhibition of outward currents, 221ff
Thyroid hormone, hair-cell development, 43ff
Tight hair bundles, 103, 363–364
Timing, synaptic role, 252–253
Tip link, 109
  development, 53ff
  and extracellular calcium, 155
  and gating springs, 112–113, 159ff
  and hair bundles, 155, 362–363
  kinocilium, 363
Tip link development, theories, 53–54
Tip link proteins, 9–10
Toadfish, cupula, 359
Tonotopy, and adaptation time constant, 177
Trachemys scripta elegans, see Red-eared turtle, Turtle
Transducer channels, activation kinetics, 65–66
  activation time constant, 158–159
  block by calcium, 163
  blocking, 162ff
  characteristic frequency and adaptation, 171, 173ff
  characteristic frequency versus conductance, 167
  cyclic adenosine monophosphate (cAMP) effect, 171
  dynamic range, 161–162
  electrical resonant frequency, 167
  genetic studies, 8–9
  hair cells, 8–9
  input resistance, 161–162
  ionic selectivity, 162
  location, 113
  molecular identity, 167ff
  motors, 337–338
  noise sources, 162
  ontogeny, 383ff
  open probability, 159ff
  photoreceptors, 8
  pore size, 163
  sensitivity, 161–162

Index 453

single-channel conductance, 165ff
tip links, 113
tonotopy, 177
Transducer currents, 157ff, 366ff
  development, 64
Transduction, 6ff, 154ff, 359ff
  channels (MET), see Transducer channels
  earliest studies, 6ff
  hair cell models, 6ff
  mammalin cochlea, 188ff
  tuning, 236–237
  utricle, 367ff
  vestibular system, 366ff
Transduction currents, see Transducer currents
Transmitters, ACh, 411–413
  γ-aminobutyric acid (GABA), 416–417
  efferent, 411ff
  hair cells, 249–250
  nitric oxide, 413–415
  peptides, 415–416
  purinergic, 416
  vestibular hair cells, 401–403
Trichannel complexes, hair cells, 229–230
Tropomyosin, hair bundles, 124
TRP channels, 3, 8–9
  cation channels, 168
Tubulin, hair bundles, 127
Tuning, hair cells, 10–11, 204ff
  ionic currents, 204ff
  role of mechanotransducer channels, 236–237
  turtle hair cell, 278
Turgor pressure, electromotility, 316–317
  and outer hair cells, 326
Turtle cochlea, receptor potentials, 205–207
Turtle hair cells, location of ionic currents, 226
Turtles, BK_{ca} channels, 230–231, 233–234
  current–voltage relationship in hair cells, 212–214
  efferent innervation, 275–276
  electrical tuning of hair cells, 207–209
  hair cell channels, 221
  hair cell transduction, 204ff
  hair cell tuning, 278

inhibition of hair cells, 278–280
IPSPs in hair cells, 276–277
splice variants in BK_{ca} channels, 233–234
Type I hair cells, bandwidth, 391
  calyceal specializations, 398ff
  calyx, 348–349
  fish, 400
  function, 358–359
  innervation, 352–353, 356–357, 403ff
  purpose, 420–421
  sensitivity to gentamicin, 400–401
  structure, 348–349
  synaptic ribbons, 396
  voltage-gated conductances, 377ff
Type I vs. type II hair cells, outwardly rectifying conductances, 379ff
Type I vs. type II physiology, 357–358
Type I-calyx synapse, transmission, 403ff
Type II hair cells, innervation, 352–353, 356–357
  receptor potentials, 387ff, 390ff
  structure, 348–349
  synaptic ribbons, 396
  voltage-gated conductances, 377ff

Unconventional myosins, 119ff
  hair-cell development, 55ff
USH1D, cadherin, 59
USH1F, cadherin, 59
Usher syndrome, myosin VIIa, 58
Utricle, 350–351
Utricular hair cells, ionic conductances, 378ff
  transduction, 367ff

Vangl2, hair-cell orientation, 51
Varitint-waddler mutant, hair bundle, 62
Vesicles, cycle in neurotransmission, 256–257
Vestibular efferent nucleus, 407–408
Vestibular efferents, ACh, 411–413
  discharge properties, 354
  GABA, 416–417
  location, 407–408
  nitrous oxide, 413–415
  peptides, 415–416
  physiological effects, 410–411

Vestibular epithelia, organization, 349ff
Vestibular hair bundle development, 48–49
Vestibular hair cells, see also Hair cells, Type I hair cells, Type II hair cells
  ACh receptors, 292–293
  α9 and α10 receptors, 292
  Ca^{2+}, 393–395
  Ca^{2+} channels, 382–383
  calcium buffering, 393–395
  calmodulin, 376
  development, 66
  development of basolateral currents, 74
  efferent inhibition, 291–293, 407, 409–410
  ephaptic transmission, 405–406
  glutamate, 401–403
  ion channels and cell location, 386–387
  K^+ channels, 379–380, 389–390
  mammalian, 348ff
  PCMA, 376
  receptors, 401–403
  synaptic morphology, 395ff
  synaptic potentials, 292
  synaptic transmission, 395ff
  transmitters, 401–403
Vestibular nerve, afferent terminal physiology, 356–358
  fiber diameter, 356–357
  physiology, 355ff
  spike rate, 355ff
Vestibular periphery, development in zebrafish, 26
  organization, 349ff
Vestibular system, adaptation, 367ff
  efferent inhibition, 291–293
  function, 349ff
  hair cell transducer adaptation, 372–373
  innervation, 351ff
  mechanoelectrical transduction, 366ff
  post-transduction processing, 377ff
  voltage-gated conductances, 377ff

Vestibular vs. auditory function, 348
Vestibular vs. cochlear hair cells, 419–420
Vezatin, hair bundles, 122
Vinculin, adhesion protein, 125
Viscous damping in the mammalian cochlea, and outer hair cell electromotility, 335
Voltage clamp vs. current clamp, hair cells, 219–221
Voltage-controlled hair bundle movements, 184
Voltage-dependent signaling, hair cell models, 10–11
Voltage-gated Ca^{2+} channels, see Ca^{2+} channels
Voltage-gated Ca^{2+} currents, see Ca^{2+} currents
Voltage-gated conductances, cochlear hair cells, 204ff
  developmental changes, 383ff
  turtle hair cells, 204ff
  vestibular hair cells, 377ff
Voltage-gated currents, see Voltage-gated Ca^{2+} channels, Voltage-gated conductances, Voltage-gated K^+ channels, Voltage-gated Na^+ channels
Voltage-gated K^+ channels, see K^+ channels
Voltage-gated Na^+ channels, see Na^+ channels

Waltzer mutation, 9–10
  and cadherin 23, 59–60
Water permeability, outer hair cell plasma membrane, 326
Wever-Bray effect, 6
Whirlin, 58
  and myosin XVa, 122

Zebrafish, mechanotransduction, 154
  vestibular organ development, 26

SPRINGER HANDBOOK OF AUDITORY RESEARCH *(continued from page ii)*

Volume 22: Evolution of the Vertebrate Auditory System
Edited by Geoffrey A. Manley, Arthur N. Popper and Richard R. Fay

Volume 23: Plasticity of the Auditory System
Edited by Thomas N. Parks, Edwin W Rubel, Arthur N. Popper, and Richard R. Fay

Volume 24: Pitch: Neural Coding and Perception
Edited by Christopher J. Plack, Andrew J. Oxenham, Richard R. Fay, and Arthur N. Popper

Volume 25: Sound Source Localization
Edited by Arthur N. Popper and Richard R. Fay

Volume 26: Development of the Inner Ear
Edited by Matthew W. Kelley, Doris K. Wu, Arthur N. Popper and Richard R. Fay

Volume 27: Vertebrate Hair Cells
Edited by Ruth Anne Eatock, Richard R. Fay and Arthur N. Popper

For more information about the series, please visit www.springer-ny.com/shar.